Molecular Biology and Biotechnology
5th Edition

Molecular Biology and Biotechnology
5th Edition

Edited by

John M Walker

School of Life Sciences, University of Hertfordshire, Hatfield, Hertfordshire AL10 9AB, UK

Ralph Raply

School of Life Sciences, University of Hertfordshire, Hatfield, Hertfordshire AL10 9AB, UK

RSCPublishing

ISBN: 978-0-85404-125-1

A catalogue record for this book is available from the British Library

Published by The Royal Society of Chemistry,
Thomas Graham House, Science Park, Milton Road,
Cambridge CB4 0WF, UK

Registered Charity Number 207890

For further information see our website at www.rsc.org

Preface

One of the exciting aspects of being involved in the field of molecular biology is the ever-accelerating rate of progress, both in the development of new methodologies and in the practical applications of these methodologies. Indeed, such developments led to the idea of the first edition of *Molecular Biology and Biotechnology* and subsequent editions have reflected the fast-moving nature of the area, not least this latest edition, which continues to reflect recent developments, with new chapters in developing areas such as genome technology, nano-biotechnology, regenerative medicine and biofuels.

The first six chapters deal with the technology used in current molecular biology and biotechnology. These deal primarily with core nucleic acid techniques and protein expression through microbial and genetic detection methods. Further chapters address the huge advances made in gene and genome analysis and update the rapid advances into yeast analysis, which is providing very exciting insights into molecular pathways. Molecular biology also continues to affect profoundly progress in biotechnology in areas such as vaccine development, use and application of monoclonal antibodies, clinical treatment and diagnosis, the production of transgenic animals, and many other areas of research relevant to the pharmaceutical industry. Chapters on all these areas have been retained and fully updated in this new edition and new chapters introduced on the applications of molecular biology in the areas of drug design and diseases, and regenerative medicine. In addition, we continue to ensure that biotechnology is not just considered at the gene level and full consideration continues to be given to applications and

Molecular Biology and Biotechnology, 5th Edition
Edited by John M Walker and Ralph Rapley
© Royal Society of Chemistry 2009
Published by the Royal Society of Chemistry, www.rsc.org

manufacturing, with chapters on downstream processing, biosensors, the applications of immobilised biocatalysts, and a new chapter on the developing area of biofuels.

Our continued intention is that this book should primarily have a teaching function. As such, this book should prove of interest both to undergraduates studying for biological or chemical qualifications and to postgraduate and other scientific workers who need a sound introduction to this ever rapidly advancing and expanding area.

John M. Walker
Ralph Rapley

Contents

Molecular Biology and Biotechnology, 5th Edition
Edited by John M Walker and Ralph Rapley
© Royal Society of Chemistry 2009
Published by the Royal Society of Chemistry, www.rsc.org

Chapter 3   Molecular Diagnostics
*Laura J. Tafe, Claudine L. Bartels, Joel A. Lefferts and
Gregory J. Tsongalis*

Chapter 4   Molecular Microbial Diagnostics
Karl-Henning Kalland

Chapter 6   The Biotechnology and Molecular Biology of Yeast
Brendan P. G. Curran and Virginia C. Bugeja

Chapter 7   Metabolic Engineering
*Stefan Kempa, Dirk Walther, Oliver Ebenhoeh and Wolfram
Weckwerth*

Chapter 8   Bionanotechnology
David W. Wright

Chapter 9   Molecular Engineering of Antibodies
James D. Marks

Chapter 10 Plant Biotechnology
Michael G. K. Jones

Chapter 10 Plant Biotechnology
Michael G. K. Jones

Chapter 11 Biotechnology-based Drug Discovery
K. K. Jain

Chapter 14 Transgenesis
Elizabeth J. Cartwright and Xin Wang

Chapter 15 Protein Engineering
John Adair and Duncan McGregor

Chapter 18 Biosensors
Martin F. Chaplin

Chapter 19 Biofuels and Biotechnology
Jonathan R. Mielenz

CHAPTER 1

Basic Molecular Biology Techniques

RALPH RAPLEY

School of Life Sciences, University of Hertfordshire, Hatfield, Hertfordshire AL10 9AB, UK

1.1 ENZYMES USED IN MOLECULAR BIOLOGY

The discovery and characterisation of a number of key enzymes have permitted the development of various techniques for the analysis and manipulation of DNA. In particular, the enzymes termed type II restriction endonucleases have come to play a key role in all aspects of molecular biology.[1] These enzymes recognise specific DNA sequences, usually 4–6 base pairs (bp) in length, and cleave them in a defined manner. The sequences recognised are palindromic or of an inverted repeat nature, that is, they read the same in both directions on each strand. When cleaved they leave a flush-ended or staggered (also termed a cohesive-ended) fragment depending on the particular enzyme used (Figure 1.1).

An important property of staggered ends is that those produced from different molecules by the same enzyme are complementary (or 'sticky') and so will anneal to each other (Table 1.1). The annealed strands are held together only by hydrogen bonding between complementary bases on opposite strands. Covalent joining of ends on each of the two strands may be brought about by the enzyme DNA ligase. This is widely exploited in molecular biology to allow the construction of recombinant DNA, *i.e.* the joining of DNA fragments from different sources.

Molecular Biology and Biotechnology, 5th Edition
Edited by John M Walker and Ralph Rapley
© Royal Society of Chemistry 2009
Published by the Royal Society of Chemistry, www.rsc.org

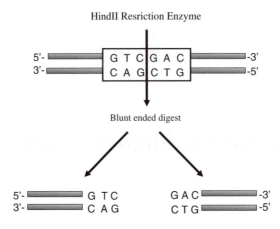

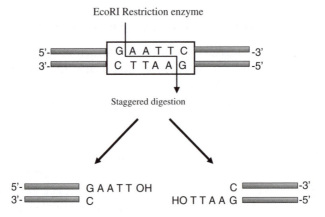

Figure 1.1 Examples of digestion of DNA by restriction endonucleases. The upper panel indicates the result of a restriction digestion forming blunt fragments with the enzyme *Hin*dIII. The bottom panel indicates the cohesive fragments produced by digestion with the enzyme *Eco*R1.

Approximately 500 restriction enzymes have been characterised that recognise over 100 different target sequences. A number of these, termed isoschizomers, recognise different target sequences but produce the same staggered ends or overhangs. A number of other enzymes have proved to be of value in the manipulation of DNA, as summarised in Table 1.2, and are indicated at appropriate points within the text.

Table 1.1 Examples of restriction endonucleases that recognise different target sequences and the resulting fragments following digestion.

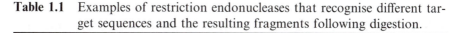

Name	Recognition Sequence	Digestion Products	
	Four Nucleotide Recognition Sequence		
*Hae*III	5'-GGCC-3' 3'-CCGG-5'	5'-GG CC-3' 3'-CC GG-5'	**Blunt End Digestion**
*Hpa*II	5'-CCGG-3' 3'-GGCC-5'	5'-C CGG-3' 3'-GGC C-5'	**Cohesive End Digestion**
	Six Nucleotide Recognition Sequence		
*Bam*HI	5'-GGATTC-3' 3'-GGCC-5'	5'-G GATCC-3' 3'-CCTAG G-5'	
*Eco*RI	5'-GAATTC-3' 3'-CTTAAG-5'	5'-G AATCC-3' 3'-CTTAA G-5'	
*Hind*III	5'-AAGCTT-3' 3'-TTCGAA-5'	5'-A AGCTT-3' 3'-TTCGA A-5'	
	Eight Nucleotide Recognition Sequence		
*Not*I	5'-GCGGCCGC-3' 3'-CGCCGGCG-5'	5'-GC GGCCGC-3' 3'-CGCCGG CG-5'	

Table 1.2 Comparison of the various labelling methods for DNA.

Labelling method	Enzyme	Probe Type	Specific Activity
5' end labelling	Alkaline Phosphatase Polynucleotide Kinase	DNA	Low
3' end labelling	Terminal Transferase	DNA	Low
Nick Translation	DNase I DNA Polymerase I	DNA	High
Random Hexamer	DNA Polymerase I	DNA	High
PCR	*Taq* DNA Polymerase	DNA	High
Riboprobes (cRNA)	RNA Polymerase	RNA	High

1.2 ISOLATION AND SEPARATION OF NUCLEIC ACIDS

1.2.1 Isolation of DNA

The use of DNA for analysis or manipulation usually requires that it is isolated and purified to a certain extent. DNA is recovered from cells by the gentlest possible method of cell rupture to prevent the DNA from fragmenting by mechanical shearing. This is usually in the presence of

EDTA which chelate the Mg^{2+} ions needed for enzymes that degrade DNA termed DNase. Ideally, cell walls, if present, should be digested enzymatically (*e.g.* lysozyme treatment of bacteria) and the cell membrane should be solubilised using detergent. If physical disruption is necessary, it should be kept to a minimum and should involve cutting or squashing of cells, rather than the use of shear forces. Cell disruption (and most subsequent steps) should be performed at 4 °C, using glassware and solutions which have been autoclaved to destroy DNase activity (Figure 1.2).

After release of nucleic acids from the cells, RNA can be removed by treatment with ribonuclease (RNase) which has been heat treated to inactivate any DNase contaminants; RNase is relatively stable to heat as a result of its disulfide bonds, which ensure rapid renaturation of the molecule on cooling. The other major contaminant, protein, is removed by shaking the solution gently with water-saturated phenol or with a

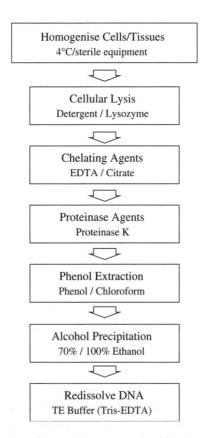

Figure 1.2   Flow diagram of the main steps involved in the extraction of DNA.

phenol–chloroform mixture, either of which will denature proteins but not nucleic acids. Centrifugation of the emulsion formed by this mixing produces a lower, organic phase, separated from the upper, aqueous phase by an interface of denatured protein. The aqueous solution is recovered and deproteinised repeatedly, until no more material is seen at the interface. Finally, the deproteinised DNA preparation is mixed with two volumes of absolute ethanol and the DNA allowed to precipitate out of solution in a freezer. After centrifugation, the DNA pellet is redissolved in a buffer containing EDTA to inactivate any DNases present. This solution can be stored at 4 °C for at least 1 month. DNA solutions can be stored frozen, although repeated freezing and thawing tend to damage long DNA molecules by shearing.

The procedure described above is suitable for total cellular DNA. If the DNA from a specific organelle or viral particle is needed, it is best to isolate the organelle or virus before extracting its DNA, since the recovery of a particular type of DNA from a mixture is usually rather difficult. Where a high degree of purity is required, DNA may be subjected to density gradient ultracentrifugation through caesium chloride, which is particularly useful for the preparation of plasmid DNA.[2] It is possible to check the integrity of the DNA by agarose gel electrophoresis and determine the concentration of the DNA by using the fact that 1 absorbance unit equates to $50 \, \mu g \, ml^{-1}$ of DNA and so

$$50 \times A_{260} = \text{concentration of DNA sample} \; (\mu g \, ml^{-1}) \qquad (1)$$

Contaminants may also be identified by scanning UV spectrophotometry from 200 to 300 nm. A ratio of 260 nm:280 nm of approximately 1.8 indicates that the sample is free of protein contamination, which absorbs strongly at 280 nm.

1.2.2 Isolation of RNA

The methods used for RNA isolation are very similar to those described above for DNA; however, RNA molecules are relatively short and therefore less easily damaged by shearing, so cell disruption can be more vigorous. RNA is, however, very vulnerable to digestion by RNases. which are present endogenously in various concentrations in certain cell types and exogenously on fingers. Gloves should therefore be worn and a strong detergent should be included in the isolation medium to denature immediately any RNases. Subsequent deproteinisation should be particularly rigorous, since RNA is often tightly associated with proteins.[3] DNase treatment can be used to remove DNA and RNA can be precipitated by ethanol. One reagent in which is commonly used in RNA extraction is guanadinium thiocyanate, which is both a strong inhibitor of RNase and a protein denaturant. It is possible to check the integrity of an RNA extract by analysing it by agarose gel electrophoresis. The most abundant RNA species are rRNA molecules, 23S and 16S for prokaryotes and 18S and 28S for eukaryotes. These appear as discrete bands on the agarose gel and indicate that the other RNA components are likely to be intact. This is usually carried out under denaturing conditions to prevent secondary structure formation in the RNA. The concentration of the RNA may be estimated by using UV spectrophotometry. At 260 nm, 1 absorbance unit equates to $40\,\mu g\,ml^{-1}$ of RNA and therefore

$$40 \times A_{260} = \text{concentration of RNA sample } (\mu g\,ml^{-1}) \qquad (2)$$

Contaminants may also be identified in the same way as for DNA by scanning UV spectrophotometry; however, in the case of RNA a 260 nm:280 nm ratio of approximately 2 would be expected for a sample containing no contamination.

In many cases, it is desirable to isolate eukaryotic mRNA, which constitutes only 2–5% of cellular RNA, from a mixture of total RNA molecules. This may be carried out by affinity chromatography on oligo (dT)-cellulose columns. At high salt concentrations, the mRNA containing poly(A) tails binds to the complementary oligo(dT) molecules of the affinity column and so mRNA will be retained; all other RNA molecules can be washed through the column with further high-salt solution. Finally, the bound mRNA can be eluted using a low concentration of salt. Nucleic acid species may also be subfractionated by more physical means such as electrophoretic or chromatographic separations based on differences in nucleic acid fragment sizes or physicochemical characteristics.[3]

1.3   ELECTROPHORESIS OF NUCLEIC ACIDS

Electrophoresis in agarose or polyacrylamide gels is the most usual way to separate DNA molecules according to size (Figure 1.3). The technique can be used analytically or preparatively and can be qualitative or quantitative. Large fragments of DNA such as chromosomes may also be separated by a modification of electrophoresis termed pulsed field gel electrophoresis (PFGE). The easiest and most widely applicable method is electrophoresis in horizontal agarose gels, followed by staining with

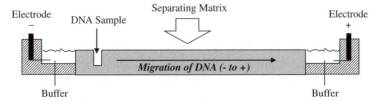

Figure 1.3 A typical setup required for agarose gel electrophoresis of DNA. The upper panel indicates a cross-section of the unit used for gel electrophoresis.

ethidium bromide. This dye binds to DNA by insertion between stacked base pairs (intercalation) and it exhibits a strong orange/red fluorescence when illuminated with ultraviolet light. Very often electrophoresis is used to check the purity and intactness of a DNA preparation or to assess the extent of a enzymatic reaction during, for example, the steps involved in the cloning of DNA. For such checks 'mini-gels' are particularly convenient, since they need little preparation, use small samples and give results quickly. Agarose gels can be used to separate molecules larger than about 100 bp. For higher resolution or for the effective separation of shorter DNA molecules, polyacrylamide gels are the preferred method.[4]

When electrophoresis is used preparatively, the piece of gel containing the desired DNA fragment is physically removed with a scalpel. The DNA may be recovered from the gel fragment in various ways. This may include crushing with a glass rod in a small volume of buffer, using agarase to digest the agarose thus leaving the DNA, or by the process of electroelution. In the latter method, the piece of gel is sealed in a length of dialysis tubing containing buffer and is then placed between two electrodes in a tank containing more buffer. Passage of an electric current between the electrodes causes DNA to migrate out of the gel piece, but it remains trapped within the dialysis tubing and can therefore be recovered easily. More commonly, commercial spin columns can be used which contain an isolating matrix used in conjunction with a bench-top microcentrifuge. The use of such standardised 'kits' in molecular biology is now commonplace. An alternative to conventional analysis of nucleic acids by electrophoresis is through the use of microfluidic systems. These are carefully manufactured chip-based units where microlitre volumes may be used and with the aid of computer analysis provide much of the data necessary for analysis. Their advantage lies in the fact that the sample volume is very small, allowing much of an extract to be used for further analysis.[5]

1.4   RESTRICTION MAPPING OF DNA FRAGMENTS

Restriction mapping involves the size analysis of restriction fragments produced by several restriction enzymes individually and in combination.[6] The principle of this mapping is illustrated, in which the restriction sites of two enzymes, A and B, are being mapped. Cleavage with A gives fragments 2 and 7 kilobases (kb) from a 9 kb molecule, hence we can position the single A site 2 kb from one end. Similarly, B gives fragments 3 and 6 kb, so it has a single site 3 kb from one end; but it is not possible at this stage to say if it is near to A's site or at the opposite end of the DNA. This can be resolved by a double digestion. If the resultant fragments are 2, 3 and 4 kb, then A and B cut at opposite ends of the molecule; if they are 1, 2 and 6 kb, the sites are near each other. Not surprisingly, the mapping of real molecules is rarely as simple as this and computer analysis of the restriction fragment lengths is usually needed to construct a map (Figure 1.4).

1.5   NUCLEIC ACID ANALYSIS METHODS

There are numerous methods for analysing DNA and RNA; however, many of them are solution based or more recently include the use of

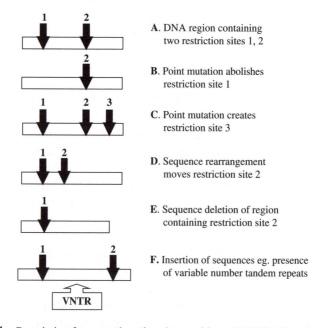

A. DNA region containing two restriction sites 1, 2

B. Point mutation abolishes restriction site 1

C. Point mutation creates restriction site 3

D. Sequence rearrangement moves restriction site 2

E. Sequence deletion of region containing restriction site 2

F. Insertion of sequences eg. presence of variable number tandem repeats

Figure 1.4   Restriction fragment length polymorphisms (RFLP). The schematic panels A–F indicate the various fragments obtained following digestion as a result of differences in the position of restriction endonuclease target sequences.

chip-based array systems. Indeed, the lab-on-a-chip approach is developing rapidly and it is possible to envisage many detection and analysis methods being developed in this format in the future.[7] However, traditional methods are still employed in many laboratories and much is still made of producing a hard copy of digested and separated single-stranded DNA fragments attached to a matrix such as nylon for analysis with an appropriate labelled probe.

1.5.1 DNA Blotting

Electrophoresis of DNA restriction fragments allows separation based on size to be carried out; however, it provides no indication as to the presence of a specific, desired fragment among the complex sample (Figure 1.5). This can be achieved by transferring the DNA from the intact gel on to a piece of nitrocellulose or nylon membrane placed in contact with it.[8] This provides a more permanent record of the sample since DNA begins to diffuse out of a gel that is left for a few hours. First the gel is soaked in alkali to render the DNA single stranded. It is then transferred to the membrane so that the DNA becomes bound to the it in exactly the same pattern as that originally on the gel. This transfer, named a Southern blot after its inventor Ed Southern, can be performed electrophoretically or by drawing large volumes of buffer through both gel and membrane, thus transferring DNA from one to the other by

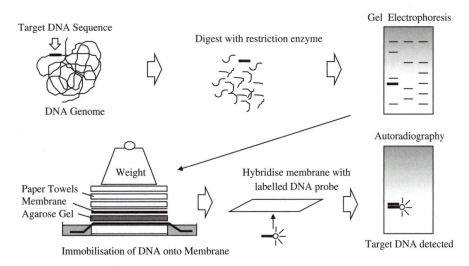

Figure 1.5   The steps involved in the production of a Southern blot and the subsequent detection of a specific DNA sequence following hybridisation with a complementary labelled gene probe.

capillary action. The point of this operation is that the membrane can now be treated with a labelled DNA molecule, for example a gene probe. This single-stranded DNA probe will hybridise under the right conditions to complementary fragments immobilised on the membrane. The conditions of hybridisation, including the temperature and salt concentration, are critical for this process to take place effectively. This is usually referred to as the stringency of the hybridisation and it is particular for each individual gene probe and for each sample of DNA. A series of washing steps with buffer are then carried out to remove any unbound probe and the membrane is developed, after which the precise location of the probe and its target may be visualised. It is also possible to analyse DNA from different species or organisms by blotting the DNA and then using a gene probe representing a protein or enzyme from one of the organisms. In this way, it is possible to search for related genes in different species. This technique is generally termed Zoo blotting.

1.5.2 RNA Blotting

The same basic process of nucleic acid blotting can be used to transfer RNA from gels on to similar membranes. This allows the identification of specific mRNA sequences of a defined length by hybridisation to a labelled gene probe and is known as Northern blotting.[9] With this technique it is not only possible to detect specific mRNA molecules but it may also be used to quantify the relative amounts of the specific mRNA. It is usual to separate the mRNA transcripts by gel electrophoresis under denaturing conditions since this improves resolution and allows a more accurate estimation of the sizes of the transcripts. The format of the blotting may be altered from transfer from a gel to direct application to slots on a specific blotting apparatus containing the nylon membrane. This is termed slot or dot blotting and provides a convenient means of measuring the abundance of specific mRNA transcripts without the need for gel electrophoresis; it does not, however, provide information regarding the size of the fragments.

A further method of RNA analysis that overcomes the problems of RNA blotting is termed the ribonuclease protection assay. Here the RNA from a sample is extracted and then mixed with a probe representing the sequence of interest in solution. The probe and the appropriate RNA fragment hybridise to form a double-stranded sequence. RNase is then added, which cleaves any single-stranded RNA present but leaves the double-stranded RNA intact. The intact RNA can then be separated by electrophoresis and an indication of the size of the fragment generated. The efficient removal of the background of RNA and

the improved sensitivity make the ribonuclease protection assay a popular choice for the analysis of specific RNA molecules.

An important step in the field of RNA analysis was the development of RNAi (RNA interference), which inhibits gene expression. Here double-stranded DNA promotes the degradation of mRNA. Double-stranded RNA in the cell is cleaved by a dicer enzyme, resulting in the formation of small 21–25 bp interfering RNAs (siRNA). The siRNA are complementary to a target RNA strand. Small RNAi proteins are guided by the siRNA to the appropriate mRNA, where the target is then cleaved and is unable to be translated. Many areas are now benefiting from the adoption of this technique in the molecular biology and biotechnology fields.[10]

1.6 GENE PROBE DERIVATION

The availability of a gene probe is essential in many molecular biology techniques, yet in many cases is one of the most difficult steps (Figure 1.6). The information needed to produce a gene probe may come from many sources, but with the development and sophistication of genetic databases this is usually one of the first stages.[11] There are a number of genetic databases throughout the world and it is possible to search these over the internet and identify particular sequences relating to a specific gene or protein. In some cases it is possible to use related proteins from the same gene family to gain information on the most useful DNA sequence. Similar proteins or DNA sequences but from different species may also provide a starting point with which to produce a so-called heterologous gene probe. Although in some cases probes have already been produced and cloned, it is possible, armed with a DNA sequence from a DNA

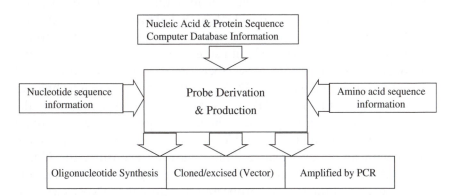

Figure 1.6 Alternative strategies for designing and producing a gene probe.

database, to synthesise chemically a single-stranded oligonucleotide probe. This is usually undertaken by computer-controlled gene synthesisers which link dNTPs together based on a desired sequence. It is essential to carry out certain checks before probe production to determine that the probe is unique, is not able to self-anneal or is self-complementary, all of which may compromise its use.[12]

Where little DNA information is available to prepare a gene probe, it is possible in some cases to use the knowledge gained from analysis of the corresponding protein. Thus it is possible to isolate and purify proteins and sequence part of the N-terminal end of the protein. From our knowledge of the genetic code, it is possible to predict the various DNA sequences that could code for the protein and then synthesise appropriate oligonucleotide sequences chemically. Due to the degeneracy of the genetic code, most amino acids are coded for by more than one codon, hence there will be more than one possible nucleotide sequence which could code for a given polypeptide. The longer the polypeptide, the greater is the number of possible oligonucleotides which must be synthesised. Fortunately, there is no need to synthesise a sequence longer than about 20 bases, since this should hybridise efficiently with any complementary sequences and should be specific for one gene. Ideally, a section of the protein should be chosen which contains as many tryptophan and methionine residues as possible, since these have unique codons and there will therefore be fewer possible base sequences which could code for that part of the protein. The synthetic oligonucleotides can then be used as probes in a number of molecular biology methods.

1.7   LABELLING DNA GENE PROBE MOLECULES

An essential feature of a gene probe is that it can be visualised by some means. In this way, a gene probe that hybridises to a complementary sequence may be detected and identify that desired sequence from a complex mixture. There are two main ways of labelling gene probes; traditionally it has been carried out using radioactive labels, but gaining in popularity are non-radioactive labels. Perhaps the most often used radioactive label is phosphorus-32 (^{32}P), although for certain techniques sulfur-35 (^{35}S) and tritium (^{3}H) are used. These may be detected by the process of autoradiography, where the labelled probe molecule, bound to sample DNA, located for example on a nylon membrane, is placed in contact with an X-ray-sensitive film. Following exposure, the film is developed and fixed just as a black and white negative and reveals the precise location of the labelled probe and therefore the DNA to which it has hybridised.

Non-radioactive labels are increasingly being used to label DNA gene probes. Until recently, radioactive labels were more sensitive than their non-radioactive counterparts. However, recent developments have led to similar sensitivities, which, when combined with their improved safety, have led to their greater acceptance.

The labelling systems are termed either direct or indirect. Direct labelling allows an enzyme reporter such as alkaline phosphatase to be coupled directly to the DNA. Although this may alter the characteristics of the DNA gene probe, they offer the advantage of rapid analysis since no intermediate steps are needed. However, indirect labelling is at present more popular. This relies on the incorporation of a nucleotide which has a label attached. At present, three of the main labels in use are biotin, fluorescein and digoxygenin. These molecules are covalently linked to nucleotides using a carbon spacer arm of 7, 14 or 21 atoms. Specific binding proteins may then be used as a bridge between the nucleotide and a reporter protein such as an enzyme. For example, biotin incorporated into a DNA fragment is recognised with a very high affinity by the protein streptavidin. This may be either coupled or conjugated to a reporter enzyme molecule such as alkaline phosphatase. This is able to convert a colourless substrate, *p*-nitrophenol phosphate (PNPP), into a yellow compound, *p*-nitrophenol (PNP), and also offers a means of signal amplification. Alternatively labels such as digoxygenin incorporated into DNA sequences may be detected by monoclonal antibodies, again conjugated to reporter molecules including alkaline phosphatase. Thus, rather than the detection system relying on autoradiography, which is necessary for radiolabels, a series of reactions resulting in either a colour or a light or chemiluminescent reaction takes place. This has important practical implications since autoradiography may take 1–3 days, whereas colour and chemiluminescent reactions take minutes.

1.7.1 End Labelling of DNA Molecules

The simplest form of labelling DNA is by 5' or 3' end labelling; 5' end labelling involves a phosphate transfer or exchange reaction where the 5' phosphate of the DNA to be used as the probe is removed and in its place a labelled phosphate, usually ^{32}P, is added. This is usually carried out by using two enzymes; the first, alkaline phosphatase, is used to remove the existing phosphate group from the DNA. Following removal of the released phosphate from the DNA, a second enzyme, polynucleotide kinase, is added, which catalyses the transfer of a phosphate group (^{32}P-labelled) to the 5' end of the DNA. The newly labelled probe

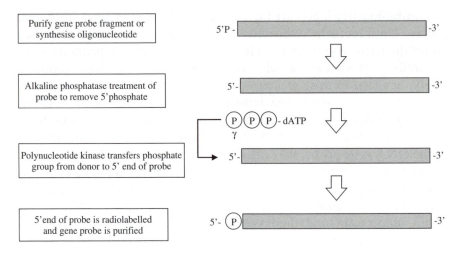

Figure 1.7 The steps involved in the production of a 5′-labelled gene probe.

is then purified, usually by chromatography through a Sephadex column, and may be used directly (Figure 1.7).

Using the other end of the DNA molecule, the 3′ end, is slightly less complex. Here a new dNTP which is labelled (*e.g.* [^{32}P]adATP or biotin-labelled dNTP) is added to the 3′ end of the DNA by the enzyme terminal transferase. Although this is a simpler reaction, a potential problem exists because a new nucleotide is added to the existing sequence and so the complete sequence of the DNA is altered, which may affect its hybridisation to its target sequence. End labelling methods also suffer from the fact that only one label is added to the DNA so they are of a lower activity in comparison with methods that incorporate label along the length of the DNA (Figure 1.8).

1.7.2   Random Primer Labelling

In random primer labelling the DNA to be labelled is first denatured and then placed under renaturing conditions in the presence of a mixture of many different random sequences of hexamers or hexanucleotides. These hexamers will, by chance, bind to the DNA sample wherever they encounter a complementary sequence and so the DNA will rapidly acquire an approximately random sprinkling of hexanucleotides annealed to it. Each of the hexamers can act as a primer for the synthesis of a fresh strand of DNA catalysed by DNA polymerase since it has an exposed 3′-hydroxyl group. The Klenow fragment of DNA polymerase is used for random primer labelling because it lacks a 5′–3′ exonuclease

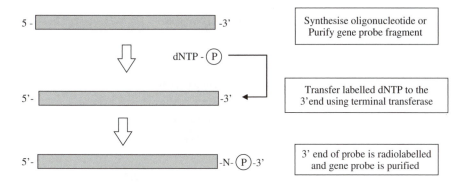

Figure 1.8 The steps involved in the production of 3′-labelled gene probe.

activity. This is prepared by cleavage of DNA polymerase with subtilisin, giving a large enzyme fragment which has no 5′ to 3′ exonuclease activity, but which still acts as a 5′ to 3′ polymerase. Thus, when the Klenow enzyme is mixed with the annealed DNA sample in the presence of dNTPs, including at least one which is labelled, many short stretches of labelled DNA will be generated (Figure 1.9). In a similar way to random primer labelling, the polymerase chain reaction may also be used to incorporate radioactive or non-radioactive labels.

1.7.3 Nick Translation

A traditional method of labelling DNA is by the process of nick translation. Low concentrations of DNase I are used to make occasional single-strand nicks in the double-stranded DNA that is to be used as the gene probe. DNA polymerase then fills in the nicks, using an appropriate deoxyribonucleoside triphosphate (dNTP), at the same time making a new nick to the 3′ side of the previous one. In this way, the nick is translated along the DNA. If labelled dNTPs are added to the reaction mixture, they will be used to fill in the nicks and so the DNA can be labelled to a very high specific activity (Figure 1.10).

1.8 THE POLYMERASE CHAIN REACTION

There have been a number of key developments in molecular biology techniques. However, one that has had the most impact in recent years has been the polymerase chain reaction (PCR). One of the reasons for the adoption of the PCR is the elegant simplicity of the reaction and relative ease of the practical manipulation steps. Frequently this is one of the first techniques to be used when analysing DNA and RNA and in

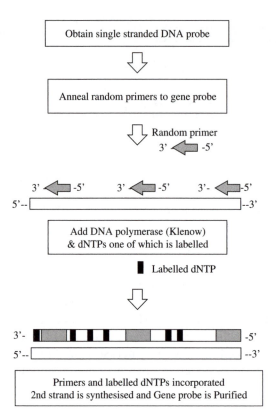

Figure 1.9 The steps involved in the production of a gene probe produced by the random hexamer method.

its quantitation it has opened up the analysis of cellular and molecular processes to those outside the field of molecular biology.

The PCR is used to amplify a precise fragment of DNA from a complex mixture of starting material, usually termed the template DNA, and in many cases requires little DNA purification. It does require the knowledge of some DNA sequence information that flanks the fragment of DNA to be amplified (target DNA). From this information, two oligonucleotide primers may be chemically synthesised, each complementary to a stretch of DNA to the 3′ side of the target DNA, one oligonucleotide for each of the two DNA strands. The result is an amplification of a specific DNA fragment which obviates the need for more time-consuming cloning procedures. The technique of the PCR is described in detail in Chapter 4. Further developments in molecular biology and biotechnology have allowed numerous genomes to be analysed and genes identified. It is not surprising that this has been aided

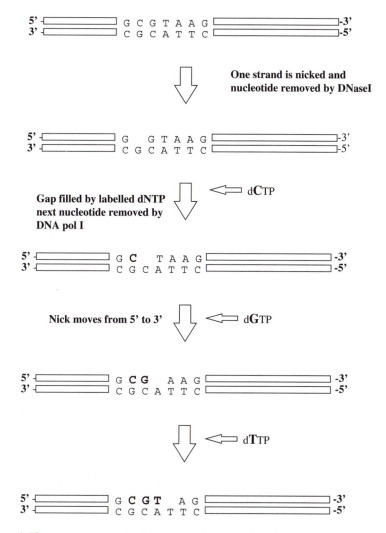

Figure 1.10 The steps involved in the production of a gene probe by the nick translation method.

by the developments in the area of bioinformatics and whole genome analysis.[13] DNA databases and other nucleic acid sequence and protein analysis software may all be accessed over the internet given the relevant software and authority (Table 1.3). This is now relatively straightforward with web browsers that provide a user-friendly graphical interface for sequence manipulation. Consequently, the new expanding and exciting areas of bioscience research are those that analyse genome and DNA sequence databases, (genomics) and also their protein counterparts

Table 1.3 Nucleic acid and protein database resources available over the Internet.

Database or Resource		URL (uniform resource locator)
General DNA Sequence Databases		
EMBL	European genetic database	http://www.ebi.ac.uk
GenBank	US genetic database	http://ncbi.nlm.nih.gov
DDBJ	Japanese genetic database	http://ddbj.nig.ac.jp
Protein Sequence Databases		
Swiss-Prot	European protein sequence database	http://expasy.hcuge.ch/ sprot/sprot-top.html
TREMBL	European protein sequence database	http://www.ebi.ac.uk/ pub/databases/trembl
PIR	US protein information resource	http://www-nbrf. gerogetown.edu/pir
Protein Structure Databases		
PDB	Brookhaven protein database	http://www.pdb.bnl.gov
NRL-3D	Protein structure database	http://www.gdb.org/ Dan/proteins/ nrl3d.html
Genome Project Databases		
Human Mapping Database Johns Hopkins, USA		http://gdbwww.gdb.org
dbEST (cDNA and partial sequences)		http://www.ncbi.nih.gov
Genethon Genetic maps based on repeat markers		http://www.genethon.fr
Whitehead Institute (YAC and physical maps)		http://www-genome. wi.mit.edu

(proteomics). This is sometimes referred to as *in silico* research, and there is no doubt that for basic and biotechnological research it is as important to have Internet and database access as it is to have equipment and reagents for laboratory molecular biology.[14]

REFERENCES

1. H. O. Smith and K. W. Wilcox, *J. Mol. Biol.*, 1970, **51**, 379.
2. K. Wilson and J. M. Walker, *Principles and Techniques of Practical Biochemistry*, Cambridge University Press, Cambridge, 2005.
3. R. E. Farrell, *RNA Methodologies. A Laboratory Guide for Isolation and Characterisation*, Academic Press, New York, 2005.
4. R. Rapley and D. L. Manning, *RNA Isolation and Characterisation Protocols*, Humana Press, Totowa, NJ, 1998.
5. P. Jones, *Gel Electrophoresis of Nucleic Acids*, Wiley, Chichester, 1995.

6. G. J. S. Jenkins, in *Molecular Biomethods Handbook*, 2nd edn, ed. R. Rapley and J. M. Walker, Humana Press, Totowa, NJ, 2008, pp. 17–27.

7. L. Chen, A. Manz and P. J. Day, *Lab Chip*, 2007, **7**, 1413.

8. E. M. Southern, *J. Mol. Biol.*, 1975, **98**, 503.

9. J. C. Alwine, D. J. Kemp and G. R. Stark, *Proc. Natl. Acad. Sci. USA*, 1977, **74**, 5350.

10. M. Aquino de Muro, in *Molecular Biomethods Handbook*, 2nd edn, ed. R. Rapley and J. M. Walker, Humana Press, Totowa, NJ, 2008, pp. 41–54.

11. G. J. Hannon, *RNAi. A Guide to Gene Silencing*, Cold Spring Harbor Laboratory Press, Cold Spring Harbor, NY, 2003.

12. T. P. McCreery and T. R. Barrette, in *Molecular Biomethods Handbook*, ed. R. Rapley and J. M. Walker, Humana Press, Totowa, NJ, 1998, pp. 73–76.

13. A. D. Baxevanis and B. F. Ouelette, *Bioinformatics. A Practical Guide to the Analysis of Genes and Proteins*, Wiley, Chichester, 2004.

14. D. W. Mount, *Bioinformatics Sequence and Genome Analysis*, Cold Spring Harbor Laboratory Press, Cold Spring Harbor, NY, 2004.

CHAPTER 2

Molecular Cloning and Protein Expression

STUART HARBRON

The Enzyme Technology Consultancy, 44 Swing Gate Lane, Berkhamsted, Hertfordshire HP4 2LL, UK

2.1 INTRODUCTION

Molecular cloning is a process for manipulating DNA that differs from cell cloning and whole animal cloning (although some of the steps involved in the process are common). Molecular cloning is used not only for gaining a better understanding of the structure, function and control of genes and their gene products, but also for commercial exploitation of proteins. Producing recombinant proteins in forms that are biologically useful is a key challenge to the pharmaceutical industry and molecular cloning leading to expression of proteins of interest is the main focus of this chapter.

Bacterial expression systems are highly attractive in this respect for a number of reasons, including:

- their rapid growth rates;
- their ability to use relatively inexpensive substrates;
- their well-characterised genetics;
- the availability of a large number of cloning vectors; and
- a variety of mutant host strains.

Molecular Biology and Biotechnology, 5th Edition
Edited by John M Walker and Ralph Rapley
© Royal Society of Chemistry 2009
Published by the Royal Society of Chemistry, www.rsc.org

Production of proteins 'requires the success of three individual factors: expression, solubility and purification'.[1] While bacterial expression systems have a number of serious drawbacks, many of these have been more or less solved over the last 20 years.[2,3] The challenge now is to produce the protein in good yield and in the right form.

2.2 HOST-RELATED ISSUES

It can be difficult to decide which host and promoter system is most suitable for heterologous protein production and the nature of the protein to be expressed is often a key factor determining successful production of the protein. A rational approach to protein expression based on the properties and provenance of the protein of interest, and then deciding on which host might be the better for its expression, remains unavailable. Rather, the approach has been in the other direction: developing approaches for molecular cloning and protein expression in commonly used host organisms. It therefore makes sense to begin with host-related considerations.

Many bacterial hosts have been optimized for heterologous protein production, partly in an attempt to identify a more or less universal system with few problems. In spite of all this work, the Gram-negative bacterium *E. coli* is the most commonly used organism for heterologous protein production, mainly because this organism is very well known and established. Hence it is no surprise that *E. coli* systems are also most commonly used for industrial and pharmaceutical protein production. Even so, 'the production of soluble proteins in *E. coli* remains a hit-or-miss affair'.[1]

The most popular hosts are *E. coli*, *B. subtilis*, yeast and cultured cells of higher eukaryotes such as insect or mammalian cells. *E. coli* is frequently used because the very large body of information available makes it relatively well understood and there are well-characterised protocols for manipulating this microbe. However, there are many proteins for which *E. coli* is not the ideal host for expression, including proteins having more than 500 amino acids, those which are highly hydrophobic, proteins having many cysteines (because the reducing environment in *E. coli* prevents the formation of disulfide bonds) and those requiring post-translational modification or other treatments.

If the protein of interest is from a eukaryotic organism, as it often is for a protein having commercial interest, then there are immediately three problems associated with expressing it in a prokaryotic system such as *E. coli*, and these problems relate to the difference in the mechanism of gene expression between the two systems.

First, bacteria are not capable of processing RNA to remove introns. Fortunately, this can be more or less easily overcome by generating double-stranded DNA copies of mRNA molecules isolated from the eukaryotic organism by using the mRNA as a template with a reverse transcriptase. This double-stranded copy, or cDNA, will not contain introns and can act as the coding sequence in expression vectors. Converting the eukaryotic sequence of interest is not entirely without its drawbacks, the most serious of which is if the mRNA is only present as a small constituent of a eukaryotic cell's mRNA population, because purification of the mRNA can be difficult. Another potential issue is that random termination of reverse transcription prior to completion of complementary strand synthesis can occur, which means that the cDNA sequence does not always include the 5' end of the gene. The problem of introns has also been addressed by synthesising fragments of the gene chemically and subsequent ligation, but this presupposes that the amino acid sequence of the protein of interest is known.[4–7]

Second, the RNA polymerase of a prokaryotic host will not bind to and transcribe the gene encoding the protein of interest unless it has an appropriate promoter sequences upstream of the coding region. Since the sequence and position of promoters are specific to each host, the choice of promoter is vital for correct and efficient transcription. Although many promoter sequences for *E. coli* are known, not a large number of them are useful. To be useful as tools for protein expression, the promoter must be strong, have a low basal expression, be easily transferred, be easily and economically induced and be unaffected by commonly used ingredients in culture media. Basal transcription, which is transcription in the absence of the inducer, can be dealt with through the use of a suitable repressor: this is especially important if the expression target introduces cellular stress, which would select for plasmid loss. Either thermal or chemical triggers can be used to initiate promoter induction and some commonly used systems are listed in Table 2.1.[8]

Finally, prokaryotic ribosomes will not bind to the mRNA produced by transcription unless there is a ribosome-binding site (RBS) on the mRNA, just before the coding region. Initiation of translation can be a significant limiting factor in expression of cloned genes.[16] Translation initiation from the translation initiation region of the transcribed messenger RNA requires an RBS and a translation initiation codon.[17] Efficiency of translation initiation is influenced by the codon following the initiation codon and abundant adenine seems to lead to highly expressed genes.[18] In addition to an initiation codon, AUG, other nucleotides, particularly in the 5' untranslated leader of the mRNA, are needed to create suitable secondary and tertiary structures in mRNA

Table 2.1 Some promoter systems for *E. coli*.[8]

Expression level	System	Induction	Cost	System
+ +	λ P_L promoter[9]	Δt	0	Invitrogen pLEX
+ → + +	*lac* promoter[10]	IPTG	+ + +	
+ +	*trc, tac* promoter[11]	IPTG	+ + +	GE Lifesciences pTrc, pGEX
+ → + + +	*araBAD* promoter (P_{BAD})[12]	L-Arabinose	+	Invitrogen pBAD
+ → + + +	rhaP_{BAD}[13]	L-Rhamnose	+ + +	
+ + → + + +	*tetA* promoter/ operator[14]	Anhydrotetracycline	+	
+ + + +	T7 RNA polymerase[15]	IPTG	+ + +	Novagen pET

and facilitate interaction between the mRNA and the ribosome. Perhaps the best known of these sequences is the Shine–Dalgarno (S–D) sequence, which is essential for translation. This sequence is located 7 ± 2 nucleotides upstream from the initiation codon, which is the canonical AUG in efficient recombinant expression systems.[19] It allows a complex to form between the mRNA and the 30S subunit of the ribosome via hydrogen bonding to the 16S rRNA. Not all *E. coli* mRNAs have an identical S–D sequence but a consensus can be identified. Optimal translation initiation is obtained from mRNAs with the SD sequence UAAGGAGG. The RBS secondary structure is highly important for translation initiation and efficiency is improved by high contents of adenine and thymine.[20]

Other sequences that increase the level of translation include translation enhancers such as the Epsilon sequence in the g10L ribosome-binding site of phage T7.[21,22]

A transcription terminator placed downstream from the sequence encoding the target gene enhances plasmid stability by preventing transcription through the origin of replication and from irrelevant promoters located in the plasmid. Transcription terminators stabilize the mRNA by forming a stem loop at the three prime end.[23]

Translation termination is preferably mediated by the stop codon UAA in *E. coli*. Increased efficiency of translation termination is achieved by insertion of consecutive stop codons or the UAAU stop codon.[24]

To obtain expression of foreign genes in *E. coli*, it is necessary to incorporate ribosome-binding motifs into the recombinant DNA molecule. Furthermore, some sequences (such as the S–D sequence) must be located at an optimal distance from the translation start codon. This is most readily achieved by construction of fusion genes where an entire untranslated leader and 5′ coding sequence from a naturally occurring gene is present. Nonetheless, all expression cassettes need to be tested thoroughly and sequences reorganized if necessary to optimise translation initiation.

2.3 VECTORS

To counter some of the issues related to the capabilities of the host, expression vectors have been developed which contain promoter and ribosome-binding sites positioned just before one or more sites for restriction endonucleases to allow the insertion of foreign DNA. These regulatory sequences, such as that from the *lac* operon of *E. coli*, are usually derived from genes which, when induced, are strongly expressed in bacteria. Since the mRNA produced from the gene is read as triplet codons, the inserted sequence must be placed so that its reading frame is in phase with the regulatory sequence. Experimentally this can be achieved by using three vectors which differ only in the number of bases between promoter and insertion site, the second and third vectors being respectively one and two bases longer than the first. When the insert is cloned using all three vectors and the resulting clones can be screened for the production of a functional foreign protein, it should be in the correct reading frame in one of them.

Expression vectors are DNA constructs that are stably maintained and propagated in a host. For a typical bacterial host, such as *E.coli* which grows and divides rapidly, the expression vector is ideally derived from a gene which, when induced, is strongly expressed. Expression vectors vary in their complexity, ease of manipulation and the length of DNA sequence they can accommodate (the insert capacity). Vectors have in general been developed from naturally occurring entities such as bacterial plasmids, bacteriophages or combinations of their constituent elements, such as cosmids. For applications to do with the expression of proteins, plasmids are the most important of these. A plasmid is an autonomously replicating, extrachromosomal circular DNA molecule, distinct from the normal bacterial genome and non-essential for cell survival under non-selective conditions. Some plasmids are capable of integrating into the host genome. Genes carried by plasmids often include those for conferring antibiotic resistance, to allow conjugation or for the metabolism of 'unusual' substrates. These are attractive candidates for modification for use as vectors, particularly if they are replicated at a high rate and are not easily 'lost' from the host in non-selective conditions. It is clear from the previous section that a

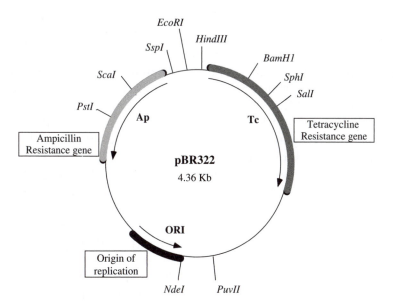

Figure 2.1 Map and important features of pBR322, including restriction sites.

number of key elements are more or less essential to the design of these vectors.[25,26]

One of the more successful plasmids is pBR322 (Figure 2.1), which has been widely used, has a number of desirable key features, which are further discussed below:

- it is small (much smaller than a natural plasmid);
- it has a relaxed origin of replication;
- two genes coding for resistance to antibiotics; and
- single recognition sites for a number of restriction enzymes at various points around the plasmid.

The small size means that it is resistant to damage by shearing and is efficiently taken up by bacteria, a process termed transformation.

A relaxed, as opposed to stringent, origin of replication means that it is not tightly linked to cell division and plasmid replication will happen far more frequently than chromosomal replication, leading to a large number of plasmid molecules per cell[25] and any vector with a replication origin in *E. coli* will replicate (together with any incorporated DNA) more or less efficiently. In stringent regulation, replication is in synchrony with cell division. The replicon may also have associated *cis* acting elements.[27] The origin of replication is most commonly ColE1, as in pBR322 (copy number 15–20) or pUC (copy number 500–700) or p15A, as in pACYC184 (copy number 10–12). These multi-copy plasmids are stably replicated and maintained under selective conditions and plasmid-free daughter cells are rare.[28] Different replicon incompatibility groups and drug resistance markers are required when multiple plasmids are employed for the co-expression of gene products. Derivatives containing ColE1 and p15A replicons are often combined in this context since they are compatible plasmids,[29] meaning that they may be stably maintained in the same cell.[30]

One of the antibiotic resistance genes allows cells that contain the plasmid to be selected: if cells are plated on medium containing an appropriate antibiotic, only those that contain plasmid will grow to form colonies. The other resistance gene can be used, as described below, for detection of those plasmids that contain inserted DNA. The most common drug resistance markers in recombinant expression plasmids confer resistance to ampicillin, kanamycin, chloramphenicol or tetracycline. Plasmid-mediated resistance to ampicillin is accomplished by expression of β-lactamase from the *bla* gene. This enzyme is secreted to the periplasm, where it catalyses hydrolysis of the β-lactam ring. Ampicillin present in the cultivation medium is especially susceptible to degradation, either by secreted β-lactamase or acidic conditions in high-density cultures. The latter effect can be alleviated by the use of analogues that are less susceptible to degradation, such as carbenicillin or by kanamycin, chloramphenicol or tetracycline, which interfere with protein synthesis by binding to critical areas of the ribosome. Kanamycin is inactivated in the periplasm by aminoglycoside phosphotransferases and chloramphenicol by the *cat* gene product, chloramphenicol acetyl transferase. Various genes confer resistance to tetracycline.[31]

Recognition sites for restriction enzymes are used to open or linearise the circular plasmid. Linearising a plasmid allows a fragment of DNA to be inserted and the circle closed. The variety of sites not only makes it easier to find a restriction enzyme which is suitable for both the vector and the foreign DNA to be inserted, but also, since some of the sites are placed within an antibiotic resistance gene, the presence of an insert can be detected by loss of resistance to that antibiotic. This is termed insertional inactivation.

The protocol utilised for using a plasmid such as pBR322 to introduce DNA encoding the protein of interest into the host cell is summarised below.

First, a fragment of DNA encoding the protein of interest and digested with *Bam*H1 is isolated and purified or produced via polymerase chain reaction (PCR). Plasmid pBR322 is also treated with *Bam*H1 and both are deproteinised to inactivate the restriction enzyme. Since *Bam*H1 cleaves to give sticky ends, the plasmid and digested DNA fragments can be ligated using T4 DNA ligase. This yields a plasmid containing a single fragment of the DNA as an insert, but the mixture will also contain products, such as plasmid which has recircularised without an insert, dimers of plasmid, fragments joined to each other and plasmid with an insert composed of more than one fragment. Most of these unwanted molecules are be eliminated during subsequent steps. The products of such reactions are usually identified by agarose gel electrophoresis.

Second, host *E. coli* is transformed using the ligated DNA plasmid. Bacteria termed competent can be induced to take up DNA from their surroundings by prior treatment with Ca^{2+} at 4 °C followed by a brief increase in temperature, termed heat shock. Plasmid DNA added to the suspension of competent host cells will thus be imported during this process. Small, circular molecules are taken up most efficiently, whereas long, linear molecules will not enter the bacteria.

Third, after a brief incubation to allow expression of the antibiotic resistance genes, the cells are plated on to medium containing the antibiotic (*e.g.* ampicillin). Any colonies that grow are obviously derived from cells that contain plasmid, since this carries the gene for resistance (to ampicillin).

Fourth (Figure 2.2), to distinguish between those colonies containing plasmids with inserts of the DNA encoding the protein of interest and those that simply contain recircularised plasmids, the colonies are replica plated, using a sterile velvet pad, on to plates containing tetracycline in their medium. The plasmid carries the tetracycline resistance gene, but the *Bam*HI site lies within this gene, which means that the plasmid will show insertional inactivation in the presence of insert, but will be intact in those plasmids that have merely recircularised. Hence colonies that grow on ampicillin but not on tetracycline must contain plasmids with inserts. Since replica plating gives an identical pattern of colonies on both sets of plates, it is straightforward to recognise the colonies with inserts and to recover them from the ampicillin plate for further growth. This illustrates the importance of a second gene for antibiotic resistance in a vector.

The fourth step can be omitted if, prior to ligation in the first step, the mixture is treated with the enzyme alkaline phosphatase, which removes 5'-phosphate groups essential for ligation. Following ligation between the 5'-phosphate of insert and the 3'-hydroxyl of plasmid, only recombinant plasmids and chains of linked DNA fragments will be formed. It does not matter that only one strand of the recombinant DNA is ligated, since the nick will be repaired by bacteria transformed by the

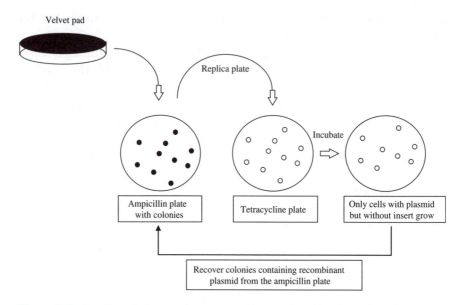

Figure 2.2   Replica plating to detect recombinant plasmids. A sterile velvet pad is
pressed on to the surface of an agar plate, picking up some cells from each
colony growing on that plate. The pad is then pressed on to a fresh agar
plate, thus inoculating it with cells in a pattern identical with that of the
original colonies. Clones of cells that fail to grow on the second plate (*e.g.*
owing to the loss of antibiotic resistance) can be recovered from their
corresponding colonies on the first plate.

modified plasmid. Including this step increases the yield of recombinant
plasmid containing inserts.

A variety of plasmids based on pBR322 have been developed, including
a series of plasmids termed pUC (Figure 2.3) and pBAD (see below). In
these, the most popular restriction sites are concentrated into a region
termed the multiple cloning site or MCS, which is part of the gene
encoding β-galactosidase. When the pUC plasmid has been used to
transform the host cell, *E. coli*, the gene is 'switched on' by adding the
inducer IPTG (isopropyl-β-D-thiogalactopyranoside) and the enzyme
β-galactosidase is produced. This enzyme hydrolyses a colourless substance called X-gal (5-bromo-4-chloro-3-indolyl-β-galactopyranoside),
leading to the precipitation of a blue insoluble material. However, disruption of the gene by the insertion of DNA encoding the protein of
interest means that X-gal is not hydrolysed. This means that a host cell
having a pUC plasmid carrying DNA encoding the protein of interest will
be white or colourless in the presence of X-gal, whereas a host cell having
an intact non-recombinant pUC plasmid will be blue since its gene is fully
functional and not disrupted. This approach, termed blue/white selection,

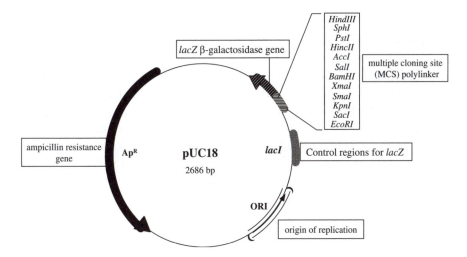

Figure 2.3 Map and important features of pUC18, including restriction sites.

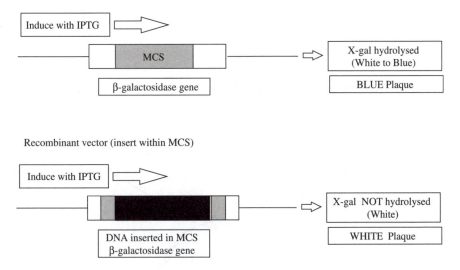

Figure 2.4 Principle of blue/white selection for the detection of recombinant vectors.

allows rapid initial identification of recombinant host cells and has been included in a number of later vector systems (Figure 2.4).

These approaches detect not only host cells containing a plasmid carrying the DNA encoding the protein of interest, but also host cells in which insertional inactivation of antibiotic resistance genes has happened as a result of the misincorporation of the DNA insert.

2.4   EXPRESSION SYSTEMS

Fortunately, not only have plasmid vectors been developed which contain promoter and ribosome-binding sites positioned just before one or more restriction sites that allow the facile insertion of the DNA encoding the protein of interest, they have also been made available commercially. Indeed, a wide range of expression systems based on different promoters is available.[3,32] Table 2.1 shows some of the commonly used systems, including the T7-based pET expression system (commercialised by Novagen) and one based on the araBAD promoter (*e.g.* Invitrogen pBAD), which are discussed below.

2.4.1   The pET Expression System

This system, which includes hybrid promoters, multiple cloning sites for the incorporation of different fusion partners and protease cleavage sites, has been developed for a variety of expression applications.[33,34] The pET plasmid, shown in Figure 2.5, is a 5.4 kb construct having the following elements:

- *lacI*   codes for the *lac* repressor protein
- *amp^R*   codes for ampicillin resistance
- *ori*   *col E1* origin of replication
- *lacO*   codes for the *lac* operon

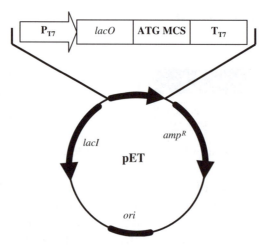

Figure 2.5   pET plasmid.

- P_{T7} codes for the T7 promoter, which is specific only for T7 RNA polymerase
- T_{T7} codes for the T7 terminator
- *MCS* multiple cloning site, has the sequence encoding the protein of interest.

A key feature is the use of P_{T7}, a 20-nucleotide sequence that is not recognized by the *E. coli* RNA polymerase, nor does T7 RNA polymerase occur in the prokaryotic genome sequence; hence in the absence of T7 RNA polymerase, P_{T7} is not activated and the protein of interest is not produced. When P_{T7} is activated, as described below, being a viral promoter it transcribes rapidly, with a maximum speed of around 230 nucleotides per second, some five times faster than *E. coli* RNA polymerase.

Expression requires a host strain lysogenised by a DE3 phage fragment, encoding the T7 RNA polymerase, under the control of the IPTG-inducible *lacUV5* promoter, and Figure 2.6 shows a schematic of the process. A copy of the *lacI* gene is present on the *E. coli* genome and on pET. *LacI* is a weakly expressed gene and a 10-fold enhancement of the repression is achieved when the overexpressing promoter mutant *LacIq* is employed.[35] The *lac* repressor protein, *LacI*, represses the *lacUV5* promoter of the host cell and the *T7/lac* hybrid promoter encoded by the expression plasmid. In the absence of an inducer then, the *lacI* tetramer binds to the *lac* operator on both the host cell genome and the plasmid. This prevents the host cell from producing T7 RNA polymerase and prevents the plasmid from producing the protein of interest.

When the inducer, typically IPTG, is introduced, it binds and triggers the release of tetrameric *lacI* from the *lac* operator on both the genome and the plasmid, which triggers the expression of T7 RNA polymerase in the host cell. Transcription of the target gene from the T7/lac hybrid promoter is thus initiated.

There can be low background expression from pET expression plasmids; this may be reduced by co-expression of T7 lysozyme, a natural inhibitor of T7 RNA polymerase, using either plasmid pLysS or pLysE. These plasmids harbour the T7 lysozyme gene in silent (pLysS) and expressed (pLysE) orientations, with respect to the cognate tetracycline responsive (Tc) promoter.[36]

Although the *lacUV5* promoter is less sensitive to regulation by the cAMP–CRP (cAMP receptor protein) complex than the *lac* promoter, incorporation of 1% glucose in the cultivation medium reduces cAMP levels and enhances repression of the promoter significantly. Host

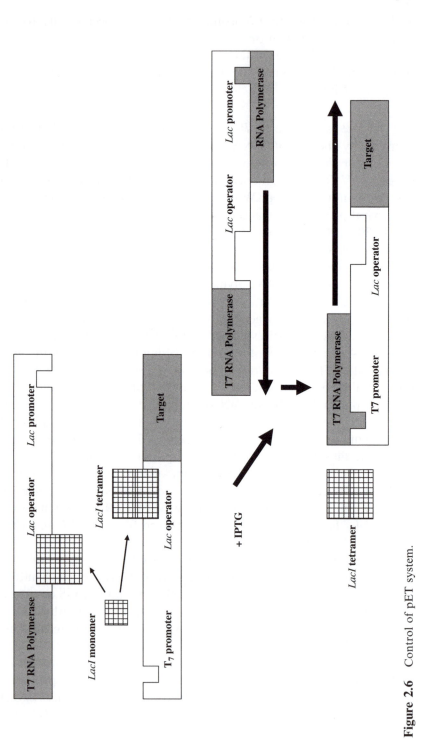

Figure 2.6 Control of pET system.

strains deficient in the *lacY* gene, which encodes lactose permease, improves control of target protein expression.[37]

2.4.2 The pBAD Expression System

Regulation of the arabinose operon in *E. coli* is directed by the product of the *araC* gene,[38] which controls the synthesis rate of the AraE, AraF and AraG proteins, required for arabinose uptake, as well as the AraB, AraA and AraD enzymes, required for its catabolism. That means that the intrinsic state of the *ara*-specific promoters is off and AraC turns them on, whereas the set state of the *lac* operon promoter is on and the *lac* repressor turns it off. In addition, pBAD is catabolite repressed, which means that growing the culture in the presence of glucose will further repress expression.

While expression of a cloned gene from plasmids containing the araBAD promoter can be modulated over several orders of magnitude in cultures grown in the presence of sub-saturating concentrations of arabinose,[12] individual cells are either fully induced or uninduced.[39] Cells having the natively controlled arabinose transport gene (araE) are either induced or uninduced, the relative fraction of which is controlled by the concentration of arabinose. The population-averaged variation in expression from pBAD as a function of inducer concentration is proportional to the percentage of cells that are fully induced (versus uninduced) rather than the level of expression in individual cells. This all-or-none phenomenon, which can have undesirable effects on the expression of heterologous genes, can be eliminated in *E. coli* by expression of araE from arabinose-independent promoters. In these arabinose-transport engineered cells, all cells in the population have approximately the same induction level.[40] Strains capable of transporting L-arabinose, but not metabolising it, such as a recA, endA strain, are therefore most suitable.

Expression plasmids based on the araBAD promoter are designed for tight control of background expression and precise control of the expression levels of the target protein.[12] This contrasts with the all-or-nothing induction achieved by most other bacterial expression systems.[41] A pBAD plasmid, derived from pBR322 and shown in Figure 2.7, is a 4.1 kb construct having the following elements:

- *araC*   ORF encoding *araC* protein
- *amp*^{R} codes for ampicillin resistance
- *ori*   *col E1* origin of replication
- *pBAD*   *araBAD* promoter

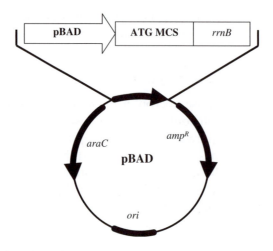

Figure 2.7 pBAD plasmid.

- *rrnB*   transcription termination region
- *MCS*   multiple cloning site, has the sequence encoding the protein of interest.

The AraC dimer binds three sites in the arabinose operon, I_1, I_2 and O_2 (Figure 2.8). In the absence of arabinose, the AraC dimer contacts the O_2 site located within the *araC* gene, 210 base pairs upstream from pBAD. The other half of the *araC* dimer contacts the I_1 site in the promoter region forming the DNA loop as shown. Transcription from pBAD and the *araC* promoter (pC) is thus inhibited by the loop. Upon binding of arabinose, the *araC* dimer changes its conformation so that it binds to the I_2 site of pBAD instead of the O_2 site. This removes the loop structure and transcription by RNA polymerase initiates. Binding of the *araC* dimer to the I_1 and I_2 sites is stimulated by cAMP receptor protein (CRP), which means that background expression from araBAD can be reduced by glucose-mediated catabolite repression.[12]

2.5 PROBLEMS

Most problems relating to expression of heterologous proteins are probably to do with the differences between codon usage in eukaryotes and *E. coli*.[8] Codon usage in *E. coli* is reflected by the level of cognate aminoacylated tRNAs available and minor or rare codons tend to be genes expressed at low levels. Codons rare in *E. coli* are often abundant in heterologous genes from eukaryotes,[42] which tend to be the proteins of

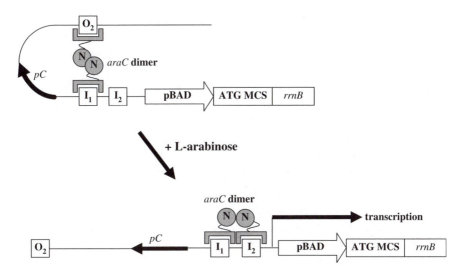

Figure 2.8 Control of pBAD system.

Table 2.2 Some rare codons in *E. coli.*

Codon	Amino acid	Frequency (per 1000) (E. coli)	tRNA gene	Frequency (per 1000) (H. sapiens)
AGG	Arg	2.1	*arg*U	12.0
CGA	Arg	2.4	*arg*W	6.2
AGA	Arg	2.4	*arg*U	12.2
CCC	Pro	2.4	*pro*L	19.8
CUA	Leu	3.4	*leu*W	7.2
AUA	Ile	5.0	*ile*X	7.5
CGG	Arg	5.0	*arg*W, *arg*X	11.4
GGA	Gly	8.2	*gly*T	16.5
AAG	Lys	8.8	*lys*	31.9

commercial interest. What this means is that attempts to express genes containing rare codons can result in translational errors due to ribosomal stalling at positions on the messenger RNA that require incorporation of amino acids coupled to minor codon tRNAs. Insufficient tRNA pools can lead to translational stalling, premature translation termination, translation frameshift and amino acid misincorporation.[43]

The most problematic codons[32] are shown in Table 2.2. A number of approaches have been used in an attempt to reduce problems associated with codon bias, such as co-transforming the host with a plasmid harbouring a gene cognate with the problematic codons.[44] Several plasmids are available for rare tRNA co-expression, most of which are based on the p15A replication origin, which permit maintenance in the presence of the ColE1 origin. Commercially available *E. coli* strains for protein expression include Rosetta 2 host strains from Novagen, which are BL21 derivatives designed to enhance the expression of eukaryotic proteins that contain codons rarely used in *E. coli*. These strains supply tRNAs for seven rare codons (AGA, AGG, AUA, CUA, GGA, CCC and CGG) on a compatible chloramphenicol-resistant plasmid. The tRNA genes are driven by their native promoters. Stratagene offer BL21-CodonPlus-RIL chemically competent cells that carry extra copies of the *argU*, *ileY* and *leuW* tRNA genes. The tRNAs encoded by these genes recognise the AGA/AGG, AUA and CUA codons, respectively. The presence of these additional tRNA genes resolves the issue of codon bias for organisms whose genome is AT-rich.

Another issue affecting efficient protein expression[32] is the stability of mRNA in *E. coli*. Exononucleases RnaseII and PNPase and the endo-nuclease RnaseE act to degrade mRNA, which can have a half-life of between seconds and 20 min. Strains are available, such as the BL21 Star strain from Invitrogen, which have a mutation in *rne131*, the gene encoding RnaseE; stability of mRNA transcripts in this strain is thus significantly improved. Another approach is to introduce 5' and 3' stabi-lising structures to reduce damage from exonuclease attack, such as by fusion to a sequence encoding green fluorescent protein.[45]

Having addressed these issues and got the protein of interest to be expressed, most frequently the cytoplasm is the first choice for hetero-logous protein production because the higher yield seems to be more attractive, even though good yields of secreted proteins are well docu-mented.[46] Particularly with the latter route, protease digestion of the protein can be problematic and although for routine protein expression *E. coli* BL21 and K12 and their derivatives are most frequently used, BL derivatives that are *lon*[47] and *ompT* protease deficient may be better.

A further problem reducing the efficiency of protein expression is inclusion body formation. Whether these form through a passive event occurring through hydrophobic interactions between unfolded poly-peptide chains or by specific clustering mechanisms is unknown.[48] However, inclusion bodies do not appear to be inert aggregates, but may be an unbalanced equilibrium between *in vivo* protein aggregation and solubilisation,[49] hence formation may be minimised by reducing the culture temperature to slow host metabolism and expression rate or addition of sorbitol, betaine, sucrose or raffinose to the culture medium. Alternatively, the DNA encoding the protein of interest may be mod-ified to include solubility-enhancing tags (see below) or by coexpression of plasmid-encoded chaperones;[26] molecular chaperones may also be used as part of a refolding strategy.[50] The idea here is that as newly synthesised polypeptide is produced on the ribosome, it associates with a trigger factor chaperone, which prevents hydrophobic patches on the polypeptide from interacting.[51] Once the trigger factor has been removed, the folding proceeds to yield the protein in a natural state. Other chaperones enhance protein degradation (DnaK and GroEL).

Overexpression of toxic proteins that are difficult or impossible to produce in bacteria can be achieved using modified strains, such as OverExpress C41 and C43 from Avidis, which are based on *E. coli* DE3.[52]

Finally, *E. coli* accumulates lipopolysaccharide (LPS) or endotoxin, which is pyrogenic in humans and other mammals; this is problematic if the protein of interest is intended for therapeutic use and it must be purified in a second step to become endotoxin free.[53]

2.6 FUSION PROTEINS

Recalling the comment that production of proteins requires the success of three individual factors, expression, solubility and purification,[1] it is clear from what has been covered so far that protein expression, although technically reasonably straightforward, does not easily yield useful protein products – the protein may be inactive, precipitated, incomplete, *etc.*

Production of the desired protein as a fusion product with another protein has become commonplace. Initially it was unavoidable, a consequence of the expression system used, but the approach has evolved onwards from merely assisting with expression to facilitate protein purification and subsequently discovering that some affinity tags also increased solubility of the protein.[54–58]

Other unlooked for benefits of using a fusion protein partner include the observation that expression levels associated with an N-terminal fusion partner can be transferred to the partner, possibly as a result of mRNA stabilisation.[45] Fusion proteins have also been used to reduce intracellular proteolysis.[59]

2.6.1 Solubility-enhancing Tags

One approach to investigate what impact a solubility-enhancing tag may have on a protein of interest is to observe the fluorescence of *E. coli* cells expressing GFP fused to a gene corresponding to the protein of interest; this has been found to correlate well with the solubility of the protein when expressed alone.[60] This screening approach was used in a directed evolution approach to produce soluble forms of three 'insoluble' proteins for genomic structural studies. The three proteins were *Pyrobaculum aerophilum* methyl transferase, tartrate dehydratase β-subunit and nucleoside diphosphate kinase, and were 50, 95 and 90% soluble, respectively.[61] The approach was also used to identify solubilising inter-action partners for 'insoluble' targets such as integration host factor β.[62]

Although some solubility-enhancing tags appear to work better than others, there is no guarantee that one of these tags will work with the protein of interest. Solubility-enhancing and purification-assisting fusion partners[1,56] that have been described or are in use are shown in Table 2.3.

Numerous solubility tags are listed in the table, but most work has utilised just a few of these, notably maltose-binding protein (MBP), N-utilization substance A (NusA), thioredoxin (Trx) and glutathione-*S*-transferase (GST). Both MBP and GST have an additional benefit in that they can function as purification-facilitating (affinity) tags; MBP

Table 2.3   Solubility-enhancing and Purification-facilitating tags.[1,56]

Tag	Protein	Source organism	Affinity matrix	Ref.
BAP	Biotin acceptor peptide		Avidin	63
CBP	Calmodulin-binding peptide		Calmodulin	64
DsbC	Disulfide bond C	*Escherichia coli*		65
FLAG	FLAG tag peptide		Anti-FLAG antibody	66
GB1	Protein G B1 domain	*Streptococcus* sp.		67
GST	Glutathione *S*-transferase	*Schistosoma japonicum*	Glutathione	68
				55
His$_6$	Hexahistidine tag		Metal chelates	69
MBP	Maltose-binding protein	*Escherichia coli*	Amylose	70
				71,72
NusA	N-Utilization substance	*Escherichia coli*		57
SET	Solubility-enhancing tag	Synthetic		73
Skp	Seventeen kilodalton protein	*Escherichia coli*		74
Strep-II	Streptavidin-binding peptide		Streptavidin	75
	SUMO	Small ubiquitin modifier	*Homo sapiens*	76
T7PK	Phage T7 protein kinase	Bacteriophage T7		74
Trx	Thioredoxin	*Escherichia coli*		38
ZZ	Protein A IgG ZZ repeat domain	*Staphylococcus aureus*	IgG	77

binds strongly to amylose resin[70] and GST binds to glutathione resin.[68] However GST is, at best, a poor solubility enhancer[78,79] in *E. coli*, while a significant body of evidence exists to show that N-terminal MBP fusions can frequently produce soluble proteins when the unfused partners are insoluble.[79–82] Although it has no independent purification-facilitating functionality, NusA protein from *E. coli* provides solubility enhancement comparable to MBP[83–85] and thioredoxin from *E. coli* has been reported in several studies to be nearly as efficient as MBP in promoting solubility,[78,83] although other studies have shown it to be less effective.[76] MBP and NusA have been used to enhance the solubilisation of ScFv antibodies in *E. coli*.[86,87] Some maltodextrin-binding proteins from other bacteria provide an even greater enhancement of solubilisation than MBP.[54] The SUMO tag, a ubiquitin-related protein, has been reported to enhance solubility and in some cases appears to be as effective as MBP.[76] A fragment of the bacteriophage T7 protein kinase gene (T7PK) not only functions as a solubility enhancer, but also appears to enhance overall levels of expression.[74]

As mentioned above, stress-induced proteins can serve to reduce inclusion body formation and thereby enhance the overall yield of an expressed protein. Two such proteins, which can be expressed in a *cis* fashion as fusion proteins, are RpoA (DNA-directed RNA polymerase α-subunit)[88] and Tsf (an elongation factor).[89] These are believed to shield active surfaces of heterologous proteins associated with non-specific protein–protein interactions that lead to the formation of inclusion bodies. The proteins were fused to nine aggregation-prone human proteins, leading to enhanced yields. They were also used for the expression of cutinase from *P. aeruginosa*.

Even with the expression systems noted above, problems still remain. Low expression yields and poor refolding efficiency of small recombinant proteins expressed in *E. coli* hinder the large-scale purification of such proteins for structural and biological investigations.[67] MBP and NusA are rather large partners; an N-terminal fragment of translation initiation factor IF2, which has a molecular weight of 17.4 kDa, has been used.[58] The advantages of using a smaller partner include the following:

- reduced energy demands on the host cell;
- diminished steric hindrance.

Recently, a His$_6$-tagged N-terminal fragment (52 residues) of *Staphylococcal nuclease* R (HR52) has been selected as a smaller efficient fusion partner for the expression of small peptides.[90] This system simplifies the purification protocol due to a one-step affinity purification procedure and dramatically increases the final yield because of the smaller size of the fusion partner. The use of HR52, which constitutes the hydrophobic core of SNase R, has its limitations, because its high hydrophobicity can interfere with the purification and refolding of hydrophobic peptides.

SET (Solubility Enhancement Tags; Stratagene),[73] having highly acidic amino acid sequences, reduce folding interference with the protein of interest. Their small size (< 30 amino acids) means that in some cases removal of the tag for structural studies may be unnecessary. Similarly, two other small protein tags, GB1 and ZZ, have been used with some success to enhance the expression and solubility of peptides and small proteins.[67,77,91] Using GB1[67] with a small cysteine-rich toxin, mutant myotoxin a (MyoP20G), the highly expressed fusion protein was refolded using an unfolding/refolding protocol, which could be monitored using heteronuclear single quantum coherence (HSQC) NMR spectroscopy. The final product yielded well-resolved NMR spectra, with a topology corresponding to the natural product. This system seems suitable for highly hydrophobic and cysteine-rich small proteins.

Even so, it is highly unlikely that there will be a single solubility-enhancing tag that works in every situation and investigators will need to keep a toolbox of solubility tags to hand, any one of which might prove the best tool for a given task. The advent of recombinational cloning and high-throughput expression techniques has made this a much easier task and data produced over the next few years should give us a better idea as to what tools to use in what circumstances; see below.

2.6.2 Purification-facilitating Tags

Purification-facilitating tags, or affinity tags, enable different proteins to be purified using a common method and obviate the need for an understanding of the dark arts associated with conventional chromatographic purification.[92] Additionally, affinity purification reduces the number of unit operations and produces high yields, imbuing the approach with a high degree of economic favourability. Some commonly use purification-enhancing tags are listed in Table 2.3.

His-tags are widely used and pET and pBAD vectors are available having a His_6-encoding sequence. A schematic for the expression and purification of a His-tagged protein of interest is shown in Figure 2.9. Purification is based on the use of metal ions complexed with a resin-immobilised chelating agent; cellular extract containing the expressed protein is applied to the column, the column is washed to remove

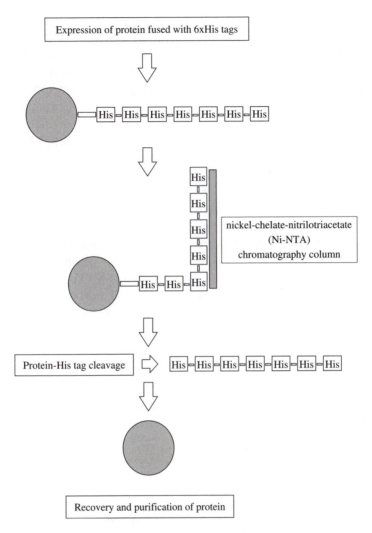

Figure 2.9 Recovery of proteins using His$_6$-tag and Ni-NTA chromatographic columns.

unbound moieties and the His-tagged protein of interest is eluted by applying a solution containing the metal ion to the column. A similar approach can be adopted for a fusion between the protein of interest and FLAG, an octapeptide recognised by the M1 monoclonal antibody. The FLAG-tagged protein may be purified on resin-immobilised M1 mAb in the presence of calcium, which is needed for binding. After subsequent washing steps, the FLAG-tagged protein of interest is eluted by applying to the column a solution containing a chelating agent.

Disadvantages of processes that use affinity tagging to purify the expressed protein include the cost of the affinity matrix that binds the tag and the consequential need for a chromatographic step. Obviously an approach that dispensed with the chromatographic step and utilised a different method for separating the tagged protein from other proteins would offer advantages. Annexin B1, a member of the annexin super-family from *Cysticercus cellulosae*, can be expressed in a soluble form in *E. coli* and purified by a Ca^{2+}-triggered precipitation step, which is fast and easy to perform, followed by resolubilisation of purified protein. Intein is a protein-intervening sequence that catalyses its efficient and precise excision from a host protein in a process mediated by pH changes or the addition of thiols and which has been engineered for protein purification. Using an annexin B1–intein hybrid as a fusion tag,[93] the expressed B1-Int-protein can be precipitated using calcium and recovered by centrifugation and the fusion partner subsequently removed by thiol-induced cleavage of the intein. The efficacy of the approach was demonstrated by expression and purification of human IL2 and single-chain plasminogen activator, urokinase-type (scuPA).

A further disadvantage is that the affinity tag may have a deleterious effect on the properties of the target protein, such as:

- a change in conformation;[94]
- lower yields;[95]
- inhibition of enzyme activity;[96,97]
- alteration in biological activity;[98]
- undesired flexibility in structural studies;[99] and
- toxicity.[100]

It is therefore desirable to remove the tag. Clearly, a strategy for the removal of the tag needs to be developed at the outset if the aim is to produce a 'native' (*i.e.* tagless) protein for human use, when it is necessary to remove not only the cleaved fusion partner but also both enzyme(s) used to cleave it. One such approach is the TAGZyme system (Qiagen), a system based on recombinant exoproteases such as DAPase, which allow the efficient and precise removal of N-terminal His-tags from proteins. One or more amino acid residues that the exoprotease is not able to remove can be included between the His-tag and the protein of interest to ensure that only the His-tag is removed. The DAPase has a C-terminal His_6-tag, which means that after treatment of the fusion protein with the enzyme, the mixture can be reapplied to the column containing the chelating agent and the cleaved protein of interest will not now bind.

Alternatively, it can be arranged that the fusion product contain a linker sequence that can be cleaved, typically by an endoprotease, such as enterokinase. Again, enterokinase-encoding regions can be included in pBAD and other expression systems.

2.6.3 HT Approaches

As noted above, although one tag may be the ideal choice for a given protein or even a given group of proteins, it does not follow that it will function equally well with every protein. Hence it remains unknown which to use for the protein of interest – there are no rules of thumb.

Fortunately, cloning technologies have been developed that have made the generation of new expression vectors relatively easy,[79,81] allowing parallel cloning into multiple vectors having different tags to become an almost routine matter. This means that the suitability of a range of solubility-enhancing tags can be assessed and compared in a single experiment.[79,80,101] Furthermore, the approach can be extended to look at larger numbers, and more diverse types, of target proteins.[79,83]

One such technology, pooled ORF expression technology (POET),[84] is a high-throughput proteomics approach that combines recombinatorial cloning of collections of sequenced ORFs with proteomic methods [two-dimensional gel electrophoreses (2DGE) and MS] to predict which ORFs in a pool will yield soluble, purified protein. The approach involves pooling hundreds of ORFs and subcloning them into a protein expression vector having a purification-facilitating tag. This yields a pool of expression plasmids, from which tagged expressed proteins are produced from a single culture of host cells and easily purified away from host proteins. Proteins in this mixture, which correspond to the original ORF pool, are separated by 2DGE, and individual proteins are identified by MS. Intensity of staining on 2DGE is roughly indicative of the likelihood of expression as a soluble protein in high yield that can be easily purified.

Several commercial systems are also available, including Invitrogen's *Gateway* system, Novagen's *Radiance* and Clontech's *In-Fusion*. The *Gateway* system not only allows exchange of ORFs between libraries, synthesised genes, PCR products or indeed many other sources, via an 'entry' clone, but the clone itself can be used to develop expression clones using a standardised procedure.[102] Furthermore, the expression clone may contain multiple ORFs in a fixed and known orthogonality. This means that it can be used to produce, in at least a semi-automated manner, permutations and combinations of ORFs, and such approaches have been used to explore metabolomic features of proteins from a particular pathway, for example. However, it may also be used to generate combinations of ORFs corresponding to proteins of interest and fusion proteins, permitting the production of a range of expression clones that can be used to produce fusion protein for simplified purification and subsequent analysis. In one study, 75 different ORFs were transferred into expression vectors in combination with four different fusion tags. The efficiency and usefulness of the exercise were evaluated in respect of yield and solubility.[103]

A similar approach has been used for high-throughput expression and screening of membrane proteins, a particular challenge hindering routine structure determination. The approach was used for 49 *E. coli* integral membrane proteins and 71% of these could be produced at sufficient levels to allow milligram amounts of protein to be relatively easily purified.[104]

In another study, 36 prokaryotic P-type transporters, a wide ensemble of modified constructs, were generated and tested for expression in *E. coli*, membrane localization, detergent extraction and homogeneity. The choice of promoter, the choice of source organism providing the cloned gene and, most importantly, the position of the affinity tag had large effects on successful production. Following the initial screening, material from nine of the 36 targets was suitable for crystallization or other structural studies.[105]

2.7 OTHER HOSTS

The focus in this chapter has been on the use of *E. coli* as the host system, but the use of bacterial expression systems is somewhat limited for eukaryotic proteins on account of the need for post-translational modifications to produce a correct glycosylation pattern. For such proteins, eukaryotic systems may be a better choice. One such system is based on a monkey COS cell line having a defective region of the SV40 genome stably integrated into the COS cell genome. Inserting an expression vector having the SV40 origin of replication and the protein of interest into the COS cell initiates viral replication, leading to a high level of expression of the protein. A disadvantage of this system is the lysis of the COS cells; another is the limited insert capacity of the vector. Other eukaryotic cloning vectors[106] for expression of the protein of interest, including *Saccharomyces cerevisiae*,[107–109] *Pichia pastoris*,[109,110], insect[111] and mammalian[107] cell lines, are available.

2.8 CELL-FREE SYSTEMS

This chapter has focused on approaches for expressing protein in a suitable host cell and the problems attendant to this process have been rehearsed in some detail; an alternative approach is the use of cell-free systems.[112] These have been known for some time,[113,114] using cell-free extracts from *E. coli*, rabbit reticulocytes and wheat embryos. Although in principle they offer high speed and accuracy, they have a number of problems, the main one being a limited life span.[115] A system based on the wheat embryo system has been developed by the Centre for Eukaryotic Structural Genomics, in cooperation with Ehime University and CellFree Sciences.[116] This automated platform for producing proteins for NMR-based structural proteomics is able to carry out as many as 384 small-scale screening reactions per week. A desktop robot is also available, which, according to the manufacturers, performs transcription, translation and batch affinity purification in around 35 h. It can run on either a six-well format or a 24-well format to express up to six or 24 genes of interest. The robot produces 2.5–3.0 mg of crude proteins per run. Other systems are available, including Expressway Cell-Free Expression Systems from Invitrogen, an *E. coli*-based *in vitro* system that is able to produce up to milligram quantities of active recombinant protein in a tube reaction format.

2.9 CONCLUSION

Molecular cloning techniques have evolved a long way from simple usage to express proteins from heterologous DNA incorporated into pBR322; now many expression systems and many hosts are available to ensure that the 'expression' part of the 'expression, solubility and purification[1]' target is met. However, there is more to protein expression than merely producing small amounts of it – a more recent challenge has been to provide a reasonable quantity of it in a state that at least approximates to the 'natural' state, for proteomic applications, for pharmaceutical research and for commercial use. This has been met by the development of solubility-enhancing and purification-facilitating tags on the one hand and of high-throughput approaches on the other. This means that is now possible quickly to screen for which of these tags might be most suitable for expressing the protein of interest, potentially rendering the purification of a protein of interest to an almost routine procedure. Molecular cloning for protein expression is thus a key technique for the future understanding and exploitation of proteomics.

REFERENCES

1. D. Esposito and D. K. Chatterjee, *Curr. Opin. Biotechnol.*, 2006, **17**, 353.

2. R. J. Slater and D. R. Williams, in *Molecular Biology and Bio-technology*, 4th edn, ed. J. M. Walker and R. Rapley, Royal Society of Chemistry, Cambridge, 2000, p. 125.
3. G. Hannig and S. C. Makrides, *Trends Biotechnol.*, 1998, **16**, 54.
4. M. H. Caruthers, S. L. Beaucage, C. Becker, W. Efcavitch, F. F. Fisher, G. Galluppi, R. oldman, P. Dettaseth, F. Martin, M. Matteucci and Y. Stabinsley, in *Genetic Engineering*, ed. J. K. Setlow and A. Hollaender, Plenum Press, New York, 1982, p. 119.
5. S. Narang, *J. Biosci.*, 1984, **6**, 739.
6. M. D. Edge, A. R. Green, G. R. Heathcliffe, P. A. Meacock, W. Shuch, D. B. Scanlon, T. C. Atkinson, C. R. Newton and A. F. Markham, *Nature*, 1981, **191**, 756.
7. D. G. Yanasura and D. J. Henner, *Methods Enzymol.*, 1990, **185**, 54.
8. K. Terpe, *Appl. Microbiol. Biotechnol.*, 2006, **72**, 211.
9. C. M. Elvin, P. R. Thompson, M. E. Argall, P. Hendry, N. P. Stamford, P. E. Lilley and N. E. Dixon, *Gene*, 1990, **87**, 123.
10. B. Gronenborn, *Mol. Gen. Genet.*, 1976, **148**, 243.
11. J. Brosius, M. Erfle and J. Storella, *J. Biol. Chem.*, 1985, **260**, 3539.
12. L. M. Guzman, D. Belin, M. J. Carson and J. Beckwith, *J. Bacteriol.*, 1995, **177**, 4121.
13. A. Haldimann, L. Daniels and B. Wanner, *J. Bacteriol.*, 1998, **180**, 1277.
14. A. Skerra, *Gene*, 1994, **151**, 131.
15. F. W. Studier and B. A. Moffatt, *J. Mol. Biol.*, 1986, **189**, 113.
16. H. A. de Boer and A. S. Hui, *Methods Enzymol.*, 1990, **185**, 103.
17. H. P. Sørensen, B. S. Laursen, K. K. Mortensen and H. U. Sperling-Petersen, *Recent Res. Dev. Biophys. Biochem.*, 2002, **2**, 243.
18. C. M. Stenstrom, H. Jin, L. L. Major, W. P. Tate and L. A. Isaksson, *Gene*, 2001, **263**, 273.
19. S. Ringquist, S. Shinedling, D. Barrick, L. Green, J. Binkley, G. D. Stormo and L. Gold, *Mol. Microbiol.*, 1992, **6**, 1219.
20. B. S. Laursen, S. Steffensen, J. Hedegaard, J. M. Moreno, K. K. Mortensen and H. U. Sperling-Petersen, *Genes Cells*, 2002, **7**, 901.
21. P. O. Olins and S. H. Rangwala, *Methods Enzymol.*, 1990, **185**, 15.
22. P. O. Olins, C. S. Devine and S. H. Rangwala, in *Expression Systems and Processes for rDNA Products*, ed. R. T. Hatch, C. Goochee, A. Moreira and Y. Alroy, American Chemical Society, Washington, DC, 1991, p. 17.
23. S. F. Newbury, N. H. Smith, E. C. Robinson, I. D. Hiles and C. F. Higgins, *Cell*, 1987, **48**, 297.

24. E. S. Poole, C. M. Brown and W. P. Tate, *EMBO J.*, 1995, **14**, 151.
25. F. Baneyx, *Curr. Opin. Biotechnol.*, 1999, **10**, 411.
26. P. Jonasson, S. Liljeqvist, P. A. Nygren and S. Stahl, *Biotechnol. Appl. Biochem.*, 2002, **35**, 91.
27. G. del Solar, R. Giraldo, M. J. Ruiz-Echevarria, M. Espinosa and R. Diaz-Orejas, *Microbiol. Mol. Biol. Rev.*, 1998, **62**, 434.
28. D. Summers, *Mol. Microbiol.*, 1998, **29**, 1137.
29. M. P. Mayer, *Gene*, 1995, **163**, 41.
30. K. G. Hardy, *Plasmids*, IRL Press, Oxford, 1987.
31. S. R. Connell, D. M. Tracz, K. H. Nierhaus and D. E. Taylor, *Antimicrob. Agents Chemother.*, 2003, **47**, 3675.
32. H. P. Sørensen and K. K. Mortensen, *J. Biotechnol.*, 2005, **115**, 113.
33. J. W. Dubendorff and F. W. Studier, *J. Mol. Biol.*, 1991, **219**, 45.
34. F. W. Studier, A. H. Rosenberg, J. J. Dunn and J. W. Dubendorff, *Methods Enzymol.*, 1990, **185**, 60.
35. M. P. Calos, *Nature*, 1978, **274**, 762.
36. F. W. Studier, *J. Mol. Biol.*, 1991, **219**, 37.
37. A. Khlebnikov and J. D. Keasling, *Biotechnol. Prog.*, 2002, **18**, 672.
38. E. Englesberg, C. Squires and F. Meronk Jr, *Proc. Natl. Acad. Sci. USA*, 1969, **62**, 1100.
39. D. A. Siegele and J. C. Hu, *Proc. Natl. Acad. Sci. USA*, 1997, **94**, 8168.
40. A. Khlebnikov, T. Skaug and J. D. Keasling, *J. Ind. Microbiol. Biotechnol.*, 2002, **29**, 34.
41. R. M. Morgan-Kiss, C. Wadler and J. E. Cronan Jr, *Proc. Natl. Acad. Sci. USA*, 2002, **99**, 7373.
42. J. F. Kane, *Curr. Opin. Biotechnol.*, 1995, **6**, 494.
43. C. Kurland and J. Gallant, *Curr. Opin. Biotechnol.*, 1996, **7**, 489.
44. G. Dieci, L. Bottarelli, A. Ballabeni and S. Ottonello, *Protein Expr. Purif.*, 2000, **18**, 346.
45. I. Arechaga, B. Miroux, M. J. Runswick and J. E. Walker, *FEBS Lett.*, 2003, **547**, 97.
46. G. Georgiou and L. Segatori, *Curr. Opin. Biotechnol.*, 2005, **16**, 538.
47. T. A. Phillips, R. A. van Bogelen and F. C. Neidhardt, *J. Bacteriol.*, 1984, **159**, 283.
48. A. Villaverde and M. M. Carrio, *Biotechnol. Lett.*, 2003, **25**, 1385.
49. M. M. Carrio, R. Cubarsi and A. Villaverde, *FEBS Lett.*, 2000, **471**, 7.
50. A. Mogk, M. P. Mayer and E. Deuerling, *ChemBioChem.*, 2002, **3**, 807.

51. E. Deuerling, H. Patzelt, S. Vorderwulbecke, T. Rauch, G. Kramer, E. Schaffitzel, A. Mogk, A. Schulze-Specking, H. Langen and B. Bukau, *Mol. Microbiol.*, 2003, **47**, 1317.
52. B. Miroux and J. E. Walker, *J. Mol. Biol.*, 1996, **260**, 289.
53. D. Petsch and F. B. Anspach, *J. Biotechnol.*, 2000, **76**, 97.
54. R. B. Kapust and D. S. Waugh, *Protein Sci.*, 1999, **8**, 1668.
55. P. A. Nygren, S. Stahl and M. Uhlen, *Trends Biotechnol.*, 1994, **12**, 184.
56. K. Terpe, *Appl. Microbiol. Biotechnol*, 2003, **60**, 523.
57. G. D. Davis, C. Elisee, D. M. Newham and R. G. Harrison, *Biotechnol. Bioeng.*, 1999, **65**, 382.
58. H. P. Sørensen, H. U. Sperling-Petersen and K. K. Mortensen, *Protein Expr. Purif.*, 2003, **32**, 252.
59. A. Jacquet, V. Daminet, M. Haumont, L. Garcia, S. Chaudoir, A. Bollen and R. Biemans, *Protein Expr. Purif.*, 1999, **17**, 392.
60. G. S. Waldo, B. M. Standish, J. Berendzen and T. C. Terwilliger, *Nat. Biotechnol.*, 1999, **17**, 691.
61. J. D. Pedelacq, E. Piltch, E. C. Liong, J. Berendzen, C. Y. Kim, B. S. Rho, M. S. Park, T. C. Terwilliger and G. S. Waldo, *Nat. Biotechnol.*, 2002, **20**, 927.
62. H. Wang and S. Chong, *Proc. Natl. Acad. Sci. USA*, 2003, **100**, 478.
63. P. J. Schatz, *Biotechnology*, 1993, **11**, 1138.
64. P. Vaillancourt, C. F. Zheng, D. Q. Hoang and L. Breister, *Methods Enzymol.*, 2000, **326**, 340.
65. Z. Zhang, Z. H. Li, F. Wang, M. Fang, C. C. Yin, Z. Y. Zhou, Q. Lin and H. L. Huang, *Protein Expr. Purif.*, 2002, **26**, 218.
66. A. Einhauer and A. Jungbauer, *J. Biochem. Biophys. Methods*, 2001, **49**, 455.
67. Y. Cheng and D. J. Patel, *Biochem. Biophys. Res. Commun.*, 2004, **317**, 401.
68. D. B. Smith and K. S. Johnson, *Gene*, 1988, **67**, 31.
69. V. Gaberc-Porekar and V. Menart, *J. Biochem. Biophys. Methods*, 2001, **49**, 335.
70. C. di Guan, P. Li, P. D. Riggs and H. Inouye, *Gene*, 1988, **67**, 21.
71. S. Nallamsetty and D. S. Waugh, *Protein Expr. Purif.*, 2006, **45**, 175.
72. J. D. Fox and D. S. Waugh, *Methods Mol. Biol.*, 2003, **205**, 99.
73. Y. B. Zhang, J. Howitt, S. McCorkle, P. Lawrence, K. Springer and P. Freimuth, *Protein Expr. Purif.*, 2004, **36**, 207.
74. D. K. Chatterjee and D. Esposito, *Protein Expr. Purif.*, 2006, **46**, 122.

75. S. Voss and A. Skerra, *Protein Eng.*, 1997, **10**, 975.
76. J. G. Marblestone, S. C. Edavettal, Y. Lim, P. Lim, X. Zuo and T. R. Butt, *Protein Sci.*, 2006, **15**, 182.
77. Y. Zhao, Y. Benita, M. Lok, B. Kuipers, P. van der Ley, W. Jiskoot, W. E. Hennink, D. J. Crommelin and R. S. Oosting, *Vaccine*, 2005, **23**, 5082.
78. M. Hammarström, N. Hellgren, S. van den Berg, H. Berglund and T. Hard, *Protein Sci.*, 2002, **11**, 313.
79. M. R. Dyson, S. P. Shadbolt, K. J. Vincent, R. L. Perera and J. McCafferty, *BMC Biotechnol.*, 2004, **4**, 32.
80. I. Kataeva, J. Chang, H. Xu, C. H. Luan, J. Zhou, V. N. Uversky, D. Lin, P. Horanyi, Z. J. Liu, L. G. Ljungdahl, J. Rose, M. Luo and B.-C. Wang, *J. Proteome Res.*, 2005, **4**, 1942.
81. D. Busso, B. Delagoutte-Busso and D. Moras, *Anal. Biochem.*, 2005, **343**, 313.
82. S. Braud, M. Moutiez, P. Belin, N. Abello, P. Drevet, S. Zinn-Justin, M. Courcon, C. Masson, J. Dassa, J. B. Charbonnier, J.-C. Boulain, A. Menez, R. Genet and M. Gondry, *J. Proteome Res.*, 2005, **4**, 2137.
83. A. Dummler, A. M. Lawrence and A. de Marco, *Microb. Cell. Fact.*, 2005, **4**, 34.
84. A. Schrodel, J. Volz and A. de Marco, *J. Biotechnol.*, 2005, **120**, 2.
85. P. Turner, O. Holst and E. N. Karlsson, *Protein Expr. Purif.*, 2005, **39**, 54.
86. H. Bach, Y. Mazor, S. Shaky, A. Shoham-Lev, Y. Berdichevsky, D. L. Gutnick and I. Benhar, *J. Mol. Biol.*, 2001, **312**, 79.
87. L. Zheng, U. Baumann and J. L. Reymond, *J. Biochem. (Tokyo)*, 2003, **133**, 577.
88. K.-Y. Ahn, J.-A. Song, K.-Y. Han, J.-S. Park, H.-S. Seo and J. Lee, *Enzyme Microb. Technol.*, 2007, **41**, 859.
89. K.-Y. Han, J.-A. Song, K.-Y. Ahn, J.-S. Park, H.-S. Seo and J. Lee, *FEMS Microbiol. Lett.*, 2007, **274**, 132.
90. Y. Cheng, D. Liu, Y. Feng and G. Jing, *Protein Pept. Lett.*, 2003, **10**, 175.
91. W. J. Bao, Y. G. Gao, Y. G. Chang, T. Y. Zhang, X. J. Lin, X. Z. Yan and H. Y. Hu, *Protein Expr. Purif.*, 2006, **47**, 599.
92. J. Arnau, C. Lauritzen, G. E. Petersen and J. Pedersen, *Protein Expr. Purif.*, 2006, **48**, 1.
93. F.-X. Ding, H.-L. Yan, Q. Mei, G. Xue, Y.-Z. Wang, Y.-J. Gao and S.-H. Sun, *Appl. Microbiol. Biotechnol.*, 2007, **77**, 483.
94. A. Chant, C. M. Kraemer-Pecore, R. Watkin and G. G. Kneale, *Protein Expr. Purif.*, 2005, **39**, 152.

95. A. Goel, D. Colcher, J. S. Koo, B. J. Booth, G. Pavlinkova and S. K. Batra, *Biochim. Biophys. Acta*, 2000, **1523**, 13.

96. K. M Kim, E. C. Yi, D. Baker and K. Y. Zhang, *Acta Crystallogr., Sect. D*, 2001, **57**, 759.

97. S. Cadel, C. Gouzy-Darmon, S. Petres, C. Piesse, V. L. Pham, M. C. Beinfeld, P. Cohen and T. Foulon, *Protein Expr. Purif.*, 2004, **36**, 19.

98. I. Fonda, M. Kenig, V. Gaberc-Porekar, P. Pristovaek and V. Menart, *Sci. World J.*, 2002, **15**, 1312.

99. D. R. Smyth, M. K. Mrozkiewicz, W. J. McGrath, P. Listwan and B. Kobe, *Protein Sci.*, 2003, **12**, 1313.

100. E. G. E. de Vries, M. N. de Hooge, J. A. Gietema and S. de Jong, *Clin. Cancer Res.*, 2003, **9**, 912.

101. R. C. Stevens, *Struct. Fold. Des.*, 2000, **8**, R177.

102. F. Katzen, *Expert Opin. Drug Discov.*, 2007, **2**, 571.

103. U. Korf, T. Kohl, H. van der Zandt, R. Zahn, S. Schleeger, B. Ueberle, S. Wandschneider, S. Bechtel, M. Schnölzer, H. Ottleben, S. Wiemann and A. Poustka, *Proteomics*, 2005, **5**, 3571.

104. S. Eshaghi, M. Hedrén, M. Ignatushchenko A. Nasser, T. Hammarberg, A. Thornell and P. Nordlund, *Protein. Sci.*, 2005, **14**, 676.

105. O. Lewinson, A. T. Lee and D. C. Rees, *J. Mol. Biol.*, 2008, **377**, 62.

106. T. Brown, *Gene Cloning and DNA Analysis*, Blackwell Publishing Ltd., Oxford, 2006.

107. L. Borsig, E. G. Berger and M. Malissard, *Biochem. Biophys. Res. Commun.*, 1997, **240**, 586.

108. L. J. Byrne, K. J. O'Callaghan and M. F. Tuite, *Methods Mol. Biol.*, 2005, **308**, 51.

109. B. Prinz, J. Schultchen, R. Rydzewski, C. Holz, M. Boettner, U. Stahl and C. Lang, *J. Struct. Funct. Genomics*, 2004, **5**, 29.

110. D. Daly and M. T. Hearn, *J. Mol. Recognit.*, 2005, **18**, 119.

111. T. A. Kost, J. P. Condreay and D. L. Jarvis, *Nat. Biotechnol.*, 2005, **23**, 567.

112. Y. Endo and T. Sawasaki, *Curr. Opin. Biotechnol.*, 2006, **17**, 373.

113. C.G. Kurland, *Cell*, 1982, **28**, 201.

114. M. Y. Pavlov and M. Ehrenberg, *Arch. Biochem. Biophys.*, 1996, **328**, 9.

115. B. E. Roberts and B. M. Paterson, *Proc. Natl. Acad. Sci. USA*, 1973, **70**, 2330.

116. D. A Vinarov and J. L. Markley, *Expert Rev. Proteomics*, 2005, **2**, 49.

CHAPTER 3

Molecular Diagnostics

LAURA J. TAFE, CLAUDINE L. BARTELS, JOEL A. LEFFERTS AND
GREGORY J. TSONGALIS

Department of Pathology, Dartmouth Medical School, Dartmouth
Hitchcock Medical Center and Norris Cotton Cancer Center, Lebanon, NH
03756, USA

3.1 INTRODUCTION

The transition of new technologies that were once thought of as
'Research Use Only' into the clinical laboratory as user developed assays
(UDA), analyte-specific reagents (ASR) or FDA-cleared diagnostic
assays has revolutionized the way that laboratory medicine is practiced.
Molecular diagnostic capabilities have spanned the entire spectrum of
what is, can and will be performed on a clinical basis in most hospital
laboratories. While the ability for clinical laboratories to detect human
genetic variation has historically been limited to a rather small number
of traditional genetic diseases where no clinical laboratory testing was
ever available, our current understanding of many disease processes at a
molecular level has expanded our testing capabilities to both human and
non-human applications. The identification of numerous new genes and
disease causing mutations, in addition to benign polymorphisms that
may influence a specific phenotype, has created a rapid demand for
molecular diagnostic testing.

Currently, there is a growing need for clinical laboratories to provide
high-quality nucleic acid-based tests within the realm of clinical

Molecular Biology and Biotechnology, 5th Edition
Edited by John M Walker and Ralph Rapley
© Royal Society of Chemistry 2009
Published by the Royal Society of Chemistry, www.rsc.org

relevance, including performance and turnaround time. This need was initially driven by the completion of the Human Genome Project that identified thousands of genes and millions of human single nucleotide polymorphisms (SNPs) and culminated in disease associations. In addition, the World Wide Web became an instant source of both medical and scientific information for the populace. Popular television programs have also incorporated the science of DNA into many of their shows, making molecular (DNA) testing common knowledge in the household. Because of this, an unprecedented demand has been placed on the clinical laboratory to provide increased diagnostic testing with unprecedented performance for rapid and accurate identification and interrogation of genomic targets.

3.2 TECHNOLOGIES

Molecular technologies first entered the clinical laboratories in the early 1980s as manual, labor-intensive procedures that required a working knowledge of chemistry and molecular biology and also an exceptional skill set. While molecular diagnostics is often described as a discipline in its infancy, the past 5 years have seen a rapid period of maturation in this field. Testing capabilities have moved very quickly away from labor-intense, highly complex and specialized procedures to more user-friendly, semi-automated procedures.[1] We have experienced this transition which began by testing for relatively high-volume infectious diseases such as Chlamydia trachomatis, HPV, HIV-1, HCV, *etc.* Much of this was championed by the availability of FDA-cleared kits and higher throughput, semi-automated instruments such as the Abbott LCX and Roche Cobas Amplicor systems.

However, in the early 1980s, the Southern blot was the method of choice for a variety of clinical applications, even though the turnaround time was in excess of 2–3 weeks. The commercial availability of restriction endonucleases and various agarose matrices helped in making this technique routine. The polymerase chain reaction (PCR) revolutionized blotting technologies and detection limits by offering increased sensitivity and the much needed shortened turnaround time for clinical result availability. Since then, many modifications to the PCR have been introduced, but none as significant as real-time capability.[2-5] The elimination of post-PCR detection systems and the ability to perform the entire assay in a closed vessel had significant advantages for the clinical laboratory. Various detection chemistries for real-time PCR were rapidly introduced and many of the older detection methods (gel electrophoresis, ASO blots, *etc.*) vanished from the laboratory.

Table 3.1 Current technologies being used in the Molecular Pathology Laboratory at the Dartmouth Hitchcock Medical Center.

Technology	*Platform/instrument*
DNA/RNA extraction	1. Manual
	2. Spin column
	3. Qiagen EZ1 robot
Fluorescence *in situ* hybridization (FISH)	1. Hybrite oven
	2. Molecular imaging system
bDNA	1. Siemens System 340
Hybrid capture	1. Digene HCII System
Real-time PCR	1. Cepheid GeneXpert
	2. ABI 7500
	3. Cepheid Smartcycler
	4. Roche Taq48
Capillary electrophoresis	1. Beckman CEQ
	2. Beckman Vidiera
Traditional PCR	1. MJ Research
	2. DNA Engines
Microarray	1. Luminex
	2. Superarray
	3. Nanosphere

The real-time PCR has become a method of choice for most molecular diagnostics laboratories. This modification of the traditional PCR allows for the simultaneous amplification and detection of amplified nucleic acid targets as it occurs. In routine clinical practice, the main advantages of real-time PCR are the speed with which samples can be analyzed, as there are no post-PCR processing steps required, and the 'closed-tube' nature of the technology. The analysis of results via amplification curve and melt curve analysis is very simple and contributes to it being a much faster method for analyzing PCR results.

However, DNA sequencing technologies, fragment sizing and LOH studies would benefit from automated capillary electrophoresis instrumentation. Several forms of microarrays are currently available as post-PCR detection mechanisms for multiplexed analysis of SNPs, gene expression and pathogen detection. Today, molecular diagnostic laboratories are armed with various instruments and technologies to address the increasing demand for clinical test results (Table 3.1).

3.3 THE INFECTIOUS DISEASE PARADIGM

Molecular infectious disease testing represents a paradigm in the application of diagnostic testing. Clinical applications of molecular

Table 3.2   Molecular infectious disease tests that are currently FDA cleared and being performed in clinical laboratories.

Bacteria	Viruses
Bacillus anthracis	Avian flu
Candida albicans	Cytomegalovirus
Chlamydia trachomatis	Enterovirus
Enterococcus faecalis	HBV (quantitative)
Francisella tularensis	Hepatitis C virus (qualitative and quantitative)
Gardnerella, Trichomonas and *Candida* spp.	HIV drug resistance
Group A streptococci	HIV (quantitative)
Group B streptococci	HBV/HCV/HIV blood screening assay
Legionella pneumophila	Human papillomavirus
Methicillin-resistant *Staphylococcus aureus*	Respiratory viral panel
Mycobacterium tuberculosis	West Nile virus
Mycobacterium spp.	
Staphylococcus aureus	

diagnostics to infectious diseases include those for qualitative testing, quantitative testing and resistance genotyping.[6–10] This paradigm spans the spectrum of current molecular capabilities and has provided significant insight to the nuances of molecular testing for other disciplines. Molecular diagnostic testing for infectious disease applications continues to be the highest volume of testing being performed in clinical laboratories (Table 3.2).

While accurate and timely diagnosis of infectious diseases is essential for proper patient management, traditional testing methods for many pathogens did not allow for rapid turnaround time. Prompt detection of the microbial pathogen allows providers to institute adequate measures to interrupt transmission to a susceptible hospital or community population. The diagnosis of an infectious diseases has typically depended upon isolation of the infective agent by a culture technique. Even though this approach is adequate for the identification of the majority of common infections, it is less than desirable for the detection of organisms that were difficult to grow *in vitro*, including long incubation times. These issues along with performance characteristics of sensitivity and specificity brought about a need for alternative techniques that would allow for direct detection of infectious agents in clinical samples while maintaining a rapid turnaround time and high performance.

Of particular importance to molecular infectious disease testing is the type of specimen for which assays have been validated. Due to the endogenous inhibitors found in some clinical specimens, there is an inherent need to check for quality of the specimen and presence of inhibitors. This can be performed by using a variety of control materials that are added to the specimen before extraction and to the nucleic acid after extraction. Proper control material for quantitative assays must also be used for accurate determination of copy number.

The rapid demand for availability of more molecular infectious disease tests led the US Food and Drug Administration (FDA) to create a regulatory category known as the analyte-specific reagent (ASR). This allowed companies to market primer and probe sets without going through a full FDA submission and hence increased the availability of reagents for laboratories to perform testing. As more laboratories desired to bring molecular testing in-house, the shortcomings of the ASR were rapidly made known. As part of an ASR, the company is not allowed to provide any instructions or recipe for use. Therefore, for laboratories with no or minimal molecular experience, a vial of probe and one of primer was difficult to deal with. Although there were many commercially available reagents now available, the laboratories' capabilities to validate and use them properly did not meet the expectations of the diagnostic community. Guidelines for proper validation of assays are available from the Clinical and Laboratory Standards Institute.

Just as we have seen a rapid trend in reagent availability, the rapid introduction of new technologies has also been responsible for pushing the molecular infectious disease applications forward. In the early 1990s, molecular techniques such as DNA probes and the PCR were introduced into the clinical microbiology laboratory. Next generations of these technologies, together with automated instrumentation are now used routinely to detect and quantify an increasing number of organisms. While real-time PCR and array platforms have made multiplex testing a reality, automated sequencing instruments have also made resistance genotyping routine in many clinical laboratories. The paradigm set forth by our expanding knowledge base from the infectious disease applications has led to new developments in other clinical applications, as will be discussed.

3.4 GENETICS

While clinical applications of molecular technologies have been developed for numerous diseases, the race to identify genes and their mutation spectra was evident early in the pre-human genome era. Current concepts of molecular mechanisms of disease that include gene associations have evolved from these early observations, as has our ability to detect routinely various genetic abnormalities. The fact that these alterations can be

Table 3.3   Genetic diseases and/or genes commonly tested for using molecular diagnostic methods

Alpha1-antitrypsin deficiency
Angelman syndrome
ApoE
Cystic fibrosis
Duchenne/Becker muscular dystrophy
Factor II (prothrombin)
Factor V Leiden
Fanconi's anemia
Fragile X syndrome
Gaucher disease
Hemochromatosis
Huntington's disease
Methylene tetrahydrofolate reductase (MTHFR)
Prader-Willi syndrome
Tay-Sachs disease

inherited through the germline or acquired has expanded our knowledge of many disease mechanisms. Molecular genetics provides a means for examining inheritance patterns at the level of nucleic acids and provides a vehicle for dissecting complex pathophysiological processes into gene defects. Many clinical laboratories are now offering molecular diagnostic testing for some of the more common diseases (Table 3.3).

Currently, there is an enormous amount of information with respect to numbers and characteristics of various genetic diseases and syndromes. Genetic diseases can be categorized into three major groups: (1) chromosomal disorders, (2) monogenic or single-gene disorders and (3) polygenic or multifactorial disorders. Chromosomal disorders are due to the loss, gain or abnormal arrangement of one or more chromosomes which results in the presence of excessive or deficient amounts of genetic material. Syndromes characterized by multiple birth defects and various forms of hematopoietic malignancy are examples of chromosomal disorders. These types of alterations usually involve large segments of DNA containing numerous genes and can be classified into four groups:

1. aneuploidy: excess or loss of one or more chromosomes;
2. deletion: breakage and/or loss of a portion of a chromosome;
3. translocation: breakage of two chromosomes with transfer of broken parts to the opposite chromosome;
4. isochromosome: splitting at the centromere during mitosis so that one arm is lost and the other duplicated to form one chromosome with identical arms.

Monogenic disorders are the result of a single mutant gene and display traditional Mendelian inheritance patterns including autosomal dominant or recessive and X-linked types. The overall population frequency of monogenic disorders is thought to be approximately 10 per 1000 live births. Polygenic or multifactorial disorders consist of chronic diseases of adulthood, congenital malformations and dysmorphic syndromes. These disorders result from multiple genetic and/or epigenetic factors which may not conform to traditional Mendelian inheritance patterns. In the two examples that follow, one disease exemplifies a single gene disorder with multiple mutations and the other a trinucleotide repeat disorder that is non-Mendelian in inheritance.

Our ability to screen individuals for many types of genetic alterations in a clinical setting is expanding at an enormous rate. Many clinical laboratories offer routine testing for several genetic diseases (Table 3.3). The discovery of the gene for cystic fibrosis (CF), known as the Cystic Fibrosis Transmembrane Regulator (CFTR) gene, led to a rapid understanding of the pathophysiology of this disease.[11] It also led to the description of more than 1000 mutations in this gene (http://www.genet.sickkids.on.ca/cftr/). Over the last several years, the American College of Medical Genetics and the American College of Obstetrics and Gynecology in an unprecedented fashion approved guidelines for a national CF screening program requiring laboratories to test for 23 mutations in the *CFTR* gene (Table 3.4).[12,13] Numerous molecular methods are commercially available for multiplex screening for these mutations (Figure 3.1).

Non-Mendelian inheritance patterns became very evident when the gene for Fragile X Syndrome (FRAX), known as FMR-1, was discovered. FRAX is the most common cause of inherited mental retardation and is due to an expansion of a trinucleotide repeat in the FMR-1 gene.[14-16] This results in methylation of an upstream CpG island and loss of FMR-1 expression (Figure 3.2). Characterization of this unstable CGG repeat and the various resulting sizes led to the identification of several allelic forms referred to as normal or common, intermediate or gray zone, permutation and full mutation. The full mutation is associated

Table 3.4 Panel of 23 recommended mutations used in screening for cystic fibrosis.

ΔF508	ΔI507	G542X	G551D	W1282X
N1303K	R553X	621+1G>T	R117H	1717-1G>A
A455E	R560T	R1162X	G85E	R334W
R347P	711+1G>T	1898+1G>A	2184delA	849+10kbC>T
2789+5G>A		3659delC		3120+1G>A

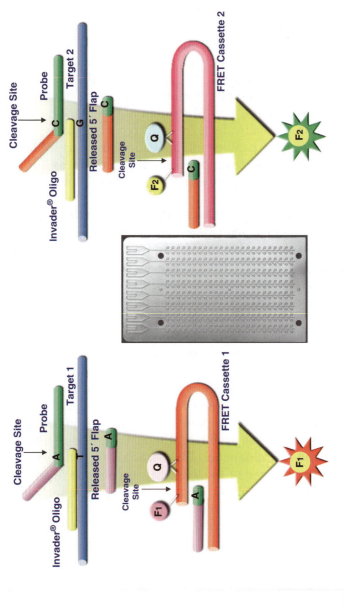

Figure 3.1 Diagram of the Invader chemistry for CF mutation testing and the microfluidic card used.

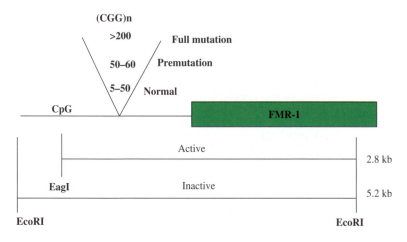

Figure 3.2 Schematic of the FMR-1 trinucleotide repeat expansion region.

with >200 CGG repeats and lack of FMR-1 expression. This mutation is X-linked and therefore males are more severely affected than females. Affected females have less severe mental retardation and can pass an expanded allele on to their offspring. Hence FRAX was one of the disorders referred to as exhibiting anticipation, the potential for the disease to increase from generation to generation. The molecular diagnostic assay for FRAX is based on Southern blot analysis and now more commonly, PCR with capillary electrophoresis is used to size individual alleles more accurately. This information can then be used in the more accurate genetic counseling of family members.

3.5 HEMATOLOGY

Applications of molecular diagnostics to hematopathology have been at the forefront of this discipline. This section will focus on malignant lymphoma as an example of the diagnostic capabilities at the molecular level. Unique biomarkers that have been identified in various hematological malignancies include gene mutations, rearrangements, translocations and gene or mRNA expression profiles,[17–21] Malignant lymphomas are a heterogeneous group of neoplasms and are categorized using the World Health Organization's Classification of Tumors of Hematopoietic and Lymphoid Tissues.[22] These neoplasms most often arise in lymphoid tissues such as lymph nodes and spleen, but may arise virtually anywhere in the body. Precise diagnosis and classification of lymphoma are important since treatment options and prognosis vary considerably. The diagnosis of lymphoma prior to 1980 was based primarily on the

Table 3.5 Common molecular diagnostic tests
applied to malignant lymphoma.

BCL-1 translocation (11;14)
BCL-2 translocation (18;14)
BCL-6 translocation (3;14)
c-Myc translocation (8;14)
Immunoglobulin heavy and light chain rearrangements
NPM-ALK translocation (2;5)
T-cell receptor gene rearrangements

histological evaluation of traditional H&E-stained slides by light microscopy. In the 1980s, our understanding of the immune system and ability to diagnose and classify non-Hodgkin and Hodgkin lymphoma improved significantly, largely due to the development of immuno-pathological methods. In the mid-1980s, the availability of molecular genetic methods further enhanced our ability to diagnose and classify lymphoid neoplasms and several tests have now become routinely available in the clinical laboratory (Table 3.5). Today, the diagnosis and classification of lymphoma requires the integration of traditional and advanced diagnostic techniques.

Non-Hodgkin lymphomas occur because of neoplastic transformation of B and T lymphocytes at different stages of normal B-cell and T-cell development. Establishing a diagnosis of non-Hodgkin lymphoma requires the determination of clonality since both B- and T-cell lymphoma represent monoclonal proliferations of B cells and T cells, respectively.[23] For B-cell neoplasms, clonality can often be determined immunopathologically by demonstrating the presence of monoclonal surface immunoglobulins or by detecting gene rearrangements in the immunoglobin gene family. In contrast, for T-cell malignancies, there is no immunopathological equivalent to monoclonal surface immunoglobulin. Thus, molecular genetic approaches for the determination of clonality in T-cell lymphoma by detecting gene rearrangements in the T-cell receptor family are especially important.[23] Other molecular genetics applications to the assessment of lymphoid malignancies include detection of chromosomal translocations in the subclassification of non-Hodgkin lymphoma. For example, in a lymph node with suspected follicular lymphoma, the detection of a translocation, t(14;18), involving the *BCL*-2 proto-oncogene, would confirm this diagnosis. Similarly, the detection of a translocation, t(11;14), involving the *BCL*-1 proto-onco-gene, would confirm a diagnosis of mantle cell lymphoma.[23,24]

The B-cell surface immunoglobulin receptor and T-cell receptor are involved in the process of antigen recognition by normal B cells and

T cells, respectively. These receptors are structurally similar in that they are heterodimer proteins linked by disulfide bonds and are composed of both variable (V) and constant (C) regions. The variable regions of these proteins are similarly involved in antigen recognition. The constant region of the immunoglobulin heavy chain protein defines the different immunoglobulin subclasses (IgG1, IgG2, IgG3, IgG4, IgA1, IgA2, IgM, IgD and IgE). The genes that code for the B- and T-cell receptors are also structurally similar and consist of a large number of exons or coding sequences. These genes undergo a process of DNA recombination or rearrangement leading eventually to transcription and translation and the production of functional receptor proteins.

The germline configuration of these genes refers to non-rearranged DNA. The exons which code for the variable regions of the immunoglobulin and T-cell receptors are referred to as variable (V) segments, diversity (D) segments and junctional (J) segments, and those which code for the constant regions are referred to as (C) segments. The process of gene rearrangement first involves the selective apposition of one D segment with one J segment by deletion of the intervening coding and non-coding DNA sequences, resulting in a DJ rearrangement. A similar process of rearrangement apposes a V segment, located in the 5' direction, to D and J to form a VDJ rearrangement. Transcription to messenger RNA (mRNA) then occurs even though the rearranged VDJ segments are not yet directly apposed to C segments, which are remotely located in the 3' direction. Subsequent splicing of the mRNA with deletion of non-coding sequences results in apposition of VDJ with C to form a VDJC mRNA, which can then be translated into a surface immunoglobulin or T-cell receptor protein (Figure 3.3).

To establish a diagnosis of B-cell or T-cell non-Hodgkin lymphoma, the ability to prove that a neoplastic population of B cells or T cells is monoclonal in origin is of central importance. A monoclonal or simply, a clonal cell population refers to a population of cells which are all derived from a single precursor cell and thus share similar characteristics. By molecular genetic methods, clonality is defined by the identification of a clonal B-cell or T-cell gene rearrangement. Traditionally this was performed by Southern blot transfer analysis. However, this technique required a large amount of high molecular weight DNA, thus eliminating the use of paraffin-embedded tissues. The PCR technique has in most instances replaced the Southern blot for evaluating for the presence or absence of B- and T-cell clonality. This methodology allows for the evaluation of minute quantities of DNA by *in vitro* amplification. Analogous to Southern blot methods, the application of PCR to detect B- and T-cell clonality involves evaluation of gene rearrangements in

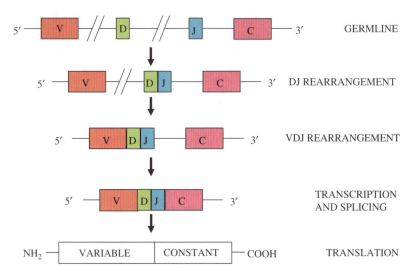

Figure 3.3 Schematic diagram of the gene rearrangement process.

those segments of DNA which code for the variable regions of the surface immunoglobulin and T-cell receptors.[23–25]

3.6 ONCOLOGY

Initial efforts in molecular diagnostics began with applications for diagnosing hematopoietic malignancies; however, applications for direct qualitative infectious disease testing far outpaced the oncology tests being performed. In part this was due to the polygenic, multifactorial complexity of the disease. It became evident very early that molecular oncology testing would not be a 'one target fits all' type of algorithm as in the identification of microbial pathogens. Interestingly, as this paradigm of being able to perform qualitative, quantitative and genotype infectious disease testing became standard practice, similar approaches to oncology were beginning to surface. The need for viral load testing emerged from the development of anti-retroviral therapeutics that warranted monitoring of viral copy numbers. Soon thereafter, genotyping efforts emerged from the need to determine viral resistance or subtypes that would better respond to therapeutics. Paralleling the applications for infectious disease testing, molecular oncology is now at a similar crossroads where applications for qualitative, quantitative and genotype testing are warranted based on novel biomarkers and therapeutics. Hence a new testing paradigm has emerged through molecular oncology testing.

Although numerous genes and mutations have been identified in many human cancer types, few have made it to the clinical laboratory as validated tests. However, this is beginning to change with our increased medical and biological knowledge of these diseases. An excellent resource for such information is the National Cancer Institute's Cancer Genome Anatomy (CGAP) website (http://cgap.nci.nih.gov). One goal of CGAP is to identify known or newly discovered single nucleotide polymorphisms (SNPs) which are of immediate importance to the study of cancer. The following two examples highlight the utility of molecular diagnostic tools in the assessment of these cancer types.

A human cancer that has made some headway with respect to molecular diagnostics is in the field of breast oncology. Higher resolution imaging, targeted drugs and improved surgical techniques have all contributed to improved patient care by detecting, diagnosing and treating breast cancer more effectively. Early and accurate detection of metastatic spread is essential for the successful management of advanced breast cancer.[26] One of the primary channels of metastatic spread of breast cancer from the primary tumor site to distant sites is the lymphatic system. Tumor cells in the breast are shed and are captured in the axillary lymph nodes (ALNs), where they can develop into metastatic lesions and then spread to other body sites. Pathological examination of the ALNs, especially sentinel lymph nodes, therefore, has become an essential and routine component in the care of breast cancer patients.[26]

New molecular techniques have been developed for detecting the presence of tumor cells in lymph nodes. Analysis of a large number of potential targets for the identification of positive nodes in a real-time RT-PCR test revealed two targets, mammaglobin and cytokeratin 19, as ideal markers for this diagnostic application.[27-31] Recently, the FDA cleared a molecular diagnostic assay for breast sentinel lymph node testing (Veridex, Warren, NJ, USA). This assay includes a standardized protocol and reagents for homogenization of lymph node tissue, isolation of RNA from the homogenized tissue and the GeneSearch Breast Lymph Node (BLN) test kit for setting up a multiplexed real-time RT-PCR assay using the SmartCycler real-time platform (Cepheid, Sunnyvale, CA, USA). The entire procedure can be completed in less than 1 h (Figure 3.4). The Intended Use statement provided by Veridex recommends this test for intra-operative or post-operative detection of nodal metastases greater than 0.2 mm. This molecular based test offers the advantage of better performance characteristics and reproducibility than many of the traditional pathology tools.

Another example of molecular biomarkers and there impact on clinical diagnostics is the FDA-cleared UroVysion Bladder Cancer Kit

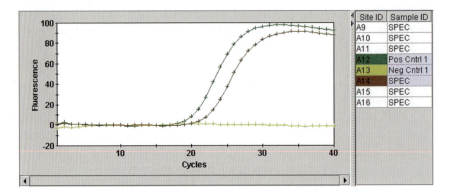

Figure 3.4   Real-time PCR amplification curve for the Breast GeneSearch assay.

(Abbott Molecular, Des Plaines, IL, USA). The diagnosis of primary and recurrent bladder cancer is a significant problem. Because this type of cancer is a chronic disease, patients need to be constantly monitored for recurrence of their disease. This assay is available for use, in conjunction with urine cytology and cystoscopy evaluation, as an aid for diagnosing and monitoring recurrence of urothelial carcinoma. The assay uses multicolor fluorescence *in situ* hybridization (FISH) to detect aneuploidy for chromosomes 3, 7 and 17 and loss of 9q21 locus (P16/ CDKN2A gene) in urine specimens, all abnormalities associated with urothelial malignancy. The combination of FISH with routine cytology evaluation is more sensitive than cytology alone at detecting urothelial carcinoma.[32–34]

Building on such applications with the advances in technology that are occurring will surely result in better management of the oncology patient. Clearly, the advantages of molecular diagnostics in the diagnosis and treatment of the cancer patient are only now becoming known. A better understanding of tumor cell biology and the pathways involved in carcinogenesis have led to novel biomarkers and therapeutics for human cancers. The diagnostic algorithms set in place for qualitative, quantitative and genotype infectious disease testing have shed light on these similar applications for oncology. In the context of patient management, not only will molecular biomarkers be responsible for the reclassification of many of these tumor types, they will also be responsible for directing therapy. Novel small-molecule therapeutics will require companion diagnostics to assess the feasibility and eligibility of a patient for a particular targeted therapy. As molecular technologies continue to improve, so will management practices.

3.7 PHARMACOGENOMICS

A rapidly growing area for molecular diagnostics is in the field of pharmacogenomics (PGX). The ability to select and dose therapeutic drugs through the use of genetic testing is becoming a reality. Although other factors will most certainly continue to play a role in drug selection and dosing, information stored in DNA can now be used to predict how an individual will respond to commonly prescribed drugs and novel targeted therapies.[35-38] The overall aim of PGX testing is to decrease adverse responses to therapy and increase efficacy by ensuring the appropriate selection and dose of therapy.[39,40]

Vogel was the first to use the term pharmacogenetics in 1959,[41] yet thousands of years earlier, in 510 BC, Pythagoras recognized that some individuals developed hemolytic anemia with fava bean consumption.[42] In 1914, Garrod expanded upon these early observations to state that enzymes detoxify foreign agents so that they may be excreted harmlessly; however, some people lack these enzymes and experience adverse effects. Hemolytic anemia due to fava bean consumption was later determined to occur in glucose-6-phosphate dehydrogenase-deficient individuals.

Over the years, the central dogma for assessing human diseases at a molecular level, DNA → RNA → protein, became the model for moving forward the newly discovered knowledge concerning pharmacogeniomics. Many genetic variants or polymorphisms in genes coding for drug-metabolizing enzymes, drug transporters and drug targets have been found to alter response to therapeutics. Technologies widely used in molecular diagnostic laboratories can be used to identify individuals with these genetic variations. These variants, such as single nucleotide polymorphisms (SNPs), can have no significant effect, change drug metabolism by >10 000-fold or alter protein binding by >20-fold. Many of these SNPs may be used to evaluate individual risk for adverse drug reactions (ADRs), so that we may decrease ADRs, select optimal therapy, increase patient compliance, develop safer and more effective drugs, revive withdrawn drugs and reduce the time and cost of clinical trials. ADRs account for more than 2 million annual hospital admissions in the USA and lead to 100 000 deaths.[39,40] Thus, the potential to minimize ADRs and improve response to therapy through genetic testing has heightened interest in the possibility of such routine clinical testing.

Most of the enzymes involved in drug metabolism are members of the cytochrome P450 (CYP) superfamily.[43-45] CYP2D6, for example, is a highly polymorphic enzyme which contains 497 amino acids.

The CYP2D6 gene is localized on chromosome 22q13.1 with two neighboring pseudogenes, CYP2D7 and CYP2D8.[46,47] More than 50 alleles of CYP2D6 have been described, of which alleles *3, *4, *5, *6, *7, *8, *11, *12, *13, *14, *15, *16, *18, *19, *20, *21, *38, *40, *42 and *44 were classified as non-functioning and *9, *10, *17, *36 and *41 were reported to have substrate-dependent decreased activity.[48] These cytochrome P450 enzymes are mainly located in the liver and gastro-intestinal tract and include > 30 isoforms.

The enzymes responsible for the majority of biotransformations and those that exhibit significant genetic polymorphisms are the CYP3A, CYP2D6, CYP2C19 and CYP2C9. Classification of these enzymes includes nomenclature, CYP2D6*1, such that the name of the enzyme (CYP) is followed by the family (CYP2), subfamily (CYP2D) and gene (CYP2D6) associated with the biotransformation. Allelic variants are indicated by a*, followed by a number (CYP2D6*1). Benign genetic variants or polymorphisms in these genes can lead to the following phenotypes: poor, intermediate, extensive and ultra-rapid metabolizers. Poor metabolizers (PM) have no detectable enzymatic activity; inter-mediate metabolizers (IM) have decreased enzymatic activity; extensive metabolizers (EM) are considered normal and have at least one copy of an active gene; ultra-rapid metabolizers (UM) contain duplicated or amplified gene copies that result in increased drug metabolism. It should also be noted that in addition to the genotype predicting a phenotype, drug-metabolizing enzyme activity can be induced or inhibited by var-ious drugs. Induction leads to the production of more enzyme within three or more days of exposure to inducers.[49] Enzyme inhibition by commonly prescribed drugs can be an issue in polypharmacy (a patient taking more than one prescribed medication) and is usually the result of competition between two drugs for metabolism by the same enzyme.

One clinical application of PGX has recently come to fruition for testing CYP polymorphisms associated with warfarin sensitivity. Oral anticoagulation with the vitamin K antagonist warfarin reduces the rate of thromboembolic events for patients in a variety of clinical settings. Warfarin therapy, however, is challenging because there are wide var-iations among patients in response to the drug and in dose require-ment.[50] Historically, to achieve and maintain an optimal warfarin dose, the prothrombin time and the international normalized ratio (INR) are monitored and doses are adjusted to maintain each patient's INR within a narrow therapeutic range. An INR of less than 2 is associated with an increased risk of thromboembolism and an INR of 4 or more is asso-ciated with an increased risk of bleeding. Polymorphisms in the gene encoding the CYP2C9 enzyme are known to contribute to variability in sensitivity to warfarin.[50,51] CYP2C9 is the enzyme primarily responsible for the metabolic clearance of the *S*-enantiomer of warfarin. Patients with certain common genetic variants of CYP2C9 require a lower dose of warfarin and a longer time to reach a stable dose (www.warfarindosing.com). They are also at higher risk for over-anticoagulation and serious bleeding.[52] Vitamin K epoxide reductase (VKORC1) is the target of anticoagulants and its common genetic variants result in altered sensitivity to warfarin. VKORC1 polymorphisms are associated with a need for lower doses of warfarin during long-term therapy. On the basis of these observations, the FDA approved a labeling change for warfarin that describes the reported effects of VKORC1 and CYP2C9 on dose requirements and calls for lower initiation doses for patients with certain genetic variations in CYP2C9 and VKORC1 enzymes.[53]

Although the genes that code for drug-metabolizing enzymes have received more attention in recent years as targets for PGX testing, the genes that code for proteins used to transport drugs across membranes also need to be considered when discussing PGX. These drug transporter proteins move substrates across cell membranes, bringing them into cells or removing them from cells. These proteins are essential in the absorption, distribution and elimination of various endogenous and exogenous substances, including pharmaceutical agents. Several groups of drug transporters that may be significant in the field of pharmacogenomics exist, including multidrug resistance proteins (MDRs), multidrug resistance-related proteins (MRPs), organic anion transporters (OATs), organic anion-transporting polypeptides (OATPs), organic cation transporters (OCTs) and peptide transporters (PepTs).[54–56]

ABCB1 is one such transporter and is a member of the ATP-binding cassette (ABC) superfamily of proteins. Also known as P-glycoprotein (P-gp) or MDR1, it is a 170 kDa glycosylated membrane protein expressed in various locations including the liver, intestines, kidney, brain and testis.[57] ABCB1 serves to eliminate metabolites and a wide range of hydrophobic foreign substances, including drugs, from cells by acting as an efflux transporter.[58] ABCB1 was first identified in cancer cells that had developed a resistance to several anticancer drugs because of an overexpression of the transporter. When expressed at normal levels in non-cancerous cells, ABCB1 has been shown to transport other classes of drugs out of cells, including cardiac drugs (digoxin), antibiotics, steroids, HIV protease inhibitors and immunosuppressants (cyclosporin A).[59] Genetic variations in the ABCB1 gene expressed in normal cells have been shown to have a role in interindividual variability in drug response.[60,61]

Another aspect of PGX testing that needs top be taken into account is the importance of drug targets, as most therapeutic drugs have targets that elicit the desired effects. These targets can include receptors, enzymes or proteins involved in various cellular events such as signal transduction and cell replication. Investigators have now identified polymorphisms in these targets that render them resistant to the particular therapeutic. Although these receptors may not show dramatic increases or decreases in activity as the drug metabolizing enzymes, biologically significant effects frequently occur.[36,43] As examples, polymorphisms have been identified in the angiotensin converting enzyme (ACE), β-adrenergic receptors and FMS-like tyrosine kinase 3 (FLT3) receptor.[62–64]

ACE is part of the angio-renin system and plays a large role in regulating cardiovascular functions such as blood pressure and cardiac output. Two genetic variants have been identified, an insertion (I-form) and deletion (D-form) of a base pair at position 287, in the gene. The prevalence of this mutation is equally distributed in the Caucasian population between the I/I (23%), I/D (49%) and D/D (28%) genotypes. Individuals with the D/D form express ACE levels 25–200% higher than I/I individuals and have a 23% increased risk of a myocardial infarction than I/D and I/I individuals.[62] Polymorphisms in the β_2-adrenoreceptor (ADRB2) gene and asthma have been extensively studied. β_2-Receptors are mainly expressed in the lung where the receptors exert their primary effect on bronchial smooth muscles, resulting in relaxation and dilation.[36] Initial studies using isoproterenol identified three single nucleotide polymorphisms that alter receptor function: Arg16Gly, Gln27Glu and Thr164Ile. Arg16Gly is associated with down-regulation of receptor expression, Gln27Glu is resistant to down-regulation and Thr164Ile displays decreased affinity along with altered coupling to the cAMP cascade.[63] FLT3 is an example of a receptor tyrosine kinase expressed and activated in most cases of acute myeloid leukemia (AML), which has a relatively high relapse rate due to acquired resistance to traditional chemotherapies.[64] An internal tandem duplication (ITD) mutation in the FLT3 gene is found in up to 30% of AML patients, while point mutations have been shown to account for approximately 5% of refractory AML. The FLT3-ITD induces activation of this receptor and results in downstream constitutive phosphorylation in STAT5, AKT and ERK pathways. This mutation in FLT3 is a negative prognostic factor in AML.

Interindividual variability has often been observed in response to chemotherapy, with most chemotherapeutic agents exhibiting up to 30% efficacy. Although many of these differences can be attributed to environmental factors, it is becoming clear that there are also genetic factors which can contribute to an individual's response to a particular therapy.

In the cancer patient, novel therapies have been developed which target specific molecules known to exacerbate the development of human tumors. These small-molecule-targeted therapies exert their actions on various known receptors such as HER2 and EGFR. One of the first and most widely used parameters for a targeted therapy is the evaluation of breast cancers for expression of the estrogen receptor (ER). ER-positive breast cancers are then treated with hormonal therapies that mimic estrogen. One of these estrogen analogues is tamoxifen (TAM), which itself has now been shown to have altered metabolism due to CYP450 genetic polymorphisms.[65] TAM is used to treat all stages of estrogen receptor-positive breast cancers.[66] TAM and its metabolites compete with estradiol for occupancy of the estrogen receptor and in doing so inhibit estrogen-mediated cellular proliferation. Conversion of TAM to its active metabolites occurs predominantly through the CYP450 system.[67] Conversion of TAM to primary and secondary metabolites is important because these metabolites can have a greater affinity for the estrogen receptor than TAM itself. For example, 4-OH-*N*-desmethyl-TAM (endoxifen) has approximately 100 times greater affinity for the estrogen receptor than tamoxifen. Activation of tamoxifen to endoxifen is primarily due to the action of CYP2D6[68] (Figure 3.5). Therefore, patients with defective *CYP2D6* alleles derive less benefit from tamoxifen therapy than patients with functional copies of CYP2D6.[69]

Another example relating PGX to oncology is the uridine diphosphate glucuronosyltransferases (UGT) superfamily of endoplasmic reticulum-bound enzymes responsible for conjugating a glucuronic acid moiety to a variety of compounds, thus allowing these compounds to be more easily eliminated. It is a member of this family that catalyzes the glucuronidation of bilirubin, allowing it to be excreted in the bile. As irinotecan therapy for advanced colorectal cancers became more widely used, it was observed that patients who had Gilbert syndrome, a mild hyperbilirubinemia, suffered severe toxicity.[70] Irinotecan is converted to SN-38 by carboxylesterase-2 and SN-38 inhibits DNA topoisomerase I activity[71,72] (Figure 3.6). SN-38 is glucuronidated by uridine diphosphate glucuronosyltransferase (UGT), forming a water-soluble metabolite, SN-38 glucuronide, which can then be eliminated.[73] The decreased glucuronidation of bilirubin and SN-38 can be attributed to polymorphisms in the *UGT1A1* gene. The 'wild-type' allele, *UGT1A1*1*, has six tandem TA repeats in the regulatory TATA box of the *UGT1A1* promoter. The most common polymorphism associated with low activity of *UGT1A1* is the *28 variant, which has seven TA repeats (Table 3.6). In August 2005, the FDA amended the irinotecan (Camptosar) package insert to recommend genotyping for the UGT1A1

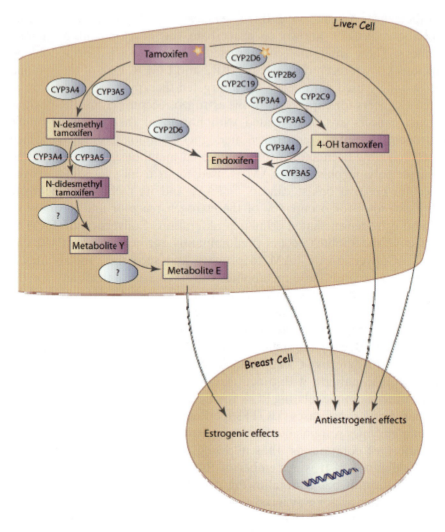

Figure 3.5   Schematic of tamoxifen metabolism.

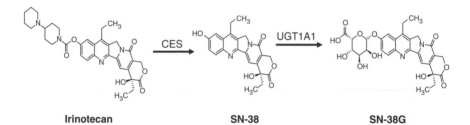

Figure 3.6   Schematic diagram of irinotecan metabolism.

Table 3.6 Frequency of UGT1A1 polymorphisms in the general Caucasian population.

TA repeat number	Allele designation	Frequency (%)
6/6	*1/*1	46
6/7	*1/*28	39
6/8	*1/*37	3
7/7	*28/*28	9
7/8	*28/*37	1.5

polymorphism and suggested a dose reduction in patients homozygous for the *28 allele.

3.8 CONCLUSION

Our knowledge base of genomics and proteomics in the clinical context of diagnostic testing is progressing at record speeds. Technology, as is typical in the clinical laboratory, has far outpaced proven clinical utility, yet routine applications of information from the 'omics era' are currently being performed. The ultimate goal of these efforts is to provide truly a 'personalized medicine' approach to patient management for the purpose of making better the clinical outcome and improve the overall well-being of the patient.

REFERENCES

1. G. J. Tsongalis and W. B. Coleman, *Clin. Chim. Acta*, 2006, **363**, 127.
2. J. Wilhelm and A. Pingoud, *ChemBioChem*, 2003, **4**, 1120.
3. C. T. Wittwer, M. G. Herrmann, A. A. Moss and R. P. Rasmussen, *Biotechniques*, 1997, **22**, 130.
4. S. B. Parks, B. W. Popovich and R. D. Press, *Am. J. Clin. Pathol.*, 2001, **115**, 439.
5. C. T. Wittwer, G. H. Reed, C. N. Gundry, J. G. Vandersteen and R. J. Pryor, *Clin. Chem.*, 2003, **49**, 853.
6. R. M. Ratcliff, G. Chang, T. Kok and T. P. Sloots, *Curr. Issues Mol. Biol.*, 2007, **9**, 87.
7. G. J. Tsongalis, *Am. J. Clin. Pathol.*, 2006, **126**, 1.
8. S. Yang and R. E. Rothman, *Lancet Infect. Dis.*, 2004, **4**, 337.
9. T. Y. Tan, *Expert Rev. Mol. Diagn.*, 2003, **3**, 93.
10. J. Versalovic and J. R. Lupski, *Trends Microbiol.*, 2002, **10**, S15.
11. M. J. Lewis, E. H. Lewis, J. A. Amos and G. J. Tsongalis, *Am. J. Clin. Pathol.*, 2003, **120** (Suppl 1), S3.

12. W. W. Grody, G. R. Cutting, K. W. Klinger, C. S. Richards, M. S. Watson and R. J. Desnick, *Genet. Med.*, 2001, **3**, 149.

13. ACOG. Preconception and prenatal carrier screening for cystic fibrosis: clinical laboratory guidelines. American College of Obstetricians and Gynecologists, Washington, D.C. 2001; 1–31.

14. S. Sherman, B. A. Pletcher and D. A. Driscoll, *Genet. Med.*, 2005, **7**, 584.

15. R. Willemsen, B. A. Oostra, G. J. Bassell and J. Dictenberg, *MRDD Res.*, 2004, **10**, 60.

16. A. Terracciano, P. Chiurazzi and G. Neri, *Am. J. Med. Genet.*, 2005, **137C**, 32.

17. D. G. Gilliland, *Semin. Hematol.*, 2002, **39**, 6.

18. H. Kiyoi, M. Yanada and K. Ozekia, *Int. J. Hematol.*, 2005, **82**, 85.

19. J. M. Scandura, *Curr. Oncol. Rep.*, 2005, **7**, 323.

20. M. C. Stubbs and S. A. Armstrong, *Curr. Drug Targets*, 2007, **8**, 703.

21. C. H. Lawrie, *Br. J. Haematol.*, 2007, **137**, 503.

22. E. S. Jaffe, N. L. Harris, H. Stein and J. W. Vardiman, *World Health Organization Pathology and Genetics of Tumours of the Haematopoietic and Lymphoid Tissues*, IARC Press, Lyon, 2001.

23. A. J. Bench, W. N. Erber, G. A. Follows and M. A. Scott, *Int. J. Lab. Hematol.*, 2007, **29**, 229.

24. Y.-L. Kwong, *Br. J. Haematol.*, 2007, **137**, 273.

25. J. J. van Dongen, A. W. Langerak, M. Brüggemann, P. A. Evans, M. Hummel, F. L. Lavender, E. Delabesse, F. Davi, E. Schuuring, R. García-Sanz, J. H. van Krieken, J. Droese, D. González, C. Bastard, H. E. White, M. Spaargaren, M. González, A. Parreira, J. L. Smith, G. J. Morgan, M. Kneba and E. A. Macintyre, *Leukemia*, 2003, **17**, 2257.

26. P. L. Fitzgibbons, D. L. Page, D. Weaver, A. D. Thor, D. C. Allred, G. M. Clark, S. G. Ruby, F. O'Malley, J. F. Simpson, J. L. Connolly, D. F. Hayes, S. B. Edge, A. Lichter and S. J. Schnitt, *Arch. Pathol. Lab. Med.*, 2000, **124**, 966.

27. J. Backus, T. Laughlin, Y. Wang, R. Belly, R. White, J. Baden, C. Justus Min, A. Mannie, L. Tafra, D. Atkins and K. M. Verbanac, *J. Mol. Diagn.*, 2005, **7**, 327.

28. E. Leygue, L. Snell, H. Dotzlaw, K. Hole, S. Troup, T. Hiller-Hitchcock, L. C. Murphy and P. H. Watson, *J. Pathol.*, 1999, **189**, 28.

29. R. J. Ouellette, D. Richard and E. Maicas, *Am. J. Clin. Pathol.*, 2004, **121**, 637.

30. M. A. Watson, S. Dintzis, C. M. Darrow, L. E. Voss, J. DiPersio, R. Jensen and T. P. Fleming, *Cancer Res.*, 1999, **59**, 3028.
31. B. K. Zehentner, *D. Carter. Clin. Biochem.*, 2004, **37**, 249.
32. K. C. Halling, W. King, I. A. Sokolova, R. G. Meyer, H. M. Burkhardt, A. C. Halling, J. C. Cheville, T. J. Sebo, S. Ramakumar, C. S. Stewart, S. Pankratz, D. J. O'Kane, S. A. Seelig, M. M. Lieber and R. B. Jenkins, *J. Urol.*, 2000, **164**, 1768.
33. M. Skacel, M. Fahmy, J. A. Brainard, J. D. Pettay, C.V. Biscotti, L. S. Liou, G. W. Procop, J. S. Jones, J. Ulchaker, C. D. Zippe and R. R. Tubbs, *J. Urol.*, 2003, **169**, 2101.
34. M. Skacel, J. D. Pettay, E. K. Tsiftsakis, G. W. Procop, C. V. Biscotti and R. R. Tubbs, *Anal. Quant. Cytol. Histol.*, 2001, **23**, 381.
35. R. M. Long, *Clin. Pharmacol. Ther.*, 2007, **81**, 450.
36. W. E. Evans and H. L. McLeod, *N. Engl. J. Med.*, 2003, **348**, 538.
37. J. Kircjjeiner, U. Fuhr and J. Brockmoller, *Nat. Rev.*, 2005, **4**, 639.
38. R. Weinshilboum and L. Wand, *Nat. Rev. Drug Discov.*, 2004, **3**, 739.
39. J. Lazarou, B. H. Pomeranz and P. N. Corey, *JAMA*, 1998, **279**, 1200.
40. B. S. Shastry, *Pharmacogenomics J.*, 2006, **6**, 16.
41. F. Vogel, *Ergeb. Inn. Med. Kinderheilkd.*, 1959, **12**, 52.
42. D. Nebert, *Clin. Genet.*, 1999, **56**, 247.
43. S. J. Gardiner and E. J. Begg, *Pharmacol. Rev.*, 2006, **58**, 521.
44. M. Ingelman-Sundberg, *Pharmacogenomics J.*, 2005, **5**, 6.
45. M. Ingelman-Sundberg, *Trends Pharmacol. Sci.*, 2004, **25**, 193.
46. M. H. heim and U. A. Meyer, *Genomics*, 1992, **14**, 49.
47. M. C. Ledesma and J. A. Agundez, *Clin. Chem.*, 2005, **51**, 939.
48. T. Andersson, D. A. Flockhart, D. B. Goldstein, S. M. Huang, D. L. Kroetz, P. M. Milos, M. J. Ratain and K. Thummel, *Clin. Pharmacol. Ther.*, 2005, **78**, 559.
49. H. K. Lee, L. D. Lewis, G. J. Tsongalis, M. McMullin, B. C. Schur, S. H. Wong and K. T. Yeo, *Clin. Chim. Acta*, 2006, **367**, 196.
50. M. R. McClain, G. E. Palomaki, M. Piper and J. E. Haddow, *Genet. Med.*, 2008, **10**, 89.
51. M. S. Wen, M. Lee, J. J. Chen, H. P. Chuang, L. S. Lu, C. H. Chen, T. H. Lee, C. T. Kuo, F. M. Sun, Y. J. Chang, P. L. Kuan, Y. F. Chen, M. J. Charng, C. Y. Ray, J. Y. Wu and Y. T. Chen, *Clin. Pharmacol. Ther.*, 2008, **84**, 83–89.
52. B. F. Gage and L. J. Lesko, *J. Thromb. Thrombolysis*, 2008, **25**, 45.
53. A. H. Wu, P. Wang, A. Smith, C. Haller, K. Drake, M. Linder and R. Valdes Jr., *Pharmacogenomics*, 2008, **9**, 169.

54. A. C. Lockhart, R. G. Tirona and R. B. Kim, *Mol. Cancer Ther.*, 2003, **2**, 685.

55. C. Marzolini, R. G. Tirona and R. B. Kim, *Pharmacogenomics*, 2004, **5**, 273.

56. B. L. Urquhart, R. G. Tirona and R. B. Kim, *J. Clin. Pharmacol.*, 2007, **47**, 566.

57. P. Borst, A. H. Schinkel, J. J. Smit, E. Wagenaar, L. Van Deemter, A. J. Smith, E. W. Eijdems, F. Baas and G. J. Zaman, *Pharmacol. Ther.*, 1993, **60**, 289.

58. S. V. Ambudkar, S. Dey and C. A. Hrycyna, *Annu. Rev. Pharmacol. Toxicol.*, 1999, **39**, 361.

59. T. Sakaeda, T. Nakamura and K. Okumura, *Biol. Pharm. Bull.*, 2002, **25**, 1391.

60. S. Hoffmeyer, O. Burk, O. Von Richter, H. P. Arnold, J. Brockmöller, A. Johne, I. Cascorbi, T. Gerloff, I. Roots, M. Eichelbaum and U. Brinkmann, *Proc. Natl. Acad. Sci. USA*, 2000, **97**, 3473.

61. K. Jamroziak and T. Robak, *Hematology*, 2004, **9**, 91.

62. N. J. SAamani, L. O'toole, K. Channer and K. L. Woods, *Circulation*, 1996, **94**, 708.

63. S. A. Green, G. Cole, M. Jacinto, M. Innis and S. B. Liggett, *J. Biol. Chem.*, 1993, **268**, 23116.

64. D. B. Shankar, J. Li, P. Tapang, J. O. McCall, L. J. Pease, Y. Dai, R. Q. Wei, D. H. Albert, J. J. Bouska, D. J. Osterling, J. Guo, P. A. Marcotte, E. F. Johnson, N. Soni, K. Hartandi, M. R. Michaelides, S. K. Davidsen, S. J. Priceman, J. C. Chang, K. Rhodes, N. Shah, T. B. Moore, K. M. Sakamoto and K. B. Glaser, *Blood*, 2007, **109**, 3400.

65. M. P. Goetz, J. M. Rae, V. J. Suman, S. L. Safgren, M. M. Ames, D. W. Visscher, C. Reynolds, F. J. Couch, W. L. Lingle, D. A. Flockhart, Z. Desta, E. A. Perez and J. N. Ingle, *J. Clin. Oncol.*, 2005, **23**, 9312.

66. Early Breast Cancer Trialists' Collaborative Group. Lancet, 1998, 351, 1451.

67. G. K. Poon, Q. Chen, Y. Teffera, J. S. Ngui, P. R. Griffin, M.P. Braun, G. A. Doss, C. Freeden, R. A. Stearns, D. C. Evans, T.A. Baillie and W. Tang, *Drug Metab. Dispos.*, 1993, **21**, 1119.

68. T. E. Klein, J. T. Chang, M. K. Cho, K. L. Easton, R. Fergerson, M. Hewett, Z. Lin, Y. Liu, S. Liu, D. E. Oliver, D. L. Rubin, F. Shafa, J. M. Stuart and R. B. Altman, *Pharmacogenomics J.*, 2001, **1**, 167.

69. Y. Jin, Z. Desta, V. Stearns, B. Ward, H. Ho, K. H. Lee, T. Skaar, A. M. Storniolo, L. Li, A. Araba, R. Blanchard, A. Nguyen,

L. Ullmer, J. Hayden, S. Lemler, R. M. Weinshilboum, J. M. Rae, D. F. Hayes and D. A. Flockhart, *J. Natl. Cancer Inst.*, 2005, **97**, 30.

70. E. Wasserman, A. Myara, F. Lokiec, F. Goldwasser, F. Trivin, M. Mahjoubi, J. L. Misset and E. Cvitkovic, *Ann. Oncol.*, 1997, **8**, 1049.
71. R. Humerickhouse, K. Lohrbach, L. Li, W. F. Bosron and M. E. Dolan, *Cancer Res.*, 2000, **60**, 1189.
72. J. T. Hartmann and H. P. Lipp, *Drug Saf.*, 2006, **29**, 209.
73. M. L. Maitland, K. Vasish and M. J. Ratain, *Trends Pharmacol. Sci.*, 2006, **27**, 432.

CHAPTER 4

Molecular Microbial Diagnostics

KARL-HENNING KALLAND

Centre for Research in Virology, The Gade Institute, University of Bergen, N-5009 Bergen, Norway

4.1 INTRODUCTION

The SARS (severe acquired respiratory syndrome) epidemic showed the power of molecular microbial diagnostics. The World Health Organization (WHO) issued a global alert in March 2003 and 1 month later the causative agent, a coronavirus, was molecularly cloned and sequenced.[1] This facilitated the design of a polymerase chain reaction (PCR) assay for the accurate detection of the SARS virus in patient materials. The sequence information was made available online and allowed diagnostic laboratories worldwide to establish their own SARS virus assays within weeks. The molecular diagnostic tools were further used to trace the origin of the epidemic back to Guangdong Province in China in November 2002. The civet cat was identified as a likely viral reservoir.

Coronaviruses are ubiquitous RNA viruses and account for a significant proportion of all upper respiratory tract infections of the common cold type. Comparison of the nucleic acid sequences of the common cold coronaviruses and the SARS virus provided clues to the increased pathogenesis of the latter.

The HIV epidemic that appeared in 1983 provides another example of the profound impact of molecular biology in the handling of emerging diseases. Indeed, effective tools are necessary since the microbiological

Molecular Biology and Biotechnology, 5th Edition
Edited by John M Walker and Ralph Rapley
© Royal Society of Chemistry 2009
Published by the Royal Society of Chemistry, www.rsc.org

world develops so quickly that we can eyewitness microbiological evolution within the course of our own lifetime. In fact, new infectious agents and new types of infectious diseases have appeared almost every year during the last 20 years (Table 4.1) and this is likely to continue. Still the causative agent cannot be identified in about 40% of respiratory diseases and more than half of all cases of encephalitis. It has been established that 20% of all human cancers are due to defined infectious agents, but indirect evidence suggests that infections may contribute to about 30% of all cancers. The role of infectious agents in autoimmune disease remains to be settled.

Table 4.1 Chronological examples of discovery of infectious agents.[a]

2005	Human retroviruses (HTLV3 and HTLV4)
	Human bocavirus
	Mimivirus
	Human coronavirus HKU1
2004	Simian foamy retroviruses
2003	SARS coronavirus
2001	Metapneumovirus
1999	Nipah virus
	West Nile virus in USA
1997	Avian influenza virus in humans (H5N1)
1996	nvCJD
	Bat lyssaviruses (Australia, Scotland)
1995	Human herpes virus 8 (Kaposi sarcoma virus)
1994	Sabia virus
	Hendra virus
1993	Sin nombre virus (hantavirus pulmonary syndrome)
1992	*Vibrio cholerae* O139
	Bartonella henselae
1991	Guanarito virus (Venezuelan haemorrhagic fever)
	Hepatitis C virus
	Ehrlichia chafeensis
	Photorhabdus asymbiotica
	Hepatitis E virus
	Human herpes virus 6
1986	*Cyclospora cayatenensis*
	Bovine spongiform encephalopathy
1985	*Enterocytozoon bieneusi*
1983	HIV-1
	Helicobacter pylori
1982	*Escherichia coli* O157:H7
	HTLV2
	Borrelia burgdorferi (Lyme disease)
1981	Toxic shock syndrome associated *Staphylococcus aureus*
1980	Human T-lymphotropic virus (HTLV1)

[a]Examples of emerging infectious agents and the year of discovery. More comprehensive information can be found at http://www.hpa.org.uk/infections/topics_az/emerging_infections/list.htm and http://www3.niaid.nih.gov/research/topics/emerging/list.htm.

4.2 CLASSICAL MICROBIOLOGICAL DIAGNOSIS

Microbiological diagnosis is involved with the detection of different classes of disease causing agents such as viruses, bacteria, fungi and many types of parasites. The mainstay of diagnosis still is the clinical history and the physical signs of the patient. However, ever since bacteria were recognised as disease-causing agents in the 1860s, clinicians have relied on supplementary laboratory tests for the confirmation of a tentative diagnosis, including microscopy, colony growth on agar plates and growth in liquid cultures supplemented with different types of chemical indicators (Table 4.2).

The focus of this chapter will be on nucleic acid-based microbiological diagnostics. Viruses are too small to be observed through the light microscope. Instead, the electron microscope is useful for the identification of viruses that cause diarrhoea and poxviruses that cause warts and other skin lesions. However, electron microscopy is hampered by its low sensitivity and the work required to prepare and examine electron micrographs. Agglutination tests are still used to detect, *e.g.*, adenovirus or rotavirus antigens in diarrhoea, but the sensitivity of this and other

Table 4.2   Classical methods in the microbiology laboratory.

A. Viral agents
Serology
 ELISA, EIA
 Complement fixation test
 Others
Cell culture
 Light microscopy
 Immunofluorescence
Electron microscopy
Immunofluorescence
Agglutination
Inoculation of egg, animals
Nucleic acid-based techniques
B. Bacteria
Light microscopy
 Of patient samples
 Of cultures and colonies
Growth in cultures
 Blood culture
 Broth culture
 Semi-solid/solid media
Fermentation indicators
Antibiotic susceptibility tests
Nucleic acid-based techniques

immunological methods is too low in most other viral diseases. Inoculation of a permissive cell culture by infected patient materials yields an efficient biological amplification of the virus, but growth takes days and different viruses require different cell types for efficient propagation, and many viruses may not grow in cell cultures at all – again limiting the speed and efficacy of diagnosis. Serological methods detect specific antibodies made by the immune system of the patient as a response to the disease causing agent or detect specific viral antigens. Serology still dominates routine diagnostics in virology. ELISA and EIA techniques have vastly improved the sensitivity, speed and accuracy of serology, but none of the serological techniques can circumvent the fact that it often takes from days to weeks following acute disease until detectable levels of antibodies are present in the circulation. In some viral diseases, usually when the incubation time is longer, such as in the case of rubella or parvovirus B19 infections, the appearance of specific antibodies coincides with the appearance of clinical signs such as skin rashes. In these cases, serology is an excellent and adequate diagnostic tool.

Nucleic acid amplification techniques, of which PCR represents the prototype, have entirely changed the basis of virological diagnosis and thereby the acute differential diagnosis of infectious diseases. The main reason is the possibility of detecting the nucleic acid of the causative agents directly, rapidly, specifically and with very high sensitivity. Nucleic acid-based techniques present, however, a number of technical challenges on their own. Due to extreme sensitivity, the amplification techniques are susceptible to contamination and their dependence on exact nucleic acid sequences may cause the tests to fail due to natural sequence variability or new mutations in critical nucleic acid sequences of the agent. The remainder of this chapter will be devoted to microbiological nucleic acid-based techniques, their principles, the applicability and remaining challenges and trends.

4.3 SAMPLE COLLECTION AND NUCLEIC ACID PURIFICATION

4.3.1 Sample Collection and Transport

Virtually all human secretions, fluids and tissue samples can be collected and examined for nucleic acids of suspected infectious agents (Table 4.3). In addition, laboratory cultures, bacterial colonies and infected cell cultures are used for nucleic acid extraction and identification. The microbiological diagnostic laboratory provides kits for taking and shipping patient samples. Some kits contain buffers that inactivate infectious

Table 4.3 Patients' samples in clinical microbiology.

Serum
EDTA blood
Plasma
Cerebrospinal fluid (CSF)
Sputum
Expectorate, nasopharyngeal aspirate
Vesicle and ulcer fluids
Urine
Faeces
Tissue biopsies and body fluids

agents, inhibit nucleases and lyse membranes to facilitate subsequent nucleic acid extraction. Whenever patient samples are collected and sent without additives, speedy delivery to the laboratory must be ensured and great care must be taken to label the sample appropriately and to avoid contamination of the outside of the collecting vessel. Several companies supply swab kits and buffer kits for stabilising samples during transport.[2] Stool transport and recovery buffer has proved useful for, *e.g.*, faecal samples (S.T.A.R.; Roche Diagnostics, Indianapolis, IN, USA). The swab extraction tube system (S.E.T.S.; Roche Diagnostics) is a simple kit for rapidly and efficiently recovering specimens attached to and absorbed into the fibres of a collection swab. IsoCode Stix (Schleicher and Schuell, Keene, NH, USA) provides a method for stabilising blood samples that are to be transported long distances for later testing by real-time PCR.[2]

4.3.2 Extraction of Nucleic Acids

Different nucleic acid extraction protocols are required for different samples and different microbiological agents. RNA is more susceptible than DNA to rapid degradation by nucleases and the best way to counteract this problem is to dissolve the sample in a guanidinium isothiocyanate-based lysis buffer as soon as possible. Viral nucleic acids usually can be extracted in the same way as host cell nucleic acids are extracted. Sometimes, *e.g.* for faecal samples, gentle vortexing followed by a low-speed centrifugation can remove debris and viral particles will remain in the supernatant. During subsequent ultracentrifugation, virus particles can be pelleted and concentrated. The thick cell wall of Gram-positive bacteria is more difficult to disrupt than the relatively thinner cell wall of Gram-negative bacteria and need special procedures.[3,4]

Haem in blood, bile in stool and components of urine and additives such as heparin may inhibit Taq polymerase and must be removed by the purification procedure prior to the amplification step. Purified nucleic acids should be eluted into a small volume of nuclease-free water or 0.1 mM EDTA. Ethanol or propan-2-ol precipitation may be used to achieve smaller volumes, but must be completely removed by pipetting and evaporation to avoid subsequent enzymatic inhibition.

4.3.3 Manual Extraction of Nucleic Acids

Several commercial manual extraction kits are available for use by clinical laboratories (see Table 2 in ref. 2). These kits vary in the method, cost and time required for extraction. Processing by manual methods involves a higher risk of contamination and switching of patient samples. US Clinical Laboratory Improvement Amendments of 1988 (CLIA) regulations provide guidelines regarding manual extraction of nucleic acids for diagnostic purposes (http://www.cms.hhs.gov/clia/).

4.3.4 Automated Extraction of Nucleic Acids

Automated extraction instruments are manufactured by a number of different companies and, like manual methods, vary in method, cost and time requirements for extraction. Additionally, these instruments vary in the specimen capacity per run and size (see Table 3 in Ref. 2 and Table 1 in Ref. 5). Comparisons of manual and automated extraction methods have found automated procedures to be equivalent and in some instances superior to manual methods (see Ref. 2 and references cited therein), although not always.[6] Automated extraction systems have certain inherent advantages over manual methods. Recovery of nucleic acids from automated instruments is consistent and reproducible. Automated extraction systems keep sample manipulation to a minimum, reducing the risk of cross-contamination of samples. They are most economical when instruments are fully loaded, although smaller and more versatile instruments have now become available. Space and cost of equipment and disposables also need to be considered.[2]

4.4 NUCLEIC ACID AMPLIFICATION TECHNIQUES

4.4.1 Polymerase Chain Reaction (PCR)

PCR represents the prototype of nucleic acid amplification techniques and its principles are described in Chapter 5, Section 3, Figure 5.7. Excellent reviews on its applicability in clinical microbiology have been published.[5,7,8] Due to its extreme sensitivity, nucleic acid amplification has revolutionised the identification of infectious agents in patient samples. Several alternative nucleic acid amplification methods have been developed for use in the clinical microbiological laboratory, such as the ligase chain reaction (LCR), isothermal strand displacement amplification (SDA) and isothermal transcription-mediated amplification (TMA/NASBA).

4.4.2 The Contamination Problem

Great care must be taken in any diagnostic laboratory to minimise the problem of contamination of samples due to the exquisite sensitivity of nucleic acid amplification techniques. Sterile filter tips must be employed for all manual pipetting steps. The logistics of the laboratory procedures are of fundamental importance. A separate room should be reserved for work with buffer components, oligonucleotide primers and probes and mastermixes. No samples or amplified products should be allowed into this room. Higher air pressure than in the surrounding rooms, restricted access, visible written hygiene guidelines on the door and use of separate clothing and shoes are advantageous. A separate room is necessary for receiving and organising patient samples and a third room for setting up the PCR reactions. Using conventional PCR, a fourth separate room will be necessary for the detection steps of the amplified products, the amplicons. Commonly used methods for the visualisation of amplicons are agarose gel electrophoresis and microplate systems with a detection probe and a streptavidin–enzyme conjugate that binds to biotinylated nucleotides incorporated into the amplicon. Following a washing step, the enzyme bound by the biotin–streptavidin interaction will change a colourless substrate into a coloured derivative in a quantitative way. Using real-time PCR (see below), the amplification and detection can be completed in a closed system and for space considerations may be performed in the same room as the reaction setup. If possible, one-way flow of persons and samples from the first to the last room is desirable. Decontamination of critical work space using UV light or 5% bleach represents an additional precaution against contamination of samples. When UTP is used in the deoxynucleotide mix of the amplification buffer of the previous setup, a preamplification step using uracil-N-glycosylase will efficiently remove amplicon contamination in the subsequent reaction.

4.4.3 Reverse PCR – cDNA Synthesis

DNA is suitable for diagnostic purposes when the infectious agent is a bacterium, a parasite or a DNA virus. A large proportion of viruses are RNA viruses, however, *i.e.* their genomes consist of RNA and are replicated in host cells as RNA. Reverse transcription of RNA to complementary DNA (cDNA) is therefore an obligatory step for the identification of RNA viruses. When PCR DNA amplification is preceded by cDNA synthesis, the combined protocol is named reverse PCR. Reverse PCR may also be useful in order to quantify transcription (mRNA synthesis) to show the activity of an infectious genome. Using enzymes such as rTth, which has both reverse transcriptase and DNA polymerase activity and is thermo-stable, reverse transcription can be achieved in one reaction and one tube. Otherwise, the reverse transcription is often carried out as a separate first step using a thermolabile reverse transcriptase (typically cloned retroviral enzymes). If random primers (hexamers or nonamers) or oligo-dT primers are used in the initial reverse transcription, then the subsequent PCR can be performed with different specific primer pairs and designed for different RNA viruses. According to the rTth reverse PCR, the same reverse primer sequence will prime both cDNA synthesis and the PCR amplification. This may increase specificity, but reduces the versatility compared with random hexamer or oligo-dT primed cDNA.

4.4.4 Nested PCR

Experience shows that whenever PCR is performed on a patient sample, the amplification products very often exhibit a smear in the following ethidium bromide-stained agarose gel electrophoresis. This is due to non-specific priming in the background of high amounts of host cell nucleic acids and the need to perform a large number of thermocycles (typically 40). Although the first PCR amplification resulted in a smear in the agarose gel, the specific nucleic acids sought will usually be highly enriched compared with the starting material. If a small amount of the first PCR is used for a second PCR amplification with a new set of internal primers, then a well-defined PCR fragment typically results if the infectious agent is present in the initial sample. The use of a second round of PCR with a new set of internal primers is called a nested PCR. If only one internal primer is used for the second round of thermocycling, it is called a semi-nested PCR. Nested and semi-nested PCR therefore increase the sensitivity as much as 1000-fold compared with conventional PCR[9] and increase specificity, but the techniques require more work and longer assay times and carry an increased risk of contamination.

4.4.5 Real-time PCR

Compared with conventional PCR, and in particular compared with nested PCR, real-time PCR has the advantage that the amplification process is recorded in real time. The risk of contamination is very much reduced since there is no need to open the reaction tube once the amplification has started, *i.e.* the post-amplification detection step is omitted. Therefore, and because more of the process can be automated, it takes much less time and labour to do real-time PCR than nested PCR. For the same reasons, the risk of sample switching is lower in real-time PCR. The sensitivity is comparable between real-time PCR and nested PCR. Real-time PCR results are digitised instantly and analysed using standardised computer algorithms, thus minimising human errors, and may provide test results within 1 h after the start of the analysis, whereas analysis based on nested PCR may take hours to complete.

4.4.6 Visualisation of Real-time PCR Amplification

The critical difference between conventional and real-time PCR is that in the latter the generation of amplification products is monitored directly in the reaction tube. Different fluorescence-based principles are employed for real-time detection. SYBR Green represents one class of DNA binding chemicals that emits much more fluorescence when intercalated into double-stranded DNA compared with the unbound chemical in solution (Figure 4.1). As a result, the SYBR Green fluorescence increases proportionally to the increase in DNA amplification products. The disadvantage of using SYBR Green is that non-specifically amplified DNA and amplification artefacts such as primer dimers will also generate increased fluorescence, thus reducing the sensitivity compared with nested PCR. Real-time PCR assays utilising an amplicon-specific probe increase the specificity substantially. A number of 'smart' assays have been invented based on three oligonucleotides,

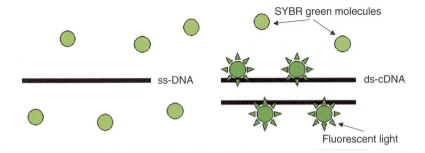

Figure 4.1   The SYBR Green molecule has high affinity to double-stranded DNA. SYBR Green fluorescence is much stronger when it is bound to DNA compared with unbound SYBR Green. ss-DNA = single-stranded DNA; ds-DNA = double-stranded DNA.

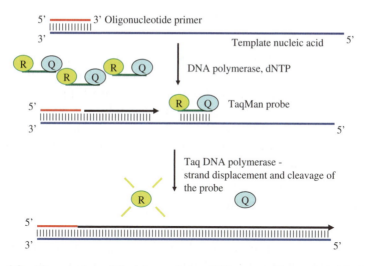

Figure 4.2 The principle of TaqMan real-time PCR assays. The probe is labelled with both a fluorescent chemical group (R) and a quencher (Q). When the Taq polymerase elongates the complementary strand from the primer, its 5′–3′ exonuclease activity cleaves any annealed probe, thus separating R from Q. Freed from its quencher, the R fluoresces when exciting light is sent in. The amount of free R, and thus fluorescence, increase exponentially along with amplicons until the PCR reaction approaches the plateau stage.

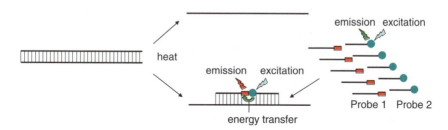

Figure 4.3 The principle of fluorescence resonance energy transfer (FRET). Two different probes are complementary to the target sequence. One probe is labelled with fluorescein and the other with LC Red fluorochrome. The hybridisation to the target sequence will bring the two fluorochromes physically close enough for the emission energy of fluorescein to excite the LC Red so that red fluorescent light is emitted. When PCR generates an increasing concentration of target, an increasing fraction of the probes can be positioned to achieve FRET. This can be monitored as increasing red fluorescence.

i.e. two primers and a probe. Several of these assays include TaqMan assays (Figure 4.2), fluorescence resonance energy transfer (FRET) (Figure 4.3) assays, Eclipse probes, Scorpion probes and molecular beacon probes.

4.4.7   Real-time PCR Equipment

Compared with conventional PCR, real-time qPCR puts extra demands on the equipment. Lasers are needed to excite fluorophores and recording devices are required to collect and digitise the fluorescent signals. Several of the available thermocyclers are listed in Table 1 in Ref. 2. The choice of thermocycler will depend on a large number of variables. Should the thermocycler be downstream of a high-throughput 96-well-format? Or is there a need for individualisation of PCR reactions that can be provided by the Smartcycler?

4.4.8   Real-time Quantitative PCR

The increasing fluorescence signal plotted on the y-axis and the increasing thermocycle numbers along the x-axis generate a sigmoidal curve. The threshold cycle is defined as the cycle number when the fluorescent signal exceeds the detection threshold. During the following 4–8 cycles, the fluorescence typically approaches an exponential increase before approaching a plateau phase.[10] During the exponential phase, the logarithm of the fluorescent signal along the y-axis and the cycle number along the x-axis generate a straight line that can be used for quantification. In this phase of the real-time PCR, the fractional threshold cycle number (C_t) is inversely related to the logarithm of the number of starting templates.[11] Parallel PCR amplification of a dilution series of known amounts of the same template therefore allows the construction of a calibration curve from which the exact amount of starting template can be calculated.[7] It must then be assumed that the amplification efficiency is equal for all samples and standards. One sign of acceptable quantification is that the slope of the calibration curve is between -3.3 and -3.4. More exact guidelines to quantification and quality control can be found in user bulletins to thermocycler software (*e.g.* ABI Prism User Bulletin No. 2, http://www.appliedbiosystems.com/).

4.4.8.1   *Absolute and Relative Quantifications*

In order to achieve absolute quantification, a known amount of reference standard is required for the dilution series and calibration curve generation. Synthetic oligonucleotides, purified PCR fragments or plasmids or *in vitro* transcribed RNA can be quantified using absorption readings at a wavelength of 260 nm (OD_{260}). Dilution series of exactly known amounts of target then allow absolute quantification. Relative quantification, relating the amount of template to an accepted standard, is often sufficient and this can be achieved using the calibration curve method or the comparative C_t method (see Ref. 7 and references cited therein or ABI Prism User Bulletin No. 2).

4.4.9 Determination of 'Viral Load' in Clinical Microbiology

'Viral load' refers to the number of viral genomes in a defined volume of a patient sample, typically in serum or plasma. Determination of viral load is becoming increasingly important with increasing use of antiviral medication. Real-time quantification of viral load is one important criterion for the start of antiviral medication against human immunodeficiency virus (HIV), hepatitis C and hepatitis B viruses, against cytomegalovirus (CMV) in immunocompromised patients and against Epstein–Barr virus (EBV) in post-transplant patients. Efficient response to antiviral therapy can be monitored as the viral load is reduced and quantification is of great value to monitor possible development of viral resistance to the medication or possible relapse after cessation of therapy. Commercial kits are widely used for quantification of viral load in clinical microbiology. A known amount of synthetic standard is typically included in each reaction to establish a competitive PCR amplification. The resulting amounts of viral and standard amplicons are compared with a standard dilution curve in order to establish the viral load. It is important to be aware that different manufacturers may have different standardisation units so that results obtained using different kits may not be comparable. Progress has recently been made to establish internationally accepted standard units for different viruses, *e.g.* for hepatitis B virus,[12] hepatitis C virus[13] and parvovirus B19.[14] The WHO provides guidelines and established standards in order to calibrate different methods to obtain comparable International Units per millilitre ($IU\,mL^{-1}$) (http://www.who.int/biologicals/reference_preparations/distribution/en/index.html).

4.4.10 Internal Controls in Microbiological Real-time qPCR

Both negative and positive controls are important in nucleic acid-based diagnosis in the microbiological laboratory.[2] The negative control consists of a parallel amplification reaction that does not contain the specific target. The most difficult decision is the number of negative controls and recommendations for a negative control for every fifth tube to control efficiently for contamination may come in conflict with available resources.[2] Positive controls can be any target sequence in a separate tube. The best positive control templates are included in the same tube as the diagnostic assay and contain flanking sequences complementary to the primers of the diagnostic assay but with a unique sequence in the region recognised by the probe. Since real-time qPCR assays are typically designed with amplicons shorter than 100 nucleotides, it is easy to obtain the desired control template sequence by oligonucleotide synthesis. A more difficult challenge is if the laboratory wants to spike the patient sample with a control sequence to control for the entire process from sample handling, via nucleic acid extraction through PCR amplification. For RNA viruses, which represent the greatest challenge, this can be achieved if a known amount of an *in vitro*-synthesised RNA is spiked into the patient sample. Any exogenous sequence will do if it can be detected separately from the sequence of the infectious agent. However, the optimal control RNA will be transcribed from a vector that can generate an RNA that contains more than 1500 nucleotides and includes a target sequence that can be amplified with the same primers as the diagnostic assay but differs in the region recognised by the probe. In all same-tube controls it is very important to titrate carefully the amount of control target. Otherwise, the control may compete out the diagnostic assay and lead to false negatives.

4.4.11 Multiplex Real-time PCR

Multiplex PCR refers to the use of more than one primer pair in one reaction tube in order to amplify more than one target sequence or to more than one primer/probe set in the case of real-time PCR. The internal control assay described above is one example of multiplex PCR. In microbiological diagnostics, the general aim of multiplex PCR is to test simultaneously for different agents in the same reaction. This requires the use of fluorophores and equipment with the ability to distinguish between the different assays. Non-specific amplification products seem to be the most limiting factor of multiplex PCR as the number of oligonucleotide primers and probes increases. Several multiplex assays for the simultaneous detection of two or more different agents are, however, well established. Examples include multiplex PCR assays for herpes simplex virus 1 and 2 (HSV-1 and HSV-2) and assays for influenza type A and B viruses.

4.4.12 Melting Curve Analysis

By taking advantage of melting curve analysis, it has been possible to increase the number of agents detected in multiplex real-time PCR, in particular to distinguish between closely related viruses such as HSV-1 and HSV-2 or influenza A and B. FRET probes (Figure 4.3) are useful for this type of analysis. When one FRET probe is designed to be complementary to one viral subtype but contains one or very few mismatches to the other viral subtype, melting curve analysis is possible due to different melting points (T_m) depending on the exact target sequence. Melting curve analysis is performed following completion of the PCR. The probe pair is allowed to anneal to the target amplicons and then the temperature is increased while the fluorescence is continuously recorded. The viral subtype with mismatch to the FRET probe will exhibit a reduced fluorescence at a lower T_m than the viral subtype with perfect homology to the FRET probe. If both subtypes are present in the same mix, this may be visualised as a curve with two different peaks when the fluorescence is plotted on the *y*-axis and temperature on the *x*-axis. A number of multiplex real-time PCR assays have been published, including simultaneous amplification and subtyping of HSV-1 and HSV2 and VZV using FRET probes[15,16] and the combination of influenza A subtypes H1N1 and H3N2, influenza B and respiratory syncytial virus (RSV) subtypes A and B.[16,17]

4.4.13 Genotyping

Viruses evolve much more rapidly than their host organisms. In particular, RNA viruses that are replicated by RNA polymerases which lack 5′ proof reading, in contrast to DNA polymerases, are prone to point mutation errors in the order of 10^{-4} per genome replication. Both DNA viruses and RNA viruses may change rapidly when there are changes in the selection pressure due to very short generation times and an enormous number of progeny in each infection. As a consequence, most viral species come in multiple serotypes and genotypes and we have the dubious possibility to eyewitness viral evolution. In many cases it is clinically relevant to distinguish between genotypes. Twelve different main genotypes are currently described for hepatitis C virus. Both prognosis and the recommended duration of PEG-interferon α-ribavirin combination therapy vary between HCV genotypes. The prevalent genotypes 1 and 4 usually require 48 weeks of combination therapy whereas 12 weeks of therapy is often sufficient for genotypes 2 and 3. Also, sustained viral response, meaning that HCV RNA cannot be detected in serum 6 months following cessation of treatment, is achieved in a higher proportion of cases for genotypes 2 and 3. More than 130 genotypes of human papillomaviruses (HPVs) are currently recorded. Less than 90% sequence homology in regions of the L1, E6 and E7 genes compared with any previously known HPV type is required to define a new HPV type. HPVs are separated into high- and low-risk types regarding their ability to cause cervical cancer. Genotyping can be

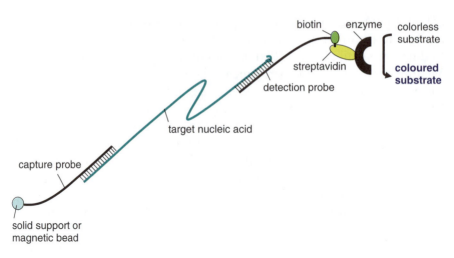

Figure 4.4   Principle of sandwich hybridisation.

achieved using DNA sequencing. Due to the current labour and cost of DNA sequencing, other methods have been developed for genotyping in the clinical microbiology laboratory such as sandwich hybridisation assays (Figure 4.4) and line probe assays (Figure 4.5).

4.5   OTHER TECHNIQUES USED IN CLINICAL MICROBIOLOGY

4.5.1   Hybridisation Techniques

The principle of complementarity, according to which A base pairs with T and C base pairs with G in a nucleic acid duplex, is the core principle of heredity, all hybridisation techniques and the PCR alike. The A–T base pairing involves two and the G–C base pairing involves three hydrogen bonds. Therefore, the relative content of A–T and G–C base pairs contributes strongly to the T_m of the nucleic acid duplex. The higher the G–C content, the higher is the T_m of the nucleic acid duplex. Different types of hybridisations were among the first techniques attempted in the laboratory detection of nucleic acids. In general, it turned out that the sensitivity of hybridisation techniques was not sufficient for routine detection of infectious agents directly in patient samples. The invention of nucleic acid amplification techniques therefore represented a breakthrough in nucleic acid-based laboratory diagnosis. However, in some cases the sensitivity of hybridisation was promising and encouraged further development of methods to detect

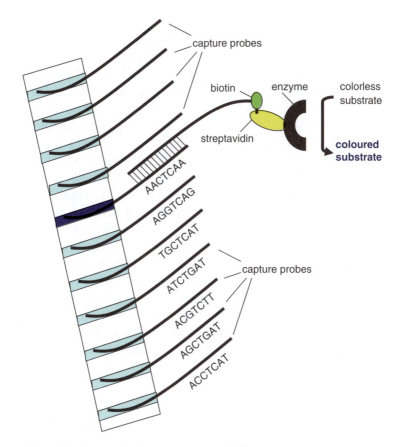

Figure 4.5 Principle of the line probe assay. Different capture probe sequences are fixed in different bands in a membrane strip. The amplification product will hybridise with the capture probe of complementary sequence. Utilising biotin-labelled PCR primers, the position of the complementary capture probe can be visualised as a coloured band following addition of streptavidin-bound enzyme that will change a non-coloured substrate to a coloured product that precipitates in the band.

nucleic acids, in particular for bacteria containing multicopy plasmids or repeated genomic sequences.[18–20]

4.5.1.1 Sandwich Hybridisation

In the classical hybridisation techniques, the nucleic acid target was usually fixed on a solid phase (nitrocellulose, nylon membrane, glass slide) and detected by a labelled probe added in the hybridisation buffer to the solid phase. These techniques, such as dot blots, colony blots, Southern blots, Northern blots and *in situ* hybridisation, usually either lacked sufficient sensitivity due to background hybridisation or were too laborious for the routine setting. The sandwich hybridisation (Figure 4.4) technique turned out, however, to have sufficiently low background and strong enough signal to be employed for some routine laboratory purposes. In this technique, the capture probe is fixed to the solid phase and available to the target in the hybridisation buffer. Following annealing between the capture probe and the specific target, unbound material is removed by washing and a new hybridisation buffer containing a detection probe is added. The detection probe is typically labelled with either biotin or an enzyme and is designed to bind to a different region of the target nucleic acid than the capture probe. The detection probe will therefore stick to the capture probe and the solid phase only if bridged by the specific target. Following a wash, retained detection probe can be visualised by addition of a colourless substrate that will change to a coloured product by the enzyme conjugated to the detection probe or by a streptavidin–enzyme complex in the case of a biotin-labelled detection probe. The solid phase may be a magnetic particle or the bottom of a 96-well plate to facilitate automation. The sandwich hybridisation technique is used in commercial tests for the detection of human papillomaviruses (HPVs) in samples from the female cervix (Digene) and in the visualisation step of conventional PCR assays (*e.g.* several Roche assays). The principle is also utilised in line probe assays.

4.5.1.2   Line Probe Assays

In line probe assays (Figure 4.5), capture probes are printed in parallel bands (lines) on a membrane. Line probe assays have been developed to distinguish strains or genotypes of infectious agents. Each band or lane contains capture probes that differ only in nucleotides that are characteristic for each strain or genotype. The target nucleic acid is typically first amplified by PCR utilizing biotin-labelled primers before addition to the line probe membrane in a hybridisation buffer. Following a wash, a streptavidin-enzyme conjugate is added and allowed to bind to biotin. Following another wash a colourless substrate is added and will be changed by the enzyme to a precipitable coloured product. The line or band corresponding to one specific genotype will then become coloured (Figure 4.5). Commercial line probe assays are used for the detection of genotypes of HPV and hepatitis C virus (HCV), strains of atypical mycobacteria and for detection of HTLV-1 (human T-cell lymphotropic virus).

4.5.1.3   Peptide Nucleic Acid Fluorescent In situ Hybridisation (PNA-FISH)

PNA-FISH is one type of *in situ* hybridisation utilising peptide nucleic acids as hybridisation probes.[21,22] Peptide nucleic acids exhibit

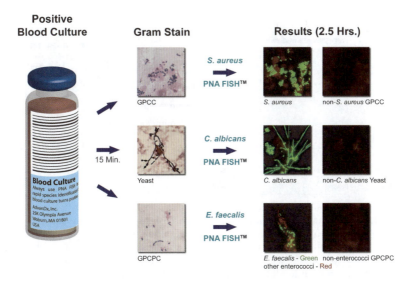

Figure 4.6 Schematics of PNA-FISH. Image provided by AdvanDx (Woburn, MA, USA).

rapid hybridisation kinetics and, when coupled to fluorescent reporter molecules, offer a rapid and sensitive verification of the infectious agent in positive blood cultures.[22] Once a blood culture turns positive, a Gram stain is performed and, based on the results, the appropriate PNA-FISH tests are selected. Identification results are available within just a few hours and can be reported to the attending physician (Figure 4.6).

4.5.2 Nucleic Acid-based Typing of Bacteria

4.5.2.1 *Restriction Endonuclease Analysis – 'DNA Fingerprinting'*
Restriction enzymes are a group of enzymes that each can recognise a short double-stranded DNA sequence of 4–9 base pairs (bp) and cleave the DNA at this site. The enzymes work by cleaving the bonds in the phosphate backbone of the DNA molecule. Depending on the exact DNA sequence of the bacterial genome, different mixtures of restriction fragments will result following digestion with one or more defined restriction enzymes. The restriction fragments can be separated using agarose gel electrophoresis and ethidium bromide staining followed by visualisation of the separated restriction fragments (DNA bands) under UV light. A 'DNA fingerprint' can thus be obtained for each species of bacterium. The method requires relatively large amounts of purified DNA and the resolution is not high.

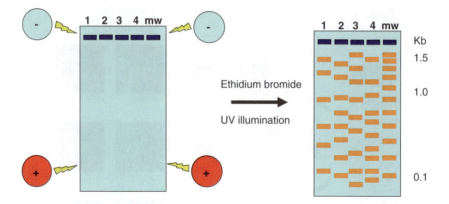

Figure 4.7   Schematics of pulse field electrophoresis. Restriction enzyme-digested
DNA is loaded into wells and separated in the agarose gel by electric fields
that alternate between the poles every 90 s for 24 h. UV illumination
visualises DNA with intercalating ethidium bromide. Different samples
have been loaded in wells 1–4 and a size marker in the rightmost lane.

4.5.2.2   Pulse Field Gel Electrophoresis (PFGE)

PFGE uses rare-cutting restriction enzymes to fragment the entire bacterial genome into large DNA fragments, which are subsequently
embedded in agarose plugs and placed in wells of the electrophoresis gel
where they become part of the gel. The large DNA fragments are separated in the agarose gel by alternate pulses of perpendicularly oriented
electric fields. Following electrophoresis and ethidium bromide staining
of the gel, characteristic DNA fragment patterns can be visualised under
UV light (Figure 4.7). The principle of PFGE is that large DNA fragments require more time to reverse direction in an electric field than do
small DNA fragments and thus are more retarded in the agarose gel. By
alternating the direction of the electric current during gel electrophoresis,
it is possible to resolve DNA fragments of 100–1000 kilobases. PFGE has
been used extensively for typing of bacterial species and is often considered to be the 'gold standard' of genomic typing methods. DNA
degradation may represent a problem.

4.5.2.3   Random Amplified Polymorphic DNA (RAPD)

In RAPD analysis, the target sequences to be amplified are unknown.
Typically, a 10-nucleotide primer of arbitrary sequence is designed, *e.g.* by
a computer program, and used for an initial PCR amplification. PCR
fragments will result only from regions of the DNA where forward (sense)
and reverse (antisense) primers are complementary to the target DNA
within a reasonable distance (up to 2000 nucleotides apart). This will

reduce the complexity and increase resolution compared with restriction endonuclease analysis. Based on differences in the genomic sequences, different PCR fragment patterns will result following agarose gel electrophoresis, and this can be used for typing and classification of bacteria.

4.5.2.4 *Amplified Fragment Length Polymorphism (AFLP)*

AFLP (Figure 4.8) combines RFLP (restriction fragment length polymorphism; see Section 5.5.2.2 and Figure 5.9) and RAPD analyses.[23] As with 'DNA fingerprinting', genomic DNA is digested with restriction enzymes. Oligonucleotide adapters are next enzymatically ligated to the ends of the DNA fragments and are designed so that the original restriction sites are not recreated following ligation. PCR primers are next designed and are complementary to the adapter sequences except that they extend one base into the intervening bacterial sequence (as in RAPD). Theoretically, the PCR will then amplify only 1/16th of all sequences with ligated flanking adapters. In a second PCR step, a new primer pair complementary to the flanks of the first PCR products but

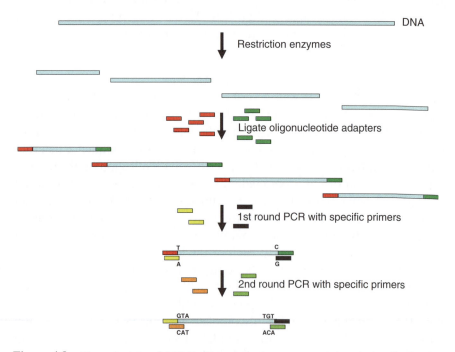

Figure 4.8 The principle of AFLP. The products are separated and visualised using gel electrophoresis and ethidium bromide and UV illumination. The 3'-terminal nucleotide sequences of PCR primers are shown.[23]

now extending 3 bp into the bacterial sequences is used to amplify a small amount of the first PCR reaction, thus theoretically reducing the sequence complexity by altogether 256-fold. The resulting fragments are then separated electrophoretically and can be visualised, *e.g.* when one of the primers of the second PCR round is fluorescently labelled.

4.5.2.5   Data Analysis of 16S Ribosomal RNA Gene Sequences – 'Ribotyping'

The 16S ribosomal RNA (rRNA) gene is present in all bacteria, often existing as a multigene family, and contains both conserved and variable regions within its about 1500 bp length. The 16S rRNA gene is for these and other reasons the most commonly used single gene for typing of bacteria.[24] Use of primers directed against the conserved regions may be used for PCR amplification followed by DNA sequencing of the amplicon. The sequences obtained may be compared with sequences in comprehensive databases available on the Internet.[24-26] Partial sequencing of the rRNA gene is often, but not always, sufficient for the identification of well-resolved bacterial species.[27] Partial sequencing of one or more additional genes may be necessary for the genetic classification and distinguishing of species.[28] It is an advantage that 16S rRNA genotyping can be done directly from patient samples without prior cultivation.

4.5.2.6   Multilocus Sequence Typing (MLST)[*]

The MLST technique is well described.[29] There are three elements to the design of a new MLST system: the choice of the isolates to be used in the initial evaluation, the choice of the genetic loci to be characterised and the design of primers for gene amplification and nucleotide sequence determination. It is advisable to assemble a diverse isolate collection on the basis of existing typing information or epidemiological data. This should comprise around 100 isolates (95 is a good number if high-throughput sequencing in 96-well microtitre plates is to be done) to ensure that the primers developed will be applicable to as many isolates as possible and to establish the levels of diversity present at each of the loci to be examined. Furthermore, the collection will ideally be representative of the bacterial population, rather than comprising a subset, such as human disease isolates. Housekeeping genes, flanked by genes of similar function, are good targets for MLST and the availability of complete genome sequences has greatly facilitated the identification of candidate loci. Experience with several bacterial species has indicated that PCR fragments of housekeeping genes

---

[*]Reprinted from reference 29 with permission from Elsevier.

around 450 bp are suitable for MLST and a nested strategy is highly recommended for sequencing.[29] The great advantage of MLST is that sequence data are unambiguous and the allelic profiles of isolates can easily be compared with those in a large central database via the Internet (in contrast to most typing procedures, which involve comparing DNA fragment sizes on gels). Allelic profiles can also be obtained from clinical material by PCR amplification of housekeeping loci directly from patient samples. Thus isolates can be precisely characterised even when they cannot be cultured from clinical material. MLST can effectively distinguish strains that possess high degrees of homology within the compared gene sequences. This technique is not as laborious as PFGE and it provides balance between sequence-based resolution and technical feasibility.

4.5.3 Pyrosequencing

Pyrosequencing (Figure 4.9) is a DNA sequencing technology based on the sequencing-by-synthesis principle. The technique is built on a four-enzyme real-time monitoring of DNA synthesis by bioluminescence using a cascade that upon nucleotide incorporation ends in a detectable light signal (bioluminescence). The detection system is based on the pyrophosphate released when a nucleotide is introduced in the DNA-strand. Thereby, the signal can be quantitatively connected to the number of bases added. Currently, the technique is limited to analysis of short DNA sequences exemplified by single-nucleotide polymorphism analysis and

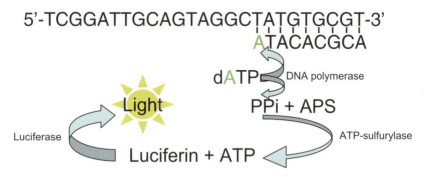

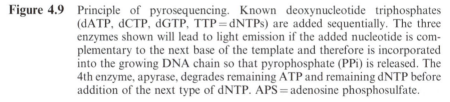

Figure 4.9 Principle of pyrosequencing. Known deoxynucleotide triphosphates (dATP, dCTP, dGTP, TTP = dNTPs) are added sequentially. The three enzymes shown will lead to light emission if the added nucleotide is complementary to the next base of the template and therefore is incorporated into the growing DNA chain so that pyrophosphate (PPi) is released. The 4th enzyme, apyrase, degrades remaining ATP and remaining dNTP before addition of the next type of dNTP. APS = adenosine phosphosulfate.

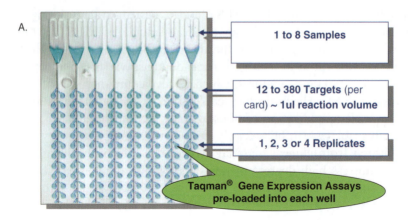

A.

1 to 8 Samples

12 to 380 Targets (per card) ~ 1ul reaction volume

1, 2, 3 or 4 Replicates

Taqman® Gene Expression Assays pre-loaded into each well

B. Load Samples

Spin

Seal

Run

Prep & Load Time 5 -10 minutes

7900HT Fast Real-Time PCR System

Figure 4.10   (A) The TaqMan low-density array (TLDA). (B) About 100 μL of master mix and hexamer primed cDNA are loaded into each well and distributed into the reaction wells by low-speed centrifugation and the card is sealed and thermocycled for about 40 cycles. Illuatration provided by courtesy of Applied Biosystems.

genotyping.[30] The technology is promising for the identification of anti-microbial resistance and bacterial strain identification.[22]

4.5.4   TaqMan Low-density Arrays (TLDAs)

The TaqMan low-density array (Figure 4.10) is a 384-well microfluidic card manufactured by Applied Biosystems where up to 384 simulta-neous real-time PCR reactions are possible without the need for liquid-handling robots or multi-channel pipettors to load samples. Each card allows for 1–8 samples to be run in parallel against 12–384 TaqMan gene expression assay targets that are preloaded into each of the 1–2 µL wells on the card. The TaqMan low-density array is customisable for human, mouse and rat genes, but human infectious agent target genes are not yet available. The TaqMan low-density array is designed for use on the Applied Biosystems 7900HT fast real-time PCR system. The TLDA format increases by one order of magnitude the number of different real-time PCR assays that are feasible in a given time and is suitable for the validation of DNA microarray gene expression results.

4.6 SELECTED EXAMPLES OF CLINICAL NUCLEIC ACID-BASED DIAGNOSIS

4.6.1 Central Nervous System (CNS) Disease

Viruses are the most common infectious agents of encephalitis (inflammation of the brain) in humans and more than 100 causative viruses are known,[31,32] followed by bacterial, parasitic, fungal and prion agents.[33] Non-infectious causes of encephalitis include autoimmune, neoplastic, metabolic and other disorders. Still, in more than half of all cases of encephalitis the aetiology remains unknown.[33] Encephalitis is not a common disease, but because early start of specific therapy may prevent serious and lasting damage to the brain, early and correct identification of the infectious agent may make a dramatic difference to each patient. Amplification of nucleic acids from cerebrospinal fluid (CSF) obtained by lumbar puncture represents a diagnostic breakthrough. It may be difficult to distinguish early symptoms of encephalitis from meningitis and meningoencephalitis most commonly caused by enteroviruses. When only an enterovirus is present in the CSF, the prognosis is usually good without specific treatment. In contrast, when a member of the herpes virus family infects the brain, early antiviral (acyclovir or gancyclovir) treatment is very important to avoid lasting damage to the brain or even death. In particular HSV-1, but also HSV-2 and VZV, accounts for a large proportion of cases with infectious encephalitis and may occur in young and otherwise healthy people. Cytomegalovirus (CMV) infection of the CNS is more commonly associated with immune suppression, immune deficiency such as AIDS or congenital CMV infection.[32] Human herpes virus 6 (HHV-6) is associated with febrile convulsions in small children and may also cause encephalitis. EBV is associated with both encephalitis and lymphomas in the brain. The very high sensitivity and specificity of real-time qPCR assays for the detection of these agents are well documented as soon as symptoms and clinical signs are apparent.[32] For therapeutic reasons, it is initially important to distinguish cases of

Mycobacterium tuberculosis, *Mycoplasma pneumoniae* and *Borrelia burgdorferi* (the cause of Lyme's disease) and pyogenic bacteria. A combination of biochemical and microscopic examination of the CSF, together with cultivation and PCR, are useful in these cases. Immuno-compromised patients, newborn and recent stays in tropical or endemic areas suggest specific diagnostic procedures in connection with CNS disease.

4.6.2 Respiratory Infections

Rhinoviruses and coronaviruses are among the most prevalent upper respiratory tract pathogens and exist in too many serotypes to make serological diagnosis possible. The common cold therefore remains a clinical diagnosis.

Lower respiratory tract infections cause a high degree of morbidity and mortality and traditional culture or serological methods often fail or take a long time. Real-time PCR and other nucleic amplification techniques have great potential for rapid and exact diagnosis. The main limitation currently is the cost compared with cheaper serological diagnoses such as the complement fixation test. In selected cases, nucleic acid extraction from expectorate followed by amplification in order to detect *Chlamydia pneumoniae*, *Mycoplasma pneumoniae*, *Bordetella pertussis*, *Legionella pneumophila*[34-38] or *Mycobacterium tuberculosis*[39,40] is most valuable. In children hospitalised with breathing difficulties due to bronchiolitis, reverse PCR tests for metapneumovirus and RSV complement the rapid immunofluorescent diagnosis available for a number of viruses in nasopharyngeal aspirates (influenza A and B viruses, parainfluenza viruses, adenovirus and RSV).[41-43] In the initial stage of new influenza epidemics, real-time PCR assays to identify and type influenza A and B viruses are replacing viral isolation as the reference standard.[16,17,44] Real-time PCR assays are available for the emergency analysis of avian influenza H5N1[45,46] and also for the SARS virus.[1,47]

4.6.3 Hepatitis

Serological tests detecting specific antibodies and antigens using ELISA and micro-EIA are adequate for the primary diagnosis of hepatitis A, B, C and E viruses (HAV, HBV, HCV and HEV). The detection of HCV RNA is a very useful complement to the standard screening and confirmation test (RIBA) to establish the diagnosis of hepatitis C. In particular, the reverse PCR is useful for the testing of the newborn with maternal anti-HCV antibodies. Detection of HCV RNA in donor blood or in pools of donor blood is implemented in many cases. However, the most important role of real-time PCR and other nucleic acid-based tests for hepatitis C and B viruses relates to therapy. Commercial HCV reverse PCR tests come in both qualitative and quantitative versions. Before PEG-interferon α–ribavirin therapy is attempted against hepatitis C, an anti-HCV positive patient should in addition test positive in the HCV RNA qualitative test. The next step is then to perform a quantitative reverse PCR-test to establish the HCV RNA in IU mL^{-1} serum. The amplification product is employed in a line probe assay in order to determine the genotype (see above). Different genotypes require different lengths of treatment and the quantitative reverse PCR test is used at intervals to monitor the response to treatment and the result may decide whether therapy is successful and should be completed or not. When therapy is completed, the HCV PCR test is used to make sure that HCV does not relapse in the next 6 months and that a sustained viral response has occurred.

Antiviral treatment is also attempted in hepatitis B in order to prevent accelerating damage to the liver and quantitative PCR to determine viral load in serum prior and during treatment is important. Only occasionally is HBV eradicated as a result of treatment. Instead, drug-resistant mutants commonly develop and are associated with specific point mutations that can be determined by DNA sequencing.

4.6.4 Gastroenteritis

Acute gastroenteritis causes almost 3 million deaths each year worldwide, mostly among children under 5 years of age in developing countries. The known causes of viral gastroenteritis include the RNA viruses rotavirus, astrovirus, sapovirus and norovirus and the DNA adenoviruses. Enteroviruses do not usually, despite their name, cause gastroenteritis, but replicate in the intestine and can be isolated from faeces in patients with meningitis, myocarditis and symptoms from many other organs. Electron microscopy has been valuable for the discovery and diagnosis of viruses in diarrhoea and the names of rotavirus ('wheel') and astrovirus ('star') refer to the electron microscopy images. Agglutination tests provide rapid diagnosis of rotavirus and adenovirus gastroenteritis. However, reverse PCR tests are increasingly coming into use for the rapid diagnosis of viral gastroenteritis. The incubation time of norovirus gastroenteritis may be less than 1 day and noroviruses commonly cause outbreaks of gastroenteritis in hospitals and in other situations where groups of people live closely together. By means of reverse PCR and DNA sequencing, the hundreds of strains of noroviruses (also known as caliciviruses/Norwalk agents) have been

identified and classified into genogroups and genetic clusters.[48] Reverse
PCR tests using nucleic acids extracted from diarrhoeal faecal specimens
have come into broad use to diagnose norovirus infection.[49,50] The
genetic variability of viruses causing gastroenteritis represents a great
challenge to molecular diagnosis and new protocols take advantage of
new chemistry and computer analysis to design multiplex PCR protocols
directed at different strains of each virus and to cover several different
viruses in one reaction.[49–52]

4.6.5 Sexually Transmitted Diseases

4.6.5.1 *Human Papillomaviruses (HPVs)*

These are among the most commonly sexually transmitted infectious
agents. Molecular cloning and DNA sequencing have been crucial for the
identification and classification of the more than 130 HPV types pre-
sently discovered. The high-risk types, including the prevalent types 16
and 18, are the cause of cervical cancer. The low-risk types cause benign
papillomas and warts and include HPV type 6 and 11 that cause genital
warts or condylomas. The cloning and recombinant expression of HPV
L1 proteins in yeast and insect cells have provided successful vaccines
against HPV 16 and 18 and one vaccine additionally protects against
HPV 6 and 11. The cytological Pap smear remains the most important
screening test for pre-stages of cervical cancer and relies on microscopic
evaluation of abnormalities in cells taken from the surface of the cervix.
Atypical cervix cells are most often caused by HPVs and detection and
typing of HPV nucleic acids are becoming increasingly used. In particular, microscopic detection of cytological changes is followed up by
HPV nucleic acid isolation and typing.[53] Several commercial nucleic
acid-based tests are on the market for this purpose. Comparison between
these tests is methodologically interesting since they comprise one
sandwich hybridisation assay (the Digene Hybrid Capture 2 assay), a
PCR-based assay (the Roche Amplicor HPV test) and a line probe assay
(the Innogenetics Inno-LiPa HPV genotyping test v2 and the Roche
Linear Array HPV genotyping test).[54–56] The additional value of focusing the diagnosis on the HPV oncogenes E6 and E7, and in particular of
monitoring their transcriptional activity, is under evaluation.[57,58]

4.6.5.2 *Chlamydia Trachomatis and Genital Mycoplasmas and Ureaplasmas*

While the diagnosis of several of the classical sexually transmitted bacterial diseases still rely on clinical experience, serology (syphilis) and
cultivation (*Neissera gonorrhoea*), nucleic acid detection methods have

revolutionised the diagnosis of genital *Chlamydia trachomatis*, one of the most ubiquitously sexually transmitted agents and associated with deep pelvic inflammation and female sterility. Efficient antibiotic treatment is available if the diagnosis is made early. Different commercial kits are used for sampling from the urethra or the cervix followed by different principles of nucleic acid amplification.[59] Self-sampling using blind vaginal swabs or urine samples may provide good alternatives to increase patient compliance.[60] Abbot's ligase chain reaction (LCR) was discontinued. Instead, Abbott has marketed a highly sensitive new test for the simultaneous detection of *Chlamydia trachomatis* and *Neissera gonorrhoeae* designed for use on the Abbott *m*2000, an automated instrument using magnetic particle and real-time PCR. The Probetec test (Beckton Dickinson) uses strand displacement amplification (SDA) and the Aptima Combo 2 test (Gen-Probe) utilises isothermal transcription mediated amplification (TMA) combined with magnetic target capture. The Cobas Amplicor PCR test (Roche) is now superseded by the Roche Cobas TaqMan48 real-time PCR test. These tests are streamlined with both positive and negative controls. The availability of several different tests provides advantages, *e.g.* in legal cases of sexual abuse or rape, when confirmation of a positive test result with a second independent test is highly recommended or necessary. A *Chlamydia trachomatis* variant that contains a 377 bp deletion in the cryptic plasmid was recently reported in Sweden. This deletion includes the targets of Cobas Amplicor, Cobas TaqMan48 and Abbott *m*2000 assays, but the mutant was readily detected by the new real-time quantitative PCR assay LightMix 480HT (TIB MOLBIOL) that targets a 136 bp fragment of the *omp*1 gene (using automatic DNA isolation with magnetic silica particles in the Roche MagNa Pure LC system combined with the LightCycler thermocycler).[61]

There is now an increasing awareness that in addition to *Chlamydia trachomatis*, several other organisms may be involved in postgonococcal urethritis, including *Mycoplasma genitalium*, *Mycoplasma hominis*, *Ureaplasma urealyticum* and *Ureaplasma parvum*, and that these organisms, if left untreated, may contribute to infertility. In-house PCR assays are now offered by many diagnostic laboratories for the specific diagnosis of these agents in urine or semen.[62,63]

4.6.6 HIV Infection and AIDS

The screening (ELISA) and confirmation (Western blot) tests for HIV-1 and HIV-2 infections provide highly reliably laboratory diagnosis of this serious infection. Nucleic acid-based amplification tests are therefore not currently important to establish the diagnosis. The drug treatment of

HIV-1 infection and AIDS has experienced remarkable progress during the last 10–15 years, in contrast with the frustrations encountered in HIV vaccine development. However, the virus still cannot be eradicated from the body and lifelong combination therapy using three different drugs is necessary to suppress disease progression. So far, 24 drugs, belonging to seven classes, have been approved for the treatment of HIV infection and inhibit defined molecular stages of the HIV life cycle, including the reverse transcriptase, the protease, the integrase and coreceptors (CCR5/CXCR4).[64] CD4$^+$ cell counts and HIV viral load in plasma must be determined prior to the start of and monitored during therapy. When therapy is effective, HIV RNA should disappear from the peripheral blood. The Roche AMPLICOR version 1.5 conventional PCR assay has been widely employed for monitoring, but the Cobas Campliprep/Cobas TaqMan system for automated nucleic acid extraction and reverse real-time PCR detection of HIV is a newer alternative.[65,66] If the HIV RNA viral load increases during therapy, it is a sign either that the patient does not take the medication properly or that resistant HIV strains have been selected. Drug resistance is associated with characteristic point mutations for each drug. DNA sequencing combined with algorithmic analysis of databases can identify which point mutations are associated with what drug resistance and be helpful in the choice of a different combination of drugs that may continue to suppress the HIV infection.

4.6.7   Bacterial Antibiotic Resistance and Virulence Factor Genes

Detection of bacterial genes that confer antibiotic resistance is important for the choice of correct treatment, to help decide which patients need to be isolated and to trace epidemics. At the same time, it may be necessary to identify virulence (disease-causing) genes. In the following, some key examples have been selected.

4.6.7.1   Methicillin-resistant Staphylococcus aureus
Staphylococcus aureus has long been a notorious problem among hospitalised patients. Whenever bacteraemia (bacteria in the blood) is suspected, the typical procedure is to identify the type of bacteria using automated blood culture systems followed by microscopy of positive blood cultures. When the microscopy shows Gram-positive cocci in clusters it is a main goal to differentiate between *S. aureus* and coagulase-negative staphylococci and to verify methicillin-resistant *S. aureus* (MRSA). MRSA strains are resistant against penicillin and all β-lactam antibiotics. A number of different nucleic acid-based techniques have been used to identify the *mecA* gene (the methicillin resistance gene) from putative MRSA colonies and at

the same time identify *S. aureus* based upon specific genes such as the *nuc* or *orfX* genes.[22,67,68] Since this verification may take more than 2 days, it is common to start treatment of patients using glycopeptide antibiotics until the results of the *mecA* test are available, and this represents a selection pressure for the increasing vancomycin resistance. Glycopeptide-resistant MRSA strains is an emerging problem.[69] One important prophylactic measure is to screen and identify asymptomatic carriers of the *mecA* gene among hospital workers and patients. Colonies grown from nasal swab samples are then analysed for *mecA* and *S. aureus* specific genes using PCR, real-time PCR or several other nucleic acid-based techniques.[22,67,68] More recently, serious MRSA infections have appeared outside hospitals; these so-called community-associated MRSA strains harbour a set of phage-carried PVL genes, in contrast to the hospital MRSA strains. It is unclear why certain MRSA strains predominate in hospitals and others in the community, but nucleic acid-based detection of PVL genes represents a very important tool for epidemiology and surveillance.[70–72] It has also been shown that different MRSA strains differ in their virulence–gene profiles and that this contributes to the complexity of defining the disease-causing abilities of *S. aureus*.[70]

4.6.7.2   *Vancomycin-resistant Enterococci*
Enterococci infections represent a serious problem in hospitals. In the same way as MRSA genes are identified in positive blood cultures, colonies of enterococci may be tested for *vanA*, *vanB* and *vanC* genes that confer resistance to glycopeptide antibiotics (vancomycin and teicoplanin) using different nucleic acid-based techniques.[22,73,74] The identification of enterococcal virulence genes in combination with vancomycin resistance may become increasingly important.[75,76] Control screening for vancomycin-resistant enterococci is based on perianal swabs from which automated DNA extraction is followed by real-time PCR. This is, however, more challenging than nasal swab screening for MRSA.[22,77] The identification of the *vanB* gene in naturally occurring gut anaerobes may be a source of confusion and calls for simultaneous detection of *vanB* and enterococcal specific genes, similar to linking *mecA* to *S. aureus* specific genes in order to avoid false positives.[22,78,79]

4.6.7.3   *Antibiotic Resistance and Virulence in Gram-negative Intestinal Bacteria*
Extended-spectrum β-lactamase (ESBL)-producing Gram-negative bacteria (such as *Escherichia coli*) are resistant against both penicillins and cephalosporins.[80] The responsible bacterial β-lactamases, including

TEM, SHV and CTX-M enzymes, can be typed by molecular techniques in order to track epidemics and monitor outbreaks. PFGE is the current 'gold standard' for molecular typing, but multilocus sequence typing (MLST) may offer advantages.[81] Real-time PCR combined with pyrosequencing may increase speed and capacity for the identification of the antibiotic resistance genes.[82,83] *Klebsiella pneumoniae* and other bacteria containing carbapenemases (KPC) is a further extension of the antibiotic resistance problem.[80,84]

Although *E. coli* is prevalent in the normal intestinal flora, some strains are highly pathogenic and cause diarrhoea or systemic disease due to acquisition of virulence genes encoded by plasmid, chromosome or bacteriophage DNA. A large number of virulence factor genes separate pathogenic *E. coli* strains into enterotoxigenic (ETEC), enteroinvasive (EIEC), enteropathogenic (EPEC), enterohaemorrhagic (EHEC) and enteroaggregative (EAEC).[85] The development of multiplex PCR and real-time PCR assays has been most useful for the survey of such virulence factors.[86–88]

4.6.7.4 *Mycobacterium tuberculosis and Atypical Mycobacteria*
Nucleic acid amplification techniques supplement direct microscopy and cultivation in the laboratory diagnosis of *M. tuberculosis*, in particular since growth in special media may take weeks. A number of genetic markers and DNA typing methods have been increasingly used to control and monitor tuberculosis.[89] Resistant and multiresistant mycobacteria represent an increasing problem. Whereas traditional susceptibility tests may take weeks to complete, DNA amplification methods can detect characteristic mutations associated with isoniazid resistance or rifampicin resistance within hours.[2] Many alternative methods have been employed for the detection of mycobacterial antibiotic resistance, such as DNA sequencing,[90] heteroduplex analysis,[91] PCR conformational polymorphism,[92] line probe assays[93] and mismatch analysis,[94] but these methods have several disadvantages. Pyrosequencing shows promise in the rapid identification of rifampicin, isoniazid and ethambutol resistance.[22,95,96]

4.7 CONCLUSION AND FUTURE CHALLENGES

Nucleic acid-based methods have revolutionised the routine laboratory diagnosis of infectious agents. In particular, the DNA amplification techniques have contributed to this by providing sufficient sensitivity for the direct examination of infected patient samples. As a result, the

clinician faced with the patient can rely much more on the microbiology laboratory in the acute differential diagnosis. Nucleic acid-based techniques have contributed to much quicker and reliable diagnosis of the individual patient, to the tracing and monitoring of epidemic outbreaks and the discovery of emerging infectious agents. There are still many problems, such as contamination during nucleic acid amplification and the ongoing biological selection and evolution in the microbiological world that without warning may cause a previously good test to fail. The potential for improvement of nucleic acid-based laboratory diagnostics is therefore vast. There is a need for backup tests directed to different regions of the target nucleic acids. At present we are seeing only the beginning of the automation and reduction of reaction volumes. PCR reactions traditionally were performed in 5–50 µL volumes, but now can already be performed in nanolitre volumes. Nanotechnology and microfluidics and electronics may become vital to saving the environment and saving money. Costs already now severely restrict the repertoire of new tests in the clinical microbiological laboratory – even in developed countries. The relative role of DNA microarray hybridisation assays and PCR amplification arrays lies in the future and we face great challenges regarding logistics and quality control of diagnostic procedures. It is therefore certain that nucleic acid-based techniques will continue to revolutionise the way in which we diagnose and treat infections.

REFERENCES

1. C. Drosten, W. Preiser, S. Gunther, H. Schmitz and H. W. Doerr, *Trends Mol. Med.*, 2003, **9**, 325.
2. M. J. Espy, J. R. Uhl, L. M. Sloan, S. P. Buckwalter, M. F. Jones, E. A. Vetter, J. D. Yao, N. L. Wengenack, J. E. Rosenblatt, F. R. Cockerill III and T. F. Smith, *Clin. Microbiol. Rev.*, 2006, **19**, 165.
3. K. Rantakokko-Jalava and J. Jalava, *J. Clin. Microbiol.*, 2002, **40**, 4211.
4. T. Schuurman, R. F. De Boer, A. M. Kooistra-Smid and A. A. Van Zwet, *J. Clin. Microbiol.*, 2004, **42**, 734.
5. K. B. Barken, J. A. Haagensen and T. Tolker-Nielsen, *Clin. Chim. Acta*, 2007, **384**, 1.
6. T. Schuurman, A. Van Breda, R. De Boer, M. Kooistra-Smid, M. Beld, P. Savelkoul and R. Boom, *J. Clin. Microbiol.*, 2005, **43**, 4616.
7. B. Kaltenboeck and C. Wang, *Adv. Clin. Chem.*, 2005, **40**, 219.
8. M. A. Espy, H. Sandin, C. Carr, C. J. Hanson, M. D. Ward and R. H. Kraus Jr, *Cytometry A*, 2006, **69**, 1132.
9. J. Ikewaki, E. Ohtsuka, R. Kawano, M. Ogata, H. Kikuchi and M. Nasu, *J. Clin. Microbiol.*, 2003, **41**, 4382.

10. J. L. Vaerman, P. Saussoy and I. Ingargiola, *J. Biol. Regul. Homeost. Agents*, 2004, **18**, 212.

11. R. G. Rutledge and C. Cote, *Nucleic Acids Res.*, 2003, **31**, e93.

12. J. C. Servoss and L. S. Friedman, *Clin. Liver Dis.*, 2004, **8**, 267.

13. J. Saldanha, N. Lelie and A. Heath, *Vox Sang.*, 1999, **76**, 149.

14. J. Saldanha, N. Lelie, M. W. Yu and A. Heath, *Vox Sang.*, 2002, **82**, 24.

15. J. Burrows, A. Nitsche, B. Bayly, E. Walker, G. Higgins and T. Kok, *BMC Microbiol.*, 2002, **2**, 12.

16. R. M. Ratcliff, G. Chang, T. Kok and T. P. Sloots, *Curr. Issues Mol. Biol.*, 2007, **9**, 87.

17. B. Stone, J. Burrows, S. Schepetiuk, G. Higgins, A. Hampson, R. Shaw and T. Kok, *J Virol. Methods*, 2004, **117**, 103.

18. K. H. Kalland, L. S. Havarstein and H. Sommerfelt, *Tidsskr. Nor. Laegeforen.*, 1987, **107**, 2510.

19. K. H. Kalland and G. Haukenes, *Tidsskr. Nor. Laegeforen.*, 1999, **119**, 802.

20. K. H. Kalland, H. Myrmel and S. A. Nordbo, *Tidsskr. Nor. Laegeforen.*, 2005, **125**, 3110.

21. M. Sogaard, H. Stender and H. C. Schonheyder, *J. Clin. Microbiol.*, 2005, **43**, 1947.

22. F. C. Tenover, *Clin. Infect. Dis.*, 2007, **44**, 418.

23. S. Bensch and M. Akesson, *Mol. Ecol.*, 2005, **14**, 2899.

24. J. M. Janda and S. L. Abbott, *J. Clin. Microbiol.*, 2007, **45**, 2761.

25. K. E. Simmon, A. C. Croft and C. A. Petti, *J. Clin. Microbiol.*, 2006, **44**, 4400.

26. D. J. Speers, *Clin. Biochem. Rev.*, 2006, **27**, 39.

27. G. E. Fox, J. D. Wisotzkey and P. Jurtshuk Jr, *Int. J. Syst. Bacteriol.*, 1992, **42**, 166.

28. B. Ghebremedhin, F. Layer, W. Konig and B. Konig, *J. Clin. Microbiol.*, 2008.

29. R. Urwin and M. C. Maiden, *Trends Microbiol.*, 2003, **11**, 479.

30. A. Ahmadian, M. Ehn and S. Hober, *Clin. Chim. Acta*, 2006, **363**, 83.

31. R. J. Whitley and J. W. Gnann, *Lancet*, 2002, **359**, 507.

32. R. L. Debiasi and K. L. Tyler, *Clin. Microbiol. Rev.*, 2004, **17**, 903.

33. C. A. Glaser, S. Honarmand, L. J. Anderson, D. P. Schnurr, B. Forghani, C. K. Cossen, F. L. Schuster, L. J. Christie and J. H. Tureen, *Clin. Infect. Dis.*, 2006, **43**, 1565.

34. S. Kumar and M. R. Hammerschlag, *Clin. Infect. Dis.*, 2007, **44**, 568.

35. M. Khanna, J. Fan, K. Pehler-Harrington, C. Waters, P. Douglass, J. Stallock, S. Kehl and K. J. Henrickson, *J. Clin. Microbiol.*, 2005, **43**, 565.

36. D. R. Murdoch, *Clin. Infect. Dis.*, 2005, **41**, 1445.
37. Y. R. Chan and A. Morris, *Curr. Opin. Infect. Dis.*, 2007, **20**, 157.
38. K. A. Beynon, S. A. Young, R. T. Laing, T. G. Harrison, T. P. Anderson and D. R. Murdoch, *Emerg. Infect. Dis.*, 2005, **11**, 639.
39. J. C. Palomino, *Curr. Opin. Pulm. Med.*, 2006, **12**, 172.
40. D. Andresen, *Paediatr. Respir. Rev.*, 2007, **8**, 221.
41. I. Lee and T. D. Barton, *Drugs*, 2007, **67**, 1411.
42. L. C. Jennings, T. P. Anderson, A. M. Werno, K. A. Beynon and D. R. Murdoch, *Pediatr. Infect. Dis. J.*, 2004, **23**, 1003.
43. C. Deffrasnes, M. E. Hamelin and G. Boivin, *Semin. Respir. Crit. Care Med.*, 2007, **28**, 213.
44. M. Petric, L. Comanor and C. A. Petti, *J. Infect. Dis.*, 2006, **194**(Suppl 2), S98.
45. M. D. Curran, J. S. Ellis, T. G. Wreghitt and M. C. Zambon, *J. Med. Microbiol.*, 2007, **56**, 1263.
46. D. L. Suarez, A. Das and E. Ellis, *Avian Dis.*, 2007, **51**, 201.
47. J. K. Louie, J. K. Hacker, J. Mark, S. S. Gavali, S. Yagi, A. Espinosa, D. P. Schnurr, C. K. Cossen, E. R. Isaacson, C. A. Glaser, M. Fischer, A. L. Reingold and D. J. Vugia, *Emerg. Infect. Dis.*, 2004, **10**, 1143.
48. R. Goodgame, *Curr. Infect. Dis. Rep.*, 2007, **9**, 102.
49. M. Hoehne and E. Schreier, *BMC Infect. Dis.*, 2006, **6**, 69.
50. W. Hymas, A. Atkinson, J. Stevenson and D. Hillyard, *J. Virol. Methods*, 2007, **142**, 10.
51. H. Yan, T. A. Nguyen, T. G. Phan, S. Okitsu, Y. Li and H. Ushijima, *Kansenshogaku Zasshi*, 2004, **78**, 699.
52. E. Royuela, A. Negredo and A. Sanchez-Fauquier, *J. Virol. Methods*, 2006, **133**, 14.
53. B. J. Morris and B. R. Rose, *Clin. Chem. Lab. Med.*, 2007, **45**, 577.
54. M. P. Stevens, S. M. Garland, E. Rudland, J. Tan, M. A. Quinn and S. N. Tabrizi, *J. Clin. Microbiol.*, 2007, **45**, 2130.
55. L. Z. Mo, S. Monnier-Benoit, B. Kantelip, A. Petitjean, D. Riethmuller, J. L. Pretet and C. Mougin, *J. Clin. Virol.*, 2008, **41**, 104.
56. M. A. De Francesco, F. Gargiulo, C. Schreiber, G. Ciravolo, F. Salinaro and N. Manca, *J. Virol. Methods*, 2008, **147**, 10.
57. T. Molden, I. Kraus, H. Skomedal, T. Nordstrom and F. Karlsen, *J. Virol. Methods*, 2007, **142**, 204.
58. A. K. Lie, B. Risberg, B. Borge, B. Sandstad, J. Delabie, R. Rimala, M. Onsrud and S. Thoresen, *Gynecol. Oncol.*, 2005, **97**, 908.
59. S. Skidmore, P. Horner and H. Mallinson, *Sex. Transm. Infect.*, 2006, **82**, 272.

60. I. J. Bakken and S. A. Nordbo, *Tidsskr. Nor. Laegeforen.*, 2007, **127**, 3202.
61. M. Unemo, P. Olcen, I. Agne-Stadling, A. Feldt, M. Jurstrand, B. Herrmann, K. Persson, P. Nilsson, T. Ripa and H. Fredlund, *Euro Surveill.*, 2007, **12**, E5.
62. S. R. Lee, J. M. Chung and Y. G. Kim, *J. Microbiol.*, 2007, **45**, 453.
63. S. Yokoi, S. Maeda, Y. Kubota, M. Tamaki, K. Mizutani, M. Yasuda, S. Ito, M. Nakano, H. Ehara and T. Deguchi, *Clin. Infect. Dis.*, 2007, **45**, 866.
64. E. De Clercq, *Nat. Rev. Drug Discov.*, 2007, **6**, 1001.
65. A. R. Oliver, S. F. Pereira and D. A. Clark, *J. Clin. Microbiol.*, 2007, **45**, 3616.
66. W. Schumacher, E. Frick, M. Kauselmann, V. Maier-Hoyle, R. Van Der Vliet and R. Babiel, *J. Clin. Virol.*, 2007, **38**, 304.
67. K. Levi and K. J. Towner, *J. Clin. Microbiol.*, 2003, **41**, 3890.
68. Y. Misawa, A. Yoshida, R. Saito, H. Yoshida, K. Okuzumi, N. Ito, M. Okada, K. Moriya and K. Koike, *J. Infect. Chemother.*, 2007, **13**, 134.
69. P. C. Appelbaum, *Int. J. Antimicrob. Agents*, 2007, **30**, 398.
70. B. A. Diep, H. A. Carleton, R. F. Chang, G. F. Sensabaugh and F. Perdreau-Remington, *J. Infect. Dis.*, 2006, **193**, 1495.
71. J. A. Mcclure, J. M. Conly, V. Lau, S. Elsayed, T. Louie, W. Hutchins and K. Zhang, *J. Clin. Microbiol.*, 2006, **44**, 1141.
72. R. R. Mcdonald, N. A. Antonishyn, T. Hansen, L. A. Snook, E. Nagle, M. R. Mulvey, P. N. Levett and G. B. Horsman, *J. Clin. Microbiol.*, 2005, **43**, 6147.
73. H. L. Young, S. A. Ballard, P. Roffey and M. L. Grayson, *J. Antimicrob. Chemother.*, 2007, **59**, 809.
74. I. Klare, C. Konstabel, D. Badstubner, G. Werner and W. Witte, *Int. J. Food Microbiol.*, 2003, **88**, 269.
75. L. J. Worth, M. A. Slavin, V. Vankerckhoven, H. Goossens, E. A. Grabsch and K. A. Thursky, *J. Hosp. Infect.*, 2008, **68**, 137.
76. I. Klare, C. Konstabel, S. Mueller-Bertling, G. Werner, B. Strommenger, C. Kettlitz, S. Borgmann, B. Schulte, D. Jonas, A. Serr, A. M. Fahr, U. Eigner and W. Witte, *Eur. J. Clin. Microbiol. Infect. Dis.*, 2005, **24**, 815.
77. L. M. Sloan, J. R. Uhl, E. A. Vetter, C. D. Schleck, W. S. Harmsen, J. Manahan, R. L. Thompson, J. E. Rosenblatt and F. R. Cockerill III, *J. Clin. Microbiol.*, 2004, **42**, 2636.
78. S. A. Ballard, K. K. Pertile, M. Lim, P. D. Johnson and M. L. Grayson, *Antimicrob. Agents Chemother.*, 2005, **49**, 1688.
79. S. A. Ballard, E. A. Grabsch, P. D. Johnson and M. L. Grayson, *Antimicrob. Agents Chemother.*, 2005, **49**, 77.

80. F. Perez, A. Endimiani, K. M. Hujer and R. A. Bonomo, *Curr. Opin. Pharmacol.*, 2007, **7**, 459.
81. L. L. Nemoy, M. Kotetishvili, J. Tigno, A. Keefer-Norris, A. D. Harris, E. N. Perencevich, J. A. Johnson, D. Torpey, A. Sulakvelidze, J. G. Morris Jr and O. C. Stine, *J. Clin. Microbiol.*, 2005, **43**, 1776.
82. C. I. Birkett, H. A. Ludlam, N. Woodford, D. F. Brown, N. M. Brown, M. T. Roberts, N. Milner and M. D. Curran, *J. Med. Microbiol.*, 2007, **56**, 52.
83. T. Naas, C. Oxacelay and P. Nordmann, *Antimicrob. Agents Chemother.*, 2007, **51**, 223.
84. L. M. Deshpande, R. N. Jones, T. R. Fritsche and H. S. Sader, *Microb. Drug Resist.*, 2006, **12**, 223.
85. M. M. Levine, *J. Infect. Dis.*, 1987, **155**, 377.
86. L. T. Brandal, B. A. Lindstedt, L. Aas, T. L. Stavnes, J. Lassen and G. Kapperud, *J. Microbiol. Methods*, 2007, **68**, 331.
87. V. K. Sharma, *Mol. Cell. Probes*, 2006, **20**, 298.
88. J. R. Yang, F. T. Wu, J. L. Tsai, J. J. Mu, L. F. Lin, K. L. Chen, S. H. Kuo, C. S. Chiang and H. S. Wu, *J. Clin. Microbiol.*, 2007, **45**, 3620.
89. K. Kremer, C. Arnold, A. Cataldi, M. C. Gutierrez, W. H. Haas, S. Panaiotov, R. A. Skuce, P. Supply, A. G. Van Der Zanden and D. Van Soolingen, *J. Clin. Microbiol.*, 2005, **43**, 5628.
90. V. Kapur, L. L. Li, S. Iordanescu, M. R. Hamrick, A. Wanger, B. N. Kreiswirth and J. M. Musser, *J. Clin. Microbiol.*, 1994, **32**, 1095.
91. D. L. Williams, L. Spring, T. P. Gillis, M. Salfinger and D. H. Persing, *Clin. Infect. Dis.*, 1998, **26**, 446.
92. M. Bobadilla-Del-Valle, A. Ponce-De-Leon, C. Arenas-Huertero, G. Vargas-Alarcon, M. Kato-Maeda, P. M. Small, P. Couary, G. M. Ruiz-Palacios and J. Sifuentes-Osornio, *Emerg. Infect. Dis.*, 2001, **7**, 1010.
93. R. Rossau, H. Traore, H. De Beenhouwer, W. Mijs, G. Jannes, P. De Rijk and F. Portaels, *Antimicrob. Agents Chemother.*, 1997, **41**, 2093.
94. K. A. Nash, A. Gaytan and C. B. Inderlied, *J. Infect. Dis.*, 1997, **176**, 533.
95. J. R. Zhao, Y. J. Bai, Y. Wang, Q. H. Zhang, M. Luo and X. J. Yan, *Int. J. Tuberc. Lung Dis.*, 2005, **9**, 328.
96. C. Arnold, L. Westland, G. Mowat, A. Underwood, J. Magee and S. Gharbia, *Clin. Microbiol. Infect.*, 2005, **11**, 122.

CHAPTER 5

Genes and Genomes

DAVID B. WHITEHOUSE

School of Life Sciences, University of Hertfordshire, College Lane, Hatfield, Hertfordshire AL10 9AB, UK

5.1 INTRODUCTION

Key scientific discoveries in the 20th century include the demonstration that genes consist of DNA, the double helical structure of DNA and the genetic code. Between them, these stimulated the development of recombinant DNA technology or 'molecular biology', which is based on the recognition that genes can be isolated, sequenced and analysed. In the 50 years between the elucidation of the DNA double helix in 1953 and the completion of the human genome sequence, dramatic progress has been made on several fronts for investigating genes and genomes, such as molecular cloning, polymorphism detection, DNA sequencing, polymerase chain reaction and genome mapping. Some of these methodologies will be addressed in this chapter.

5.1.1 Background

Genomic analysis has far-reaching consequences because it potentially offers an unprecedented perspective into the nature of the genetic contribution to species biology. Genomics and genetics pervade all areas of basic biology, biotechnology and medicine, where in many cases there

Molecular Biology and Biotechnology, 5th Edition
Edited by John M Walker and Ralph Rapley
© Royal Society of Chemistry 2009
Published by the Royal Society of Chemistry, www.rsc.org

are clear-cut and immediate benefits such as the diagnosis of genetic disease. Genomics and genetics are also of growing value to several other fields, including anthropology, archaeology, biotechnology, ecology and forensic science, and also to commerce, where there are already ethical debates concerning the availability of personal genetic information for employment and insurance purposes.

5.1.1.1 Eukaryote Genomes

Cellular genomes (*i.e.* non-viral) vary tremendously in size and complexity. The smallest known from a prokaryote (bacterium) contains about 600 000 DNA base pairs; in contrast, eukaryote genomes such as human and mouse genomes have some 3 billion DNA base pairs (Table 5.1). In eukaryotes, genomic analysis is frequently complicated by their polyploid nature. In humans and the majority of the animal kingdom, including 'model species' such as mice and fruit flies, the DNA is packaged into two sets of chromosomes, one set being inherited from each parent. The human diploid genome is organised into 23 pairs of linear chromosomes, one set being inherited from each parent. All pairs of metaphase chromosomes appear identical in normal karyotypes apart from the sex chromosomes, X and Y, both of which are present in male cells. Sometimes gene pairs differ in their DNA sequences to a small degree and these 'allelic' differences can have a marked effect on the expression of the 'phenotype'. The effects of such allelic differences, which could be as small as the replacement of a single nucleotide within the amino acid coding portion of the gene or in the associated non-coding DNA that regulates gene action, give rise to mendelian genetic phenomena referred to as 'dominant', 'recessive' or 'codominant'. Detecting and characterising the molecular features that differentiate alleles and their expression have remained an important quest for genome scientists and geneticists.

Table 5.1 Approximate haploid genome sizes in various organisms

Organism		Genome size (Mb)
Bacteria	*Mycoplasma genitalium*	0.6
	Escherichia coli	4.6
Yeast	*Saccharomyces cerevisiae*	14
Nematode worm	*Caenorhabditis elegans*	100
Flowering plant	*Arabidopsis thalania*	115
Fruit fly	*Drosophilia melanogaster*	165
Puffer fish	*Fugu rubripes rubripes*	400
Human	*Homo sapiens*	3300

5.1.1.2   Gene Identification

In contrast to prokaryotes, gene identification in eukaryotes is further complicated because most chromosomal DNA is non-coding. For example, it has been estimated that less than 2% of human nuclear DNA consists of the 20 000–25 000 or so genes that constitute our genomes.[1] At first sight, this number of genes appears to be too low to underpin the genetic blueprint for sophisticated organisms such as vertebrates. However, it is clear that because of post-transcriptional processing, the number of messenger RNA (mRNA) transcripts vastly exceeds the actual number of genes. Further heterogeneity is introduced by post-translational modification during protein synthesis and maturation. Thus a relatively small number of genes appear to be capable of producing several different sequences of mRNA (transcripts) through various mechanisms such as alternative promoter usage, splicing and polyadenylation.[2,3] The non-genic portion of the genome, *i.e.* the remainder of the 6 000 000 000 DNA base pairs in the diploid human genome, is often referred to as 'junk DNA' because its function is not yet fully understood.

5.1.1.3   Genomic Analysis

In prokaryotes and viruses, genomic analysis is vastly simplified because the genomes consist of comparatively small single chromosomes with a considerably reduced complexity of gene organisation. Prokaryote genes consist of uninterrupted arrays of amino acid coding sequences that extend from the start site to the finish of the coding region with relatively little intergenic non-coding DNA. In eukaryotes the position is more complex: the DNA is arranged into several chromosomes and typical genes are not continuous stretches of coding sequence but rather are broken up into exons, which encode amino acids and non-coding introns.

The first genome sequence to be completed was that of the virus-like bacteriophage PhiX174 reported by Sanger and colleagues in 1977.[4,5] PhiX174 consisted of approximately 5375 nucleotides and 11 genes with no intergenic or intronic sequences. The publication of the PhiX174 sequence predated the publication of the full sequence of the human genome by some 25 years. Sanger's key discovery of the 'chain termination' DNA sequencing method enabled the nucleotide base order to be read in comparatively long stretches of DNA. This is the basis for modern DNA sequencing, which led to the development of the high-throughput protocols that allowed the completion the draft human genome sequence in 2001.[6,7]

5.2 KEY DNA TECHNOLOGIES

Key technologies relevant to the analysis of genes and genomes include the discovery of restriction endonucleases and cloning vectors, which led to molecular cloning, DNA sequencing methods and, more recently, the polymerase chain reaction.

5.2.1 Molecular Cloning Outline

Molecular cloning provides a convenient way of breaking very large segments of DNA associated with genomes into smaller pieces that are suited to detailed analysis. Cloning involves the amplification of specific segments of DNA leading to the accumulation of large amounts of identical DNA fragments that can be used for various purposes. Classical molecular cloning employs bespoke DNA cloning vectors, which are often derived from circular DNA molecules called bacterial plasmids and bacteriophage; such recombinant vectors replicate efficiently in specialised strains of bacterial host cells, resulting in the production of multiple copies of the recombinant DNA molecule that was inserted into the vector.[8] Before a DNA fragment can be inserted into a cloning vector, it must be prepared so that the precise DNA sequences of its ends are compatible with the ends of the vector molecule DNA sequence. This is achieved using commercially available enzymes that cleave double-stranded DNA at specific sequences; these are known as restriction enzymes or restriction endonucleases. In Nature, restriction enzymes constitute part of the bacterial defence system by digesting double-stranded DNA from harmful bacteriophage and other foreign DNA as it enters the cell. To date, more than 3500 restriction enzymes have been identified, some of which are commercially available and used for routine molecular biology applications.[9] If both the vector and the fragments of 'foreign' DNA are cut with the same restriction enzyme, such as *Eco*RI, which cuts at the sequence 5'GAATTC3', it is a relatively straightforward procedure to join, or ligate, the two molecules because their matching ends have small overhangs with complementary base pairs (Figure 5.1). The resulting molecule is a recombinant vector which is capable of growing in the host cell.

5.2.2 Cloning Vectors

Modern cloning vectors are engineered so that they contain 'multiple cloning sites' or 'polylinkers'. These are segments of DNA that consist of several unique restriction enzyme recognition sites for the most widely

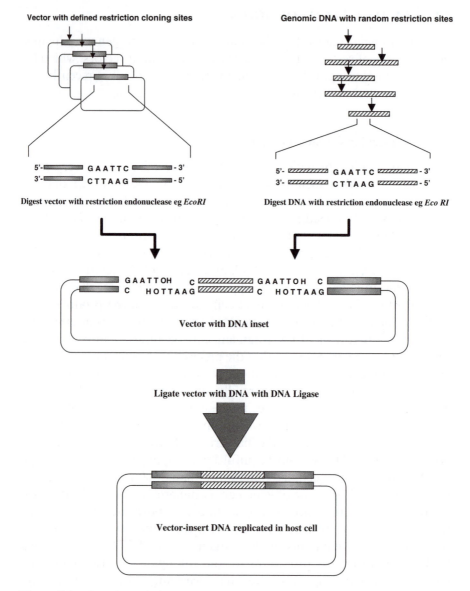

Figure 5.1 Overview of the restriction endonuclease digestion and DNA ligation process in DNA cloning.

available restriction enzymes (Figure 5.2). Thus, if a circular vector such as pUC18 is cut with any one of the polylinker specific restriction enzymes, it will become linearised but unfragmented. Under the correct conditions, the linear vector is able to ligate to a fragment of DNA that has been generated using the same restriction enzyme. Upon successful

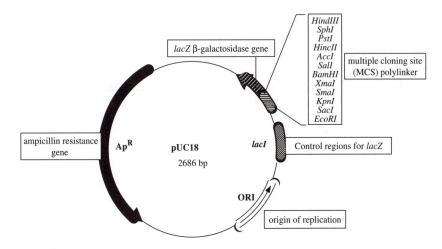

Figure 5.2 Map indicating the important features of the plasmid cloning vector pUC18.

Table 5.2 Examples of vectors generally available for cloning DNA fragments

Vector	Host Cell	Vector Structure	Insert Range (kb)
M13	*E. coli*	Circular virus	1–4
Plasmid	*E. coli*	Circular plasmid	1–5
Phage λ	*E. coli*	Linear virus	2–25
Cosmids	*E. coli*	Circular plasmid	5–45
BACs[a]	*E. coli*	Circular plasmid	50–500
YACs[b]	*S. cerevisiae*	Linear chromosome	100–2000

[a]BAC: bacterial artificial chromosome
[b]YAC: yeast artificial chromosome

ligation using the enzyme DNA ligase, the vector molecule containing the recombinant insert DNA is re-circularised. Other necessary features of cloning vectors include an origin of DNA replication and one or more selectable marker genes; typically these are antibiotic resistance genes.[8]

Many types of vector and host cells are employed in cloning experiments. The choice of vector depends on the size of the insert to be cloned and the purpose of the experiment (Table 5.2). Plasmids and lambda bacteriophage vectors are used for cloning small inserts of up to a few kilobases of DNA; cosmid vectors, on the other hand, can replicate efficiently with up to about 50 kb of DNA insert; bacterial artificial chromosomes (BACs) and yeast artificial chromosomes (YACs) are capable of replicating significantly larger inserts of up to about 500 kb of DNA (BACs) and more than 1000 kb of DNA (YACs). The last two vectors have been vital tools in the eukaryotic genome projects, where vast lengths of DNA had to be mapped and arranged into a series of overlapping clones or contigs. Experience showed that whereas YACs were capable of holding more 1 megabase of insert DNA, they tended to become unstable, causing the inserts to rearrange, whereas BACs were found to be far more stable and have become the mainstay for many genome projects.

5.2.3   The Cloning Process

The recombinant vector molecule must be introduced into its 'matching' host cell in order to replicate and produce multiple copies. The process by which DNA is introduced into the host cell is known as bacterial transformation. Since 'naked' vector DNA is hydrophilic and the bacterial cell wall is normally impermeable to such molecules, the host cell must be made 'competent' by treatment with calcium chloride in the early log phase of growth. This causes the cell to become permeable to chloride ions. When competent cells are mixed with DNA and heat shocked at 42 °C, the swollen cells are able to take up the naked DNA molecules. It is believed that only a single DNA molecule is permitted to enter any single cell. Thus individual colonies of transformed bacteria grow single recombinant vector molecules. The bacterial cells are usually grown on selective media so that only transformants survive to form colonies. Thus, for example, if the vector contains an ampicillin resistance gene and the cells are grown on ampicillin-containing media, only the cells containing the vector will form colonies.

5.2.3.1   Library Screening

There are several methods of screening for transformed colonies that contain recombinant vectors. For example, where the cloning site in the vector lies within an antibiotic resistance gene, successful integration of the insert will lead to inactivation of the resistance gene and recombinant colonies can be identified by a technique known as replica plating (Figure 5.3). In this method, the pattern of colonies in the original Petri dish is printed on to a nutrient agar plate containing the selective antibiotic. The position of the recombinant colonies, *i.e.* those that fail to grow on the selective antibiotic, is noted so that they can be picked from the master plate. In other vectors, a method called blue/white selection can be performed. In blue/white selection, successful integration of a foreign DNA molecule in the vector destroys an enzyme gene (the *LacZ* gene of β-galactosidase) that otherwise forms a blue product when the

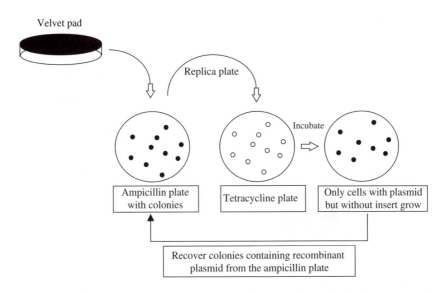

Figure 5.3   Replica plating to detect recombinant plasmids. A sterile velvet pad is
pressed on to the surface of an agar plate, picking up some cells from each
colony growing on that plate. The pad is then pressed on to a fresh agar
plate containing the selective antibiotic, thus inoculating it with cells in a
pattern identical with that of the original colonies. Clones of cells that fail
to grow on the second plate (owing to the loss of antibiotic resistance) can
be recovered from their corresponding colonies on the first plate.

transformed colonies are exposed to the substrate X-gal. Thus recombinant colonies are white and non-recombinants are blue (Figure 5.4).

5.2.3.2   Identifying Clones

Bacterial clones containing the sought after recombinant vectors can be
identified by hybridisation with specific radioactively labelled or
enzyme-labelled cDNA or genomic DNA probes or alternatively by the
immunodetection of protein products (using specialised expression
vectors which allow a cloned foreign cDNA to be transcribed to express
its protein product). Both approaches are technically straightforward.
Both involve the transfer of bacterial colonies from a master agar plate
on to carefully orientated nitrocellulose or nylon membranes. The cells
are then lysed and the DNA (or protein) from the lysed colonies is
immobilised on the membrane, which is used for the probing step.
Recombinant colonies can be detected as spots on X-ray film either by
autoradiography or enzyme-generated chemiluminescence. The spots on
the X-ray film can then be aligned with the agar master plate allowing
the correct colonies to be picked (Figure 5.5). For protein detection in

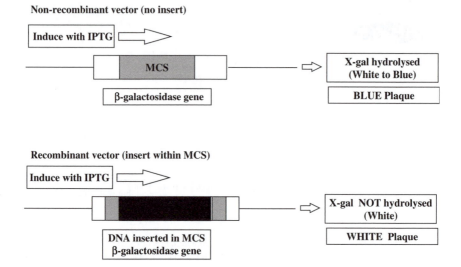

Figure 5.4   Principle of blue/white selection for the detection of recombinant vectors. In the presence of the inducer IPTG the β-galactosidase (*LacZ*) gene is transcribed. The recombinant DNA insert disrupts the expression of the *LacZ* gene which encompasses the multiple cloning site (MCS), hence the substrate X-gal is not hydrolysed and the recombinant colonies remain white.

expression vectors, antibody probes are employed and in a manner analogous to an ELISA test. The antibody probe is conjugated to an enzyme such as horseradish peroxidase or alkaline phosphatase. It is the activity of the bound enzyme on its chemiluminescence or chromogenic substrate that reveals the position of the recombinant colonies.

5.2.4   DNA Libraries

'DNA library' is the term used to describe a collection of recombinant clones or DNA molecules generated from a specific source of DNA. There are two main types of DNA library which are very distinct in their origin and purpose. DNA from a nucleated cell, whatever the tissue source, from a specific organism is used to make a 'genomic' library. The idea of a general genomic library is to produce a set of clones that contain enough DNA fragments so that the entire genome of the organism is represented. Variations of genomic DNA libraries such as chromosome-specific libraries that were prepared from chromosomes sorted by flow cytometry[10] were employed in the Human Genome Project in an attempt to shorten the path between the starting DNA and generation of the genome map. The second type is the cDNA library,

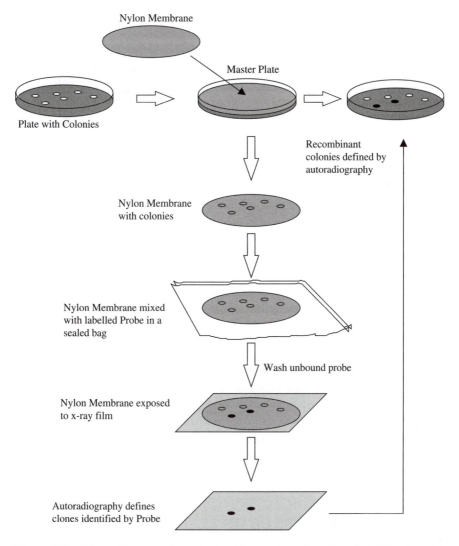

Figure 5.5 Method for detection of recombinant clones by colony hybridisation with labelled gene probes. The bacteria immobilised on the nylon (or nitrocellulose) membrane are lysed in alkaline conditions to make the plasmid DNA accessible to the probe.

which is made from mRNA that has been reverse transcribed by the enzyme reverse transcriptase. Reverse transcriptase produces complementary DNA (or cDNA) fragments which are then cloned into a vector. Therefore, unlike a genomic DNA library, a cDNA library is representative of the expressed genes in a particular cell or tissue type. Thus a skeletal muscle cDNA library contains sequences expressed in

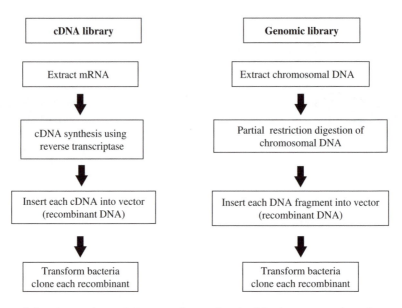

Figure 5.6   Comparison of the general steps involved in the construction of genomic and cDNA libraries.

the muscle tissue at the time the mRNA was harvested. cDNA libraries are particularly useful for cloning sequences where there is biological information. For example, it is known that mammalian skeletal muscle produces high levels of phosphoglucomutase (PGM1) enzyme activity, hence PGM1 cDNA clones are expected (and found) to be well represented in skeletal muscle cDNA libraries.[11] The essential features of both types of library are illustrated in Figure 5.6.

5.3   THE POLYMERASE CHAIN REACTION (PCR)

In some respects, the PCR can be regarded as a form of molecular cloning, since it is a technique analogous to the DNA replication process that takes place in cells and the outcome is the same, namely the generation of new DNA molecules based exactly upon the sequence of the existing ones. PCR is a laboratory technique that is currently a mainstay of molecular biology. One of the reasons for the global adoption of the PCR is the elegant simplicity of the reaction and relative ease of the practical manipulation steps. Indeed, combined with the relevant bioinformatics resources for the design of oligonucleotide primers and for determination of the required experimental conditions, it provides a rapid means for DNA identification and analysis.[12]

One problem with early PCR reactions was that the temperature needed to denature the DNA also denatured the DNA polymerase. However, the availability of a thermostable DNA polymerase enzyme isolated from the thermophilic bacterium *Thermus aquaticus*, found in hot springs, provided the means to automate the reaction. *Taq* DNA polymerase has a temperature optimum of 72 °C and survives prolonged exposure to temperatures as high as 96 °C and so is still active after each of the denaturation steps.[13]

The PCR is often used to amplify a fragment of DNA from a complex mixture of starting material usually termed the template DNA. However, in contrast to conventional cell-based cloning, PCR does require knowledge of the DNA sequences which flank the fragment of DNA to be amplified (target DNA). From this sequence information, two oligonucleotide primers are chemically synthesised, each complementary to a stretch of DNA to the 3' side of the target DNA, one oligonucleotide for each of the two DNA strands. For many applications PCR has replaced the traditional DNA cloning methods as it fulfils the same function, the production of large amounts of DNA from limited starting material; however, this is achieved in a fraction of the time needed to clone a DNA fragment. Although not without its drawbacks, the PCR is a remarkable development which has changed the approach of many scientists to the analysis of nucleic acids and continues to have a profound impact on core genomic and genetic analysis.

5.3.1 Steps in the PCR

The PCR consists of three well-defined times and temperatures termed steps: (i) denaturation at high temperature, (ii) annealing of primer and target DNA and (iii) extension in the presence of a thermostable DNA polymerase.[13] A single round of denaturation, annealing and extension is termed a 'cycle' (Figure 5.7). A typical PCR experiment consists of 30–40 cycles. In the first cycle, the double-stranded 'high molecular weight' template DNA is (i) denatured by heating the reaction mix to 95 °C. Within the complex mass of DNA strands, the region to be specifically amplified (target) is thus made accessible to the primers. The temperature is then (ii) decreased to between 40 and 60 °C to allow the hybridisation of the two oligonucleotide primers, which are present in excess, to bind to their complementary sites that flank the target DNA. The annealed oligonucleotides act as primers for DNA synthesis, since they provide a free 3'-hydroxyl group for DNA polymerase. The DNA synthesis step (iii) is termed extension and carried out at 72 °C by a thermostable DNA polymerase, most commonly *Taq* DNA polymerase.

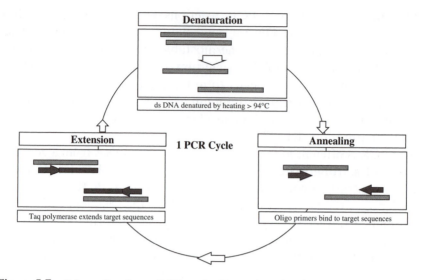

Figure 5.7   Schematic of one PCR cycle illustrating the three steps: denaturation, annealing and extension.

DNA synthesis proceeds from both of the primers until the new strands have been extended along and beyond the target DNA to be amplified. It is important to note that, since the new strands extend beyond the target DNA, they will contain a region near their 3′ ends which is complementary to the other primer. Hence, if another round of DNA synthesis is allowed to take place, not only the original strands will be used as templates but also the new strands. The products obtained from the new strands will have a precise length, delimited exactly by the two regions complementary to the primers. As the system is taken through successive cycles of denaturation, annealing and extension, all the new strands will act as templates and so there will be an exponential increase in the amount of DNA produced. The net effect is to amplify selectively the target DNA and the primer regions flanking it, leading to the production of millions of effectively identical copies.

For efficient annealing of the primers, the precise temperature at which the annealing occurs is critical and each PCR system has to be defined and optimised by the user. One useful technique for optimisation of annealing temperature is called 'Touchdown' PCR, where a programmable thermocycler machine is used to decrease the temperature incrementally until the optimum annealing is reached.[14] Reactions that are not optimised may give rise to other DNA products in addition to the specific target or may not produce any amplified products at all. Another approach to reducing spurious non-specific amplification during the early stages of the reaction is termed 'Hot Start' PCR. In this method, the reaction components are heated to the melting temperature before adding the polymerase. At one time Hot Start was achieved by introducing a physical wax barrier between the *Taq* polymerase and the remainder of the reaction components which melted at the denaturation temperature, allowing the *Taq* polymerase access to the reaction mix. More recently, modified polymerase enzyme systems have been developed that inhibit polymerisation at ambient temperature, either by the binding ligands such as antibodies or by the presence of bound inhibitors that dissociate only after a high-temperature activation step is performed.[15] Such modified polymerases are routinely used in PCR.

5.3.2 PCR Primer Design and Bioinformatics

The specificity of the PCR lies in the design of the two oligonucleotide primers. These not only have to be complementary to sequences flanking the target DNA but also must not be self-complementary or bind each other to form dimers, since both prevent authentic DNA amplification. They also have to be matched in their GC content and have similar annealing temperatures and be incapable of amplifying unwanted genomic sequences. Manual design of primers is time consuming and often hit or miss, although equations such as the following are still used to derive the annealing temperature (T_a) for each primer:

$$4(G + C) + 2(A + T) = T_m$$

where T_m is the melting temperature of the primer/target duplex and G, C, A and T are the numbers of the respective bases in the primer. In general, T_a is set 3–5 °C lower than T_m. On occasions, secondary or primer dimer bands may be observed on the electrophoresis gel in addition to the authentic PCR product. In such situations, Touchdown or Hot start regimes may help. Alternatively, raising T_a closer to T_m can enhance the specificity of the reaction.

The increasing use of bioinformatics resources such as Oligo, Generunner and Primer Design Assistant[16] in the design of primers makes the design and the selection of reaction conditions much more straightforward. These computer-based resources allow the sequences to be amplified, primer length, product size, GC content, *etc.*, to be input and following analysis, provide a choice of matched primer sequences. Indeed, the initial selection and design of primers without the aid of bioinformatics would now be unnecessarily time consuming. Finally, before ordering or synthesising the primers, it is wise to submit proposed sequences to a nucleotide sequence search program such as BLAST,[17] which can be used to interrogate GenBank or other comprehensive public DNA sequence databases to increase confidence that the reaction will be specific for the intended target sequence.

5.3.3 Reverse Transcriptase PCR (RT-PCR)

RT-PCR is an extremely useful variation of the standard PCR which permits the amplification of specific mRNA transcripts from very small biological samples without the need for the rigorous extraction procedures associated with mRNA purification for conventional cloning purposes. Conveniently, the dNTPs, buffer, *Taq* polymerase, oligonucleotide primers, reverse transcriptase (RT) and the RNA template are added together to the reaction tube. The reaction is heated to 37 °C which allows the RT to work and permits the production of a cDNA copy of the RNA strands that anneal to one of the primers in the mix. Following 'first strand synthesis', a normal PCR is carried out to amplify the cDNA product, resulting in 'second strand synthesis', and subsequently a dsDNA product is amplified as usual. The choice of primer for the first strand synthesis depends on the experiment. If amplification of all mRNAs in the cell extract is required, then an oligo dT primer that would anneal to all the polyA tails can be used. If a specific cDNA is sought, then a coding region-specific primer can be used with success, otherwise a random primer could be used. The method is fast, accurate and simple to perform. It has many applications, such as the assessment of transcript levels in different cells and tissues (when combined with Q-PCR; see Section 5.3.4). When combined with allele-specific primers, it also allows the amplification of cDNA from single chromosomes. RT-PCR is widely used as a diagnostic tool in microbiology and virology.[18–20]

5.3.4 Quantitative or Real-time PCR

Another useful PCR development is quantitative PCR (Q-PCR) or real-time PCR (confusingly also referred to as RT-PCR). Quantitative PCR is suited to many applications because of the rapidity of the method compared with conventional PCR, while simultaneously providing a lower limit of detection and greater dynamic range. Another advantage is that Q-PCR permits a rigorous analysis of PCR problems as they arise. Early quantitative PCR methods involved the comparison of a standard or control DNA template amplified with separate primers at the same time as the specific target DNA.[21] These types of quantification rely on the reaction being exponential and so any factors affecting this may also affect the result. Other methods involve the incorporation of a radiolabel through the primers or nucleotides and their subsequent detection following purification of the amplicon. An alternative automated real-time PCR method is the 5′ fluorogenic exonuclease detection system or TaqMan.[22] In its simplest form, a DNA-binding dye such as SYBR Green is included in the reaction. As amplicons accumulate, SYBR Green binds the dsDNA proportionally. Fluorescence emission of the dye is detected following excitation. The binding of SYBR Green is non-specific. Therefore, in order to detect specific amplicons, an oligonucleotide probe labelled with a fluorescent reporter and quencher molecule at either end is included in the reaction in the place of SYBR Green. When the oligonucleotide probe binds to the target sequence, the 5′ exonuclease activity of *Taq* polymerase degrades and releases the reporter from the quencher (Figure 5.8). A signal is generated which increases in direct proportion to the number of starting molecules. Hence the detection system is able to induce and detect fluorescence in real time as the PCR proceeds. In addition to quantification of the

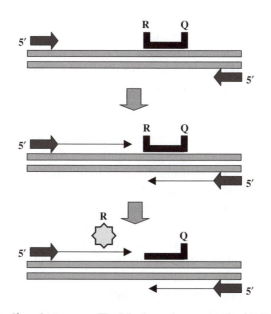

Figure 5.8 The 5′ nuclease assay (TaqMan) employs a standard PCR together with a small oligonucleotide probe. The probe is labelled with a fluorescent reporter (R) at one and a quencher (Q) at the other. The close proximity of the R–Q quenches fluorescence. The R–Q probe binds to its complementary PCR fragment until cleaved by the 5′ nuclease activity of Taq polymerase. Once the R group is released an increase in fluorescence is detected. Thus the performance of the PCR is measured in real time.

reaction, real-time PCR may also be used for genotyping single nucleotide polymorphisms and for accurate determination of amplicon melting temperature using curve analysis. Other probe-based PCR systems have been devised, such as the use of scorpion probes.[23]

5.4 DNA SEQUENCING

The determination of the order or sequence of nucleotide bases along a length of DNA is one of the central techniques in molecular biology and has played the key role in genome mapping and sequencing projects.

Two basic techniques have been developed for efficient DNA sequencing, one based on an enzymatic method frequently termed Sanger sequencing, after its developer, and a chemical method, Maxam and Gilbert sequencing, named for the same reason.[4,24] For large-scale DNA analysis, Sanger sequencing and its variants are by far the most effective methods and many commercial kits are available for its use. However, there are certain occasions, such as the sequencing of short oligonucleotides, where the Maxam and Gilbert method is still more appropriate.[25]

One absolute requirement for Sanger sequencing is that the DNA to be sequenced is in a single-stranded form. Traditionally this demanded that the DNA fragment of interest be cloned into the specialised bacteriophage vector M13, which is naturally single stranded.[26,27] Although M13 is still widely used, the advent of the PCR has provided a rapid means to amplify a region of any genome or cDNA for which primer sequences are available and generate the corresponding nucleotide sequence. This has led to an explosion in DNA sequence information and has provided much impetus for polymorphism discovery by resequencing regions of the genome from individuals.[1]

The Sanger method is simple and elegant and in many ways mimics the natural ability of DNA polymerase to extend a growing nucleotide chain based on an existing template. Initially the DNA to be sequenced is allowed to hybridise with an oligonucleotide primer, which is complementary to a sequence adjacent to the $3'$ side of DNA within a vector such as M13 (or within an amplicon in the case of PCR). The oligonucleotide will then act as a primer for synthesis of a second strand of DNA, catalysed by DNA polymerase. Since the new strand is synthesised from its $5'$ end, virtually the first DNA to be made will be complementary to the DNA to be sequenced. One of the deoxyribonucleoside triphosphates (dNTPs) which must be provided for DNA synthesis is radioactively labelled with ^{33}P or ^{35}S and so the newly synthesised strand will be radiolabelled.

5.4.1 Dideoxynucleotide Chain Terminators

The reaction mixture is then divided into four aliquots, representing the four dNTPs, A, C, G and T. Using the adenine (A) tube as an example, in addition to all of the dNTPs being present in the mix, an analogue of dATP is added [2′,3′-dideoxyadenosine triphosphate (ddATP)], which is similar to A except that it has no 3′-hydroxyl group. Since a 5′ to 3′ phosphodiester linkage cannot be formed without a 3′-hydroxyl group, the presence of the ddATP will terminate the growing chain. The situation for tube C is identical except that ddCTP is added; similarly, the G and T tubes contain ddGTP and ddTTP, respectively.

Since the incorporation of a ddNTP rather than a dNTP is a random event, the reaction will produce new molecules varying widely in length, but all terminating at the same type of base. Thus four sets of DNA sequence are generated, each terminating at a different type of base, but all having a common 5′ end (the primer). The four labelled and chain-terminated samples are then denatured by heating and loaded next to each other on a polyacrylamide gel for electrophoresis. Electrophoresis is performed at approximately 70 °C in the presence of urea, to prevent renaturation of the DNA, since even partial renaturation alters the rates of migration of DNA fragments. Very thin, long electrophoresis gels are used for maximum resolution over a wide range of fragment lengths. After electrophoresis, the positions of radioactive DNA bands on the gel are determined by autoradiography. Since every band in the lane from the ddATP sample must contain molecules which terminate at adenine and that those in the ddCTP terminate at cytosine, *etc.*, it is possible to read the sequence of the newly synthesised strand from the auto-radiogram. Under ideal conditions, sequences of approximately 300 DNA bases can be read from one gel.

5.4.2 Sequencing Double-stranded DNA

It is also possible to undertake direct DNA sequencing from double-stranded molecules such as plasmid cloning vectors and PCR amplicons. The double-stranded DNA must be denatured prior to annealing with primer. In the case of plasmids, an alkaline denaturation step is suffi-cient. However, for PCR amplicons this is more problematic. Unlike plasmids, amplicons are short and reanneal rapidly. Denaturants such as formamide and dimethyl sulfoxide have been used to prevent the reannealing of PCR strands following their separation. Another strategy is to bias the amplification towards one strand by using a primer ratio of 100:1, which also overcomes this problem to a certain extent.

It is possible physically to separate and retain one PCR product strand by incorporating a molecule such as biotin into one of the primers. Following PCR, the strand that contains the biotinylated primer may be removed by affinity chromatography with streptavidin-coated magnetic beads, leaving the complementary PCR strand. This magnetic affinity purification provides single-stranded DNA derived from the PCR amplicon and, although somewhat time consuming, it does provide high-quality single-stranded DNA for sequencing.[28]

5.4.3 PCR Cycle Sequencing

One of the most useful methods of sequencing PCR amplicons is termed PCR cycle sequencing. This is not strictly a PCR, since it involves linear amplification with a single primer. Approximately 20 cycles of denaturation, annealing and extension take place. Radiolabelled or fluorescently labelled dideoxynucleotides are then introduced in the final stages of the reaction to generate the chain-terminated extension products. Automated direct PCR sequencing is increasingly being refined, allowing greater lengths of DNA to be analysed in a single sequencing run.[29]

5.4.4 Automated DNA Sequencing

Advances in fluorescent labelling chemistry have led to the development of high-throughput automated sequencing techniques. Essentially, most systems involve the use of dideoxynucleotides labelled with different fluorochromes (often referred to as dye terminators). The advantage of this modification is that since a different label is incorporated with each ddNTP, it is unnecessary to perform four separate reactions. Therefore, the four chain-terminated products are run on the same track of a denaturing electrophoresis gel. Each product with its base-specific dye is excited by a laser and the dye then emits light at its characteristic wavelength. A diffraction grating separates the emissions which are detected by a charge-coupled device (CCD) and the sequence is interpreted by a computer. The advantages of the techniques include real-time detection of the sequence. In addition, the lengths of sequence that may be analysed are in excess of 500 bp. Capillary electrophoresis is increasingly being used for the detection of sequencing products. This is where liquid polymers in thin capillary tubes are used, obviating the need to pour sequencing gels and requiring little manual operation. This substantially reduces the electrophoresis run times and allows high throughput to be achieved. A number of large-scale sequence facilities are now fully automated, allowing the rapid acquisition of sequence data. Automated sequencing for genome projects is usually based on cycle sequencing using instruments such as the ABI PRISM 3700 DNA Analyzer. This can be formatted to produce simultaneous reads from 384-well cycle sequencing reaction plates. The derived nt sequences are downloaded automatically to databases and manipulated using a variety of bioinformatics resources.

5.4.5 Pyrosequencing

Rapid PCR sequencing has also been made possible by the use of pyrosequencing, which can be regarded as a second-generation sequencing method without the need for cloning in *E. coli* or any host cell. In one format of pyrosequencing, a PCR template is hybridised to an oligonucleotide and incubated with DNA polymerase, ATP sulfurylase, luciferase and apyrase. During the reaction, the first of the four dNTPs are added and, if incorporated, release pyrophosphate (PPi), hence the name 'pyrosequencing'. The ATP sulfurylase converts the PPi to ATP, which drives the luciferase-mediated conversion of luciferin to oxyluciferin to generate light. Apyrase degrades the resulting component dNTPs and ATP. This is followed by another round of dNTP addition. A resulting pyrogram provides an output of the sequence. The method provides short reads very quickly and is especially useful for the determination of mutations or discovery of single nucleotide polymorphisms (or SNPs; see Section 5.5.2.4).[30] Another pyrosequencing format involves direct analysis of DNA fragments and this system allows the rapid sequencing of entire genomes by the 'shotgun' approach. First genomic DNA is randomly sheared and ligated to linker sequences that permit individual molecules captured on the surface of a bead to be amplified while isolated within an emulsion droplet.[31] A very large collection of such beads is arrayed in the 1.6 million wells of a fibre-optic slide. The micro-array is presented sequentially with each of the four dNTPs and the amount of incorporation is monitored by luminometric detection as before. The second-generation Roche 454 Genome Sequencer FLX is reportedly able to produce 100 Mb of sequence with 99.5% accuracy for individual reads averaging over 250 bases in length. Once again, the derived sequences can be downloaded automatically to databases and manipulated using a variety of bioinformatics resources.

5.5 GENOME ANALYSIS

5.5.1 Mapping and Identifying Genes

There are two main mapping approaches to identifying genes on chromosomes: genetic mapping and physical mapping. Genetic mapping relies on observing the recombination frequencies between pairs of polymorphic markers segregating within families. A genetic map is constructed by analysing several pairs of polymorphic genetic markers. The genetic distance between the markers is determined by their recombination frequency and measured in centimorgans (cM), where 1 cM equates to 1% recombination, rather than number of nucleotides. For any pair of marker loci, the smaller the recombination frequency, the shorter is the genetic distance between the markers. The process of constructing a genetic map is called linkage analysis; this aims to determine whether or not a pair of gene loci tend to be co-inherited or separated by recombination.[32] At one time, linkage analysis in humans was a slow process because of the dearth of polymorphic markers. However, the discovery of variable numbers of tandem repeat poly-morphisms (VNTRs) and the production of genome wide panels of polymorphic microsatellites in the 1990s rendered linkage analysis one of the major tools in the Human Genome Project. Additionally, a number of physical mapping techniques were developed that enabled genes to be identified and localised purely on the basis of their physical positions along the chromosomes; this became known as 'reverse genetics' and later 'positional cloning'.

5.5.2 Tools for Genetic Mapping

5.5.2.1 Genetic Polymorphism

Some molecular variants are extremely common in populations; these commonly occurring variants are named polymorphisms and they have become one of the indispensable tools in genetics and genomics. Genetic polymorphism is a precise term which describes commonly occurring molecular variation at a specific locus such that at least 2% of the population will be heterozygous. 'Locus' means a DNA site in the genome, but not necessarily a functional gene. Most genetic poly-morphisms have absolutely no effect on the phenotype of the organism, they represent 'normal variation'.

5.5.2.2 RFLPs and Minisatellites

The first DNA polymorphisms were detected using Southern blotting techniques of genomic DNA digested with restriction enzymes.[33] This gave rise to restriction fragment length polymorphisms (RFLPs), which are due to the cleavage of certain alleles and the lack of cleavage of alternative alleles, resulting from the presence or absence of specific restriction enzyme recognition sites in the DNA. Southern blotting involves the transfer of DNA fragments generated by restriction enzyme digestion from an electrophoresis gel to an immobilising membrane such as nitrocellulose or nylon. The immobilised DNA is then probed with a radioactively or biotin-labelled DNA probe that could have originated from a cDNA clone or a genomic DNA clone or a PCR product. When the autoradiograph of the probed DNA digests from a set of unrelated individuals was examined, person to person variation in the resulting pattern of the bands identified an RFLP.[34] The majority of RFLPs consist of two alleles (cut and uncut) and are found to result from single nucleotide replacements (Figure 5.9). Southern blotting remains one of the simplest methods to analyse gene dosage. The amount of the labelled probe that hybridises to the filter bound 'target' is proportional to the amount of target on the filter. Hence, when comparing the band patterns of samples that have been very carefully loaded on to the electrophoresis gel, it is possible to make a quantitative assessment using densitometry to measure the intensity of the bands on the X-ray film.

Restriction enzyme analysis of human DNA gave rise to a further class RFLPs known as VNTRs, where the variation is in the length of the DNA rather than the presence or absence of restriction enzyme sites. VNTRs are blocks of repetitive DNA sequences called minisatellites that occur throughout the genome, but that tend to be found towards the ends of human chromosomes. Hypervariable minisatellites were discovered by Jeffreys *et al.* in 1985 and were detected by hybridisation of a minisatellites probe to Southern blots of DNA digested with a restriction enzyme, usually *Hin*F1, that cuts the DNA either side of the minisatellites.[35] DNA probes such as 33.15 recognise a common core sequence of about 10–15 base pairs which was shared between many different minisatellite loci. These multilocus probes revealed tremendous person-to-person variations in the complex patterns of bands which became known as DNA fingerprints.[36] DNA fingerprints were of great utility to forensic medicine and other fields where it was necessary to identify individuals unambiguously; however, DNA fingerprints were not amenable to mendelian segregation analysis or gene mapping because it was not possible to assign alleles within the complexity of bands displayed on the X-ray film.

A considerable advance with respect to gene mapping was made following the discovery of single-locus VNTR probes. Single-locus probes (SLP) were developed from specific cloned minisatellites and are thus able to detect individual minisatellite loci. They produce simple patterns on Southern blots in which heterozygous individuals display two bands and homozygotes a single band. Hence SLPs were useful for gene mapping since the bands could be tracked through families (Figure 5.10).

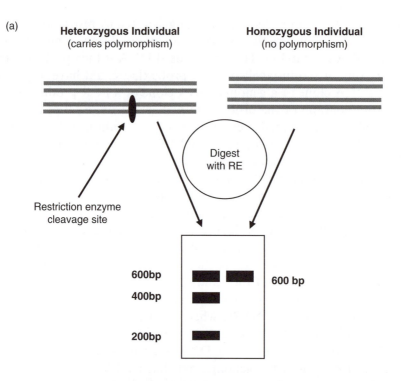

(a)

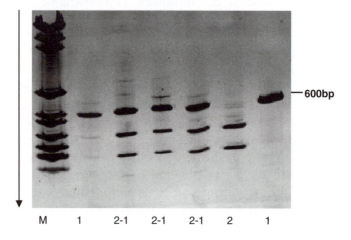

(b)

Figure 5.9 RFLP analysis of a 600 bp PCR product. (a) Schematic analysis of products from heterozygous and homozygous individuals digested with a restriction endonuclease. Three bands arise from the heterozygous individual, one which represents an uncleaved fragment and the other a cleaved fragment. (b) An *Sfa*II RFLP revealed in six individuals following digestion of a 600 bp PCR product.

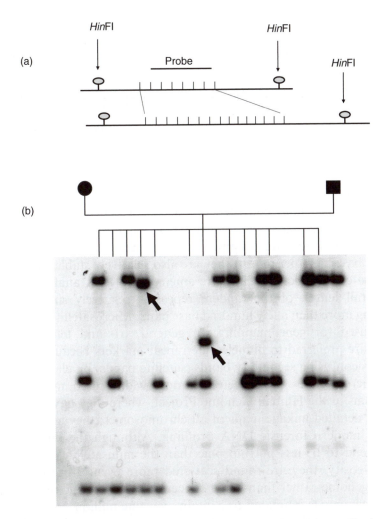

Figure 5.10 (a) Schematic illustration of a heterozygous VNTR locus. The invariant
*Hinf*I sites reveal a length polymorphism resulting from the repeat
number variation in the two minisatellite alleles. (b) Southern blot analysis of hypervariable minisatellite B6.7 using a single locus probe on
*Hinf*I digests of genomic DNA from a large family. The arrows indicate
new mutations not found in either parent.

The average heterozygosity of polymorphic VNTR loci is about 70%;
those loci with very high heterozygosities, such as B6.7 in Figure 5.10,
tend to exhibit considerable mutability and instability.[37] It is estimated
that there are between 15 000 and 20 000 VNTR loci in the human
genome. However, as they tend to cluster preferentially near the ends of
chromosomes, they are of limited use for genome mapping purposes.

Interestingly, it has since been discovered that the repeat blocks in many VNTR loci vary slightly within the sequences of the repeats, giving rise to repeat types along the length of the two alleles. These variants are detectable by PCR-based methods and they are the basis of the minisatellite variant repeats (MVR) system. This is a highly informative and elegant means of producing individual genetic barcodes which could be of tremendous use in forensic analysis of individuals in humans and many other species.[38]

5.5.2.3 *Microsatellites*

Whereas minisatellites consisted of repeats of between 10 and 70, or more, DNA base pairs per repeat block, microsatellites are tandem repeats which have a much simpler structure, consisting of between two and five base pairs per repeat; they are also relatively stable. In contrast to minisatellites, microsatellites are amenable to PCR analysis because the overall length of the tandem repeat is relatively small, seldom exceeding more than a few hundred base pairs. Microsatellites have been widely used for gene mapping. In general, tri- and tetranucleotide repeats are preferred to dinucleotide repeat markers because they produce less confusing secondary bands (satellite bands) on electrophoresis gels. Microsatellites have been the most widely used marker polymorphisms in the Human Genome Project, where they have been used to build detailed linkage maps of all chromosomes; they also form the basis of modern forensic DNA profiling. Although the microsatellite blocks tend to be less polymorphic than the minisatellites in terms of heterozygosity, they are found scattered more evenly throughout the human genome rather than being concentrated in specific regions. The human genome is estimated to contain over one million such loci.[6]

Large-scale gene mapping studies frequently involve the analysis of numerous microsatellite loci by gel or capillary electrophoresis. This type of approach can yield several hundred genotypes from a single experiment. Capillary electrophoresis is amenable to automation such that the products of a PCR reaction can be introduced robotically to the electrophoretic instruments. In this case, the microsatellites are amplified using fluorophore-labelled PCR primers and the reaction products are detected by laser-stimulated fluorescence as they pass by a detector that records the electrophoresed DNA fragments. Automated techniques such as capillary electrophoresis can be carried out with relatively little technical support in comparison with the demands of gel electrophoresis. In practical terms, microsatellite analysis can be conducted using instruments such as the Applied Biosystems Genescan 3100. The dye-labelled PCR products are loaded on to the instrument and electrophoresed. Specialised software is then used to process and determine the allele sizes of the products and thus the genotypes.

Interestingly, some triplet DNA repeat microsatellite sequences have been found to play a key role in the aetiology of some human genetic disorders, particularly those associated with certain inherited neurological conditions such as Huntington disease (HD, OMIM 143100), myotonic dystrophy (DM, OMIM 160900) and fragile X syndrome (FMR1, OMIM 309550). In these cases, the onset and severity of the disorders can be related to the increase in the number of trinucleotide repeats from generation to generation. For example, in HD, an autosomal dominant disease that gives rise to progressive and neural cell death, the onset of disease is associated with increases in the increasing length of a CAG triplet repeat located within the Huntington gene on human chromosome 4. The CAG triplet encodes the amino acid glutamine, which gives rise to a polyglutamine tract within the Huntington gene. It has been noted that in healthy individuals the number of CAG repeats ranges between 10 and 36, whereas in HD patients the number ranges between 37 and 100.[39]

The molecular mechanisms leading to microsatellite variation are not completely understood, although microsatellite loci certainly exhibit higher mutation rates compared with single nucleotide substitutions in genes.[40] It is generally accepted that replication slippage is the most common mutational mechanism leading to the gain or loss of one or more repeat units.[41] Other mutational mechanisms such as those based on unequal crossing over or duplication events have also been considered.[42] The tendency to mutate might also depend on the chromosomal environment of a particular locus and whether or not it is transcribed.[43,44]

5.5.2.4 *Single Nucleotide Polymorphisms (SNPs)*

By far the most widespread type of genetic variants is the single nucleotide polymorphism. Commonly referred to as SNPs (or Snips), these variant sites are found scattered throughout the genomes of most species. An SNP is caused by a single nucleotide replacement such as GAATTC to GTATTC that occurs with a frequency of at least 1% in the population. SNP is a new name for an old phenomenon. SNPs are the cause of most RFLPs and indeed the type of mutation that was indirectly detected in isozyme analysis by zone elctrophoresis before the advent of DNA technology.[45] Detailed studies of human DNA suggest that SNPs are extremely abundant: a recent estimate is that there are 10 million SNPs in the human genome,[46] which gives an average of one

SNP per 300 base pairs. In a recent mapping effort, more than 3 million human SNPs have been analysed in more that 200 individuals.[47] However, SNPs are not distributed evenly throughout the genome and the frequency can vary up to 10-fold.[48] Unsurprisingly, because of selection pressure, SNPs are less abundant in the coding regions than non-coding regions; other factors that affect their distribution are local recombination and mutation rates.[49] The majority of SNPs are located in repeat regions which are difficult to analyse; even so, SNPs are the best choice of polymorphic markers that cover the whole genome. Thus it has been estimated that there are 5.3 million common SNPs, each with a frequency of 10–15%, accounting for the bulk of DNA differences. Such SNPs present themselves on average once every 600 base pairs within the human genome in non-coding DNA,[50] hence the level of heterozygosity in the human genome resulting from SNPs is extraordinarily high. Many of the current the SNP detection methods can equally detect small insertions and deletions in the DNA. Although SNPs have a use for classical gene mapping by linkage analysis, they have attracted great interest as tools for large-scale association studies with the purpose of associating specific sequences and mutations with specific phenotypes.[51] SNPs can be used simply as genetic markers to identify disease genes by linkage or association analysis, or alternatively the SNPs themselves may on rare occasions be the actual variants in DNA sequences that cause the phenotype. SNPs within genes, especially coding regions that lead to missense mutations, may well change the function, regulation or expression of a protein.[52] SNPs have also been discovered in the splice sites of genes and these can result in variant protein products with different exon arrays.[53] A number of SNPs have been discovered in the controlling region of genes and some of these are reported to affect expression and regulation of proteins.[54,55] It should be emphasised that most SNPs occur in non-coding sequences far away from functional genes are therefore unlikely to have any functional significance, unless by chance they were to affect the configuration of a significant motif such as a remote locus control region.

SNP analysis is currently seen as the most likely approach to discovering genes and mutations that contribute to disease. The US National Center for Biotechnology Information (NCBI) curates and maintains an SNP database (dbSNP) which contains sequence details of several million SNPs, together with short insertions and deletions, from a variety of species.

The structural simplicity of SNPs compared with microsatellite renders them ideal for full-scale automation and unambiguous scoring of alleles. However, in contrast with the microsatellite loci, because most

SNPs represent single base changes or small insertions or deletions, they can have only two alleles and thus their heterozygosity cannot exceed 0.5, whereas microsatellite in contrast may have a vast number of alleles at individual loci with heterozygosities sometimes approaching 1.0. It should be noted that the low polymorphism information content of SNPs compared with microsatellites is more than offset by the fact that SNPs are present in such great abundance throughout the genome.

5.5.3 Mutation Detection

Since the early days of gene product analysis, electrophoresis-based methods have prevailed as the most common means for mutation detection. With the completion of the Human Genome Project and the production of the reference DNA sequence of the genome, there has been an increasing demand for faster and improved detection methods of polymorphic markers in the effort to map and to establish the function of all genes.

Although the role of classical DNA sequencing has remained a prominent weapon in the armoury for SNP detection,[56] a variety of alternative laboratory approaches (Table 5.3) have been developed to screen for, and type, new and existing SNPs.[57] These methods include single-strand conformation polymorphism (SSCP), denaturing high-performance

Table 5.3 Semi-quantitative comparison of some methods for discovering new SNPs (detection) and for typing known SNPs (screening) in PCR products

Technology	Detection	Screening
Direct DNA sequencing	*****	****
SSCP (single-strand conformation polymorphism)	****	*****
DGGE (denaturing gradient gel electrophoresis)	****	***
Gel-based heteroduplex analysis	***	**
Chemical cleavage mismatch	****	**
dHPLC, ~100% efficient	*****	****
Dot blot methods (ASOs)	NA	*****
Oligonucleotide ligation assay	NA	***
RFLP	*	***
Real-time PCR, *e.g.* TaqMan, Applied Biosystems	NA	*****
Primer extension minisequencing	?	****
DNA microarrays	?	*****
Allele-specific PCR	NA	***** (useful for direct haplotyping)

liquid chromatography (dHPLC) and microarray technology, which perhaps offers the greatest promise for the future.

5.5.3.1 *Single-stranded Conformation Polymorphism (SSCP)*

Originally described by Orita *et al.* in 1989,[58] SSCP has been widely applied in the field of human genetics, where it has been used both for SNP detection and for SNP discovery,[59] although it is generally preferred for the former application. The essence is to denature PCR products to make them single stranded and then to separate them by gel electrophoresis under non-denaturing conditions. In the absence of strong denaturants, a single-stranded fragment of DNA will adopt a unique and specific three-dimensional conformation as it attempts to fold into the most stable structure. The mutated form will adopt a different conformation to its wild-type counterpart. The differences in the conformers can be assessed on electrophoresis gels, where a heterozygous sample normally displays four bands, one for each denatured strand, and homozygotes normally display two bands (Figure 5.11). For mutation detection, the resolving power of the technique is improved by scanning PCR fragments that are less than 300 base pairs in length.[60] Provided that PCR-SSCP analysis is conducted under appropriate conditions, for example by repeating each electrophoresis experiment at two running temperatures or by varying the amount of mild denaturant included in the gel (such as formamide), it has been shown to be an efficient approach to discovering new mutations.[61–63] Initially SSCPs were detected using autoradiography of radiolabelled PCR products, followed later by silver staining to visualise unlabelled DNA fragments. A semi-automated method, PLACE-SSCP, has been developed in which the products of the PCR are labelled with fluorescent dyes and analysed by capillary electrophoresis under SSCP conditions.[64] As with many other mutation detection techniques, SSCP analysis is amenable to multiplexed PCR formats where several PCR products labelled with different fluorophores are analysed in the same electrophoretic lane.[65]

5.5.3.2 *Denaturing High-performance Liquid Chromatography (DHPLC)*

DHPLC has been shown to be a fast and reliable method for SNP detection and discovery.[66–68] The technique is based on the analysis of homoduplexes and heteroduplexes formed between reference and mutant DNA molecules by iron-pair reversed-phase high-performance liquid chromatography under partially denaturing conditions. For DNA fragments of between about 100 and 1500 base pairs, DHPLC has been

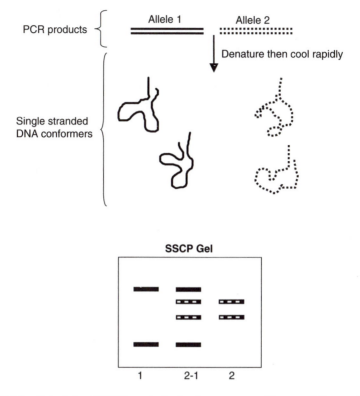

Figure 5.11 Principle of SSCP analysis. In the presence of heat and formamide, the dsDNA PCR products from a heterozygote are denatured for several minutes to form single-stranded DNA (ssDNA). Immediately before loading on to a non-denaturing electrophoresis gel, the samples are cooled on ice to encourage the formation of ssDNA conformers. Each conformer has a unique electrophoretic mobility as shown on the gel diagram.

shown to be capable of detecting all single base substitutions and also small insertions and deletions.

To perform the analysis, DNA fragments amplified by PCR from reference and test chromosomes are mixed together, fully denatured at 95 °C and allowed to re-anneal by reducing the temperature slowly to about 25 °C. This results in three classes of duplexes: homoduplexes of the reference DNA and mutant DNAs and, third, mismatched hetero-duplexes formed between reference and mutant DNA. Mismatched heteroduplexes and homoduplexes are then introduced to an ion pair of the reversed-phase HPLC flow path where they bind to the HPLC capillary. The temperature is then increased to a threshold level, usually between 50 and 60 °C, whereupon the DNA duplexes become *partially denatured.* Initially the duplexes are retained in the HPLC capillary through ionic binding interactions between the negatively charged phosphate backbone of the partially denatured DNA fragments and the beads in the cartridge, which are coated with the positively charged ion-pair reagent TEAA. The column is then eluted with a TEAA–acetonitrile gradient under conditions such that the heteroduplexes with a mismatched base pair(s) elute before the more stable homoduplexes. Note that for SNP detection purposes, the two homoduplexes are indistinguishable. The eluted fragments then pass through a UV detector and the absorbance is measured and the data are analysed by computer. Two sets of peaks are observed for each duplex (Figure 5.12). The method is rapid, taking on average 7 min to analyse a genotype, and amenable to multiplexing through the use of fluorescence primer detection analogous to high-throughput DNA sequencing strategies.[69] From a practical perspective, the method requires the use of *Pfu* polymerase instead of *Taq* polymerase for generating PCR products. *Pfu* has proof-reading properties which minimise the introduction of confusing PCR induced mutations.

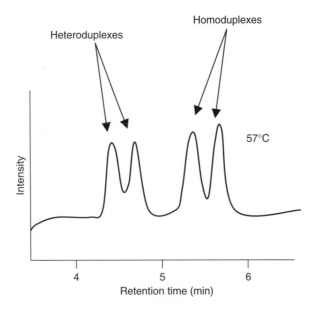

Figure 5.12   Schematic of dHPLC analysis of PCR products showing that the homoduplex and heteroduplex DNA species are differentially eluted when the column is held at the T_m of the PCR fragment, in this case 57 °C. N.B. to detect homozygous point mutations, the mutant PCR product is mixed with reference wild-type DNA.

5.5.3.3 DNA Microarrays

The need for rapid, high-quality, comprehensive gene-related data for a variety of purposes has stimulated the development of DNA microarrays or biochips. These provide a radically different approach to large-scale characterisation of genes and gene expression.[70] DNA microarrays are a departure from electrophoresis-based approaches and currently represent the extreme end of miniaturisation. A microarray is an ordered array of many thousands of DNA oligonucleotides, either single-stranded oligonucleotides or double-stranded cDNAs, attached (either by direct printing or *in situ* synthesis of short oligonucleotides) to a glass or silicon 'chip' that is about the size of a microscope coverslip. The bound DNA can then be hybridised with test DNA or RNA which has been labelled with one or two fluorescent dyes depending on the experimental design. Following an incubation step, the unhybridised material is washed away and the result is recorded using a confocal laser scanner. The data are collected and displayed automatically using dedicated computer programs.

Initially microarrays were designed to measure mRNA transcripts from thousands of genes in a single experiment.[71] This enabled the physiological state of cells and overall gene expression pattern to be correlated. For example, transcriptional profiles have been obtained for many types of human cancer and the accumulated data promise to lead the way to a better understanding of neoplasia and new therapeutic targets.[72]

DNA microarrays are not limited to gene expression studies. The first genotyping biochips were devised to identify key mutations in highly variable medically important genes and genomes such as the tumour suppressor gene *TP53* (OMIM 191170)[73] and the human immunodeficiency virus (HIV).[74] In both cases, successful genotyping of clinical material is achieved using the microarray approach. More recently, genotyping chips capable of assigning SNP alleles on a 'whole genome' basis have been devised, which have applications in a variety of post-genomic projects such as the HapMap initiative to map all human variations in different population groups. Three commercially available microarrays are available that make it possible to analyse simultaneously many thousands of SNPs in an individual's DNA. Two microarrays from Affymetrix, with roughly 100 000 and 500 000 SNPs, respectively, and a microarray from Illumina, with more than 300 000 SNPs, enable a large sample of the genetic variation of an individual to be assessed in a single experiment.[75] Such tools should improve the opportunities for correlating common diseases with genetic differences in statistically high-powered association studies. Commercially available microarrays have now been developed to monitor SNP profiles in economically or scientifically important non-humans such as cattle and mice. However, the current estimate is that there are more than 10 million SNP sites in the human genome and considerable future developments will probably be required to give the required coverage of SNPs in all human populations sufficient for genome-wide association studies.

5.6   GENOME PROJECTS BACKGROUND

The Human Genome Project was conceived in the mid-1980s, but was officially launched in 1990. This led to an upsurge in DNA sequencing activity and an avalanche of sequence data in a variety of species. By 1995, the first cellular (as opposed to viral) genome of a bacterium had been completed and this was followed by a stream of sequenced genomes, from prokaryotes initially and then eukaryotes. Two independent drafts of the human genome euchromatic DNA sequence were published simultaneously in 2001 as a result of a race between by the publicly funded International Human Genome Project and the privately funded initiative led by Celera Genomics.[6,7] These groups estimated that the human genome consists of between 30 500 and 35 500[6] and between 26 000 and 38 000[7] expressed genes. Today this is considered an overestimate, the actual number being closer to 20 000–25 000 genes, very much lower than the pre-genome project estimates of 80 000–140 000 genes. Rapid and powerful advances in cloning and sequencing technologies and computational biology have transformed genome sequencing initiatives from massive long-term endeavours to relatively quick and much less expensive undertakings. By mid-2008, complete genome sequences had been generated for 809 species, including 94 complex eukaryotic genomes.[76] There are currently more than 2500 genomes under study and the number is set to grow as the precision and speed of genome sequencing technologies improve and the cost is reduced.[76] It is estimated that the financial cost of the Human Genome Project, which involved a huge international effort over more than 10 years, was US $500 million, but it is predicted that the cost of resequencing any individual's entire DNA sequence could soon be just $1000. This would bring human genomic analysis into the realms of personal medicine.

5.6.1   Mapping and Sequencing Strategies

The Human Genome Project aimed to produce four types of map: physical, genetic, DNA sequence and gene. Physical and genetic maps provide essential anchor points and frameworks to align DNA sequences and assign genes. A high-resolution physical map based on the analysis of overlapping DNA clones represents the actual distance in DNA base pairs between genetic markers and other landmarks. However, the ultimate physical map is the DNA sequence itself. Low-resolution physical maps are generated from techniques such as somatic cell hybridisation and fluorescence *in situ* hybridisation; these methods are also applicable for assigning genes to chromosomes and will be addressed in the Section 5.7.

There are two approaches to genome sequencing: whole genome shotgun sequencing (WGS) and the more labour-intensive hierarchical shotgun sequencing (HS). In simple organisms such as bacteria and viruses, where the chromosomes are haploid and very little repeat sequence occurs, or for sequencing individual human genes, WGS works well.[77,78] In contrast, for eukaryotic genomes, where repeat sequences often abound, including the human genome (> 50% repeats), and there is considerable heterozygosity, it has been argued that HS offers advantages over WGS, and this was the approach adopted by the publicly funded International Human Genome Sequencing Consortium.[6]

5.6.1.1 Hierarchical Sequencing

This approach relies on the production of a set of large-insert clones (typically 100–200 kb each) that cover the entire genome.[6] Since the clones are all independent and ultimately positioned on a physical map in an order that represents each chromosome, repeated sequences are far less troublesome, leading to fewer gaps than encountered by the WGS approach. For many genome projects, high-capacity vectors such as BACs are advantageously used for generating large-insert clones since they are less likely to rearrange than alternatives such as YACs. Long-range physical maps are generated from the production of 'contigs'. Contigs are a set of overlapping DNA fragments that have been obtained from independent clones and positioned relative to one another so that they form a contiguous array. To obtain contigs, genomic libraries must prepared from high molecular weight DNA that has either been partially digested with restriction enzymes or randomly sheared. Partial digestion or random shearing leads to the production of a set of overlapping clones, whereas complete digestion would produce a set of fragments with no overlaps (Figure 5.13). Partial digestion ensures that when each DNA fragment is cloned into a vector, it has ends that will overlap with other clones. Thus, when the overlaps are identified, the clones can be positioned or ordered, so that a physical map is produced.

(a) Partial DNA digestion at restriction enzyme sites (E)

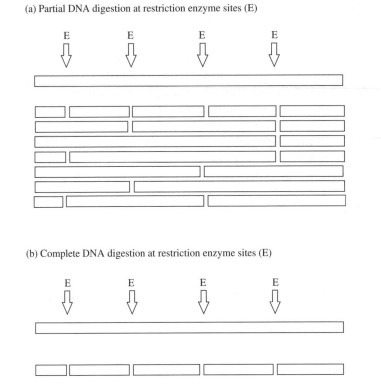

(b) Complete DNA digestion at restriction enzyme sites (E)

Figure 5.13 Comparison of partial and complete digestion of DNA molecules at restriction enzymes sites (E).

Large-insert clones are broken down further into sets of smaller overlapping subclones and sequenced using the shotgun sequencing approach. In order to position the overlapping ends into a contig representing the large insert clone, it is preferable to undertake DNA sequencing of both ends of the individual subclones (double-barrelled shotgun sequencing). Eventually, the entire DNA sequence of the large-insert clone is obtained by computer-based alignment of individual subclone DNA end sequences. In order to minimise the overlaps and identify the large-insert clones for further sequencing, restriction enzyme mapping can be undertaken to produce a 'fingerprint clone contig'. In the Human Genome Project, fingerprint clone contigs were mapped to human chromosomal locations by each chromosome workgroup using resources such as panels of human radiation hybrids (RH), fluorescence *in situ* hybridisation (FISH) with human chromosomes and existing genetic maps. Radiation hybrids are panels of human–hamster cell hybrids formed by fusing human cells containing radiation-generated fragments of human chromosomes with hamster cells. Panels of radiation hybrids that contain characterized fragments from all human chromosomes can be used for constructing genetic maps that are complementary to both recombination maps and physical maps based on contigs.[79]

In order to define a common way for all research laboratories to order clones and connect physical maps together, an arbitrary molecular technique based on the PCR has been developed to generate sequence-tagged sites (STS). These are small, unique sequences between 200 and 300 base pairs that are amplified by PCR.[80] The uniqueness of the STS is defined by the PCR primers that flank the STS. If the PCR results in amplification, then the STS is present in the clone being tested. In this way, defining STS markers that lie approximately 100 kb apart along a contig map allows the ordering of those contigs. Thus, all groups working with clones have publicly available defined landmarks with which to order clones produced in their DNA libraries (Figure 5.14). STSs may also be generated from polymorphic markers that may be traced through families along with other DNA markers and located on a genetic linkage map. These polymorphic STSs may thus serve as markers on both a physical map and a genetic linkage map for each chromosome and therefore provide a useful means for aligning the two types of map.

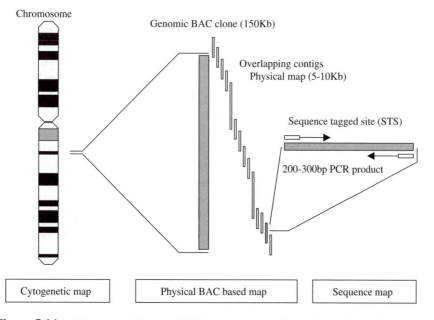

Figure 5.14   Schematic of the use of STS markers in the hierarchical physical mapping a human chromosome using BAC clones.

In addition to the human genome, the hierarchical sequencing approach has been used to sequence several genomes, including those of the yeast *Saccharomyces cerevisiae* and the nematode worm *Caenorhabditis elegans*.[81,82] High-quality BAC clone-based physical genome maps in one species can be of great value to genome projects in other species where some conservation of genomic sequence and gene order might be expected. For example, the outputs of the Human Genome Project have provided anchors for ordering BAC clones generated in several other species such as mouse, rat and cattle, allowing simplification of clone alignments and physical map building in addition to the generation of comparative maps in the respective species.[83–85] The International Human Genome Sequencing Consortium sequence was reported as finished in 2004 (Build 35) and contains 2.85 billion nucleotides interrupted by 341 gaps. It covers approximately 99% of the euchromatic genome.[86]

5.6.1.2   Whole Genome Shotgun Sequencing (WGS)

In contrast to the hierarchical BAC by BAC approach, which relies on the availability of genetic and physical maps for success, WGS is based on the strategy of sequencing a vast number of random genomic clones followed by intensive computer-based analysis of the DNA sequences which identifies matching sequences in different clones. This permits the assembly of a chromosomal DNA sequence, in principle without other map resources. As with the HS approach, overlapping clones are required, but since the clones are destined for direct sequence analysis, only vectors that contain small to medium inserts are normally used. Hence once the overlaps have been identified, the entire sequence is assembled. WGS was the approach adopted by the privately funded human genome initiative.[7]

Although WGS remains somewhat controversial for sequencing complex genomes of 'higher' organisms because of the problems associated with repeat sequences and heterozygosity, it is a widely used approach. The number of complex genomes sequenced by this WGS is increasing and includes the fruit fly *Drosophila*, mosquito (anopheles), mouse, puffer fish, dog and grapevine.[87–92] However, in some cases, such as the silk worm genome project, the WGS method has resulted in many seemingly irresolvable gaps in the genome and so the BAC-based hierarchical ordering of clones was used to close the gaps.[93] Advances in computational analysis of WGS sequences suggest that the problems caused by repeat sequences could be overcome, hence the approach can be expected to gain more ground in future genome projects.[94]

5.7 GENE DISCOVERY AND LOCALISATION

A major goal of all genome projects is to identify, map and characterise the genes. These objectives are prerequisites for advances in biotechnology, medicine and genetics.

5.7.1 Laboratory Approaches

Some well-established methods for mapping genes predated the DNA era, including somatic cell hybridisation and family-based linkage analysis. Somatic cell hybridisation was a forerunner of the radiation hybrid mapping panels that played a significant role in the Human Genome Project for high-resolution gene mapping. Somatic cell mapping panels are derived from human–rodent cell hybrids that are formed by fusing human cells with rodent cells. The hybrids preferentially lose human chromosomes at random, which allows the establishment of a panel of clones that retain different human chromosomes. This permits the assignment of a DNA probe, or an enzyme detected by its activity, to a particular human chromosome.[95] Both somatic cell hybrid analysis and linkage analysis require knowledge of the gene to be mapped and therefore cannot be used alone for gene discovery. However, both approaches can be used to provide low-resolution localisation for a gene: in the case of the somatic cell hybrids, to a particular chromosome, and for linkage analysis, to a linkage group, which may be associated with a chromosome in some instances.[32] A limitation of linkage analysis is that the gene being mapped and the marker genes must be heterozygous in key parent individuals; this was frequently a stumbling block before the availability of polymorphic DNA markers such as microsatellites and SNPs.

Several gene mapping and identification techniques have been developed for use with cloned DNA. A gene can be localised to a chromosome by fluorescence *in situ* hybridisation (FISH). FISH is carried out on R-banded spreads of metaphase chromosomes using a labelled genomic clone. The clone is hybridised to its complementary sequence on a chromosome and its position revealed by fluorescence confocal microscopy. By alternating the microscope between the hybridised and banded display, it is possible to assign a gene to a particular chromosome band to a resolution of a few megabases.[11]

A high-resolution development of FISH, sometimes called Fibre FISH, can be used to order cloned genes on DNA fibres prepared from chromatin from interphase nuclei. The fibres are made to extend and using the approach it is possible to hybridise three genomic clones simultaneously enabling their relative positions and order to be determined.[96] This is approach can resolve the order of clones that are only about 2–7 kb apart.

Identifying genes *de novo* in cloned DNA presents a different type of challenge. Several strategies have been developed with varying degrees of success. In some species, expressed genes are associated with upstream unmethylated GpC dinucleotide-rich sequences. This has allowed the use of the restriction endonuclease *Hpa*II to scan cloned DNA for CpG dinucleotide-rich regions. Because this enzyme only cleaves unmethylated cytosine of the CpG dinucleotide, cleavage at such sites may indicate the presence of an expressed gene. Upon digestion with *Hpa*II, these regions form tiny fragments and are known as *Hpa*II tiny fragments (HTF) or CpG islands.[97]

Considerable progress in gene localisation and discovery has been made with a PCR-based method known as EST mapping. An EST or Expressed Sequence Tag is a small PCR product that has been generated from a cDNA sequence, thus reflecting an expressed gene in the cell or tissue from which the mRNA was prepared. Public domain ESTs are available from the dbEST database.[98] Release 053008 of dbEST contains more than 50 million ESTs from a large number of species. EST mapping used with large insert genomic clones leads to discovery and provides information on genome organisation, including gene density and localisation.[99] ESTs in many species reflect genes of unknown function. For example, a recent study of the brown planthopper, a serious pest of rice plants, generated a library of more than 37 000 nuclear genome ESTs some of which were unrelated to any gene sequences in the databases. These and others could be used to search for genes of relevance to understanding the biology of the pest species and identifying potential target genes for developing novel insecticides against planthoppers.[100,101]

A further gene isolation system, which uses adapted vectors termed exon trapping or exon amplification, may be used to identify exon sequences. Exon trapping requires the use of a specialised expression vector that will accept fragments of genomic DNA containing sequences for splicing reactions to take place. Following transfection of a eukaryotic cell line, a transcript is produced that may be detected by using specific primers in an RT-PCR. This indicates the nature of the foreign DNA by virtue of the splicing sequences present.[102]

5.7.2 Bioinformatics Approaches

Parallel developments in bioinformatics and biocomputing have been essential for the success of the genome projects. In the closing stages of the Human Genome Project, some institutions were generating up to 175 000 individual DNA sequence reads per day, information which needed to analysed, submitted to databases, annotated and checked for quality. A multiplicity of DNA and protein sequences from different species is submitted daily by laboratories worldwide to databases such as GenBank (USA, DNA sequences), EMBL-BANK (Europe, DNA sequences), DDBJ (Japan, DNA sequences) and SwissProt (Europe, protein sequences). All of these ever-expanding and annotated databases are available to the public and the three major DNA sequence databases share information on a daily basis. These massive primary sequence databases have stimulated the emergence of a large number of specialised genomic and protein databases dedicated to various subjects such as a particular species group, biological features such as disease markers, protein families, protein domains, *etc.* The Nucleic Acids Research online Molecular Biology Database Collection is a public repository that lists more than 1000 databases.[103]

Many algorithms such as BLAST[17] have been developed to search databases for matching sequences of nucleotides or amino acids and there are now unparalleled opportunities for characterising genes and for studying groups of related genes *in silico*. It has been estimated that the majority of new cDNA and EST sequences will show similarity to proteins of known function and that many of the sequences will show similarity to each other. Further precision for identifying gene families is afforded by automated approaches to access databases such as Pfam to search for protein domains.[104]

Other computational approaches have been devised to predict features of genes in otherwise anonymous genomic sequences. Grail is a gene finder program which predicts exons, genes, promoter regions, polyA tails and other features associated with expressed DNA. 'Cloning' genes using database resources and dedicated algorithms is now a reality and biocomputing has become an applied science which enables investigators dealing with sequences held in data repositories to plot a course that may seldom require the use of conventional laboratory equipment. Organising and synthesising the exponentially expanding nucleotide data delivered by the second-generation sequencing projects is perhaps the major challenge for genomics.

In addition to the DNA and protein sequence databases, there has been an accompanying development of catalogue databases. Two examples from human genetics are OMIM (Online Mendelian Inheritance in Man), which catalogues salient details and literature references for more than 18 000 human genetic traits (the majority of which are known at the sequence or molecular basis levels), and the Human

Genome Nomenclature Database (HGNC), which assigns names to all known human genes and curates a searchable database. There are many others serving various research communities and biological systems.

5.8 FUTURE DIRECTIONS

In medicine, a major challenge is to unravel the genetic basis of common diseases, which at some stage during life may affect the majority of individuals in all populations. The Human Genome Project has produced the resources to tackle these multifactorial disorders such as cardiovascular disease, cancer, diabetes, bipolar disorder, asthma, multiple sclerosis, *etc.* These complex traits do not exhibit mendelian inheritance patterns often observed with single gene disorders such as cystic fibrosis. Instead, the action of several genes, each with a small effect and environmental influences, can be expected to modify the risk of disease. Various types of association studies are used in an attempt to define the genes that underlie the propensity for some individuals to develop such diseases. Population- and family-based association studies such as whole genome association studies and affected sib pair analysis make use of the gene maps, polymorphic marker resources and genome technologies in the attempt to pinpoint small regions of chromosomes associated with the disease phenotype. Chromosomal regions showing positive association with disease can then be subjected to detailed analysis by efficient resequencing strategies[1] and functional studies in the attempt to identify the causative mutations.[105] As the genome becomes better understood, it is expected that more complex trait genes and mutations will be identified, leading to an improved understanding of molecular pathology and, it is hoped, new effective therapies.

Interestingly, except for a few promising cases,[106] the optimistic predictions for gene therapy that were made more than 20 years ago have not yet been fulfilled. Understanding the expression patterns of transcription of disease predisposing genes (from microarray data, for example) seems to offer more potential for therapeutic developments. For instance, there have been considerable advances in the field of RNA silencing or transcriptional inactivation, which fosters the belief that small interfering RNA (siRNA) molecules could one day have a therapeutic role in common diseases, including cancers, by 'switching off' deleterious gene expression.[107]

The genome projects and gene analysis in other species can be expected to produce enormous scientific and economic benefits. The information from the genome projects of model organisms such as mouse, fruit fly, yeast and the nematode worm has been integrated with the human data to provide insights into evolution, refining maps, *etc.* Animal and plant genome projects are of great importance to agriculture and biotechnology. Transgenesis or genetic modification (GM) of food organisms, both plant and animal, is likely to grow in importance, particularly as world food supplies become more stretched. Genetically modified crops, principally corn, soybean, rapeseed and cotton, have already been produced with a range of GM characters such as herbicide resistance, insect resistance, virus resistance, delayed fruit ripening (tomato), altered oil content, *etc.* In the next decades, transgenic animals that produce increased yields of meat, improved health, disease resistance, optimised fat content and ability to thrive in different environments could be produced. Transgenics also holds great promise for the low-cost production of pharmaceuticals such as therapeutic proteins expressed in milk and plants. The growing effort of genome sequencing and gene discovery promises to overcome the previously encumbering difficulties associated with locating genes for species improvement, new antibiotic targets, *etc.* However, before implementation of these new GM strategies can occur, a full risk assessment is necessary to understand the possible impacts on human health and the environment.

REFERENCES

1. S. Levy, G. Sutton, P. C. Ng, L. Feuk, A. L. Halpern, B. P. Walenz, N. Axelrod, J. Huang, E. F. Kirkness, G. Denisov and Y. Lin, *et al., PLoS Biol.* 2007, **5**, e254.
2. P. Carninci, T. Kasukawa, S. Katayama, J. Gough, M. C. Frith, N. Maeda, R. Oyama, T. Ravasi, B. Lenhard and C. Wells, *et al., Science*, 2005, **309**, 1559.
3. P. Carninci, *Trends Genet.*, 2006, **22**, 501.
4. F. Sanger, S. Nicklen and A. R. Coulson, *Proc. Natl. Acad. Sci. USA*, 1977, **74**, 5463.
5. F. Sanger, G. M. Air, B. G. Barrell, N. L. Brown, A. R. Coulson, J. C. Fiddes, C. A. Hutchison III, P. M. Slocombe and M. Smith, *Nature*, 1977, **265**, 687.
6. International Human Genome Sequencing Consortium, *Nature*, 2001, **409**, 860.
7. J. C. Venter, M. D. Adams, E. W. Myers, P. W. Li, R. J. Mural, G. G. Sutton, H. O. Smith, M. Yandell, C. A. Evans and R. A. Holt, *et al., Science*, 2001, **291**, 1304.
8. J. Sambrook and P. MacCallum, *Molecular Cloning: a Laboratory Manual*, 3rd edn, Cold Spring Harbor Laboratory Press, Cold Spring Harbor, NY, 2001.

9. R. J. Roberts, T. Vincze, J. Posfai and D. Macelis, *Nucleic Acids Res.*, 2007, **35**, Database Issue, D3–D4.

10. K. E. Davies, B. D. Young, R. G. Elles, M. E. Hill and R. Williamson, *Nature*, 1981, **293**, 374.

11. D. B. Whitehouse, W. Putt, J. U. Lovegrove, K. Morrison, M. Hollyoake, M. F. Fox, D. A. Hopkinson and Y. H. Edwards, *Proc. Natl. Acad. Sci. USA*, 1992, **89**, 411.

12. R. K. Saiki, S. Scharf, F. Faloona, K. B. Mullis, G. T. Horn, H. A. Erlich and N. Arnheim, *Science*, 1985, **230**, 1350.

13. R. K. Saiki, D. H. Gelfand, S. Stoffel, S. J. Scharf, R. Higuchi, G. T. Horn, K. B. Mullis and H. A. Erlich, *Science*, 1988, **239**, 487.

14. R. H. Don, P. T. Cox, B. J. Wainwright, K. Baker and J. S. Mattick, *Nucleic Acids Res.*, 1991, **19**, 4008.

15. D. J. Sharkey, E. R. Scalice, K. G. Christy Jr, S. M. Atwood and J. L. Daiss, *Biotechnology*, 1994, **12**, 506.

16. S. H. Chen, C. Y. Lin, C. S. Cho, C. Z. Lo and C. A. Hsiung, *Nucleic Acids Res.*, 2003, **31**, 3751.

17. S. F. Altschul, T. L. Madden, Al. A. Schaffer, J. Zhang, Z. Zhang, W. Miller and D. J. Lipman, *Nucleic Acids Res.*, 1997, **25**, 3389.

18. J.-P. Lai, J.-H. Yang, S. D. Douglas, X. Wang, E. Riedel and W.-Z. Ho, *Clin. Diagn. La.b Immunol.*, 2003, **10**, 1123.

19. S. Corbet, J. Bukh, A. Heinsen and A. Fomsgaard, *J. Clin. Microbiol.*, 2003, **41**, 1091.

20. B. S. Coulson, J. R. Gentsch, B. K. Das, M. K. Bhan and R. I. Glass, *J. Clin. Microbiol.*, 1999, **37**, 3187.

21. R. Higuchi, C. Fockler, G. Dollinger and R. Watson, *Biotechnology*, 1993, **11**, 1026.

22. P. M. Holland, R. D. Abramson, R. Watson and D. H. Gelfand, *Proc. Natl. Acad. Sci. USA*, 1991, **88**, 7276.

23. A. Solinas, L. J. Brown, C. McKeen, J. M. Mellor, J. Nicol, N. Thelwell and T. Brown, *Nucleic Acids Res.*, 2001, **29**, E96.

24. A. M. Maxam and W. Gilbert, *Proc. Natl. Acad. Sci. USA*, 1977, **74**, 560.

25. J. Kieleczawa, *J. Biomol. Tech.*, 2005, **16**, 220.

26. J. Messing, B. Gronenborn, B. Muller-Hill and P. Hans Hopschneider, *Proc. Natl. Acad. Sci. USA*, 1977, **7**, 3642.

27. B. Gronenborn and J. Messing, *Nature*, 1978, **272**, 375.

28. R. E. March, W. Putt, M. Hollyoake, J. H. Ives, J. U. Lovegrove, D. A. Hopkinson, Y. H. Edwards and D. B. Whitehouse, *Proc. Natl. Acad. Sci. USA*, 1993, **90**, 10730.

29. H. M. Pang and E. S. Yeung, *Nucleic Acids Res.*, 2000, **28**, E73.

30. M. Ronaghi, S. Shokralla and B. Gharizadeh, *Pharmacogenomics*, 2007, **8**, 1437.
31. D. Dressman, H. Yan, G. Traverso, K. W. Kinzler and B. Vogelstein, *Proc. Natl. Acad. Sci. USA*, 2003, **100**, 8817.
32. A. J. F. Griffiths, J. H. Miller, D. T. Suzuki, R. C. Lewontin and W. M. Gelbart (Eds.), *An Introduction to Genetic Analysis*, Freeman, New York, 1993.
33. E. M. Southern, *Trends Biochem. Sci.*, 2000, **25**, 585.
34. R. Leach, R. DeMars, S. Hasstedt and R. White, *Proc. Natl. Acad. Sc.i USA*, 1986, **83**, 3909.
35. A. J. Jeffreys, V. Wilson and S. L. Thein, *Nature*, 1985, **314**, 67.
36. A. J. Jeffreys, V. Wilson and S. L. Thein, *Nature*, 1985, **316**, 76.
37. K. Tamaki, C. A. May, Y. E. Dubrova and A. J. Jeffreys, *Hum. Mol. Genet.*, 1999, **8**, 879.
38. K. Tamaki, C. H. Brenner and A. J. Jeffreys, *Forensic Sci. Int.*, 2000, **113**, 55.
39. J. F. Gusella and M. E. MacDonald, *Curr. Opin. Neurobiol.*, 1995, **5**, 656.
40. Y. C. Li, A. B. Korol, T. Fahima, A. Beiles and E. Nevo, *Mol. Ecol.*, 2002, **11**, 2453.
41. C. Schlötterer and D. Tautz, *Nucleic Acids Res.*, 1992, **20**, 211.
42. J. M. Hancock, in *Microsatellites: Evolution and Applications*, D. Goldstein and C. Schlötterer (Eds.), Oxford University Press, New York, 1999, pp. 1–9.
43. H. Ellegren, *Nat. Genet.*, 2000, **24**, 400.
44. J. D. Hawk, L. Stefanovic, J. C. Boyer, T. D. Petes and R. A. Farber, *Proc. Natl. Acad. Sci. USA*, 2005, **102**, 8639.
45. H. Harris and D. A. Hopkinson, *Handbook of Enzyme Electrophoresis in Human Genetics (with Supplements)*, Oxford American Publishing, New York, 1976.
46. E. Lai, *Genome Res.*, 2006, **11**, 927.
47. International HapMap Consortium, *Nature*, 2007, **449**, 851.
48. International SNP Map Working Group, *Nature*, 2001, **409**, 928.
49. D. E. Reich, S. F. Schaffner, M. J. Daly, G. McVean, J. C. Mullikin, J. M. Higgins, D. J. Richter, E. S. Lander and D. Altshuler, *Nat. Genet.*, 2002, **32**, 135.
50. L. Kruglyak and D. A. Nickerson, *Nat. Genet.*, 2001, **27**, 234.
51. G. A. Thorisson and L. D. Stein, *Nucleic Acids Res.*, 2003, **31**, 124.
52. F. J. de Serres, *Environ. Health Perspect.*, 2003, **111**, 1851.
53. M. Krawczak, J. Reiss and D. N. Cooper, *Hum. Genet.*, 1992, **90**, 41.

54. E. M. El-Omar, M. Carrington, W. H. Chow, K. E. McColl, J. H. Bream, H. A. Young, J. Herrera, J. Lissowska, C. C. Yuan and N. Rothman, *et al.*, *Nature*, 2000, **404**, 398.

55. A. Ligers, N. Teleshova, T. Masterman, W. X. Huang and J. Hillert, *Genes Immun.*, 2001, **2**, 45.

56. P. Taillon-Miller, E. E. Piernot and P.-Y. Kwok, *Genome Res.*, 1999, **9**, 499.

57. P.-Y. Kwok (Ed.), *Single Nucleotide Polymorphisms, Methods and Protocols*, Humana Press, Totowa, NJ, 2003.

58. M. Orita, H. Iwahana, H. Kanazawa, K. Hayashi and T. Sekiya, *Proc. Natl. Acad. Sci. USA*, 1989, **86**, 2766.

59. T. Tozaki, N.-H. Choi-Miura, M. Taniyama, M. Kurosawa and M. Tomita, *BMC Med. Genet.*, 2002, **3**, 6.

60. K. Hayashi and D. W. Yandell, *Hum. Mutat.*, 1993, **2**, 338.

61. S. P. Yip, W. Putt, D. A. Hopkinson and D. B. Whitehouse, *Ann. Hum. Genet.*, 1999, **63**, 129.

62. C. Mroske, J. Muci, J. Wang, K. Li, W. Song, J. Yan, J. Feng, Q. Liu and S. S. Sommer, *Anal. Biochem.*, 2007, **368**, 250.

63. S. P. Yip, D. A. Hopkinson and D. B. Whitehouse, *Biotechniques*, 1999, **27**, 20.

64. T. Tahira, A. Suzuki, Y. Kukita and K. Hayashi, *Methods Mol. Biol.*, 2003, **212**, 37.

65. J. S. Ellison, *Mol. Biotechnol.*, 1996, **5**, 1.

66. D. Guzowski, A. Chandrasekaran, C. Gawel, J. Palma, J. Koenig, X. P. Wang, M. Dosik, M. Kaplan, C. C. Chu and S. Chavan, *et al.*, *J. Biomol. Tech.*, 2005, **16**, 154.

67. D. Liu, Y. Zhang, Y. Du, G. Yang and X. Zhang, *DNA Seq.*, 2007, **18**, 220.

68. M. Shi, Y. Hou, J. Yan, R. Bai and X. Yu, *J. Forensic Sci.*, 2007, **52**, 235.

69. A. Premstaller, W. Xiao, H. Oberacher, M. O'Keefe, D. Stern, T. Willis, C. G. Huber and P. J. Oefner, *Genome Res.*, 2001, **11**, 1944.

70. E. M. Southern, *Trends Genet.*, 1996, **12**, 110.

71. V. Trevino, F. Falciani and H. A. Barrera-Saldaña, *Mol. Med.*, 2007, **13**, 527.

72. D. R. Rhodes, S. Kalyana-Sundaram, V. Mahavisno, R. Varambally, J. Yu, B. B. Briggs, T. R. Barrette, M. J. Anstet, C. Kincead-Beal and P. Kulkarni, *et al.*, *Neoplasia*, 2007, **9**, 166.

73. S. A. Ahrendt, S. Halachmi, J. T. Chow, L. Wu, N. Halachmi, S. C. Yang, S. Wehage, J. Jen and D. Sidransky, *Proc. Natl. Acad. Sci. USA*, 1999, **96**, 7382.

74. R. Gonzalez, B. Masquelier, H. Fleury, B. Lacroix, A. Troesch, G. Vernet and J. N. Telles, *J. Clin. Microbiol.*, 2004, **42**, 2907.

75. I. Pe'er, P. I. de Bakker, J. Maller, R. Yelensky, D. Altshuler and M. J. Daly, *Nat. Genet.*, 2006, **38**, 605.

76. Genomes OnLine Database, http://www.genomesonline.org/index.htm, 2008.

77. R. C. Gardner, A. J. Howarth, P. Hahn, M. Brown-Luedi, R. J. Shepherd and J. Messing, *Nucleic Acids Res.*, 1981, **9**, 2871.

78. S. L. Chissoe, A. Bodenteich, Y. F. Wang, Y. P. Wang, D. Burian, S. W. Clifton, J. Crabtree, A. Freeman, K. Iyer and L. Jian, *et al.*, *Genomics*, 1995, **27**, 67.

79. G. Gyapay, K. Schmitt, C. Fizames, H. Jones, N. Vega-Czarny, D. Spillett, D. Muselet, J. F. Prud'homme, C. Dib and C. Auffray, *et al., Hum. Mol. Genet.*, 1996, **5**, 339.

80. E. A. Stewart, K. B. McKusick, A. Aggarwal, E. Bajorek, S. Brady, A. Chu, N. Fang, D. Hadley, M. Harris and S. Hussain, *et al., Genome Res.*, 1997, **7**, 422.

81. H. W. Mewes, K. Albermann, M. Bahr, D. Frishman, A. Gleissner, J. Hani, K. Heumann, K. Kleine, A. Maierl and S. G. Oliver, *et al., Nature*, 1997, **387** (6632 Suppl), 7.

82. *C. elegans* Sequencing Consortium, *Science*, 1998, **282**, 2012.

83. S. G. Gregory, M. Sekhon, J. Schein, S. Zhao, K. Osoegawa, C. E. Scott, R. S. Evans, P. W. Burridge, T. V. Cox and C. A. Fox, *et al., Nature*, 2002, **418**, 743.

84. M. Krzywinski, J. Wallis, C. Gosele, I. Bosdet, R. Chiu, T. Graves, O. Hummel, D. Layman, C. Mathewson and N. Wye, *et al., Genome Res.*, 2004, **14**, 766.

85. International Bovine BAC Mapping Consortium, *Genome Biol.*, 2007, **8**, R165.

86. International Human Genome Sequencing Consortium, *Nature*. 2004, **431**, 931.

87. M. D. Adams, S. E. Celniker, R. A. Holt, C. A. Evans, J. D. Gocayne, P. G. Amanatides, S. E. Scherer, P. W. Li, R. A. Hoskins and R. F. Galle, *et al., Science*, 2000, **287**, 2185.

88. R. A. Holt, G. M. Subramanian, A. Halpern, G. G. Sutton, R. Charlab, D. R. Nusskern, P. Wincker, A. G. Clark, J. M. C. Ribeiro and R. Wides, *et al., Science*, 2002, **298**, 129.

89. Mouse Genome Sequencing Consortium, *Nature*, 2002, **420**, 520.

90. S. Aparicio, J. Chapman, E. Stupka, N. Putnam, J. M. Chia, P. Dehal, A. Christoffels, S. Rash, S. Hoon and A. Smit, *et al., Science*, 2002, **297**, 1301.

91. E. F. Kirkness, V. Bafna, A. L. Halpern, S. Levy, K. Remington, D. B. Rusch, A. L. Delcher, M. Pop, W. Wang and C. M. Fraser, *et al., Science*, 2003, **301**, 1898.

92. R. Velasco, A. Zharkikh, M. Troggio, D. A. Cartwright, A. Cestaro, D. Pruss, M. Pindo, L. M. Fitzgerald, S. Vezzulli and J. Reid, *et al., PLoS ONE*, 2007, **2**, e1326.

93. K. Yamamoto, J. Nohata, K. Kadono-Okuda, J. Narukawa, M. Sasanuma, S. I. Sasanuma, H. Minami, M. Shimomura, Y. Suetsugu and Y. Banno, *et al., Genome Biol.*, 2008, **9**, R21.

94. S. Istrail, G. G. Sutton, L. Florea, A. L. Halpern, C. M. Mobarry, R. Lippert, B. Walenz, H. Shatkay, I. Dew and J. R. Miller, *et al., Proc. Natl. Acad. Sci. USA*, 2004, **101**, 1916.

95. H. N. Bu-Ghanim, C. M. Casimir, S. Povey and A. W. Segal, *Genomics*, 1990, **8**, 56.

96. N. Hornigold, M. van Slegtenhorst, J. Nahmias, R. Ekong, S. Rousseaux, C. Hermans, D. Halley, S. Povey and J. Wolfe, *Genomics*, 1997, **41**, 385.

97. C. A. Sargent, I. Dunham and R. D. Campbell, *EMBO J.*, 1989, **8**, 2305.

98. M. S. Boguski, T. M. Lowe and C. M. Tolstoshev, *Nat. Genet.*, 1993, **4**, 332.

99. L. L. Qi, B. Echalier, S. Chao, G. R. Lazo, G. E. Butler, O. D. Anderson, E. D. Akhunov, J. Dvořák, A. M. Linkiewicz, A. Ratnasiri and J. Dubcovsky, *et al., Genetics*, 2004, **168**, 701.

100. H. Noda, S. Kawai, Y. Koizumi, K. Matsui, Q. Zhang, S. Furukawa, M. Shimomura and K. Mita, *BMC Genomics*, 2008, **9**, 117.

101. D. R. Price, H. S. Wilkinson and J. A. Gatehouse, *Insect Biochem. Mol. Biol.*, 2007, **3**, 1138.

102. G. M. Duyk, S. W. Kim, R. M. Myers and D. R. Cox, *Proc. Natl. Acad. Sci. USA*, 1990, **87**, 8995.

103. M. Y. Galperin, *Nucleic Acids Res.*, 2008, **36** (Database Issue), D2–D4.

104. R. D. Finn, J. Mistry, B. Schuster-Böckler, S. Griffiths-Jones, V. Hollich, T. Lassmann, S. Moxon, M. Marshall, A. Khanna and R. Durbin, *et al., Nucleic Acids Res.*, 2006, **34** (Database Issue), D247–D251.

105. W. Bodmer and C. Bonilla, *Nat Genet.*, 2008, **40**, 695.

106. J. W. Bainbridge, M. H. Tan and R. R. Ali, *Gene Ther.*, 2006, **13**, 119.

107. S. Walchli and M. Sioud, *Front. Biosci.*, 2008, **13**, 3488.

CHAPTER 6

The Biotechnology and Molecular Biology of Yeast

BRENDAN P. G. CURRAN[a] AND VIRGINIA C. BUGEJA[b]

[a] School of Biological and Chemical Sciences, Queen Mary, University of London, Mile End Road, London E1 4NS, UK; [b] School of Life Sciences, University of Hertfordshire, College Lane, Hatfield, Hertfordshire AL10 9AB, UK

6.1 INTRODUCTION

The yeast *Saccharomyces cerevisiae* plays a central role in both biotechnology, the profitable exploitation of biological systems by humans, and molecular biology, which can be described as studying and manipulating biological systems at the molecular level. This is because it is the most successfully exploited microorganism, with a record of exploitation reaching back over 2000 years, and as the most molecularly characterised eukaryotic organism it is currently providing an unparalleled insight into cellular structure and function at the molecular level.

Despite a complete ignorance of the underlying mechanisms, the biotechnological exploitation of yeast began thousands of years ago with the production of beer, wine and bread. Today we still use yeast to make bread, wine and beer but can also manipulate molecules of DNA to direct yeast cells to produce vast quantities of molecules with a

Molecular Biology and Biotechnology, 5th Edition
Edited by John M Walker and Ralph Rapley
© Royal Society of Chemistry 2009
Published by the Royal Society of Chemistry, www.rsc.org

staggering array of functions. Therefore, in addition to its traditional roles in biotechnology, the baker's yeast *Saccharomyces cerevisiae* plays a major role in the biotechnology of heterologous protein production. So too does its non-fermentative cousin *Pichia pastoris*. These include a multitude of molecular medicines, mostly non-native (heterologous) proteins (Table 6.1). Both are also being used in pathway engineering where the basic heterologous protein production technologies provide the tools to develop much more sophisticated recombinant cloning technologies which essentially re-engineer the yeast cell's biochemistry. Finally, the post-genomic analysis of the molecular biology of *S. cerevisiae* provides a deep insight into how these cells metabolise, grow and divide, thereby providing a paradigm for the eukaryotic cell – the basic building block of every living multicellular organism on the face of this planet. The biotechnology and molecular biology of yeast therefore covers: the production of heterologous protein products for exploitation by humans, the development of novel pathway engineering and post-genomic analyses that facilitate a deeper understanding of cellular biology.

Table 6.1 Heterologous protein production in the yeasts *Saccharomyces cerevisiae* and *Pichia pastoris*.

Recombinant protein	Therapeutic use
Produced in *Saccharomyces cerevisiae*[64]	
Hirudan	Anticoagulant
Insulin	Diabetes mellitus
Somatropin	Growth disturbance
Glucagon	Hypoglycaemia
Platelet-derived growth factor (PDGF)	Lower-extremity diabetic neuropathic ulcers
Hepatitis B surface antigen (HBsAg)	Hepatitis B vaccination
Major capsid protein from four Human papillomovirus (HPV) types	Vaccination against diseases caused by HPV
Urate oxidase	Hyperuricaemia
Granulocyte-macrophage colony stimulating factor (GM-CSF)	Chemotherapy-induced neutropenia
Products under development in *Pichia pastoris*[65]	
Angiostatin	Antiangiogenic factor
Endostatin	Antiangiogenic factor
Epidermal growth factor (EGF) analogue	Diabetes
Elastase inhibitor	Cystic fibrosis
Human serum albumin	Blood volume stabiliser in burns treatment
Insulin-like growth factor-1	Insulin-like growth factor-1 deficiency

6.2 THE PRODUCTION OF HETEROLOGOUS PROTEINS BY YEAST

Heterologous, or recombinant, proteins are produced when recombinant DNA technology is used to ensure the expression of a gene product in an organism in which it would not normally be made. It is very challenging to assemble all of the genetic components required to ensure that a specific DNA sequence is transcribed into an mRNA molecule, which can subsequently be translated into a functional protein by a cell. Moreover, due to the bewildering array of molecular processes (*e.g.* glycosylation, acetylation, myristelation, *etc.*) that impact on protein processing, the final protein structure is dependent upon the type of host cell in which the expressed protein is synthesised. Thus, although gene expression vector technology was originally developed in *E. coli*, the production of recombinant proteins, especially if they are for *human use*, often requires an appropriately engineered DNA molecule to be transformed into a *eukaryotic* host cell. Therefore, shortly after the first heterologous proteins was expressed in *E. coli* in 1977,[1] the yeast *S. cerevisiae* was developed as the first eukaryote host system. Many other eukaryotes have been developed as host systems since then, and one of these is another yeast called *Pichia pastoris*.

6.2.1 The Yeast Hosts

Yeasts are the simplest eukaryotes and they share many of the attributes of bacteria: they are unicellular, grow rapidly, can be transformed with DNA and can form colonies on a plate.

S. cerevisiae was developed as the first eukaryotic host cell to express heterologous proteins, not only because it shared a number of useful attributes with *E. coli* but also because it had a long, safe history of use in commercial fermentation processes. This made it particularly suitable for approval by regulatory bodies charged with the responsibility of ensuring the safe production of medically important heterologous proteins. Moreover, it also has its own autonomously replicating plasmid, can carry out post-translational modifications of expressed proteins and secretes a small number of proteins into the growth medium. This, as we shall see, can be exploited both to produce glycosylated proteins and to simplify the purification of heterologous proteins. It has a very primitive glycosylation pathway, however, which makes it unsuitable for many human proteins. That is one of the main reasons why its cousin *P. pastoris* was developed as an alternative yeast host cell. In addition, this yeast is a methylotroph, one of a small number of yeast species that share a biochemical pathway allowing the cells to utilise methanol as a sole carbon source. They grow to much higher cell densities in fermenters due to the absence of toxic levels of ethanol and, importantly, they have a glycosylation pathway which produces more authentic patterns for human heterologous protein products than *S. cerevisiae*.

6.2.2 Assembling and Transforming Appropriate DNA Constructs into the Hosts

6.2.2.1 The DNA Constructs

In order for the yeast host cells to be able to express the foreign protein, appropriate yeast expression vectors need to be engineered. These are generally shuttle vectors, which consist of DNA sequences that will ensure appropriate protein expression in the yeast host cells and a section of bacterial DNA that allows the molecule to be assembled and engineered in *E. coli*. Typical vectors can be seen in Figure 6.1. They consist of a backbone of bacterial DNA including an origin of replication for *E. coli*, a selectable marker such as ampicillin resistance and a number of appropriate restriction sites into which the various yeast-specific DNA sequences can be inserted. These latter include a yeast selectable marker, a strong yeast promoter and terminator to control mRNA production, appropriate DNA sequences to direct translation when the mRNA has been generated and a polycloning site for insertion of the DNA sequence encoding the heterologous protein. The heterologous DNA is inserted between the promoter and terminator. The heterologous DNA is almost always a cDNA of the desired protein coding sequence because yeast cannot normally recognise regulatory sequences or excise introns in DNA sequences from higher eukaryotes such as humans.

There are two main types of expression vectors used in *S. cerevisiae* biotechnology. YEp (yeast episomal plasmid) vectors are based on the ARS (autonomously replicating sequence) sequence from the endogenous yeast 2 μm plasmid which contains genetic information for its own replication and segregation (for a review, see Ref. 2). They are capable of autonomous replication, are present at 20–200 copies per cell and under selective conditions are found in 60–95% of the cell population (Figure 6.1A). The addition of a centromeric sequence converts a YEp into a yeast centromeric plasmid (YCp) (Figure 6.1B). YCps are normally present at one copy per cell, can replicate without integration

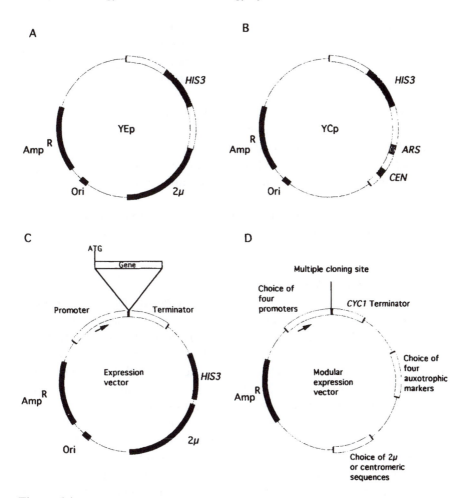

Figure 6.1   Schematic diagrams of yeast cloning vectors. A yeast episomal (YEp)
vector (A) consists of a prokaryotic gene for resistance to ampicillin
(Amp^R), a prokaryotic replication origin (ori), a yeast auxotrophic marker
(HIS3) and a yeast 2μ DNA sequence. A yeast centromeric (YCp) vector
(B) contains a yeast centromere (CEN) and an autonomous relpication
sequence (ARS) instead of the 2μ DNA sequence. The addition of yeast
promoter and terminator sequences generates a yeast expression vector
(C); transcription initiation of heterologous genes cloned into a unique
cloning site is indicated by the arrow. A extremely versatile series of
modular expression vectors (D) provides a choice of promoters (CYC1,
ADH, TEF, and GPD); a choice of selectable marker genes (HIS3,LEU2,
TRP1 and URA3); a choice of copy number (centromeric or 2μ plasmid)
and contains a multiple cloning site between the promoter and terminator
sequences (see reference 7).

into a chromosome and are stably maintained during cell division even in the absence of selection. An extremely versatile family of expression vectors providing alternative selectable markers, a range of copy numbers and differing promoter strengths (see Figure 6.1d) has been produced by Mumberg *et al.*[3]

S. cerevisiae vectors are usually transformed as circular molecules into the host cells. On the other hand, *P. pastoris* expression vectors are designed to be cut and linearised before transformation. This allows the constructs to integrate into the host chromosome and thereafter there is no need for further selection in order to maintain the construct in the host cells.

There are now a number of commercially available vectors for both *S. cerevisiae* and *P. pastoris*.

6.2.2.2   Transformation and Selection

There are several procedures that achieve efficient transformation of yeast. Exponentially growing cells can be enzymatically treated to remove their cell walls and the resulting sphaeroplasts exposed to DNA in the presence of calcium ions and polyethylene glycol, before being embedded in hypertonic selective agar to facilitate cell wall regeneration and subsequent colony formation. Frequencies of up to 10^6 transformants per μg DNA can be obtained using this technique.[4,5]

Intact yeast cells can be transformed by treating them with alkali metal cations (usually lithium), in a procedure analogous to *E. coli* transformation[6] or by electroporation,[7] which involves using a brief voltage pulse to facilitate entry of DNA molecules into the cells. Transformation efficiencies obtained using whole cells can be lower than those obtained using the protoplasting method, but transformed cells can be spread directly onto selective plates (rather than embedded in agar) thus facilitating procedures that require the use of colony screens. These are now the methods of choice both for *S. cerevisiae* and *P. pastoris*.

Transformants are identified by selection. Unlike the dominant antibiotic resistance markers used in *E. coli* transformations, many yeast selectable markers are genes which complement a specific auxotrophy (*e.g.* Leu, His, Trp) and thus require the host cell to contain a recessive, non-reverting mutation. The most widely used selectable markers and their chromosomal counterparts are listed in Table 6.2. Further details of these and other recessive selectable markers can be found on the Internet.[8] The most widely used dominant selectable marker systems are also included in Table 6.2.

Table 6.2 Selectable markers for yeast transformation

Auxotrophic markers		
Gene	*Chromosomal mutation*	*Ref.*
HIS3	*his3-Δ1*	66
LEU2	*leu2–3,leu2–112*	5
TRP1	*trp1–289*	67
URA3	*ura3–52*	66
Dominant markers		
Gene	*Selection*	*Ref.*
CUP1	Copper resistance	68
G418^R	G418 resistance (kanamycin phosphotransferase)	52,69
TUN^R	Tunicamycin resistance	70

6.2.3 Ensuring Optimal Expression of the Desired Protein

6.2.3.1 Ensuring High Levels of mRNA

The first step in expressing heterologous proteins is the production of mRNA from the DNA construct. The overall level of heterologous mRNA in the cell is a balance between the production of mRNA and its stability in the cytoplasm. The former is determined by the copy number of the expression vector and the strength of the promoter; the latter by the specific mRNA sequence.

In the case of *S. cerevisiae*, expression vectors based on YEp technology have a high copy number but require selective conditions to ensure their stable inheritance. Plasmid instability can be prevented by introducing a centromere into the vector, but at the cost of reducing the plasmid copy number to 1–2 copies per cell.

High-level mRNA production is also dependent on the type of promoter chosen to drive expression. The major prerequisite for an expression vector promoter is that it is a strong one. Those most frequently encountered are based on promoters from genes encoding glycolytic enzymes, *e.g.* phosphoglycerate kinase (*PGK*), alcohol dehydrogenase 1 (*ADH*1) and glyceraldehyde-3-phosphate dehydrogenase (*GAPDH*), all of which facilitate high-level constitutive mRNA production. Constitutive expression can be disadvantageous when the foreign protein has a toxic effect on the cells, but this can be circumvented by using a regulatable promoter to induce heterologous gene expression after cells have grown to maximum biomass. The most commonly used one, based on the promoter of the galactokinase gene (*GAL*1), is induced when glucose is replaced by galactose in the medium, but a number of others are also available. Transcript stability is also vitally important in maintaining high levels of mRNA in the cell. Despite the fact that 'instability elements' have been identified,[9] mRNA half-lives cannot be accurately predicted from primary structural information so they must be empirically determined; unstable transcripts curtail high-level expression. *S. cerevisiae* often fails to recognise heterologous transcription termination signals because its own genes lack typical eukaryotic terminator elements. This can result in the production of abnormally long mRNA molecules which are often unstable.[10] As this can result in a dramatic drop in heterologous protein yield,[11] expression vectors frequently contain the 3' terminator region from a yeast gene (*e.g. CYC1, PGK* or *ADH1*) to ensure efficient mRNA termination (Figure 6.1d).

In the case of *P. pastoris*, the cells can be grown on either on methanol using an alternative alcohol oxidase locus, in which case the heterologous protein is continuously expressed, or on glucose, in which case the heterologous gene is repressed until induced by methanol. The tight level of regulation allows for extremely precise control of the expression of the heterologous gene. This easily regulated promoter has practical advantages over the more cumbersome galactose-inducible ones used to regulate heterologous expression in *S. cerevisiae*.

6.2.3.2 Ensuring High Levels of Protein
A high level of stable heterologous mRNA does not necessarily guarantee a high level of protein production. The protein level depends on the efficiency with which the mRNA is translated and the stability of the protein after it has been produced.

The site of translation initiation in 95% of yeast mRNA molecules corresponds to the first AUG codon at the 5' end of the message. It is advisable to eliminate regions of dyad symmetry and upstream AUG triplets in the heterologous mRNA leader sequence to ensure efficient initiation of translation. The overall context of the sequence on either side of the AUG codon (with the exception of an A nucleotide at the −3 position) and the leader length do not appear to affect the level of translation (for a review, see Ref. 12).

Achieving high-level transcription and translation of a heterologous gene in any expression system does not necessarily guarantee the recovery of large amounts of heterologous gene product: Some proteins are degraded during cell breakage and subsequent purification; others are rapidly turned over in the cell. The powerful tools provided by a detailed knowledge of yeast molecular biology can be exploited to minimise this. Genes for proteolytic enzymes can be inactivated either by mutation or the presence of protease inhibitors during extraction. A more elegant route is to exploit the yeast secretory pathway to smuggle heterologous proteins out of the cell into the culture medium where protease levels are low. This not only minimises the exposure of heterologous proteins to intracellular protease activity but also facilitates their recovery and purification due to the very low levels of native yeast proteins normally present in culture media.

Entry into the secretory pathway is determined by the presence of a short hydrophobic 'signal' sequence on the N-terminal end of secreted proteins. The 'signal' sequences from *S. cerevisiae*'s four major secretion products have been attached (by gene manipulation) to the N-terminus of heterologous proteins and used with varying degrees of success to direct their secretion: Invertase[13] and acid phosphatase[14] signal sequences target proteins to the periplasmic space whereas α factor[15] and killer toxin[16] signals target the proteins to the culture medium. A typical secretion vector is shown in Figure 6.2. Secretion can also be used to produce proteins that have an amino acid other than methionine at their N-terminus. If a secretory signal is spliced onto the heterologous gene at the appropriate amino acid (normally the penultimate one), then the N-terminal methionine which is obligatory for translation initiation will be on the secretory signal. Proteolytic cleavage of this signal from

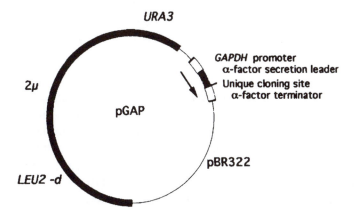

Figure 6.2   Schematic representation of the yeast secretion vector pGAP (J. Travis, M. Owen, P. George, R. Carrell, S. Rosenberg, R. A. Hallewell, and P. J. Barr, *J. Biol. Chem.*, 1985, **260**, 4384.). The vector contains LEU2-d and URA3 yeast selectable marker genes, pBR322 sequences for amplification in *E. Coli* and 2μ sequences for autonomous replication in yeast. The expression 'cassette' contains a unique cloning site flanked by GAPDH promoter, α-factor secretion leader and α-factor terminator sequences. Transcription iniation is indicated by the arrow.

Signal sequence   Heterologous protein

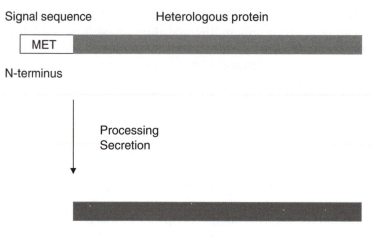

N-terminus

Processing
Secretion

Mature protein

Figure 6.3 Schematic diagram showing the secretion of a heterologous protein using a signal sequence. Cleavage of the secretory signal in the endoplasmic reticulum removes the N-terminal methionine, thereby generating a heterologous protein with an authentic N-terminal amino acid.

the heterologous protein in the endoplasmic reticulum (ER) will generate an authentic N-terminal amino acid (Figure 6.3).

Heterologous proteins can also be secreted from *P. pastoris* and the most widely used secretion signal sequences include the *S. cerevisiae* α factor prepro sequence and the signal sequence from *Pichia*'s own acid phosphatase gene.

6.2.3.3 Obtaining the Appropriate Protein Structure and Function

The objective of heterologous gene expression is the high-level production of biologically active, authentic protein molecules. It is therefore important to consider the nature of the final product when choosing the expression system. The protein size, hydrophobicity, normal cellular location, needed for post-translational modification(s) and ultimate use, must be assessed before an appropriate expression system is chosen.

The secretory pathway is often chosen for heterologous protein production because apart from the fact that it enhances protein stability and can generate proteins lacking an N-terminal methionine, secretion facilitates the accurate folding of large proteins and contains the machinery for post-translational modification. A direct comparison between the intracellular production and extracellular secretion of prochymosin and human serum albumin resulted in the recovery of small quantities of mostly insoluble, inactive protein when they were produced intracellularly, but the recovery of soluble, correctly folded, fully active protein when they were secreted.[17,18]

The biological activity and/or stability of heterologous proteins can also be affected by the post-translational addition of carbohydrate molecules to specific amino acid residues. Glycosylation in yeast is of both the N-linked (via an asparagine amide) and O-linked (via a serine or threonine hydroxyl) types, occurring at the sequences Asn–X–Ser/Thr and Thr/Ser, respectively. Inner core N-linked glycosylation occurs in the ER and outer core glycosylation in the Golgi apparatus. However, it is important to note that the number and type of outer core carbohydrates attached to glycosylated proteins in yeast are different to those found on mammalian proteins. In many cases these differences can be tolerated, but if the protein is being produced for therapeutic purposes they may cause unacceptable immunogenicity problems. One approach to overcoming this problem is to remove the glycosylation recognition site by site-directed mutagenesis. This strategy was successfully used to produce urokinase type plasminogen activator.[19]

Despite the advantages that secretion offers for the production of heterologous proteins in yeast, higher overall levels of protein production are often possible using intracellular expression. Some proteins form insoluble complexes when expressed intracellularly in *S. cerevisiae* but many others do not. Human superoxide dismutase was recovered as a soluble active protein after expression in yeast. It was also efficiently acetylated at the N-terminus to produce a protein identical with that found in human tissue.[20] Other proteins can be produced as denatured, intracellular complexes which can be disaggregated and renatured after harvesting. The first recombinant DNA product to reach the market was a hepatitis B vaccine produced in this way.[21]

P. pastoris is regarded as a more efficient and more faithful glycosylator of secreted proteins. Many proteins of therapeutic importance have been successfully made using both intracellular and extracellular production in *Pichia*. These include proteins involved in the prevention and treatment of clots, peptide hormones and cytokines and protein vaccines including a hepatitis B vaccine, which is already on the market.

However, many proteins of interest require complex post-translational modification and yeast is a primitive eukaryote unable to effect those molecular changes. However, as will be seen in the following section, molecular biologists are currently developing a whole series of yeast strains that have been genetically manipulated so that they have modified and/or completely new biochemical pathways.

6.3   FROM RE-ENGINEERING GENOMES TO CONSTRUCTING NOVEL SIGNAL AND BIOCHEMICAL PATHWAYS

However, the contribution of yeast to biotechnology and molecular biology extends far beyond the production of heterologous proteins. The ease with which yeast cells can be genetically manipulated, coupled with yeast's homologous recombination system, makes it one of the most versatile eukaryotic cloning systems, thereby facilitating the *large-scale manipulation of genetic material* needed in order to produce re-engineered mammals and even synthetic life forms. Moreover, as a simple unicellular eukaryote, *S. cerevisiae* has many of the basic biochemical pathways found in complex multicellular organisms including cell to cell signalling. Therefore, in addition to use as a host cell for the manipulating extra-large DNA molecules, the cell itself can be genetically modified so that it can generate novel signal and biochemical pathways. These include genetically engineering *novel biological reporter systems* that facilitate the analysis of human hormone and brain receptors and the *in vivo* analysis of protein–protein interactions from a huge variety of organisms, and re-engineering numerous metabolic pathways for the production *of novel biochemical products*, including 'humanised' glycosylation of hetero-logous proteins.

6.3.1   Large-scale Manipulation of Mammalian and Bacterial DNA

Our detailed knowledge of yeast molecular biology, coupled with the versatility of its genetic manipulation, led to the development of novel and indeed extremely powerful cloning vectors.

6.3.1.1   Cloning Extra-large Segments of DNA

Yeast artificial chromosomes (YACs) are specialised vectors capable of accommodating extremely large fragments of DNA (100–1000 kb).[22] Schematic diagrams of a YAC and its use as a cloning system are shown in Figure 6.4. YACs contain a centromere, an autonomously replicating sequence, two telomeres and two yeast selectable markers separated by a unique restriction site. They also contain sequences

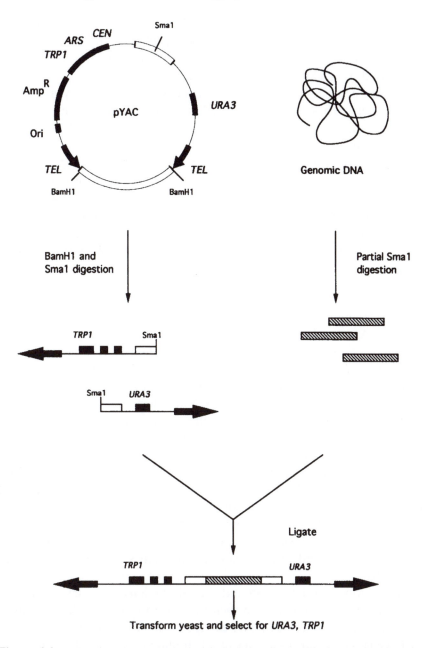

Figure 6.4   Schematic diagram of a YAC cloning vector, indicating prokaryotic gene for resistance to amplicillin (Amp^R), prokaryotic replication origin(ori), yeast auxotrophic markers (URA3, TRP1), autonomous replication sequence(ARS), yeast centromere (CEN) and telomeres (TEL).

for replication and selection in *E. coli*. YACs are linear molecules when propagated in yeast but must be circularised by a short DNA sequence between the tips of the telomeres for propagation in bacteria. When used as a cloning vehicle, the YAC is cleaved with restriction enzymes to generate two telomeric arms carrying different yeast selectable markers. These arms are then ligated to suitably digested DNA fragments, transformed into a yeast host and maintained as a mini-chromosome.

YACs have become indispensable tools for mapping complex genomes such as the human genome[23] because they accommodate much larger fragments of DNA than bacteriophage or cosmid cloning systems, thus simplifying the ordering of the human genome library. The complete library can be contained in approximately 10 000 clones, cutting by a factor of five the number of clones required by other vector systems.

6.3.1.2 *Manipulating Mammals*

Once a YAC has been successfully transformed into a yeast cell, the highly efficient homologous recombination system of *S. cerevisiae* can be exploited *in vivo* to manipulate extensively both YAC vector sequences (retrofitting) and their inserts.[24] For example, homologous recombination can be used to retrofit mammalian selectable markers into the vector arms and/or introduce specific mutations into any genomic sequence carried in a YAC, thus generating artificial chromosomes that can be used in the production of transgenic mice (Figure 6.5). A linear DNA fragment consisting of neomycin resistance and *LYS2* genes sandwiched between the 5' and 3' ends of the *URA3* gene can be targeted into the *URA3* locus on the right arm of the YAC. Homologous recombination at this locus, which can be selected for by selecting for transformants on a lysine-deficient medium, generates a useful uracil minus phenotype and introduces a mammalian selectable marker into the construct.

Homologous recombination and the selectable/counter selectable nature of the uracil phenotype can then be used to introduce desired mutations into the heterologous DNA carried in the YAC (Figure 6.6A,B): a *URA3* gene is inserted close to a suitably mutated site in a sub-clone of the relevant section of the gene of interest. This is then cut from the plasmid and used to transform yeast. Homologous recombination of the mutated version of the gene into the DNA inserted into the YAC insert is selected for by simply growing the yeast cells on medium lacking uracil. Removal

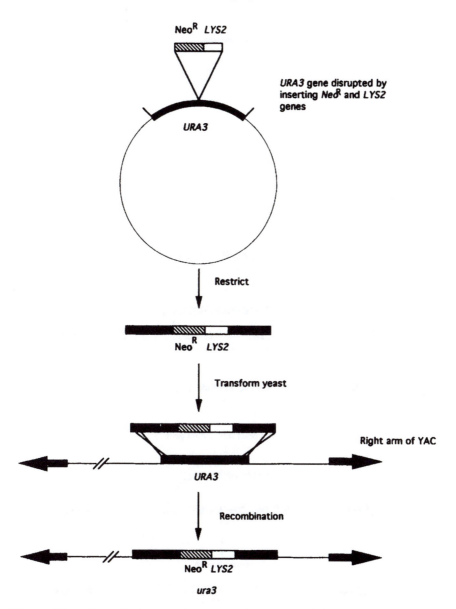

Figure 6.5 Schematic diagram showing the integration of linear yeast DNA into a homologous region of chromosomal DNA carried in a YAC. A linear fragment of DNA carrying the neomycin resistance gene (Neo^R) and the LYS2 gene flanked by URA3 gene sequences is isolated from plasmid DNA and transformed into the YAC carrying yeast strain. Homologous recombination events which result in the replacement of the wild-type (URA3) gene on the right arm of the YAC with this linear fragment are selected for by growth on lysine deficient medium.

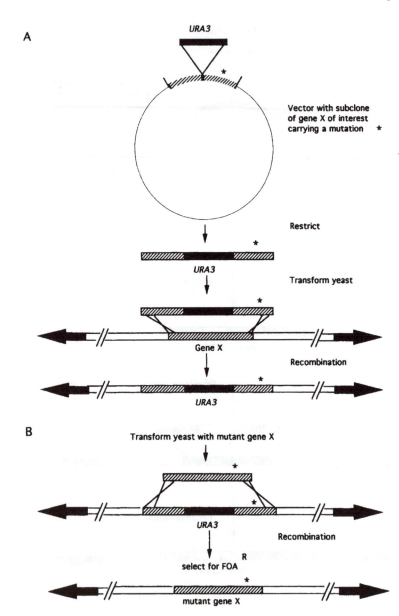

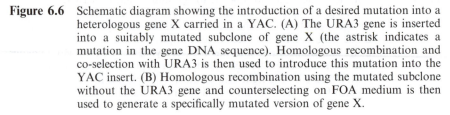

Figure 6.6   Schematic diagram showing the introduction of a desired mutation into a
heterologous gene X carried in a YAC. (A) The URA3 gene is inserted
into a suitably mutated subclone of gene X (the astrisk indicates a
mutation in the gene DNA sequence). Homologous recombination and
co-selection with URA3 is then used to introduce this mutation into the
YAC insert. (B) Homologous recombination using the mutated subclone
without the URA3 gene and counterselecting on FOA medium is then
used to generate a specifically mutated version of gene X.

of the *URA3* gene by a second homologous recombination event, selected for by growth on media containing 5-fluoroorotic acid (FOA),[25] generates a YAC vector containing a specifically mutated version of the gene of interest. A simplified diagram of this procedure is shown in Figures 6.6A and B and precise details can be found elsewhere.[24] The YAC containing the manipulated genes can then be used to create transgenic mice, thus allowing the analysis of large genes or multigenic loci *in vivo*. Many such murine and human genes have been introduced into mice and display correct stage- and tissue-specific expression.[24]

6.3.1.3 Building Synthetic Organisms

The homologous recombination mechanism of yeast can also be used to generate an artificial chromosome that can be maintained in yeast but is very different to the types of YAC described above. This process of transformation-associated recombination (TAR) exploits the fact that short, specific non-yeast sequences at the ends of a linear DNA will recombine exclusively with the correct sequence in DNA molecules within the cell. In Figure 6.7, the vector has been linearised such that it carries specific human DNA sequences on both ends of the fragment. These will specifically recombine with and circularize with the target sequence. The vector also has a centromere and a selectable marker. However, it lacks an autonomous replication sequence (ARS) and so, even if it could circularize on to itself, will not survive because it cannot replicate in yeast. However, human ARS are recognised in yeast cells and therefore vectors carrying one will survive. However, this is only possible if the vector can undergo homologous recombination with the specific human target sequences on the chromosomal DNA. Therefore, rather than generate an entire library of genomic sequences in order to isolate a specific chromosomal segment, TAR cloning allows the direct isolation of specific targets.[26,27] A quick PCR of part of the desired segment confirms if it has been successfully isolated. There have been a number of extremely useful developments of this type of TAR technology to join and isolate all types of DNA targets. These include isolation of a functional copy of the human *BRCA1* gene.[28] This technology has also recently been exploited by Venter and co-workers in the cloning of an entirely synthetic chromosome for the bacterium *Mycoplasma* genitalium.[29] This is the first time an entirely synthetic chromosome has been engineered and it was made possible by the incredibly accurate homologous recombination system in yeast.

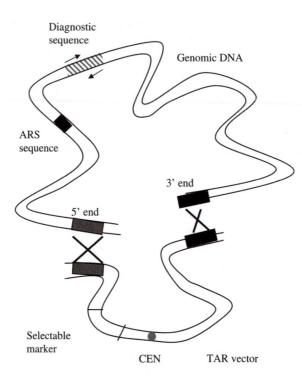

Figure 6.7   Isolation of a specific fragment of geonomic DNA by exploiting Transformation-Associated Recombination (TAR). Random segments of
genomic DNA are co-transformed into yeast with a linearised vector
which lacks an autonomous replication sequence (ARS). The linearised
vector carries a centromere, selectable marker and two specific targets
sequences. Site specific recombination creates an artificial chromosome
consisting of the yeast vector and the targeted fragment of genomic DNA,
which supplies an ARS thereby allowing re-circularised vectors carrying
an insert to replicate in the yeast cell. A simple PCR reaction is then used
to confirm that the appropriate segment has been successfully recombined
into the linear vector.

6.3.2   Novel Biological Reporter Systems

Our extensive understanding of the biochemistry and molecular biology
of yeast cells has resulted in the development of a number of imaginative
manipulations of promoter elements, transcription factors and signal
cascade proteins to transduce molecular interactions into easily scoreable yeast phenotypes.

6.3.2.1   *A Yeast-based Reporter for Human Brain Receptor Activity*
Yeast cells have been manipulated to provide a 'readout' of heterologous
G-protein-coupled receptor (GPCR) activity: The well-characterised

mating signal transduction pathway has been extensively re-engineered to replace a cell arrest phenotype with a HIS + phenotype[30] (Figure 6.8). The membrane-bound α factor receptor can be deleted and replaced with a heterologous one. The regulatory subunit of the yeast G-protein can then be manipulated to ensure activation of the mating pathway when the agonist binds to the foreign receptor. Deletion of the gene responsible for initiating cell cycle arrest and the fusion of a *HIS*3 gene to a mating pathway activated promoter effectively transduces the receptor-agonist interaction to a scoreable HIS + phenotype. As GPCRs represent the targets for the majority of currently prescribed pharmaceutical drugs, this system has exciting potential for the development of high-throughput screening technology.[31]

6.3.2.2 A Yeast-based Reporter for Human Steroid Receptor Activity

The yeast cell is insensitive to oestrogen, yet when the human oestrogen receptor was expressed in yeast cells and the oestrogen receptor element (ERE) was cloned into a disabled *CYC1* promoter fused to a β-galactosidase gene, oestrogen induced β-galactosidase enzyme production[32] (Figure 6.9). This reporter system was sufficiently sensitive to analyse the effect of site specific mutations on hormone binding efficiency and to measure the effectiveness of agonists and antagonists on hormone action.

6.3.2.3 A Yeast-based Reporter for Detecting Protein–Protein Interactions

The interaction of two proteins X and Y can be transduced into a scoreable phenotype by subjecting them to analysis by the extensively used two-hybrid system[33] (Figure 6.10). The interaction of X and Y depicted in Figure 6.10A results in the expression of a reporter gene because X has been genetically fused to the DNA binding site (DBS) of a transcription factor and in a separate construction Y has been fused to the activation domain (AD). Specific X–Y interactions bring the two transcription factor domains sufficiently close together to drive expression of the reporter gene. However, when two proteins do not interact (Figure 6.10B), expression of the reporter gene is not induced. Further extensions of this procedure are continually being developed in yeast to address the complex interplay of a wide variety of molecules in the cell.[34]

Yeast vectors are available in which the DBS (commonly derived from *GAL4* or *Lex* A) and the AD (commonly derived from *GAL4* or the viral activator VP16) are on separate 2 μm-based expression plasmids carrying different selectable markers (for a review, see Ref. 35). The genes for

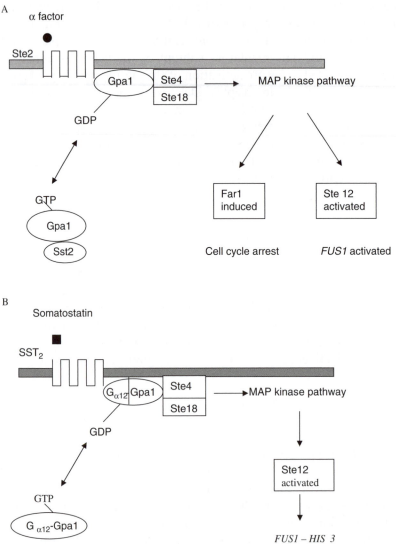

Figure 6.8   Heterologous receptor analysis using a re-engineered yeast pathway. The yeast mating pathway (A) was re-engineered (B). Now the yeast α-factor receptor is replaced by a heterologous human receptor protein (SST_2). Binding of the appropriate human ligand (somatostatin) results in activation of the mating pathway cascade. However, by deleting *Far1* the cells fail to arrest their cell cycle and by fusing the *HIS3* gene to an appropriate promoter (*FUS1*) the cascade triggers the expression of *HIS3*. The ligand–receptor interactions can therefore be assayed by the number of cells with a *HIS3+* phenotype.

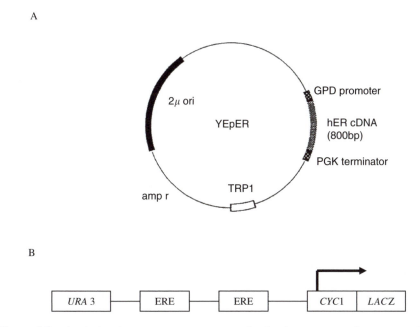

Figure 6.9 Analysis of oestrogen receptor protein. In the presence of oestrogen the receptor protein expressed from YEpER (A) binds to the oestrogen response elements in the promoter of the *LACZ* gene (B). The level of β-galactosidase activity can therefore be used to assay the effect of molecules that interfere with/accentuate the hormone–receptor interaction.

the proteins under analysis are inserted into these specific vectors and the plasmids are co-transformed into a suitable yeast host. Expression from a reporter gene (*e.g. LacZ, HIS*3, *LEU*2 or *URA*3) is used to test for a proposed protein interaction; to establish the effect of site-directed mutagenesis on previously characterised interacting proteins, or X may be used as a bait to trap unknown interacting Ys from an expression library fused to the AD. The two hybrid system has been successfully applied to study a broad spectrum of protein–protein interactions from a wide variety of different species.[36]

6.3.3 Novel Biochemical Products Include Humanised EPO

Not only does the homologous recombination system allow for extensive re-engineering of YAC vectors and their inserts, it also provides a mechanism for the targeted deletion of individual genes within the genome of the yeast host cells. This allows scientists to modify existing biochemical pathways by deleting native genes, and transforming in

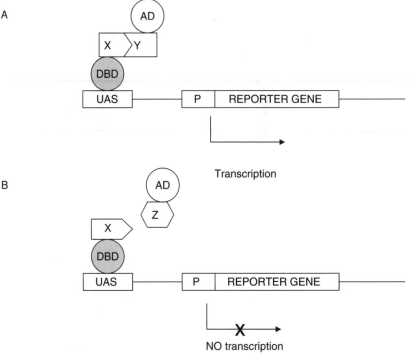

UAS   Upstream Activator sequence
P   Promoter
DBD   DNA Binding Domain
AD   Activation Domain

Figure 6.10   The yeast two-hybrid system. Protein X is fused to a DNA binding domain protein (DBD) and protein Y is fused to an activation domain (AD) of a transcription factor. Specific interaction of proteins X and Y brings the DBD and AD together, thus driving reporter gene transcription (A). In (B) there is no induction of expression because although protein X is bound to the DNA the activation domain remains separate from the DNA binding domain because its fusion partner Z is unable to interact with X.

either foreign versions of the deleted ones, or novel genetic material encoding a new biochemical function. In essence, biochemical pathway engineering projects allow biotechnologists to convert yeast cells into surrogate plant or animal cells. In the former case this usually involves hijacking the central metabolism of yeast to produce secondary metabolites normally synthesised in plants. In the latter case our detailed knowledge of yeast biochemistry and cell biology is exploited to re-engineer the cell such that it can synthesize the complex sugar moieties needed to produce glycoproteins normally produced in mammalian cells.

6.3.3.1 Genetically Engineering the Production of Secondary Metabolites

In addition to primary metabolism, which comprises all of the pathways necessary for the survival of a cell secondary metabolism occurs in a wide variety of organisms. The products of these pathways, although useful to the cell, are not essential. There is a huge array of these natural products, including many medicines, fragrances and flavours that are commonly isolated from plants, microbes and marine organisms. Unsurprising, these compounds are normally produced in rather limited amounts in the organisms in which they occur and purifying them is often difficult. Moreover their intrinsic complexity makes them very difficult to chemically synthesis. An alternative is to transfer the genes encoding the enzymes required for the biosynthesis of these compounds into cells from whence the products can be easily extracted. The yeast *S. cerevisiae* is increasingly being used in this endeavour (for reviews, see Refs 37 and 38). There are many examples of yeast cells being re-engineered into surrogate plant cells by transforming plant genes into yeast such that the central metabolic pathway is hijacked to generate the basic building blocks required for the synthesis of the complex molecules in question. Two very exciting examples are presented in Figure 6.11.

Here the central biochemical pathway in yeast is the pathway leading from the products of glycolysis, through multiple steps including Isopentenyl pyrophosphate (IPP) geranyl diphosphate (GPP) and farnesyl diphosphate (FPP), to squalene, and thence to ergosterol, the central steroid component of the membrane. This pathway can be hijacked into secondary metabolism by expressing the appropriate plant enzymes. Ro and co-workers cloned the amorphadiene synthase gene from the plant *Artemisia annua* and expressed it in yeast.[39] This hijacked this central biosynthetic pathway at FPP (see Figure 6.11) and converted FPP into the cyclical molecule amorphadiene. The further cloning and coexpression of a cognate cytochrome P450 hydroxylase and cytochrome P450 reductase then mediated three oxidation steps to generate Artemisinic acid, which was extracted and treated chemically to generate Aritemisinin, a highly effective product in the treatment of the malaria causing parasite *Plasmodium falciparum*. Alternatively the same pathway can be 'hijacked' at an earlier point by the expression of GGPPS

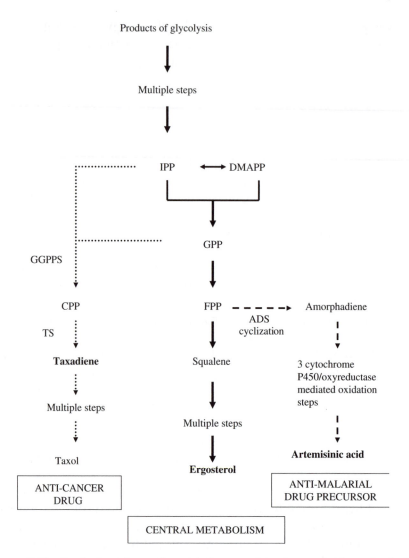

Figure 6.11   Producing plant cell metabolites, used in the production of valuable drugs, by hijacking the central metabolism of yeast. Expression of a plant amorphadiene synthase enzyme converts the yeast metabolite FPP into Amorphadiene. The further coexpression of a cognate cytochrome P450 hydroxylase and cytochrome P450 reductase generates Artemisinic, a precursor to the anti-malarial drug Artemisinin. Alternatively the same pathway can be by 'hijacked' at an earlier point by expressing Genranylgenranyl diphosphate synthase (GGPPS) and Taxadiene synthase (TS) from yew trees. These convert the yeast central metabolites IPP and GPP via CPP, into a plant cell metabolite called Taxadiene. This is a precursor in the production of the anticancer drug Taxol.

and TS from the yew tree such that the cell hijacked into making Taxadeine a metabolic precursor of [39] the anticancer drug Taxol.[40]

6.3.3.2 Humanising the Glycosylation of Heterologous Proteins in P. pastoris

If the production of secondary metabolites requires detailed knowledge of yeast biochemistry, the production of complex glycoproteins requires knowledge of both biochemistry and cellular biology. This is because of the enzymes involved in this pathway: some are in the cytoplasm, where they synthesize monosaccharide precursors and assemble them into the oligosaccharide required by the pathway; others reside in the ER membrane, where they flip back and forth as they transport oligosaccharides from the cytoplasm into the ER; others, such as the sialic acids transporter enzyme, are embedded as transporters in the Golgi membrane; and yet others tethered by hydrophobic tails sit in the membrane of the ER or Golgi, where they either transfer sugar residues on to the glycoprotein (glycotransferases) or trim sugar residues (glycosidases) as the complex oligosaccharides emerges. Moreover sialic acid a complex monosaccharide required for most human glycoproteins, is not even synthesised in yeast.

Because they are eukaryotic cells, yeasts share the early sections of the basic glycosylation pathway with mammalian cells. They share many of the same basic sugar residues used in the assembly of the sugar coating of proteins and indeed have identical biochemistry in the first (so-called *core glycosylation*) stages in the process. In *core* glycosylation the sugars that are added to proteins are initially enzymatically 'activated' as sugar nucleotides in the cytoplasm. Then two UDP-*N*-acetyl glucosamine followed by one GDP-mannose are sequentially added to the phosphate group of dolichol phosphate (a long lipid molecule in the ER with a terminal phosphate group in the cytoplasm). A further four GDP-mannose units are added, generating a small oligosaccharide (2 Glc Nac 5 Man) tethered to dolichol by its phosphate group. This entire structure is then flipped into the lumen of the ER, where enzyme-bound enzymes add further oligosaccharides and the entire structure is transferred on to a growing polypeptide chain via specific asparagine residues within specific recognition amino acid sequences. This oligosaccharide is further processed before being exported into the Golgi complex, where *terminal glycosylation* occurs. In the Golgi complex, N-linked oligosaccharides are further processed, and the addition of sugars to appropriate −OH residues of amino acids (*O*-linked glycosylation) also commences. It is in the Golgi that yeast and higher organisms diverge in their biochemistry (Figure 6.12): Yeast simply add further mannose residues; higher organisms trim the arriving oligosaccharide and then build up a complex array of sugar moieties onto the exposed sugars. In particular: the cytoplasmic biosynthetic pathway for the synthesis of CMP-Sialic acid, Golgi transporters for UDP-galactose and UDP-GlcNAc and CMP-sialic acid and the enzymes required to transfer these critically important saccharides onto the growing oligosaccharide are completely absent from yeast cells.

Thus in order to transform a yeast cell into a surrogate animal cell, it was necessary to re-engineer the cells such that:

1. Extra mannose residues were no longer added to the core oligosaccharide the Golgi complex.
2. Transporters for the nucleotide-sugars UDP-GalNAc and CMP-sialic acid, which are absent in yeast but are required for human-type oligosaccharide assembly, were expressed and targeted to the Golgi complex.
3. A selection of glycosidase and glycotransferase genes were expressed and targeted to the Golgi in order to assemble the correct type and sequence of sugar moieties.
4. The genes encoding the biochemical pathway for the production of CMP-sialic acid – a sugar which is an absolute requirement for humanised therapeutic proteins were expressed.

In a series of genetic engineering steps of increasing sophistication and complexity (for a review, see Ref. 41), each of these biological obstacles was overcome. In brief:

1. The addition of mannose in the Golgi complex was prevented by using gene knockout technology to delete *OCH1*, the enzyme responsible for transferring mannose sugars onto the core oligosaccharide in the Golgi complex.[42]
2. Genes for the appropriate transporter proteins were isolated from a range of organisms transformed into *P.pastoris* and successfully targeted into the Golgi membrane.[42–44]
3. A ground-breaking procedure was developed in which one of a range of N-terminal ER (or early Golgi) localisation signals from either *S. cerevisiae* or *P. pastoris* was fused in frame to one of a range of mannosidase or transferase catalytic domains from a whole range of organisms. In one such study, more than 600 different combinations were tested in order to identify a mannosidase capable of efficient high-level trimming of mannose

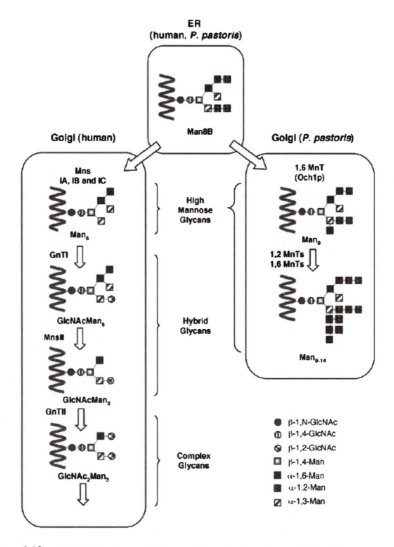

Figure 6.12 N-Linked glycosylation pathway in humans and in *P. pastoris*. This comparison highlights the differences between human and Pichia glycosylation pathways. From Stephen R. Hamilton, *et al. Science*, 2003, **301**, 1244–1246. Reprinted with permission from AAAS. Yeast simply add further mannose residues; higher organisms trim the arriving oligosaccharide and then build up a complex array of sugar moieties onto the exposed sugars. The production of High Mannose Glyucans was prevented in by deleting genes from the Pichia host. The expression of a series of non-yeast enzymes then converted the host cell's glycosylation pathway into a mammalian surrogate.

residues.[42] In the same study, a transfer enzyme (GnT1) capable of adding UDP-GlcNAc to oligosaccharides in the Golgi complex was isolated by screening a series of 67 different protein fusions engineered between fungal Golgi localisation signal peptides and GnT1catalytic domains from a range of higher organisms.[42] The same group later developed a more complex glycosylation pathway by adding a further two eukaryotic enzymes using the same strategy of creating chimeric proteins consisting of fungal leader sequences fused to catalytic domains.[43]

4. Finally, the same group expressed the genes for four human enzymes which modify nucleotide-sugars UDP-GlcNAc into the human specific CMP-sialic acid in the cytoplasm and then transport it into the Golgi complex. They finally ensured the addition of the terminal sialic acid residues on to the mature glycoprotein by successfully targeting a sialyltransferase/yeast-leader chimeric protein into the Golgi.[44]

This generated a yeast cell which has been genetically engineered by the deletion of native genes and the addition of 14 new genes, with the latter being generated using DNA sequences from nine different organisms including three yeast species, the fruit fly, mouse, rat and human, in order to generate a fully humanised glycosylation pattern of heterologous human proteins. It has already been used to express EPO, which has been shown to have the same biological effects as EPO of human origin.[44] This yeast has also been used to express human antibodies with extremely accurate and reproducible glycosylation patterns[45] – much more consistent in fact than mammalian cell cultures, which produce natural variability in the glycans on expressed proteins due to multiple enzymes competing for the same transient glycan structure. Not only do these yeast cells outshine mammalian cells in terms of expressing human proteins with reproducible glycosylation patterns but, because different yeast strains have been engineered with different combinations of glycosylation enzymes, work can now begin to study how these sugar moieties, which are major determinants in the effector function of these complex molecules, mediate a whole range of immune responses.

Finally, following the successful assembly of this complex humanised N-linked glycosylation pathway, the less well characterised O-linked glycosylation pathway is currently being re-engineered[46] in *S. cerevisiae*. Although this has just recently begun, giant strides have been made in that direction already.

6.4 YEAST AS A PARADIGM OF EUKARYOTIC CELLULAR BIOLOGY

Most fundamental cellular functions are conserved from yeast to humans. Therefore, not surprisingly, as the first eukaryotic organism to have its entire genome sequenced (for a review, see Ref. 47), *S. cerevisiae* is currently in the process of providing a unique insight into eukaryotic cell biology. Shortly after the genome was sequenced, yeast geneticists set about using homologous recombination to knock out each individual reading frame to identify what each gene did. These gene knockouts provided unparalleled insight into basic eukaryotic cellular biology – the most dramatic revelation being that many gene deletions could be well tolerated in the haploid yeast cell revealing the robustness of biological systems. *S. cerevisiae* was thereafter used as the model eukaryote in which to develop methods of globally monitoring cellular mRNAs, proteins, metabolites and the myriad interactions of these cellular molecules. Currently it is the organism in which all of this information is being combined with decades of insight gleaned from the literature to generate a holistic counterweight to this reductionism – systems biology.

6.4.1 Genomic Insights

Once the yeast genome sequence became available, it rapidly became apparent that many human disease genes have homologues in this simple eukaryotic cell. Comparative sequence analysis between *S. cerevisiae* and human genes has already provided valuable insight into human cellular metabolism. A variety of examples are given in Table 6.3, of which perhaps the most dramatic to date is the analysis of an autosomal recessively inherited disease, Rhizomelic chondrodysplasia punctata, which presents with symptoms of severe growth and mental retardation. Patients with this condition were found to carry mutations leading to defects in peroxisome biogenesis which are functionally equivalent to the yeast peroxisome targeting mutants (*pex5* and *pex7*).[48,49] However, this serves as an excellent model for many other human diseases, including cancer, neurodegenerative diseases such as Alzheimer's, Parkinson's and CJD,[50] and various metabolic disorders. Moreover, yeast also facilitates fundamental studies of molecular mechanisms such as apoptosis – a key to embryonic development.

However, more than simply facilitating homology searches, the sequencing of the genome has paved the way to using the powerful homologous recombination system in yeast to produce a bar-coded set of yeast strains, each one carrying a precisely deleted single open reading

Table 6.3   Yeast genes homologous to positionally cloned human genes

Human disease	Human gene	Yeast gene	Yeast gene function	Yeast phenotype
Hereditary nonpolyposis cancer	MSH2	MSH2	DNA repair protein	Increased mutation frequency
Cystic fibrosis protein	CFTR	YCFl	Membrane transport	Cadmium sensitivity
Wilson's disease	WND	CCC2	Copper transport atpase	Iron uptake deficiency
Glycerol kinase deficiency	GK	GUT1	Defective glycerol kinase	Defective glycerol utilization
Rhizomelic chon- drodysplasia punctata	PEX.5, PEX7	PEX.5, PEX7	Peroxisome targeting mutants	Peroxisome dysfunction
Ataxia telangiectasia	ATM	TEL1	Phosphoinositol3- kinase	Short telomeres

frame. A PCR-based gene disrupted procedure[51] has been used to disrupt each open reading frame (ORF), affording analysis of any resultant physiological effects (Figure 6.13). Long PCR primers, homologous to a section of the DNA sequence under investigation at the 5' end and homologous to a selectable marker (frequently the *KanR* gene) at their 3' terminus, are used to generate a PCR fragment which consisting of two short target gene segments on either side of a selectable marker. The target gene is then disrupted by homologous recombination and the event selected for by growth on appropriate selective medium. By incorporating one shared 18mer and one uniquely identifiable 20mer sequence into the 5' primer between the ORF and the *KanR* sequences (Figure 6.13),[52] each of the 6000 ORFs in the yeast genome has been successfully deleted. Not only does this allow insight into the effect of each deletion, but also PCR can be used to monitor the population dynamics of these differently tagged yeast strains growing together under competitive growth conditions.

6.4.2   Transcriptomes, Proteomes and Metabolomes and Drug Development

Not only was it the first eukaryote to have its genome sequenced, but it is also the organism in which most subsequent high-throughput genomics technologies were developed: microarray technology[53] provides exquisitely sensitive information on transcript levels in the cell for every gene at any given moment in time. This transcriptomic information is complemented by extensive information on protein type and levels (the proteome). The 2-hybrid technology mentioned earlier and complementary 'pull down' technology (where one protein is tagged and isolated and other attached proteins identified) have been used to

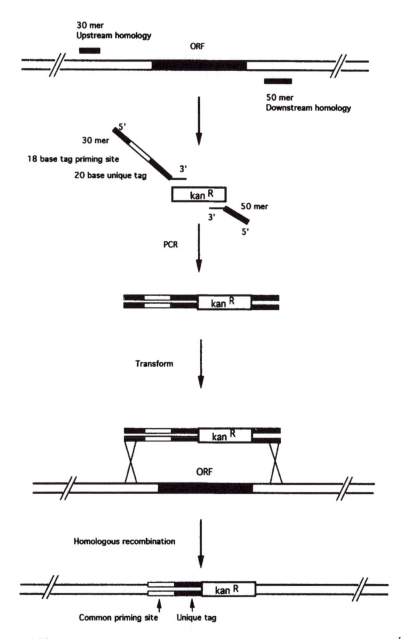

Figure 6.13   PCR gene deletion strategy. The Kanamycin resistance gene (Kan^R) is amplified using an 86 mer forward primer that contains 30 bases of upstream homology to the yeast gene of interest, an 18 base tag priming site, a uniquely indentifiable 20 base sequence tag and 18 bases of homology to the Kan^R gene. The reverse primer is a 68 mer that contains 50 bases of downstream homology to the yeast gene of interest and 18 bases of homology to the Kan^R gene. The PCR products are transformed into a haploid yeast strain and selected for on G-418 containing medium. Homologous recombination replaces the targeted yeast ORF with a common 18 mer priming site, a unique 20 base tag and the Kan^R gene.

provide information on protein–protein[54-56] interaction (the interactome[57]). A catalogue of all cellular metabolites constitutes the metabolome.[58] All of these provide global insights which can be used to
ascertain how this eukaryotic cell responds to genetic and/or environmental perturbation. By altering either (or both), biotechnologists can
now monitor eukaryotic cell responses to a variety of cellular insults and
novel therapeutic agents. This is the basis of a whole range of yeast-
based technologies for the identification of drug targets[59,60] for use in
humans because they are composed of eukaryotic cells with so much in
common with yeast.

6.4.3 Systems Biology

The extraordinary research efforts which produced the current post-
genomic era have generated mind-numbing amounts of data about
individual cellular constituents. However, cells are dynamic entities and
the next challenge is to obtain insight into the dynamic fluxes within the
cell with a view to producing mathematical and computer models
of cells and eventually multicellular organisms. These goals are a long
way off, but yeast is once again leading the way. Systems biology is
currently being applied to many biological systems and consists of the
same core principles: a computer model of the system is designed using
all of the data available; the model is then used to predict how cells (or
groups of cells) will respond based on how the computer model
responds to altering a parameter *in silico*; if the model fails to predict
the outcome in the laboratory then any discrepancy creates new
information, which can then be used to inform and improve the model.
Hence the current objectives of systems biology are to develop biologically meaningful quantitative descriptions of living cells and to generate new computational tools to facilitate this. Once again yeast
researchers have been pioneers in this area. *S. cerevisiae* has been
intensively studied for decades and, with the additional data from post-
genomic high-throughput analyses, provides an ideal model organism
in which to undertake the reiterative process of model building, prediction, perturbation, analysis and new model. Indeed, the first yeast
study on systems biology investigated the relatively simple and hugely
characterised galactose regulon,[61] and it was here that the types of
challenges that lie ahead were revealed: despite 40 years of publications
in this area and all of the tools available to test the system, the best
model was still incorrect. This gives a foretaste of just how much
information will be needed in order to generate biologically-meaningful

models of even the simplest modules in a cell. However, the same paper[61] revealed the power and exciting possibilities offered by of this type of approach: the inconsistency between the prediction and biological reality revealed a regulatory parameter that had heretofore been unknown. Systems biology may well be in its infancy in yeast, but it promises to provide unparalleled insight into basic cell functions[62] and of course how cells respond to changes in substrates, genetic changes and environmental insults.

6.5 FUTURE PROSPECTS

Yeast biotechnological and molecular biology have had an extremely successful and synergistic history. Its central importance in alcohol production made yeast an organism of research value – the more one knew, the better one could control it. This in turn made it the organism of choice when a eukaryotic cell was needed in which to express heterologous protein. The technologies that arose subsequently provided even more insight into its molecular biology, allowing us to re-engineer it as surrogate animal and plant cells. At the same time, its homologous recombination system allows it to be used to engineer entire chromosomes for synthetic biology. We know more about it than any other eukaryotic cell and therefore it is set to lead us into the future as the model organism for the development of systems biology.

The near future will see further developments in metabolic pathway cloning, with a detailed systems analysis of metabolic flux[63] facilitating the optimisation of secondary metabolite production. At the same time, glycosylation-specific strains will generate a whole new host of heterologous proteins and allow glycobiologists to probe the exciting and elusive world of glycan structure and function in complex organisms. TAR cloning will facilitate the study of the organisation and evolution of complex genomes and indeed is also set to allow the creation of novel life forms with the ability to address global problems of energy limitations and climate change.

In short, yeast biotechnology and molecular biology have just come of age and the best is yet to come.

REFERENCES

1. K. Itakura, T. Hirose, R. Crea, A. D. Riggs, H. L. Heyneker, F. Bolivar and W. H. Boyer, *Science*, 1977, **198**, 1056.
2. A. B. Futcher, *Yeast*, 1988, **4**, 27.

3. D. Mumberg, R. Muller and M. Funk, *Gene*, 1995, **156**, 119.
4. A. Hinnen, J. B. Hicks and G. R. Fink, *Proc. Natl. Acad. Sci. USA*, 1978, **75**, 1929.
5. J. D. Beggs, *Nature*, 1978, **275**, 104.
6. H. Ito, Y. Fukada, K. Murata and A. Kimura, *J. Bacteriol.*, 1983, **153**, 163.
7. E. Meilhoc, J.-M. Masson and J. Teissie, *Bio/Technology*, 1990, **8**, 223.
8. http://genome-www.stanford.edu/Saccharomyces/alleletable.html.
9. R. Parker and A. Jacobson, *Proc. Natl. Acad. Sci. USA*, 1990, **87**, 2780.
10. K. S. Zaret and F. Sherman, *J. Mol. Biol.*, 1984, **177**, 107.
11. J. Mellor, M. J. Dobson, N. A. Roberts, A. J. Kingsman and S. M. Kingsman, *Gene*, 1985, **33**, 215.
12. T. F. Donahue and A. M. Cigan, *Methods Enzymol.*, 1990, **185**, 366.
13. D. T. Moir and D. R. Dumais, *Gene*, 1987, **56**, 209.
14. A. Hinnen, B. Meyhack and R. Tsapis, *in Gene Expression in Yeast*, Eds. M. Kornola, E. Väisänen Kauppakirjapaino, Helsinki, 1983, p.157.
15. A. J. Brake, J. P. Merryweather, D. G. Coit, U. A. Heberlein, F. R. Masiarz, G. T. Mullenbach, M. S. Urdea, P. Valenzuela and P. J. Barr, *Proc. Natl. Acad. Sci. USA*, 1984, **81**, 4642.
16. N. Skiper, M. Sutherland, R. W. Davies, D. Kilburn, R. C. Miller, A. Warren and R. Wong, *Science*, 1985, **230**, 958.
17. R. A. Smith, M. J. Duncan and D. T. Moir, *Science*, 1985, **229**, 1219.
18. T. Etcheverry, W. Forrester and R. Hitzeman, *Bio/Technology*, 1986, **4**, 726.
19. L. M. Melnick, B. G. Turner, P. Puma, B. Price-Tillotson, K. A. Salvato, D. R. Dumais, D. T. Moir, R. J. Broeze and G. C. Avgerinos, *J. Biol. Chem.*, 1990, **265**, 801.
20. R. A. Hallewell, R. Mills, P. Tekamp-Olsen, R. Blacker, S. Rosenberg, F. Otting, F. R. Masiarz and C. J. Scandella, *Bio/Technology*, 1987, **5**, 363.
21. D. E. Wampler, E. D. Lehman, J. Boger, W. J. McAleer and E. M. Scolnick, *Proc. Natl. Acad. Sci. USA*, 1985, **82**, 6830.
22. D. T. Burke, G. F. Carle and M. V. Olsen, *Science*, 1987, **236**, 806.
23. P. Sudbery, *in Human Molecular Genetics*, Addison Wesley Longman, England, 1998, p. 209.
24. K. R. Peterson, C. H. Clegg, Q. Li and G. Stamatoyannopoulos, *Trends Genet.*, 1997, **13**, 61.

25. J. D. Boeke, F. LaCroute and G. R. Fink, *Mol. Gen. Genet.*, 1984, **197**, 345.
26. N. Kouprina and L. Vladimir, *FEMS Microbiol. Rev.*, 2003, **27**, 629.
27. N. Kouprina and L. Vladimir, *Nat. Protocols*, 2008, **3**, 371.
28. L. A. Annab, N. Kouprina, G. Solomon, P. L. Cable, D. E. Hill, J. C. Barrett, V. Larionov and C. A. Afshari, *Gene*, 2000, **250**, 201.
29. D. G. Gibson, G. A. Benders, C. Andrews-Pfannkock, E. A. Denisova, H. Baden-Tillson, J. Zaveri, T. B. Stockwell, A. Brownley, D. W. Thomas, M. A. Algire, C. Merryman, L. Young, V. L. Noskov, J. I. Glass, J. C. Venter, C. A. Hutchinson III and O. Smith, *Science*, 2008, **319**, 1215.
30. M. H. Pausch, *Tibtech*, 1997, **15**, 487.
31. J. R. Broach and J. Thorner, *Nature*, 1996, **384**, 14.
32. C. K. Wrenn and B. S. Katzenellenbogen, *J. Biol. Chem.*, 1993, **268**, 24089.
33. S. Fields and O. Song, *Nature*, 1989, **340**, 245.
34. R. K. Brachmann and J. D. Boeke, *Curr Opin. Biotechnol.*, 1997, **8**, 561.
35. R. D. Gietz, B. Triggsraine, A. Robbins, K. C. Grahame and R. A. Woods, *Mol. Cell. Biochem.*, 1997, **172**, 67.
36. K. H. Young, *Biol. Reprod.*, 1998, **58**, 302.
37. B. Huang, J. Guo, B. Yi, X. Yu, L. Sun and W. Chen, *Biotechnol. Lett.*, 2008, 9663.
38. J. A. Chemler, Y. Yan and M. A. G. Koffas, *Microb. Cell Factories*, 2006, **5**, 20.
39. D.-K. Ro, E. M. Paradise, M. Ouellet, K. J. Fisher, K. L. Newman, J. M. Ndungu, K. A. Ho, R. A. Eachus, T. S. Ham, J. Kirby, M. C. Y. Chang, S. T. Withers, Y. Shiba, R. Sarpong and J. D. Keasling, *Nature*, 2006, **440**, 940.
40. J. M. DeJong, Y. Liu, A. P. Bollon, R. M. Long, S. Jennewein, D. Williams and R. B. Croteau, *Biotechnol. Bioeng.*, 2006, **93**, 212.
41. S. R. Hamilton and T. U Gerngross, *Curr. Opin. Biotechnol.*, 2007, **18**, 387.
42. B. K. Choi, P. Bobrowicz, R. C. Davidson, S. R. Hamilton, D. H. Kung, H. Li, R. G. Miele, J. H. Nett, S. Wildt and T. U. Gerngross, *Proc. Natl. Acad. Sci. USA*, 2003, **100**, 5022.
43. S. R. Hamilton, P. Bobrowicz, B. Bobrowicz, R. C. Davidson, H. Li, T. Mitchell, J. H. Nett, S. Rausch, T. A. Stadheim, H. Wischnewski, S. Wildt and T. U. Gerngross, *Science*, 2003, **301**, 1244.

44. S. R. Hamilton, R. C. Davidson, N. Sethuraman, J. H. Nett, Y. Jiang, S. Rios, P. Bobrowicz, T. A. Stadheim, H. Li, B. K. Choi, D. Hopkins, H. Wischnewski, J. Roser, T. Mitchell, R. R. Strawbridge, J. Hoopes, S. Wildt and T. U. Gerngross, *Science*, 2006, **313**, 1441.

45. H. Li, N. Sethuraman, T. A. Stadheim, D. Zha, B. Prinz, N. Ballew, P. Bobrowicz, B. K. Choi, W. J. Cook and M. Cukan, *Nat. Biotechnol.*, 2006, **24**, 210.

46. K. Amano, Y. Chiba, Y. Kasahara, Y. Kato, M. K. Kaneko, A. Kuno, H. Ito, K. Kobayashi, J. Hirabayashi, Y. Jigami and H. Narimatsu, *Proc. Natl. Acad. Sci. USA*, 2008, **105**, 3232.

47. A. Goffeau, B. G. Barrell, H. Bussey, R. W. Davis, B. Dujon, H. Feldmann, F. Galibert, J. D. Hoheisel, C. Jacq, M. Johnston, E. J. Louis, H. W. Mewes, Y. Murakami, P. Philippsen, H. Tettelin and S. G. Oliver, *Science*, 1996, **274**, 546.

48. P. E. Purdue, J. W. Zhang, M. Skoneczny and P. B. Lazarow, *Nat. Genet.*, 1997, **15**, 381.

49. A. M. Motley, *et al., Nat. Genet.*, 1997, **15**, 377.

50. M. W. Walberg, *Arch Neurol.*, 2000, **57**, 1129.

51. A. Baudin, O. Ozier-Kalogeropoulos, A. Denouel, F. Lacroute and C. Cullin, *Nucleic Acids Res.*, 1993, **21**, 3329.

52. D. D. Shoemaker, D. A. Lashkari, D. Morris, M. Mittmann and R. W. Davis, *Nat. Genet*, 1996, **14**, 450.

53. J. L. DeRisi, V. R. Iyer and P. O. Brown, *Science*, 1997, **278**, 680.

54. P. Uetz, L. Giot and G. Cagney, *et al., Nature*, 2000, **403**, 623.

55. A. C. Gavin, M. Bosche and R. Krause, *et al., Nature*, 2002, **415**, 41.

56. Y. Ho, A. Gruhler and A. Heilbut, *et al., Nature*, 2002, **415**, 180.

57. A. H. Y. Tong, G. Lesage and G. D. Bader, *et al., Science*, 2004, **303**, 808.

58. J. I. Castrillo, A. Hayes, S. Mohammed, S. J. Gaskelland and S. G. Oliver, *Phytochemistry*, 2003, **62**, 929.

59. N. Bharucha and A. Kumar, *Comb. Chem. High Throughput Screening*, 2007, **10**, 618.

60. A. B. Parsons, R. Geyer, T. R. Hughes and C. Boone, *Prog. Cell Cycle Res.*, 2003, **5**, 159.

61. T. Ideker, V. Thorsson and J. A. Ranish, *et al., Science*, 2001, **292**, 929.

62. R. Mustacchi, S. Hohmann and J. Nielsen, *Yeast*, 2006, **23**, 227.

63. J. Nielsen and M. C. Jewett, *FEMS Yeast Res*, 2008, **8**, 122.

64. G. Walsh, *Nat. Biotechnol.*, 2006, **24**, 769.

65. B. U. Gerngross, *Nat. Biotechnol.*, 2004, **22**, 1409.

66. K. Struhl, D. T. Stinchcomb, S. Scherer and R. W. Davis, *Proc. Natl. Acad. Sci. USA*, 1979, **76**, 1035.
67. G. Tschumper and J. Carbon, *Gene*, 1980, **10**, 157.
68. S. Fogel and J. Welch, *Proc. Natl. Acad. Sci. USA*, 1982, **79**, 5342.
69. T. D. Webster and R. C. Dickson, *Gene*, 1983, **26**, 243.
70. J. Rine, W. Hansen, E. Hardeman and R. W. Davis, *Proc. Natl. Acad. Sci. USA*, 1983, **80**, 6750.

CHAPTER 7

Metabolic Engineering

STEFAN KEMPA,[a] DIRK WALTHER,[b] OLIVER EBENHOEH[a] AND
WOLFRAM WECKWERTH[a]

[a]GoForSys, University of Potsdam, Institute of Biochemistry and Biology,
c/o Max-Planck-Institute for Molecular Plant Physiology, Am Mühlenberg 1,
14424 Potsdam, Germany; [b] Max-Planck-Institute for Molecular Plant
Physiology, Am Mühlenberg 1, 14424 Potsdam, Germany

7.1 INTRODUCTION

Making use of the metabolic properties of a living organism has a long
tradition in human civilisation, with applications in production of food
and fermentation of food for conservation, taste and salubriousness.

Since the introduction of recombinant DNA technology, metabolic
engineering has made great progress in the manipulation and optimisation of new cellular processes in bacteria, yeast, filamentous fungi,
plants and mammalian cells.[1]

In the past, improvements in productivity and yield were achieved
mainly by classical breeding methods or mutagenesis combined with
large-scale phenotype screening. With increasing knowledge about
genomic structure and gene expression and the application of directed
genetic engineering, more targeted manipulation strategies came up for
modifying the properties of organisms.

However, the knowledge of the specific biochemical properties of
those organisms is ultimately required for successful establishment or
improvement of biotechnological applications. The regulatory processes

Molecular Biology and Biotechnology, 5th Edition
Edited by John M Walker and Ralph Rapley
© Royal Society of Chemistry 2009
Published by the Royal Society of Chemistry, www.rsc.org

controlling metabolism are complex and the degree of complexity increases dramatically first from the transition from prokaryotes to eukaryotes and also from single-cellular to multi-cellular organisms with specialised organs. Interestingly, the metabolic complexity does not increase to the same level as the regulatory complexity. The organisation of the central metabolism is similar in a wide range of organisms. Thus, the use of unicellular systems such as bacteria and yeast is one of the major focuses of molecular engineering. The situation changes if the production of proteins or complex chemical molecules is desired, which involves a series of modification steps more pronounced in complex eukaryotic systems. In the case of proteins, post-translational modifications such as glycosylation and protein folding might be impaired in simpler organisms. The same holds true for more complicated chemical molecules such as alkaloids.

A complete new avenue to metabolic engineering was opened with the onset of genome sequencing (see http://www.genomesonline.org/gold. cgi and Figure 7.1). Further novel bioanalytical techniques evolved for genome-wide molecular analyses and a new comprehensive description of metabolic regulatory networks opens new perspectives for metabolic engineering. The understanding of metabolic pathways and their regulation at a systems level raises the opportunity to identify targets for manipulation which will have the strongest influence on desired metabolic process. It helps also to evaluate the limits of the metabolic system and to identify competing pathways which impair the efficiency of the desired process (see Sections 7.2.2.2 and 7.2.2.3). Thus increased knowledge in systems biology and recent technological developments serve as a basis for modern bioengineering.

7.2 THEORETICAL APPROACHES FOR METABOLIC NETWORKS

The initial phase of pathway manipulation requires an understanding of the targeted metabolic network or, better, the complete picture of metabolism in the targeted biological system. Structural analysis of metabolic networks identifies entry points for metabolic engineering.[2,3]

A plethora of methods exists to address structural modelling, kinetic modelling and control of metabolism, namely elementary flux modes (EFM) and flux balance analysis (FBA), rate equation-driven modelling using complex differential equations and kinetic constants and metabolic control analysis (MCA). In the following sections we will introduce briefly the basic principles of these techniques and present some applications.

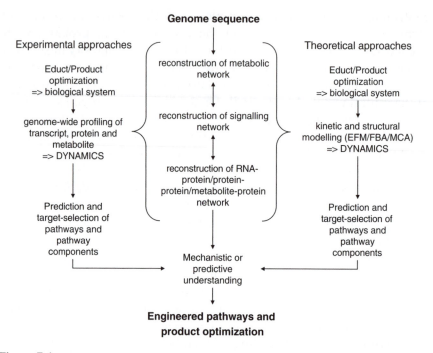

Figure 7.1   Overall scheme summarizing all different areas of systems biology. The consequent integration of theoretical and experimental approaches and the iterative improvement of predictability will necessarily lead to a better understanding of life.

7.2.1   Kinetic Modelling

Metabolic systems are characterized by different types of quantities. These are variables such as the concentrations of metabolites, which may change in a short time-scale, and system parameters such as kinetic constants, which are fairly constant during the life span of an organism. Although in most cells the levels of enzyme concentrations determining the maximal activities of a certain reaction may change by adaptation to environmental conditions, the special ordering of enzymatic reactions, that is, the stoichiometry of metabolic pathways, was fixed during evolution.

The classical approach to modelling metabolic systems is mainly concerned with the simulation of the time-dependent behaviour of the variables at fixed values of the parameters by using systems of ordinary differential equations. This approach dates back to the pioneering work of Garfinkel and Hess[4] on glycolysis. Later, this metabolic chain and related pathways of cellular energy metabolism were favoured subjects of successful mathematical modelling.[5–11]

In such an approach, the variables represent the concentrations of the different chemical species and their temporal evolution is determined by production and consumption rates describing the enzyme activities. In its most general form, the system of ordinary differential equations can be written as

$$\frac{d\vec{x}}{dt} = \mathbf{N} \cdot \vec{v}(\vec{x}; \vec{p}) \tag{1}$$

where $\vec{x}$ is the vector containing the concentrations of the various chemical species and the vector $\vec{v}$ contains expressions for the enzymatic rates of all reactions, which in general depend on the concentrations $\vec{x}$ and other parameters, comprised in $\vec{p}$, describing the enzyme's kinetic behaviour. The time derivative and the reaction rates are linked by the so-called stoichiometry matrix $\mathbf{N}$. An entry n_{ij} represents the stoichiometric coefficient of metabolite i in reaction j and is negative if i is consumed by reaction j and positive if it is produced. This matrix is determined by the topology of the metabolic system and therefore remains constant independent of external influences or changes in the kinetic parameters.

The investigation of a metabolic system with ordinary differential equations is helpful to predict a system's dynamic behaviour under certain external conditions or parameter changes, which might represent, for example, the knock-down or knock-out of a gene or the targeted inhibition of some specific enzymes. For the formulation of the equation system (1), detailed knowledge on the enzymatic rates of every participating reaction is necessary. Correct rate expressions reflect the enzyme mechanisms including allosteric regulations, such as the inhibition or activation due to the presence of other metabolites. The experimental determination of enzymatic rate laws and their parameters is a labour-intensive and time-consuming process and it is therefore not surprising that our knowledge of these parameters is still rather limited. However, for relatively small and well-studied systems, such as single central pathways, the required information is often available, allowing for the mathematical description of the enzymatic rates. Commonly used rate laws are mass-action kinetics, describing the enzymatic rate as a multilinear function in the substrate concentrations, and the well-known Michaelis–Menten rate law, which reflects that enzymatic rates saturate with high substrate concentrations as a result of limited free enzyme.

We will outline the general approach by a simple illustrative example, which was first studied by Higgins[101] and Selkov.[12] We consider the

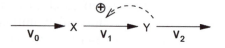

Figure 7.2   Higgins–Selkov oscillator.

reaction system depicted in Figure 7.2 of three consecutive reactions with two intermediates X and Y. The first substrate, X, is supplied at a rate v_0, then metabolised to intermediate Y at a reaction rate v_1, and subsequently Y is degraded at a rate v_2. In particular, we consider a product activation of the reaction converting X into Y (v_1). This mechanism is observed, for example, in the enzyme phosphofructo-kinase which is activated by its own product, fructose-1,6-bisphosphate.

The general form of the differential equations describing the dynamic behaviour of the system reads

$$\frac{d}{dt}\begin{pmatrix} X \\ Y \end{pmatrix} = \begin{pmatrix} v_0 - v_1 \\ v_1 - v_2 \end{pmatrix} \tag{2}$$

or, in a form equivalent to Eq. (1):

$$\frac{d\vec{x}}{dt} = \mathbf{N} \cdot \vec{v} \text{ with } \vec{x} = \begin{pmatrix} X \\ Y \end{pmatrix}, \mathbf{N} = \begin{pmatrix} 1 & -1 & 0 \\ 0 & 1 & -1 \end{pmatrix} \text{ and } \vec{v} = \begin{pmatrix} v_0 \\ v_1 \\ v_2 \end{pmatrix} \tag{3}$$

To solve this system, expressions for the rates as functions of substrate concentrations and other parameters need to be known. We will now make a few simplifying assumptions allowing us to formulate the required rate expressions. First, we assume the influx to be constant, v_0= constant. Second, we assume that degradation of Y occurs under non-saturating conditions and can be approximated by a mass-action rate law:

$$v_2 = k_2 \cdot Y. \tag{4}$$

Last, we need an expression for rate v_1. This rate expression should reflect that product Y acts as an activator, which in particular means that an increase in Y will result in an increase of rate v_1. The exact form of such an expression for a realistic system would have to be derived from the detailed enzymatic mechanism and subsequently the para-meters would have to be fitted to experimental data or, alternatively, if the detailed mechanism is not known, a heuristic equation must be derived and fitted against measured values. For our case study, we assume that the expression has a particularly simple form, namely

$$v_1 = k_1 \cdot X \cdot Y^2. \tag{5}$$

The resulting system of coupled differential equations reads

$$\frac{\mathrm{d}X}{\mathrm{d}t} = v_0 - k_1 \cdot X \cdot Y^2 \tag{6}$$

$$\frac{\mathrm{d}Y}{\mathrm{d}t} = k_1 \cdot X \cdot Y^2 - k_2 \cdot Y. \tag{7}$$

were v_0, k_1 and k_2 are system parameters that have to be provided.

Solving the system numerically allows one to plot the dynamics of the variables as functions of time. Figure 7.3 depicts the time course for the variable X for two different choices of parameters. It can be seen that the system exhibits drastically different modes of behaviour when the parameter v_0 is varied. For $v_0 > 1$ (dashed line), the system rapidly

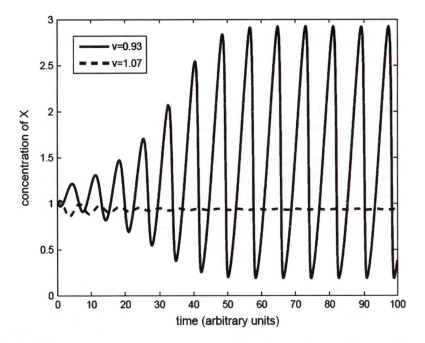

Figure 7.3 Temporal dynamics of the variable X in the Higgins–Selkov oscillator. Shown are two time courses determined for different influx rates v_0 Whereas for values $v_0 > 1$ the system quickly adapts a stationary phase (dashed line), for values $v_0 < 1$ the system might show oscillatory behaviour (solid line).

assumes a steady state and the concentrations remain constant. In contrast, for $v_0 < 1$ (solid line), the variables display an oscillatory behaviour and, in fact, the system approaches a stable limit cycle.

In general, the phenomenon that small changes in some parameters can result in a qualitative change of the system's behaviour is termed *bifurcation* and is commonly observed in larger systems. Such a behaviour is not understandable from pure observations alone, but an underlying theory is required to explain how the same system may exhibit fundamentally different modes of action. Model calculations can help in identifying which key parameters are critical for the system's behaviour, a knowledge that is crucial to predict theoretically which particular biotechnological modifications are most promising to yield the desired effect.

7.2.2 Metabolic Control Analysis (MCA), Elementary Flux Modes (EFM) and Flux-balance Analysis (FBA)

7.2.2.1 *Metabolic Control Analysis (MCA)*

Intuitively, to engineer a metabolic system with the goal of increasing the yield of the desired compound (metabolite) or flux through a metabolic pathway system, one would proceed by identifying and manipulating so-called rate-limiting-steps, *i.e.* particular steps within pathways that exert the most control on the system as a whole. However, despite the plausibility of this approach, most experimental attempts to boost pathway fluxes by overexpressing single candidate rate-limiting step enzymes have largely met with failure.[13] Apparently, the control of the system as a whole is much more distributed across the system's components than appreciated.

The study of how global yield parameters are distributed across system components is the domain of sensitivity analysis. Sensitivity analysis has found many applications across many engineering disciplines as well as in economics. The application of the general concepts of sensitivity analysis to metabolic pathway systems by Kacser *et al.*[14] (1973) and Heinrich and Rapoport[15] (1974) laid the foundation of metabolic control analysis (MCA). MCA provides a rigorous mathematical framework to study both qualitatively and quantitatively how the control of metabolic fluxes and concentrations of intermediate compounds is distributed among the participating pathway enzymes, thereby allowing the identification of enzymatic steps with the strongest effects on metabolite levels and fluxes and also the mutual dependencies between them.[13,16]

Central to MCA are so-called *control coefficients* as unit-free relative measures of the magnitude of change in system variables, such as fluxes,

J, or metabolite concentrations, S, in response to local perturbations (enzyme concentrations or rates). In their general form, control coefficients are defined as

$$C_i^A = \frac{\partial A/A}{\partial v_i/v_i} = \frac{\partial \ln A}{\partial \ln v_i} \tag{8}$$

where A is the variable of interest (flux J or metabolite concentration S), i refers to the enzymatic step being perturbed and v_i is the associated enzymatic rate at the steady state. Because enzymatic rates can be considered to depend linearly on enzyme concentrations, E, and thus $v = k[E]$), Eq. (8) can be re-stated as

$$C_i^A = \frac{\partial \ln A}{\partial \ln[E_i]} \tag{9}$$

At steady state, it could be shown that the sum over all flux control coefficients in a fully connected metabolic pathway system must add up to one, and all concentration control coefficients add up to zero:

$$\sum_i C_i^J = 1 \tag{10}$$

for the flux control coefficient, and

$$\sum_i C_i^{[S_j]} = 0 \tag{11}$$

for the concentration control coefficient, where the summation is made over all steps i and j denotes a particular metabolite of interest.

The interpretation of these so-called *summation theorems* has far-reaching consequences as they capture global system properties. As the sum over all control coefficients is constant, increases in one control coefficient are necessarily accompanied by corresponding decreases of control coefficients associated with other steps in the pathway system. Thus, all control coefficients are tightly linked and control is shared between all enzymes to different degrees. A strictly rate-limiting enzyme would be signified by a flux-control coefficient of 1, while all other enzymes have zero flux control coefficients. Although not impossible, this is highly unlikely to occur in reality.

While the summation theorems deal with systemic properties and the influence of individual enzymatic steps on the overall flux, a wealth of experimental data are available for more local enzyme characteristics such as dependency of enzymatic rates on substrate concentrations or temperature or other parameters and data on detailed enzymatic mechanisms are available. In a general form, the dependence of the local enzymatic rate, v, of enzyme i in the pathway on the parameter p (*e.g.* temperature), is described by *elasticity coefficients*:

$$\varepsilon_p^i = \frac{\partial v_i/v_i}{\partial p/p} = \frac{\partial \ln v_i}{\partial \ln p} \tag{12}$$

Every enzyme has as many elasticity coefficients as parameters influencing it.

Control coefficients describe global properties, elasticity coefficients capture properties of the isolated enzyme. As a particularly powerful aspect of MCA, the *connectivity theorems* allow integration of the global and local properties into a unified view:

$$\sum_i C_i^J \varepsilon_{[S]}^i = 0 \tag{13}$$

The sum of the products of the flux-control coefficient of all (i) steps affected by S and its elasticity coefficients towards S is zero. And for the concentration control coefficient:

$$\sum_i C_i^{[A]} \varepsilon_{[S]}^i = 0, \; A \neq S \tag{14}$$

(the reference metabolite, A, is different to the perturbed metabolite, S); and

$$\sum_i C_i^{[S]} \varepsilon_{[S]}^i = -1 \tag{15}$$

if both reference and perturbed metabolite are the same. The connectivity theorems describe how perturbations on metabolites of a pathway propagate throughout the entire metabolic network.

Experimental approaches to measure individual control coefficients require the introduction of small perturbations.[16]. The variable of interest (flux or the concentration of a particular metabolite) is then measured after a new steady state has been reached. Such perturbations can be introduced, for example, by changing enzyme concentrations (knock-down/-out) or by applying enzyme-specific inhibitors. A number of MCA-related software suites are also available, such as *Gepasi* and *SCAMP*.[17,18]

7.2.2.2 Elementary Flux Mode (EFM) Analysis

Engineering metabolic systems to boost the yield of a desired metabolite or to synthesize a certain compound, given a set of starting compounds, naturally leads to the concept of metabolic pathways thought of as molecular assembly lines connecting substrates to their products. To achieve the desired outcome, the identified production pathway needs to be targeted by bioengineering means, possibly alternative pathways identified or other, competing pathways suppressed. Because many known metabolic reaction routes are not linear reaction chains, but are branched or even cyclical, and many metabolites participate in several different metabolic reactions, especially so-called pool metabolites such as water and ATP, an unambiguous pathway assignment form a given metabolic map proves challenging and the term 'pathway' conceptually difficult.

Elementary flux mode analysis (EFM) offers a rigorous definition of pathways by defining them as minimal sets of enzymes that can operate at the steady state with consideration of their directionality (irreversibility of reactions).[2,19,20] 'Minimal' means that deleting any one enzyme in a given mode would lead to the interruption of the flux carried by this mode. The combination of elementary modes – which can partly overlap – reconstructs the entire metabolic network and describes all possible fluxes and states (flux patterns) that the system can be in. Elementary modes cannot be decomposed into smaller sub-networks as those sub-paths would not be able to carry a steady-state flux. Elementary modes connect via *internal* metabolites starting and end nodes that are considered *external*, *i.e.* metabolites whose concentrations are buffered and are provided as a reservoir and can be considered sources or sinks, while internal metabolites are compounds participating in the biochemical reactions and their production and consumption are balanced at the steady state. It is therefore obvious that any EFM analysis first starts by defining external and internal metabolites, thus establishing the operational system boundaries. In addition, all directions need to be defined (irreversible reaction steps). Under the appropriate definition, an elementary mode describes a flux that can be sustained even though all other enzymes are rendered non-functional (*e.g.* knock-out mutants). Hence potentially wasteful routes can be switched off without affecting the yield of the desired product. Likewise, EFM provides an understanding of what consequences gene deletions or gene additions would have on the metabolic network of an organism. Furthermore, alternative, but perhaps sub-optimal, pathways leading to the same desired product can be identified in cases where the best pathway (mode) may be not easily amenable to biotechnological interventions. EFM has also been used to identify maximum conversion yields.[3,21] Software tools for the analysis of EFM have also been developed and made available.[22]

7.2.2.3   Flux-balance Analysis (FBA)

The perhaps ultimate goal of systems biology can be formulated as given the components of a system being able to describe mathematically and, therefore, being able to simulate the behaviour of the system under varying external conditions and to understand its emerging properties such as stability and regulation. Attaining this goal would also provide the ideal prerequisite for rational metabolic engineering approaches as the metabolic system can then be simulated and theoretically optimized with a desired objective (e.g. yield of a particular metabolite) in mind. The technological advances achieved by the various 'omics' technologies have put one aspect of this goal within reach. For many organisms, we have already deciphered essentially all genes and therefore, in principle, all encoded proteins and largely along with their functional annotations. Therefore, the enzymatic potential of an organism – all possible metabolic reactions – can be considered known. This knowledge is further augmented by actual measurements of metabolite and protein levels and functional characterisations of proteins. However, describing all possible reactions in a mathematical framework using established metabolic modelling techniques proved challenging as many required kinetic parameters reflecting the actual enzymatic mechanism remain undetermined.

Flux balance analysis (FBA) provides a framework that still allows a description of the metabolic capabilities of an organism without the need for detailed kinetic modelling (for reviews, see Refs 23–25). The basic concept of FBA is to reduce all metabolic reactions and metabolic exchange processes to a minimal description that allows the application of conservation principles that must be fulfilled, in particular conservation of mass reflected by the stoichiometry the chemical reactions and other processes that change the concentration of metabolites, and then to infer the most likely distribution of metabolic fluxes according to a reasonable optimisation criterion, such as biomass production. Based on the reconstructed metabolic network for a particular organism derived from its genome sequence and bioinformatics (gene predictions, functional assignments based on homology) and experimental annotation (e.g. EST sequences, proteomics measurements) and imposing mass balance, it is relatively straightforward to write down all reactions and processes (unified treated as fluxes) that alter the concentrations of a metabolite based on the stoichiometry of the metabolic reactions (see also the previous section above and Figure 7.1). Those metabolic reactions participate in transport processes (export and import of compounds across the system's boundaries) and also growth processes that effectively use up a given metabolite, *i.e.* compounds being removed from the system and contributing towards its growth:

$$\frac{dX_i}{dt} = V_{synthesis} - V_{degradation} - V_{growth/use} \pm V_{transport} \tag{16}$$

And generalized for all m metabolites, n fluxes and using matrix notation:

$$\frac{dX}{dt} = NV \tag{17}$$

where N is the $m \times n$ stoichiometric matrix (including the stoichiometric relationships for transport and usage processes) and V is the vector including all fluxes (metabolic, transport as well as usage fluxes).

At the steady state, Eq. (18) applies:

$$\frac{dX}{dt} = NV = 0 \tag{18}$$

Hence the solutions for the flux distributions at the steady state are contained in the null space of the stoichiometric matrix, N. Since generally $m < n$, *i.e.* the number of metabolites is smaller than the number of fluxes operating on them, the system of equations is underdetermined and can only be solved by assuming additional boundary conditions (for example, constrained ranges of fluxes). Furthermore, by introducing objective functions that are to be maximized or minimized (*e.g.* maximum flux leading to maximum growth or maximum yield of a certain metabolite or minimum overall flux rates), all possible solutions contained in Eq. (18), the so-called flux cone, are reduced to the flux distribution of the system under constrained optimisation and can be derived by suitable techniques such as linear programming.

For simple organisms such as bacteria, it is plausible to assume that the natural flux distribution has evolved under the principle of growth maximisation. Evidently, for higher organisms, such optimisation principles are less obvious. As the introduced optimisation principles are to a large extent hypothetical, consideration of more physical constraints such as thermodynamic realisability based on Gibbs free energy have been proposed recently.[26]

FBA also provides a powerful methodological framework to assess biotechnological objective functions such as the yield maximisation of a desired metabolite and furthermore to assess the effects of changes of the underlying metabolic network, by deleting or adding certain enzymes, and then to re-evaluate the metabolic capabilities of the system and to determine to optimal distribution of fluxes.[27]

7.3 EXPERIMENTAL APPROACHES FOR METABOLIC ENGINEERING

7.3.1 Tools for Metabolic Engineering

Engineering of metabolic pathways can be achieved by manipulating the expression level, kinetic parameters or activation state of metabolic enzymes.

Selection of wanted properties was for a long time the strategy to achieve improvements in the targeted processes. The productivity was monitored and strains or individuals with the best performance were selected for propagation. Over time, this strategy was very effective and led to organisms with very specific properties optimised for the desired processes. These selection procedures might have changed metabolic regulation at all levels. New techniques evolved for molecular analyses and molecular biology, directing targeted manipulations by recombinant DNA technologies. Using molecular biological techniques, modification of metabolism at all levels can be achieved. For example, mutagenesis as an undirected approach or targeted mutagenesis of genomes/genes is used to modify the desired properties of the organism.

The expression level of a metabolic enzyme can be changed by over-expression of target genes or the use of antisense RNAi and its expression can be terminated by gene insertion knock-outs or, if possible, replacement. Regulatory sites and allosteric regulation substrate specificity can be manipulated specifically by site-directed mutagenesis of the gene coding sequence. Also, complete new pathways can be introduced by expressing enzymes of foreign origin in the target organism.

For targeted manipulation of the metabolism of organisms, a deep knowledge of metabolic regulation is crucial. This forced the development of technologies to monitor the metabolism of the organisms.

Methods for metabolite measurement have been highly developed for decades. However, most of those methods were targeted to a subset of metabolites for identification of a specific compound class. The recent development of metabolomics techniques changed the view on metabolism and serves as a cornerstone for systems biology.

7.3.2 Metabolomics

'Omic' technologies measure many variables simultaneously in a biological sample.[28] These measurements represent snapshots of the system, enabling a methodical search for correlations among the variables (environmental parameters, metabolites, proteins, transcripts and others) and thus describing the system.[29] The systematic and especially the time-dependent description of living systems requires a substantial sample throughput in parallel with comprehensive analysis of as many constituents as possible. In view of the chemical and physical diversity of small biological molecules, the challenge remains to develop one or more protocols to gather the whole 'metabolome' (all small molecules present in a sample). No single technique is suitable for the analysis of the different types of molecules, which is why a mixture of techniques has to be used.[28] In proteomics and transcriptomics, problems arise from the sheer number of dynamically fluctuating transcripts and proteins and also post-transcriptional and post-translational regulation and modification. In the field of metabolomics, the general estimation of the size and the dynamic range of a species-specific metabolome is at a preliminary stage.

Metabolic fingerprinting and metabolomics with high sample throughput but decreased dynamic range and deconvolution of individual components achieve a global view on the *in vivo* dynamics of metabolic networks. Here, the reader's attention is directed to excellent reviews covering this topic, including NMR, direct infusion mass spectrometry and infrared spectroscopy.[30–35] A lower sample throughput but unassailable identification and quantitation of individual compounds in a complex sample are achieved by GC–MS and LC–MS technology.

Owing to major steps forward in these technologies, it is possible to match specific demands with specific instruments and novel developments in the performance of mass analysers. However, it is important to know that each type of technology exhibits a bias towards certain compound classes, mostly due to ionisation techniques, chromatography and detector capabilities. GC–MS has evolved as an essential technique for metabolomics owing to its comprehensiveness and sensitivity.[29,36–46] The coupling of GC with time-of-flight (TOF) mass analysers is an emerging technology. High scan rates provide accurate peak deconvolution of complex samples.[47–51] GC–TOF-MS capabilities provide an improvement over conventional GC–MS analysis with respect to the analysis of ultra-complex samples, which is particularly important for the metabolomics approach.[29,40,44] Samples of ultra-complexity contain hundreds of coeluting compounds varying in their abundance by several orders of magnitude. Hence accurate mass spectral deconvolution and a broad linear dynamic range represent an indispensable prerequisite for high-quality spectra and peak shapes. Modern GC–TOF-MS applications and incorporated mass spectral deconvolution algorithms fulfil these requirements.

Other promising technologies are CE–MS,[52–56] particularly for the analysis of polar and thermolabile compounds, and electrochemical detection in parallel with LC–MS for the analysis of redox-active compounds.[57–59] The coupling of LC and MS is the most established technique to address targeted identification and quantitation of specific metabolites in complex mixtures.[60–63] In contrast, for the non-targeted analysis of all compounds in a complex sample,[64–67] novel deconvolution algorithms have to be implemented taking into account the differences of data acquisition capacities of specific mass analysers, for instance quadrupole time-of flight instruments or ion traps.[64,66,68,69]

7.3.3   Metabolomics in the Context of Metabolic Engineering

The metabolomics technology allows for genome-wide analysis of metabolite dynamics and therefore permits a comparison of genome-wide mathematical models of metabolism and their metabolite level predictions (see Figure 7.1). Other examples are the combination of metabolite concentration measurements –metabolomics – and FBA (see above) to reveal constraints for structural modelling.[26] Thus, a combination of modelling and experimental approaches for metabolic networks will constantly improve our knowledge about predictive behaviour to identify the control points as targets for genetic manipulation.

Another aspect for the conclusive understanding of a metabolic network in a sequenced species is genome annotation, thus the functional assignment of the genes. Here, the application of metabolomics and proteomics techniques help to improve genome annotation, as recently demonstrated in *Chlamydomonas reinhardtii*.[70] The high-throughput character of metabolomics techniques allows the fast detection of metabolites from all ranges of metabolism (see above). A projection of these detected metabolites into a theoretical metabolic network of the respective organism identifies gaps which need to be filled otherwise the network would not function. Based on these discrepancies, one can go back to the draft genome sequence and search for corresponding proteins or protein domains which fulfil the missing reactions.[70] The systematic development of this species-specific draft metabolic network in combination with EFM and FBA techniques (see above) will then allow the identification of key points in the network for targeted manipulation, *e.g.* higher flux into lipids during oil plant seed filling processes.[71]

A related consideration in metabolic engineering is the experimental determination of fluxes of metabolites within the cell.[72] This determination provides an unambiguous description of metabolism before and/or after engineering interventions. The elucidation of metabolic fluxes is important for a number of reasons. First, the set of fluxes through a cell's metabolic pathways is a key descriptor of its physiology and for evaluating mechanisms for metabolic engineering.[1,73] While gene transcripts, protein levels and metabolite concentrations tend to vary in a seemingly unpredictable fashion, metabolic fluxes might be more invariant though capturing effectively a cell's metabolic state. They provide a further means for pinpointing the effect of engineering interventions (*e.g.* knock-outs/-ins, up-/down-regulations) on cellular metabolism alluding to their effectiveness and suggesting additional engineering strategies.

7.4 EXAMPLES IN METABOLIC ENGINEERING

7.4.1 Metabolic Engineering of Plants

It is very attractive to use the production apparatus of plants for the synthesis of complicated chemical structures, especially if stereochemical properties are considered. Also, the use of solar energy and inorganic nutrients as a main requirement for plant growth might be a reason why metabolic engineering of plants is employed in a broader range.

Alkaloids are a structurally diverse class of nitrogenous compounds that are found in many plants and often exhibit physiological activity. Throughout history, plants that produce alkaloids and their extracts have been exploited for their medicinal and toxic properties. Modern examples of widely used plant-derived alkaloids include analgesics (morphine and codeine), stimulants (caffeine and nicotine) and chemotherapeutics (vincristine, vinblastine, camptothecin derivatives and paclitaxel). While most alkaloids are formed from amino acids such as phenylalanine, tyrosine, tryptophan, ornithine and arginine, they can be derived from a number of substrates (*e.g.* purines for caffeine). In plants, over 12 000 alkaloid structures have already been elucidated,[74] providing drug companies with a diverse set of structures valuable for pharmacological screening.[75]

Successful characterisation efforts focused on the following: nicotine of *Nicotiana*; the tropane alkaloids of *Hyoscyamus*, *Datura* and *Atropa*; the isoquinoline alkaloids of *Coptis* and *Eschscholtzia californica*; and the terpenoid indole alkaloids of *Catharanthus roseus*. As more genes are cloned, it becomes feasible to engineer plant systems for the production of valuable secondary metabolites.

Most of the efforts, however, have not been focused on genetically engineered cell cultures, tissue cultures or plants, but rather on non-engineered cell lines studied under elicitation, with precursor feeding or in optimized media. Although these classical methods have hastened characterisation efforts, metabolic engineering offers the most promising method for improved product composition and increased alkaloid yields of plants and cultured cell systems. Chemical synthesis remains prohibitively expensive due to the existence of multiple chiral centres, and the length and incomplete characterisation of the pathways required for alkaloid production make the use of alternative systems such as microorganisms impossible (for a review, see Ref. 76).

Amino acids represent important targets for metabolic engineering for a number of reasons. Some amino acids (such as proline, arginine, methionine and glutamate) are directly or indirectly involved in the regulation of plant responses to various environmental signals, including light and mineral availability and also biotic and abiotic stresses. Other amino acids (such as lysine, threonine, methionine and tryptophan) contribute significantly to the nutritional quality of plant-based foods. These 'essential' amino acids cannot be synthesized by humans and have to be supplied in the diet. Although the regulation of amino acid metabolism in higher plants may be analogous to that in microorganisms, the multicellular and multiorgan nature of higher plants introduces additional levels of complexity that render metabolic fluxes much more difficult to predict and engineer. Understanding the regulation of metabolic fluxes is also of particular importance because in many cases metabolic engineering needs to be targeted to specific organs (for a review, see Ref. 77).

Stress resistance of plants is also considered a topic of metabolic engineering. Drought and salinity, for example, are major limitations on crop productivity and quality globally. Increasing the resistance of crops to these osmotic stresses was one of the first objectives of plant metabolic engineering and remains a major goal today. One way in which many plants and other organisms cope with osmotic stress is to synthesize and accumulate compounds termed osmoprotectants (or compatible solutes). These are small, electrically neutral molecules that are non-toxic at molar concentrations and stabilize proteins and membranes against the denaturing effect of high concentrations of salts and other harmful substances.

In dry or saline environments, osmoprotectants can serve both to raise the cellular osmotic pressure and to protect cell constituents. Their protective effects also extend to temperature extremes and other stresses. Chemically, osmoprotectants are of three types: betaines (fully *N*-methylated amino acid derivatives) and related compounds such as dimethyl sulfoniopropionate (DMSP) and choline-*O*-sulfate, certain amino acids such as proline and ectoine and polyols and non-reducing sugars such as trehalose and raffinose. To investigate the basic mechanisms of such abiotic stress responses, a plethora of plant mutants exists, paving the way for metabolic engineering or stress-resistant plants.[78]

7.4.2 Acetate Metabolism and Recombinant Protein Synthesis in *E. coli* – a Test Case for Metabolic Engineering

Single-cellular organisms are the most common used in biotechnological processes, they are easy to cultivate and an increase in biomass can easily be achieved. New pathways have been introduced and established metabolic pathways modified; those studies will not be reviewed in greater detail here.

Whereas on the one hand it is desirable to modify metabolic pathways for the production of target compounds, it is also useful to modulate the production of compounds which compete for internal recourses.

The culture of *E. coli* for the commercial production of recombinant proteins has increased significantly in recent years.[79] Acetate as a by-product of fermentation is undesirable because it retards growth even at concentrations as low as $0.5\,\mathrm{g\,L^{-1}}$,[80] and it inhibits protein formation. Moreover, acetate production represents a diversion of carbon that might otherwise have generated biomass or protein product.[81] There have been a number of attempts to apply metabolic engineering to reduce carbon flow to acetate-producing pathways.[82–84] The reduction of acetate production in *E. coli* led to a more than 10-fold increase in recombinant protein production.[85] This example demonstrates that knowledge of the metabolic properties of the target organism is the key for improving biotechnological processes.

7.4.3 Metabolic Flux Analysis and a Bioartificial Liver

In the light of the tremendous shortage of donor livers for transplantation, the bioartificial liver (BAL) assist device provides a viable alternative for treating patients with fulminant hepatic failure.[86]

This device employs primary hepatocytes or hepatoma cell lines to provide a whole complement of anabolic, catabolic, detoxifying and secretory liver-specific functions.[87,88]

Since the cells in these systems are exposed to plasma during clinical use,[89] understanding the effect of plasma on their metabolic state is essential to optimise effectively and rationally the BAL.

One method that can help gain a better insight into cellular functions and metabolism is metabolic flux analysis (MFA), which is related to FBA (see above). MFA relates experimental measurements of substrate inputs and product outputs to a steady-state stoichiometric model to determine intracellular flux distributions between metabolite pools. An MFA model has previously been developed to investigate flux distributions in hepatocyte cultures.[90]

7.5 OMICS TECHNOLOGIES OPEN NEW PERSPECTIVES FOR METABOLIC ENGINEERING

The complexity of genome organisation – structural diversity, gene duplication and redundancy – inherently implies that molecular phenotypes are not a phenomenon that can be understood in the context of single gene expression, but rather as the output of gene interaction networks.[91] Consequently, interaction networks are best determined by multiparallel measurement of transcripts, proteins and metabolites. These interactions can be viewed as correlation networks. However, correlations *per se* contain no information on causality.[92] Nevertheless, correlation of gene and protein expression analysis and the resulting metabolic phenotype correspond well with our understanding of causality, for instance the genotype–phenotype relationship.[93] From the statements above, it is evident that co-regulation and causal connectivity can be best described if variables of different levels are analysed in an integrative data matrix.

The comprehensive profiling of biological samples requires both statistical and novel data-mining tools to reveal significant correlations. It is further enhanced by profound studies on theoretical metabolic networks (see above).[94–96,2,14] Most of these approaches can be divided into the following classes: (i) studies on network topology and properties based on genome-predicted theoretical reaction pathways and/or regulatory gene networks (see Figure 7.1), (ii) measuring biochemical networks such as protein association, gene, protein and metabolite correlation and co-regulation,[97] and finally (iii) combining experimental data with theoretical modelling (see Figure 7.1). System structures are defined with reference to gene annotation or pathway, gene and protein databases. Comprehensive invasive investigations such as MS-based protein–protein inter-association analyses are also used. The modelling of metabolic pathways is complicated by inherently complex cellular and regulatory structures and the gap of knowledge concerning genome organisation.[98] Not all possible pathways and enzymatic reactions are currently known and it will take years to elucidate functions of unknown and putative proteins in genomes. Consequently, the models are fragmentary. The presence and absence of pathways under various conditions has to be considered as a major question.[99,100]

Many experimental and modelling approaches are conclusive for accessible systems such as *E. coli* and yeast, but not easily applied in more complex systems such as plants. The hope is that results from these studies can help to elucidate gene functions in other species based on sequence homology and conserved protein domain structures and finally lead to a better understanding of metabolic network regulation as a groundwork for pathway manipulation.[32]

7.6 ACKNOWLEDGEMENT

The section on MCA was written by also consulting the very clear and concise description of MCA available at http://dbkgroup.org/mca_home.htm.

REFERENCES

1. J. E. Bailey, *Science*, 1991, **252**, 1668–1675.
2. S. Schuster, D. A. Fell and T. Dandekar, *Nat. Biotechnol.*, 2000, **18**, 326–332.
3. S. Schuster, S. Klamt, W. Weckwerth, F. Moldenhauer and T. Pfeiffer, *Bioprocess Biosyst. Eng.*, 2002, **24**, 363–372.
4. D. Garfinkel and B. Hess, *J. Biol. Chem.*, 1964, **239**, 971.
5. T. A. Rapoport, R. Heinrich and S. M. Rapoport, *Biochem. J.*, 1976, **154**, 449–469.
6. A. Werner and R. Heinrich, *Biomed. Biochim. Acta*, 1985, **44**, 185–212.
7. A. Joshi and B. O. Palsson, *J. Theor. Biol.*, 1989, **141**, 515–528.
8. A. Joshi and B. O. Palsson, *J. Theor. Biol.*, 1990, **42**, 41–68.
9. M. Rizzi, M. Baltes, U. Theobald and M. Reuss, *Biotechnol. Bioeng.*, 1997, **55**, 592–608.
10. P. J. Mulquiney, W. A. Bubb and P. W. Kuchel, *Biochem. J.*, 1999, **342**, 567–580.

11. P. J. Mulquiney and P. W. Kuchel, *Biochem. J.*, 1999, **342**, 597–604.

12. E. E. Selkov, *Eur. J. Biochem.*, 1968, **4**, 79.

13. D. A. Fell, *Biotechnol. Bioeng.*, 1998, **58**, 121–124.

14. H. Kacser, J. A. Burns and D. A. Fell, *Biochem. Soc. Trans.*, 1995, **23**, 341–366.

15. R. Heinrich and T. A. Rapoport, *Eur. J. Biochem.*, 1974, **42**, 89–95.

16. D. A. Fell, *Biochem. J.*, 1992, **286**, 313–330.

17. H. M. Sauro and D. A. Fell, *Math. Comput. Modell.*, 1991, **15**, 15–28.

18. P. Mendes, *Computer Appl. Biosci.*, 1993, **9**, 563–571.

19. S. Schuster and C. Hilgetag, *J. Phys. Chem.*, 1995, **99**, 8017–8023.

20. S. Schuster and H. V. Westerhoff, *Biosystems*, 1999, **49**, 1–15.

21. S. Schuster, T. Dandekar and D. A. Fell, *Trends Biotechnol.*, 1999, **17**, 53–60.

22. T. Pfeiffer, I. Sanchez-Valdenebro, J. C. Nuno, F. Montero and S. Schuster, *Bioinformatics*, 1999, **15**, 251–257.

23. A. Varma and B. O. Palsson, *Appl. Environ. Microbiol.*, 1994, **60**, 3724–3731.

24. K. J. Kauffman, P. Prakash and J. S. Edwards, *Curr. Opin. Biotechnol.*, 2003, **14**, 491–496.

25. J. M. Lee, E. P. Gianchandani and J. A. Papin, *Brief. Bioinf.*, 2006, **7**, 140–150.

26. A. Hoppe, S. Hoffmann and H. G. Holzhutter, *BMC Syst. Biol.* 2007, **1**, 23–28.

27. C. H. Schilling, J. S. Edwards, D. Letscher and B. O. Palsson, *Biotechnol. Bioeng.*, 2000, **71**, 286–306.

28. W. Weckwerth, *Annu. Rev. Plant Biol.*, 2003, **54**, 669–689.

29. W. Weckwerth, K. Wenzel and O. Fiehn, *Proteomics*, 2004, **4**, 78–83.

30. J. K. Nicholson, J. C. Lindon and E. Holmes, *Xenobiotica*, 1999, **29**, 1181–1189.

31. J. K. Nicholson, J. Connelly, J. C. Lindon and E. Holmes, *Nat. Rev. Drug Discov.*, 2002, **1**, 153–161.

32. J. O. Castrillo and S. G. Oliver, *J. Biochem. Mol. Biol.*, 2004, **37**, 93–106.

33. R. Goodacre, S. Vaidyanathan, W. B. Dunn, G. G. Harrigan and D. B. Kell, *Trends Biotechnol.*, 2004, **22**, 245–252.

34. D. B. Kell, *Curr. Opin. Microbiol.*, 2004, **7**, 296–307.

35. W. B. Dunn and D. I. Ellis, *Trends Anal. Chem.*, 2005, **24**, 285–294.

36. H. Sauter, M. Lauer and H. Fritsch, *Abstr. Pap. Am. Chem. Soc.*, 1988, **195**, 129–AGRO.

37. O. Fiehn, J. Kopka, P. Dormann, T. Altmann, R. N. Trethewey and L. Willmitzer, *Nat. Biotechnol.*, 2000, **18**, 1157–1161.

38. U. Roessner, C. Wagner, J. Kopka, R. N. Trethewey and L. Willmitzer, *Plant J.*, 2000, **23**, 131–142.
39. U. Roessner, A. Luedemann, D. Brust, O. Fiehn, T. Linke, L. Willmitzer and A. R. Fernie, *Plant Cell*, 2001, **13**, 11–29.
40. W. Weckwerth, V. Tolstikov and O. Fiehn, *in Proceedings of the 49th ASMS Conference on Mass Spectrometry and Allied Topics*, American Society of Mass Spectrometry, Chicago, 2001, pp. 1–2.
41. C. Wagner, M. Sefkow and J. Kopka, *Phytochemistry*, 2003, **62**, 887–900.
42. C. D. Broeckling, D. V. Huhman, M. A. Farag, J. T. Smith, G. D. May, P. Mendes, R. A. Dixon and L. W. Sumner, *J. Exp. Bot.*, 2005, **56**, 323–336.
43. J. W. Webb, S. C. Gates, J. P. Comiskey and D. F. Weber, *Abstr. Pap. Am. Chem. Soc.*, 1986, **191**, 70–ANYL.
44. W. Weckwerth, M. E. Loureiro, K. Wenzel and O. Fiehn, *Proc. Natl. Acad. Sci. USA*, 2004, **101**, 7809–7814.
45. P. Jonsson, J. Gullberg, A. Nordstrom, M. Kusano, M. Kowalczyk, M. Sjostrom and T. Moritz, *Anal. Chem.*, 2004, **76**, 1738–1745.
46. O. Fiehn, J. Kopka, R. N. Trethewey and L. Willmitzer, *Anal. Chem.*, 2000, **72**, 3573–3580.
47. J. T. Watson, G. A. Schultz, R. E. Tecklenburg and J. Allison, *J. Chromatogr.*, 1990, **518**, 283–295.
48. T. Veriotti and R. Sacks, *Anal. Chem.*, 2001, **73**, 4395–4402.
49. S. E. Stein and D. R. Scott, *J. Am. Soc. Mass Spectrom.*, 1994, **5**, 859–866.
50. S. E. Stein, *J. Am. Soc. Mass Spectrom.*, 1999, **10**, 770–781.
51. C. S. Tong and K. C. Cheng, *Chemom. Intell. Lab. Syst.*, 1999, **49**, 135–150.
52. P. Schmitt-Kopplin and M. Frommberger, *Electrophoresis*, 2003, **24**, 3837–3867.
53. T. Soga, Y. Ohashi, Y. Ueno, H. Naraoka, M. Tomita and T. Nishioka, *J. Proteome Res.*, 2003, **2**, 488–494.
54. T. Soga, Y. Ueno, H. Naraoka, K. Matsuda, M. Tomita and T. Nishioka, *Anal. Chem.*, 2002, **74**, 6224–6229.
55. T. Soga, Y. Ueno, H. Naraoka, Y. Ohashi, M. Tomita and T. Nishioka, *Anal. Chem.*, 2002, **74**, 2233–2239.
56. S. Sato, T. Soga, T. Nishioka and M. Tomita, *Plant J.*, 2004, **40**, 151–163.
57. P. H. Gamache, D. F. Meyer, M. C. Granger and I. N. Acworth, *J. Am. Soc. Mass Spectrom.*, 2004, **15**, 1717–1726.

Chapter 7

58. B. S. Kristal, K. E. Vigneau-Callahan and W. R. Matson, *Anal. Biochem.*, 1998, **263**, 18–25.
59. K. Kaddurah-Daouk, Beecher, Kristal, Matson, Bogdanov, and Asa, *Pharmagenomics*, January, 2006, **1**, 46–52.
60. J. L. Josephs and M. Sanders, *Rapid Commun. Mass Spectrom.*, 2004, **18**, 743–759.
61. D. V. Huhman and L. W. Sumner, *Phytochemistry*, 2002, **59**, 347–360.
62. J. P. Shockcor, A. Nichols, H. Antti, R. S. Plumb, J. M. Castro-Perez, H. Major and S. Preece, *Drug Metab. Rev.*, 2003, **35**, 1–1.
63. L. Yang, M. Amad, W. M. Winnik, A. E. Schoen, H. Schwein-gruber, I. Mylchreest and P. J. Rudewicz, *Rapid Commun. Mass Spectrom.*, 2002, **16**, 2060–2066.
64. V. V. Tolstikov, A. Lommen, K. Nakanishi, N. Tanaka and O. Fiehn, *Anal. Chem.*, 2003, **75**, 6737–6740.
65. V. V. Tolstikov and O. Fiehn, *Anal. Biochem.*, 2002, **301**, 298–307.
66. E. von Roepenack-Lahaye, T. Degenkolb, M. Zerjeski, M. Franz, U. Roth, L. Wessjohann, J. Schmidt, D. Scheel and S. Clemens, *Plant Physiol.*, 2004, **134**, 548–559.
67. G. J. Dear, J. Ayrton, R. Plumb and I. J. Fraser, *Rapid Commun. Mass Spectrom.*, 1999, **13**, 456–463.
68. A. L. Duran, J. Yang, L. J. Wang and L. W. Sumner, *Bioinformatics*, 2003, **19**, 2283–2293.
69. B. Kenney and J. P. Shockcor, *Pharmagenomics*, 2003, **3**, 56–63.
70. P. May, S. Wienkoop, S. Kempa, B. Usadel, N. Christian, J. Rupprecht, J. Weiss, L. Recuendo-Munoz, O. Ebenhöh, W. Weckwerth, and D. Walther, *Genetics*, 2008, **179**, 157–166.
71. J. Schwender, F. Goffman, J. B. Ohlrogge and Y. Shachar-Hill, *Nature*, 2004, **432**, 779–782.
72. J. Schwender, J. Ohlrogge and Y. Shachar-Hill, *Curr. Opin. Plant Biol.*, 2004, **7**, 309–317.
73. G. Stephanopoulos and J. J. Vallino, *Science*, 1991, **252**, 1675–1681.
74. A. El-Shazly, M. Abdel-All, A. Tei and M. Wink, *Z. Naturforsch., Teil C*, 1999, **54**, 295–300.
75. R. Verpoorte, R. van der Heijden and J. Memelink, *Transgenic Res.*, 2000, **9**, 323–343.
76. E. H. Hughes and J. V. Shanks, *Metab. Eng.*, 2002, **4**, 41–48.
77. G. Galili and R. Hofgen, *Metab. Eng.*, 2002, **4**, 3–11.
78. A. Grover, A. Pareek, S. L. Singla, D. Minhas, S. Katiyar, S. Ghawana, H. Dubey, M. Agarwal, G. U. Rao and J. Rathee, *et al. Curr. Sci.*, 1998, **75**, 689–696.

79. M. Arbabi-Ghahroudi, J. Tanha and R. MacKenzie, *Cancer Metastasis Rev.*, 2005, **24**, 501–519.
80. K. Nakano, M. Rischke, S. Sato and H. Markl, *Appl. Microbiol. Biot.*, 1997, **48**, 597–601.
81. J. C. March, M. A. Eiteman and E. Altman, *Appl. Environ. Microbiol.*, 2002, **68**, 5620–5624.
82. J. Delgado and J. C. Liao, *Biotechnol. Prog.*, 1997, **13**, 361–367.
83. K. Hosono, H. Kakuda and S. Ichihara, *Biosci. Biotechnol. Biochem.*, 1995, **59**, 256–261.
84. K. Y. San, G. N. Bennett, A. A. Aristidou and C. H. Chou, *Recombinant DNA Technol.*, 1994, **721**, 257–267.
85. M. S. Wong, S. Wu, T. B. Causey, G. N. Bennett and K. Y. San, *Metab. Eng.*, 2008, **10**, 97–108.
86. J. H. Hoofnagle, R. L. Carithers, C. Shapiro and N. Ascher, *Hepatology (Baltimore, MD)*, 1995, **21**, 240–252.
87. F. D. Watanabe, C. J. P. Mullon, W. R. Hewitt, N. Arkadopoulos, E. Kahaku, S. Eguchi, T. Khalili, W. Arnaout, C. R. Shackleton and J. Rozga, *et al. Ann. Surg.*, 1997, **225**, 484–491.
88. M. L. Yarmush, M. Toner, J. C. Y. Dunn, A. Rotem, A. Hubel and R. G. Tompkins, *Ann. N. Y. Acad. Sci.*, 1992, **665**, 238–252.
89. S. L. Nyberg and S. P. Misra, *Mayo Clin. Proc.*, 1998, **73**, 765–771.
90. C. Chan, F. Berthiaume, K. Lee and M. L. Yarmush, *Metab. Eng.*, 2003, **5**, 1–15.
91. A. Wagner, *Nonlinearity*, 1996, **9**, 607–629.
92. A. Wagner, *Biol. Philos.*, 1997, **14**, 83–101.
93. M. Glinski and W. Weckwerth, *Mass Spectrom. Rev.*, 2006, **25**, 173–214.
94. J. A. Papin, N. D. Price, S. J. Wiback, D. A. Fell and B. O. Palsson, *Trends Biochem. Sci.*, 2003, **28**, 250–258.
95. E. Ravasz and A. L. Barabasi, *Phys. Rev., Feb.*, 2003, **67**, Issue 2.
96. R. Steuer, J. Kurths, O. Fiehn and W. Weckwerth, *Bioinformatics*, 2003, **19**, 1019–1026.
97. W. Weckwerth, *Physiol. Plant.*, 2008, **132**, 176–189.
98. W. Weckwerth and O. Fiehn, *Curr. Opin. Biotechnol.*, 2002, **13**, 156–160.
99. E. M. Marcotte, *Nat. Biotechnol.*, 2001, **19**, 626–627.
100. J. Ihmels, R. Levy and N. Barkai, *Nat. Biotechnol.*, 2004, **22**, 86–92.
101. J. Higgins, *Ind. Eng. Chem.*, 1967, **59**, 19–62.

CHAPTER 8
Bionanotechnology

DAVID W. WRIGHT

Department of Chemistry, Vanderbilt University, Nashville TN 37235, USA

8.1 INTRODUCTION

Today, researchers find themselves involved in cross-disciplinary science that a decade ago was simply inconceivable. Nowhere is this more apparent than at the cusp of two rapidly developing fields, nanoscience and biotechnology. But what is nanotechnology? The prefix nano means a billionth (1×10^{-9}). Nanotechnology is the study and application of unique structures having dimensions on the order of a billionth of a meter that exhibit novel size-controlled electronic, optical or catalytic properties. These structures can be metallic, semiconductor or magnetic nanoparticles, nanowires or nanotubes. The underlying basis for such nanoscale effects is that every property of a material has a characteristic and critical length associated with it. The fundamental physics and chemistry of a material will change when the dimensions of a solid become comparable to one or more of these characteristic lengths, many of which exist at the nanometer length scale.

One of the challenges of understanding nanotechnology is the vocabulary. The nanoscale is populated by a diverse group of players. If only one length of a three-dimensional structure is of a nanodimension, the structure is known as a quantum well. When a material exists on the nanoscale along two sides, the structure is referred to as a nanowire. A quantum dot has all three dimensions in the nanometer range. The

Molecular Biology and Biotechnology, 5th Edition
Edited by John M Walker and Ralph Rapley
© Royal Society of Chemistry 2009
Published by the Royal Society of Chemistry, www.rsc.org

assembly of these structures into hierarchical assemblies relies on both physical (*e.g.* lithography, scanning probe microscopy, electrophoretic strategies, ball milling or Langmuir–Blodgett films) and chemical methods (*e.g.* interparticle electrostatic interactions, covalent coordination, template recognition with subsequent crosslinking or crystal engineering).[1] Although effective for the preparation of certain nano-scale architectures, many of the physical methods are limited because they tend to be slow, have a large infrastructure cost and do not lend themselves to preparing designed nanostructures which span the macroscopic dimension. In contrast, the advantages of chemical methods are that building blocks may be linked in a massively parallel fashion. This is particularly useful for the rapid construction of two- or three-dimensional structures. Unfortunately, current chemical methodologies, especially when compared with the above physical methods, are difficult to control. Increasingly, there has been interest in the use of biomolecules to overcome this challenge.[2]

It is not surprising, then, to find that nanotechnology is interested in the realm of biology, for they share the nanoscale (Figure 8.1). Consider the tendon, whose function it is to attach muscle to bone. The principal building block of the tendon is the assemblage of amino acids (~ 0.6 nm) that form the gelatin-like protein collagen (~ 1 nm), that coils into a left-handed triple helix (~ 2 nm). These individual helical proteins then assemble into a fibrillar nanostructure in which collagen assembles to form microfibrils (~ 3.5 nm), subfibrils (10–20 nm) and fibrils (50–500 nm). These fibers then form clusters of mesoscopic fibers called a fascicle (50–300 μm) and, finally, the macroscopic tendon itself (10–50 cm).[3] A number of research groups are actively focusing on the use of biomolecules to direct the formation of nanostructures and extended nanoscale assemblies due to the inherent molecular recognition of such molecules.

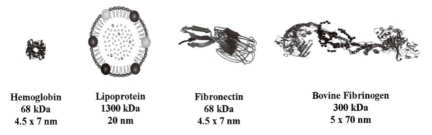

Hemoglobin	Lipoprotein	Fibronectin	Bovine Fibrinogen
68 kDa	1300 kDa	68 kDa	300 kDa
4.5 x 7 nm	20 nm	4.5 x 7 nm	5 x 70 nm

Figure 8.1 Many proteins have dimensions and molecular weights that place them in the nano-regime.

In an emergent field such as nanobiotechnology, it is difficult to predict the ultimate achievements of the field. Today, the discipline is making significant progress in four broad areas: separations, imaging/diagnostics, drug delivery and synthesis of new materials. As a testament to the true interdisciplinary nature of these endeavors, each of these areas is being driven by rapid advances in the other. Consequently, the resulting hybrid of bionanotechnology holds the promise of providing revolutionary insight into the many aspects of biology. However, it also represents a substantial challenge. Biological systems have been making functional nanoscale devices since the beginning of life and there is much to learn from biology about how to build nanostructured materials. Yet, how can a life scientist who is trying to develop new gene transfer methods, but who does not know the difference between a buckyball and a II–VI semiconductor core–shell quantum dot, gain entrée into the field of nanotechnology? The purpose of this chapter is to provide such an entrée – a glimpse into the potential impact of these developments. Another obstacle to overcome is how to present the numerous, almost daily changing aspects of this field. This chapter will introduce the field of bionanotechnology through one of its most successful elements, biomolecular nanoparticle (NP) systems. Thus, while the specifics of bionanotechnology go far beyond the examples discussed here, the themes of nanoscale effects, biomolecular interfaces and emerging real-world applications are all well represented.

8.2 SEMICONDUCTOR QUANTUM DOTS

8.2.1 Quantum Confinement Effects

Quantum dots (qdots) are nanometer-sized crystalline clusters (1–10 nm) made from a variety of semiconductor materials. Qdots are characterized by large absorption spectra, but narrow and very symmetric emission bands with full widths at half-maximum of 25–35 nm. The emission bands can span the spectrum from the ultraviolet to the infrared (365–1350 nm) depending on the size of the dot. These materials typically have a large absorption cross-section and long fluorescence lifetimes (> 10 ns). Further, they possess a photostability many orders of magnitude greater than conventional organic fluorophores. Given their advantages relative to organic fluorophores, qdots have emerged as a new class of fluorescent probes for biological applications.[4]

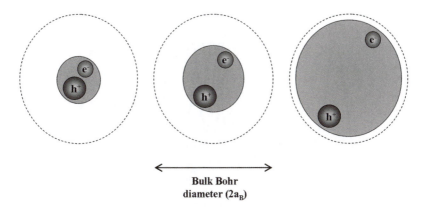

**Bulk Bohr
diameter (2a_B)**

Figure 8.2 Quantum confinement. As the size of a nanocrystal increases the bandgap
decreases, similar to the classic particle in a box.

The origin of the unique size-dependent optical and electronic properties of semiconductor nanoparticles arises from the particle's nanoscale structure. When a photon of sufficient energy $[hv > E_g$ (bandgap)] is
absorbed by a semiconductor, an electron is excited out of the valence
band into the conduction band, creating an electron–hole pair. In bulk
semiconductors, the average distance between the electron and the hole
is called the bulk Bohr radius, a_B. As the size of the crystal decreases to
approach the bulk Bohr radius, the energy levels become discrete and
boundary conditions are imposed on the wavefunctions of the two
charges. This condition results in the phenomena of quantum confinement (Figure 8.2).

To a first approximation, quantum confinement exhibited in nano-
crystals follows behavior similar to that of the classic particle-in-a-box
problem. For a particle in a box, the energy difference between the
conduction and the valence bands is related to the inverse square of the
nanocrystal radius:

$$E_n \approx \frac{h^2 n^2}{\varepsilon m R^2}, \ n = 1, \ 2, \ 3 \tag{1}$$

where h is Planck's constant (6.626×10^{-34} J s), m is mass and R is the
nanocrystal radius. As the size of the nanocrystal increases, the bandgap
narrows and both the absorption and fluorescence wavelengths shift
towards the red (Figure 8.3).[5] Capitalizing on this quantum confinement
permits exceptional tunability of absorption and emission wavelengths
by simply changing the size of the nanocrystal.

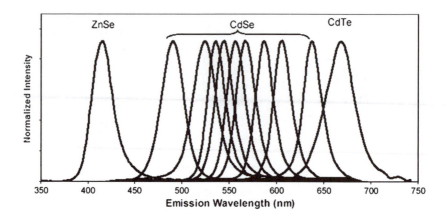

Figure 8.3   Emission spectra of semiconducting quantum dots. The variable CdSe spectra illustrate the blue to red shift that occurs with increasing particle size.

8.2.2   Biotechnological Applications of Fluorescent Semiconductor Quantum Dots

With commercial sources of qdots available and a growing variety of surface chemistry strategies for functionalization, the successful application of qdots to a diverse range of imaging challenges has grown. These include such applications as immunofluorescence assays, biotechnology detection, live cell imaging, single-molecule biophysics and *in vivo* animal studies. Although qdots represent a powerful tool for some imaging applications, it must be remembered that a square peg will not fit into a round hole. The examples below highlight some of the most suitable applications of qdots and discuss some of the challenges that researchers are currently facing.

The broadband adsorption and narrow, symmetrical and size-tunable emission bands of qdots facilitate their use for multiplexed detection of signals. The important consequence of this broad absorption is that one wavelength of light can excite multiple qdots, each with different emission maxima. Highly robust commercial preparations of CdSe/ZnS qdots, emitting in the visible spectrum, match the detection range of many typical imaging devices. Further, with the elimination of many chromatic aberration and alignment issues encountered with standard fluorescence microscopy, colocalization studies are possible.[6] A recent example of dynamic multicolor imaging can be seen in the visualization of viral proteins in the membrane of an infected host cell.

Benzten *et al.* directly labeled both the F (fusion) and G (attachment) proteins of respiratory syncytical virus (RSV) using qdots with the primary antibody attached covalently to the surface. Using confocal laser scanning microscopy, the F and G proteins were co-localized on the surface of the infected cells. The sensitivity of detection of the F protein with the qdots as a function of plaque-forming units (PFUs) showed a linear response at 18 h over a range of 35–110 PFU per well (multiplicity of infection 0.0032). After longer periods of time (360–42 h), lower rates of infection could be detected as a result of the inherent viral replication in culture. In terms of absolute sensitivity, the F and G proteins could be detected as early as 1 h post-infection, on a par with the most sensitive RT-PCR methods.[7]

There are several potential applications arising from the increased photostability of qdots relative to organic dyes. High fluorescence photostability allows for the repeated imaging of immunostained samples that will maintain their crisp, high-resolution three-dimensional definition. For example, pathology samples or culture assays can be readily reviewed without worry about loss of signal that is a limiting factor with organic dyes. The photostability is also a clear advantage in live cell imaging experiments, where cells or single molecules need to be monitored over hours or days. Dubertret *et al.* demonstrated cell lineage tracing by injecting qdots in a single *Xenopus* frog cell during an early embryonic stage and followed their fate during subsequent development over days.[8] Similarly, the lateral dynamics of qdot-labeled glycine receptors has been monitored in neuronal membranes as a function of time.[9]

Qdots have also been surprisingly successful for *in vivo* animal imaging in a number of species. These deep tissue studies have imaged lymph nodes, vascular markers, blood vessels and grafted tumors.[6] The utilization of a polyethylene glycol (PEG) co-ligand is a common approach in all of these studies to enhance circulation time and reduce non-specific binding. With the advent of NIR CdTe/CdSe qdots, lymph nodes were imaged 1 cm deep in tissue.[10] As synthetic advances in the production and functionalization of both NIR and IR qdots improve, the properties of these probes combined with time-gated microscopy methods, to reduce background autofluorescence, may result in detection sensitivities that rival those of radiolabeled probes.

Although the above examples highlight the promise of qdot applications as biological probes, it is important also to mention some of their challenges. Foremost, is the difficulty of targeting these probes to the cytoplasm. Despite some interesting attempts using membrane translocating peptides, electroporation or transfection reagents, qdots tend to accumulate in vesicles or appear non-homogeneously distributed in the cytoplasm.[11,12] Currently, there has been no real success in overcoming this technique for the convenient imaging of cytosolic targets.

Another application that has not met with complete success due to the inherent properties of qdots is the development of FRET sensors. The narrow and tunable emission characteristics of qdots can be exploited to customize donor emission in fluorescence (Föster) resonance energy transfer (FRET) assays between qdot donors and fluorescent organic dye acceptors.[13] Such assays convey only qualitative information about molecular associations in ensemble measurements. Unfortunately, there are several hurdles to be cleared before more quantitative measurements can be made by FRET. The first challenge is that minor variations in the surface defects of individual qdots can introduce significant spectral heterogeneity.[14] The second challenge is the environment-dependent intermittency of qdot fluorescence – the so-called blinking phenomenon. Qdot blinking is associated with charge trapping and untrapping at surface defect sites resulting in bright and dark states.[15] Such blinking results in the random loss of distance information on all time scales and can impact the energy transfer efficiency. With improvements in synthetic strategies to eliminate blinking and improve spectral homogeneity, qdots may eventually become effective FRET nanoscale biosensors.

8.3   MAGNETIC NANOPARTICLES

8.3.1   Nanoscaling Laws and Magnetism

Biological examples of magnetic nanoparticles were first discovered in magnetotactic bacteria. These unique iron oxide magnetites are used by these organisms to orient themselves into an optimal environment. Magnetic nanoparticles have also been found in the nasal capsules of salmon. These particles are believed to respond to the geomagnetic field of the Earth, providing the salmon with assistance in reaching their spawning grounds. Similar nanostructures have also been found in the brains of homing pigeons, sea turtles and even humans.[16]

Scientists have developed artificial magnetic nanoparticles through chemical synthesis. Recent advances have provided new types of magnetic nanoparticles with precisely tuned compositions, shapes and size. As with many nanostructures, a number of interesting phenomena have been observed in these particles that are distinct from their bulk properties. In the bulk, the fundamental magnetic properties of coercivity (H_c) and susceptibility (χ) are determined principally by the parameters of composition, crystallographic structure, magnetic anisotropy and vacancies and defects. Yet, when magnetic materials are reduced in size to the nanoscale regime, these fundamental properties are no longer

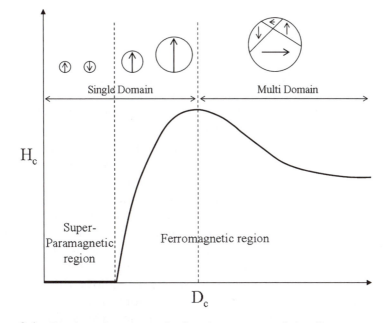

Figure 8.4 Size-dependent magnetic domain structures. Spin alignment occurs in single domain particles having sizes below a critical size (D_c). Reproduced by permission of The Royal Society of Chemistry.

permanent. Furthermore, the size, shape and compositional structure become important determinants of the magnetic properties.[17]

One of the interesting size-dependent phenomena of magnetism on the nanoscale is the observed changes in magnetic coercivity. Whereas bulk magnets contain multiple magnetic domain structures, nanoparticles possess single domain magnetic structures below a specific critical size (D_c). Below this size, all of the magnetic spins in the nanoparticles align unidirectionally (Figure 8.4). The magnetic coercivity has been shown to increase as the size of the nanoparticles increases with the relationship

$$H_c = \frac{2K_u}{m_s}\left[1 - 5\left(\frac{kT}{K_u V}\right)^{\frac{1}{2}}\right] \qquad (2)$$

where m_s is the saturation magnetization.[18]

The saturation magnetization of the particles themselves is also strongly dependent on the size. In bulk materials, the intrinsically disordered magnetic spin layers near the surface are negligible, since the surface layer is minimal compared with the entire volume of the magnet. Yet, on the nanoscale, the disordered surface effects can be dramatic as they now represent a significantly larger fraction of the total volume.

This size effect follows the relationship described as

$$m_s = M_s \left(r - \frac{d}{r} \right)^3 \qquad (3)$$

where r is the size, M_s is the saturation magnetization of bulk material and d is the thickness of the disordered surface layer.[19]

This effect can be seen in the case of magnetism-engineered iron oxide (MEIO), Fe_3O_4 nanoparticles. As the size of the MEIO nanoparticles increases over a range 4–12 nm, the mass magnetization values increase linearly as a plot of $m_s^{1/3}$ *versus* r^{-1}. Such size-dependent properties directly affect their magnetic resonance (MR) signal enhancement capabilities for molecular imaging methodologies.[20]

The crystallinity of the magnetic nanoparticles plays an important role in determining its magnetic coercivity.[21] This is not surprising when one considers how the degree of crystallinity and the organization that the crystal lattice must impart on the interactions of spin within the nanoparticles. This can be seen in magnetic nanoalloys with anisotropic crystalline structures.[22] Co–Pt core–shell nanoparticles composed of an isotropically structured face-centered cubic (fcc) Co core and a non-magnetic Pt shell displays superparamagnetic behavior with zero coercivity at room temperature. When the particle is annealed, the resulting CoPt nanoalloy adopts a crystalline face-centered tetragonal (fct) structure with room temperature ferromagnetic behavior and a coercivity value of 5300 Oe.

The use of magnetic dopants to modify the composition of nanoparticles can provide access to tunable magnetization.[23] MEIO particles, Fe_3O_4, have a ferrimagnetic spin structure. In the fcc-packed oxygen lattice, Fe^{2+} and Fe^{3+} ions occupying octahedral (O_h) sites have spin that aligns parallel to an external magnetic field (Figure 8.5), whereas Fe^{3+} ions occupying tetrahedral (T_d) sites have spins that align antiparallel to the field. Since high-spin Fe^{3+} and Fe^{2+} have a d^5 and d^6 electron count, respectively, the total magnetic moment per unit $(Fe^{3+})T_d$ $(Fe^{2+}Fe^{3+})O_hO_4$ is approximately $4\,\mu_B$. Incorporation of an M^{2+} (M = Mn, Co, Ni) magnetic dopant with electronic configurations of d^5, d^4 and d^3, respectively, at the O_h Fe^{2+} site result in a predictable change in the net magnetization to 5, 3 and $2\,\mu_B$, respectively (Figure 8.6).

8.3.2 Biotechnological Applications of Magnetic Nanoparticles

Magnetic particles have shown considerable promise as probes for magnetic resonance imaging, as they provide strong contrast effects to

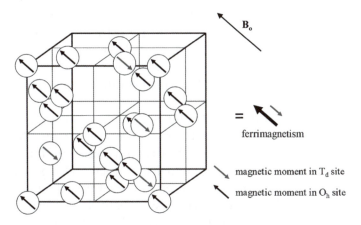

Figure 8.5 Ferrimagnetic spin structure of fcc-packed Fe_3O_4 lattices. Reprinted with permission from Jun *et al.*, *J. Am. Chem. Soc.*, 2008, **41**, 179. Copyright 2008 American Chemical Society.

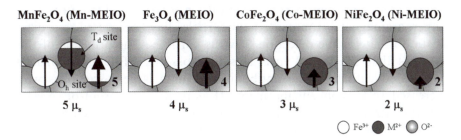

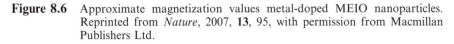

Figure 8.6 Approximate magnetization values metal-doped MEIO nanoparticles. Reprinted from *Nature*, 2007, **13**, 95, with permission from Macmillan Publishers Ltd.

surrounding tissues. These effects can be understood in terms of their influence on the spin–spin relaxation times (T_2) of surrounding water molecules. Conventional iron oxide contrast agents such as the super-paramagnetic iron oxides or the cross-linked iron oxides are of limited utility due to their poor magnetic contrast effects. New nanoparticles systems such as MEIO nanoparticles have high and tunable mass magnetization values capable of enhancing T_2 relaxation times.

The importance of nanoscaling laws in designing optimal MEIO systems can be seen in the size dependence of the relaxivity coefficient (r_2), a direct indication of contrast enhancement. In 4 nm MEIO particles, the relaxivity coefficient is $78 \, mM^{-1} \, s^{-1}$, but increases to 106, 130 and $218 \, mM^{-1} \, s^{-1}$ for 6, 9 and 12 nm nanoparticles, respectively.[20] Using dopants, the efficacy of these particles can also be tuned by changes in their composition. A 12 nm Mn-doped MEIO with the highest magnetization values of 110 emu g^{-1} (Mn + Fe) exhibits the best MR contrast with an r_2 of 358 mM^{-1} s^{-1}. Other metal-doped MEIOs have values of 101 emu g^{-1} (Fe) for all Fe, 99 emu g^{-1} (Co + Fe) for Co-doped and 85 emu g^{-1} (Ni + Fe) for Ni-substituted with r_2 values of 218, 172 and 152 mM^{-1} s^{-1}, respectively.[23]

Given that the r_2 of the Mn–MEIO particle is six times higher that of most conventional molecular MR contrast imaging agents, these properties have been translated into effective enhancements for the ultra-sensitive detection of *in vivo* biological targets. Using Mn–MEIO particles conjugated with Herceptin and injected into the tail vein of a mouse, a small HER2/nue cancer was selectively detected by MRI imaging. In contrast, the same tumor was undetectable using convention cross-linked iron oxide–Herceptin conjugate.[23]

Although magnetic beads have long been widely used for magnetic-based sensing and separations, problems still exist with these methods due to low magnetic susceptibility and considerable magnetic inhomo-geneity. Magnetic nanoparticles can offer a solution to these problems. Superparamagnetic iron oxide nanoparticles have been used in a diag-nostic format for the detection of virus particles. The principle of this assay is that the target virus acts to cross-link SPIO nanoparticles derivatized with targeting antibodies, forming larger assemblies with increased relaxation enhancements (Figure 8.7). In proof-of-concept experiments, the formation of the virus aggregated nanoassemblies was confirmed by light scattering experiments. When the samples were examined by MRI, the virus particles were detected in concentrations as low as five particles in 10 μl of biological sample. It is likely that

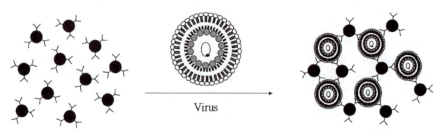

Magnetic viral nanosensor   Viral-induced nanoassembly

Figure 8.7   Antibody-functionalized nanoparticles induce the formation of nano-assemblies in the presence of a virus. This allows for viral detection by measuring the change in spin–spin relaxation times of surrounding molecules brought about nanoassembly formation.

detection limits could be further enhanced with a shift to MEIO particles and/or greater magnetic field strength.[24]

Magnetic nanoparticles can also enhance magnetophoretic separations. When a magnetic field is applied perpendicular to the flow direction of a microfluidic channel, magnetic particles experience a magnetic force which drives their lateral movement to a given velocity. The velocity of that magnetophoretic lateral movement (v_{lat}) is proportional to the magnetic susceptibility of the particle and the square of the particle radius. The magnetic susceptibility of the nanoparticles is a key component of velocity control and consequently the achievement of good magnetophoretic separations.[25]

Allergy tests require the quantitative detection of allergen-specific antibodies (IgE) in the serum of a patient suffering from allergies. Unfortunately, the IgE concentrations in allergy patients are usually low, driving the need for highly sensitive detection methods. To demonstrate the efficacy of magnetophoretic separation and sensing of dust mite IgE antibodies, microbeads coated with mite allergen from *Dermotophagiodes farina* were first mixed with target IgE (Figure 8.8). To this solution, secondary anti-human IgE-coated MEIO nanoparticles were added. The resulting solution was injected into the microchannel of a magnetophoretic separator. At high concentrations of the target IgE, significant lateral movement of the beads was achieved ($v_{lat} = 15\,\mu\text{m s}^{-1}$). At lower concentrations of target, reduced ($v_{lat} = 2\,\mu\text{m s}^{-1}$) or negligible lateral movement was detected. This is consistent with the idea that in this nanohybrid sandwich assay, higher concentrations of specific IgE result in more MEIO nanoparticles binding on to the microbeads. In sera, quantitative detection of target IgEs was achieved at the sub-picomolar levels ($\sim 500\,\text{fM}$) when using a target IgE concentration *versus* lateral velocity calibration curve.[26]

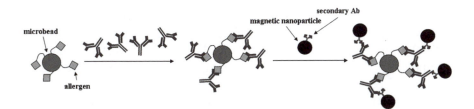

Figure 8.8 Sandwich assay demonstrating the detection and separation of specific allergen antibodies using antibody-functionalized magnetic nanoparticles. Reprinted with permission from Jun *et al.*, *J. Am. Chem. Soc.*, 2008, **41**, 179. Copyright 2008 American Chemical Society.

Researchers are beginning to achieve a better understanding of the nanoscaling laws for magnetism. From these studies, it is apparent that size, shape and composition have a tremendous impact on magnetic parameters such as coercivity and magnetization values. Using these tunable properties, magnetic particles have begun to find their way into a variety of biotechnology applications. Research and discovery will lead to continued improvements in MRI, biosensing and magnetic separations and new advances in next-generation drug delivery and hyperthermia treatments.

8.4 ZEROVALENT NOBLE METAL NANOPARTICLES

8.4.1 Nanoscale Properties of Zerovalent Nobel Metal Nanoparticles

For centuries before the terms nanotechnology and nanoparticles were coined, zerovalent noble metal nanoparticles had been known and studied. Many of the beautiful colors of medieval stained-glass windows are the result of small metal oxide nanoparticles in the glass. Particles of different size scatter different wavelengths of light, imparting the different colors to the glass.[27] The photoreduction of silver salts to small colloidal particles has long been a part of the process of image formation in photography.[28] With the growth of bionanotechnology, noble metal nanoparticles have been found to be indispensable building blocks in a variety of systems, ranging from scaffolds for antigen mimics, electron transfer mediators in enzyme-based biosensors to integral components in DNA detection.

AuNPs and AgNPs demonstrate a variety of interesting nanoscale size- and shape-dependent tailoring of their physical and chemical properties making them highly tunable components in catalytic, sensor and biological imaging applications. There has been considerable interest in resonant Rayleigh scattering from AuNPs and AgNPs. Specifically, light scattering from silver nanoparticles made by lithographic or colloidal techniques has been used extensively for biological and chemical analysis.[29] The utility of these plasmon-resonant particles (PRPs) stems from the sensitivity of the localized surface plasmon resonance (LSPR) to local chemical environment and refractive index. These surface plasmon waves are surface electromagnetic waves that propagate parallel along a metal/dielectric interface. Since the wave is on the boundary of the metal and the external medium (air or water for example), these oscillations are very sensitive to any change of this boundary, such as the adsorption of molecules to the metal surface.

Light scattering from PRPs is 10^6 times more intense than fluorescence emitted from commercially available fluorophores and PRPs do not suffer from photodegradation.[30] Moreover, intense scattering facilitates the detection of particles, typically of 30–100 nm diameter, using commercially available optical microscopes, which adds to the popularity of this field.

In discussing shape- and size-dependent optical properties of metal nanoparticles, it is important to note that nanoparticle extinction has both light absorption and scattering components, which are size and shape dependent and may or may not have the same wavelength dependence.[31] Theoretical experiments by Schatz and co-workers[32] show the same wavelength dependence for both extinction and scattering for small spherical gold nanoparticles (30 nm diameter). For larger diameter nanoparticles (*e.g.* 100 nm), scattering is enhanced, relative to absorption in the red region of the spectrum.

Size and shape dependence on light scattering for AgNPs has been reported. In general, as silver nanoparticles increase in size, their scattering spectra are red shifted.[33] Moreover, changes to the shape of AgNPs has a significant effect on their observed light scattering spectrum. Silver spheres have a maximum scattering peak at ~400 nm and pentagons at ~500 nm, and triangles have a peak maximum further red shifted to 750 nm, all of which are size dependent. In addition, the observed light scattering spectra for triangular-shaped particles are blue shifted as the corners of the triangles go from sharp and well-defined to truncated or rounded.[33,34] With these tunable properties, AuNPs and AgNPs are highly functional components of optical labeling schemes for a wide array of biological systems.

On the nanoscale, noble metal nanoparticles have unique electrical properties that can be used in biosensing applications.[35] The ability of AuNPs to store charge as nanoscale capacitors was originally demonstrated by Murray's group by observing quantized double-layer (QDL) charging peaks for monodisperse AuNPs referred to as gold monolayer protected clusters (MPCs).[36] QDL charging results from a single electron transfer into or out of the AuNP metallic core through the thiol shell. QDL peaks can be observed for monodisperse MPCs when this single electron transfer causes a change in the MPC potential. Improvements in techniques for obtaining monodisperse MPCs have enhanced the ability to observe sequential QDL peaks, especially for hexanethiol gold MPCs.[37] It is very likely that this same quantitated behavior can be explained, in concept, by a similar 'particle-in-a-box' model as has been done for semiconductor quantum dots.

8.4.2   Bionanotechnology Application of Zerovalent Noble Metal Nanoparticles

The 'wiring' of redox enzymes to electrodes is the basis of many amperometric biosensors.[38] Despite many proof-of-concept demonstrations, the inherent insulation of the protein shell prevents direct electric contact between the enzyme active site and the electrode, limiting efficient electron transfer and ultimately sensitivity. Gold nanoparticles (AuNP) can be used as nanoelectrodes effectively to shorten electron transfer distances and mediate charge transport. In such a scheme, an AuNP is connected to the electrode with a dithiol bridge moiety and a cofactor from the target enzyme. When incubated with an apoenzyme, the apoenzyme binds the cofactor-functionalized nanoparticle forming a haloenzyme oriented electrode. Such an approach was first demonstrated by Willner's group in the design of an amperometric sensor for glucose (Figure 8.9).[39]

The enzyme glucose oxidase (GOx) drives the conversion of glucose to gluconic acid in an FAD cofactor-dependent reaction. The AuNP sensor was constructed by linking an FAD-functionalized AuNP to the electrode surface through a dithiol bridge moiety. The alignment of GOx on the particles via the reconstitution of the haloenzyme and the shortening of the enzyme–electrode electron transfer distance allowed the bioelectrocatalytic oxidation of glucose. Analysis of the characterized electrode revealed $k_{et} = 5000\,s^{-1}$, approximately seven-fold higher than the rate of electron transfer to the native electron acceptor, O_2.[39] The creation of such a hybrid NP–enzyme system not only achieves the goals of efficient amperometric sensors, but also suggests a general approach for the tailoring of effective electrode surfaces for other applications such as biofuel cells.

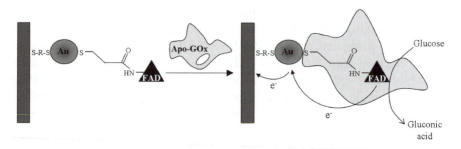

Figure 8.9   Gold nanoparticle enabled biosensor. The gold nanoparticles help to shorten electron transport distances and mediate the charge transfer. Interaction with GOx allows the detection of glucose by inducing bioelectrocatalytic oxidation.

Nanotechnology approaches to the development of sensitive non-isotopic approaches to the detection of DNA have had a significant impact on the field. Initial nucleic acid detection efforts utilizing functionalized gold nanoparticles were based on the formation of nanoparticle networks induced by the presence of target DNA.[40] In these experiments, one solution of 13 nm AuNP was functionalized with DNA complementary to the 5′ end of the target DNA. A second solution of 13 nm AuNP was functionalized with DNA complementary to the 3′ end of the target DNA. When these solutions were mixed they remained a pink–red color characteristic of 13 nm nanoparticles. However, the addition of target DNA caused the agglomeration of a nanoparticle networks (Figure 8.10). As a result, the formation of these networks caused a red shift in the plasmon resonance of the AuNP causing the solution to shift from a pink–red color to purple.

More advanced detection schemes have utilized capture DNA bound to a glass slide, allowing for multiplexed detection.[41] After the glass slide has been functionalized with capture DNA, it is incubated in hybridization buffer containing both target DNA, part of which will hybridize to the capture DNA and nanoparticle probes functionalized with DNA that

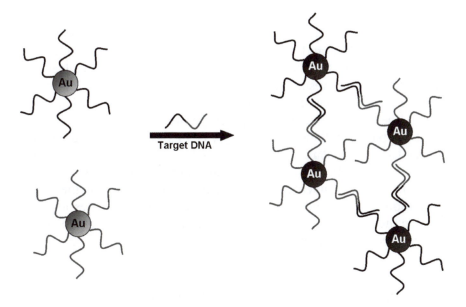

Figure 8.10 A depiction of target DNA inducing the formation of nanoparticle networks of functionalized gold nanoparticles. This aggregation causes a shift in the plasmon resonance of the nanoparticles resulting in a red to purple color change of the solution.

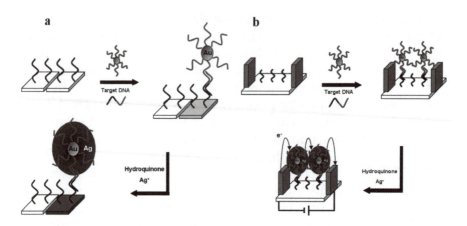

Figure 8.11   (a) Schematic of the 'sandwich' formed between the functionalized glass
surface and the functionalized gold nanoparticles in the presence of
target DNA. This 'sandwich' formation can then be visually enhanced
through the reduction of Ag(I) by hydroquinone mediated by the surface
of the gold nanoparticles. (b) Electrical detection of target-induced
'sandwich' formation. Following sandwich formation and silver
enhancement, a bridge is formed between the two electrodes. This bridge
creates a change in resistance which in turn is measured to determine the
presence of target DNA.

will hybridize to a different part of the target DNA. In the presence of
target DNA, a 'sandwich' is formed between target DNA (the middle)
and capture DNA and nanoparticles (the ends) (Figure 8.11a). To
amplify the signal, the slide can subsequently be exposed to silver ions
and hydroquinone, resulting in the reduction of the silver by hydroquinone at the surface of the gold nanoparticles. If the 'sandwich' is not
formed, the glass will remain transparent; however, if it has been formed,
it will be opaque so the slide can simply be imaged on a flatbed scanner.

This detection scheme was later adapted to utilize electrical detection
(Figure 8.11b).[42] In this scheme, capture DNA was bound to a silica
wafer between two microelectrodes fixed on the surface of the wafer.
Formation of the 'sandwich' and silver enhancement creates a bridge
between the two electrodes resulting in a change in resistance which can
be measured with a multimeter.

8.5   MAKING NANOSCALE STRUCTURES USING
BIOTECHNOLOGY

Biological processes produce an expansive array of complex materials
including laminate composites and ceramics such as bone, teeth and

shells; magnetic materials, such as the forms of magnetite found in magnetobacteria and the brains of homing pigeons; novel silver nanoclusters produced as a result of heavy metal detoxification mechanisms by the bacteria *Pseudomonas stutzeri*; and arrays of precisely fabricated diffracting architectures resulting in the multitude of intense colors observed in insects and birds. An essential ingredient to understanding these biomaterials is an examination of the molecular interactions at inorganic–organic interfaces which result in the controlled nucleation and growth of these novel materials. The biomaterials often represent unique crystal forms extending over several size domains and are synthesized in aqueous solutions at room temperature and standard pressure.[16] In recent years, researchers have begun to mimic mechanisms ascribed to natural production and stabilization of nano-scale materials.

Diatoms are unicellular, eukaryotic algae that form a diverse array of nanopatterned silica structures[43] on the scale of 16 gigatons per year. Silaffins are highly post-translationally modified peptides derived from the Sil1 protein of the diatom *Cylindrotheca fusiformis* that have been implicated in the biosilicification process.[44] Within the native peptide, all of the lysines have been modified to long-chain polyamine moieties and the serines have been post-translationally phosphorylated. Silaffins from a variety of diatoms, and also other long-chain polyamines, have been shown to promote silica condensation from a solution of monosilicic acid.[45] The post-translationally modified silaffins self-assemble into supramolecular structures, providing a template for silicic acid poly-condensation. Whereas silaffins effectively precipitate silica nanospheres under mildly acidic conditions, the non-modified R5 peptide (H_2N–SSKKSGSYSGSKGSKRRIL–CO_2H) is capable of precipitating silica at neutral pH. A synthetic site-directed mutagenesis study of the full-length non-post-translationally modified R5 peptide suggests that the C-terminus RRIL motif serves as an organizing element that permits the formation of a supramolecular assembly of peptides creating a locally high concentration of primary amine side-chain residues that drives *in vitro* silica precipitation.[46]

Recent work with a variety of polyamines such as poly-L-lysine, pentapropylenehexamine and polyallylamine hydrochloride have shown that they too can act to form silica nanospheres, most likely in an aggregation-dependent manner.[47] This biomimetic approach has been extremely successful in the synthesis of bio-like metal oxides in the laboratory. Dendrimers represent unique unimolecular polymer templates whose functionality can be tuned through the judicious choice of branching elements and terminal groups. They have been used not only to drive the condensation of silica,[48] but also as functional elements for the creation of silica composites.[49]

Peptides can also serve as templates for the formation of semiconductor nanoparticles. In the presence of high levels of cadmium salts, a variety of plants (*e.g.* tomatoes and carrots) and algae protect themselves through the synthesis of peptides, known as phytochelatins.[50] Structurally similar to glutathione, phytochelatins have the general structure (γ-Glu–Cys)$_n$–Gly (n = number of dipeptide repeats). These peptides are capable of nucleating the formation of fluorescent CdS quantum dots for the efficient storage of these toxic heavy metal ions.[51] With these systems as inspiration, several groups have developed platforms for the discovery of novel semiconductor nucleating peptide templates.

There are several platforms for the presentation of combinatorial peptide libraries, including synthetic libraries, cell-surface display and the example highlighted here, phage display. Libraries of 10^{11} members of combinatorial peptides can be displayed as fusions to the capsid proteins of one of several bacteriophage. The first such libraries were constructed using the filamentous phage M13 (Figure 8.12). Although everything from peptides to antibody fragments to engineered proteins have been displayed the on all of the five capsid proteins of M13, the majority of libraries utilize capsid protein III or VIII. The M13 viral particle carries five copies of pIII at one of its ends and 2700 copies of pVIII, which covers the filamentous virus along its length ($\sim 1\,\mu$m).[52]

In a typical affinity selection experiment, the inorganic surface is exposed to a phage-displayed combinatorial peptide library for several hours. After the unbound phage have been washed away, phage particles displaying binding sequences are eluted from the surface with an

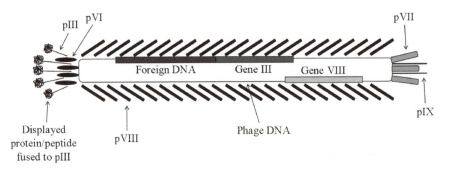

Figure 8.12   Schematic of M13 bacteriophage. Proteins and peptides are typically displayed fused to the pIII or pVIII proteins.

Figure 8.13   Biopanning. Phage displayed peptides are exposed to an inorganic surface. Unbound phage are washed away. Acid elution frees the bound
phage. Multiple panning rounds are generally employed to identify the
strongest binding peptides. DNA analysis then allows for the identification of peptide sequence responsible for binding.

acidic buffer wash. The recovered binding phage are amplified by
infecting a culture of *Escherichia coli* and the output phage used in
subsequent selection rounds. Typically, it can take three or more rounds
of enrichment to isolate binding phage (Figure 8.13). In searching
for peptide ligands to nanoscale materials, the emphasis has largely
been on strong interactions that have the potential to mediate the
nucleation of inorganic structures or that can promote the assembly of
heterocomponent systems.

One of the most stunning successes in this approach has been work by
Belcher's group. They have selected a number of phage (M13) displayed
peptides capable of nucleating ZnS quantum dots.[53] These phage, when
exposed to a ZnS precursor solution, assembled into a self-supporting
hybrid film composed of phage particles aligned along their length to
form sheets. Each phage had a ZnS nanocrystal attached to the presented protein III peptide nucleating domain. Furthermore, these viruses at high concentrations behave as functionalized liquid crystals. They
also demonstrated a method for the construction of novel hetero bio-
inorganic structures by expressing ZnS and/or CdS nucleating peptides
as fusions to the coat protein VIII. The resulting phage nucleated ZnS
and/or CdS along their length, creating a semiconductor nanowire
showing preferential orientation of their nanocrystal component.[54]
Phage that expressed both the ZnS and CdS nucleating peptides were
also used to form heterostructured nanowires. The long filamentous,
programmable virus suggests a highly manipulable template for the
synthesis of nanostructures.

A different approach to the synthesis of nanoscale structures is to use
biomolecular templates. Nanoscale fibers and tubes serve as effective
templates for the construction of 1D nanowires due to their inherent 1D
organization. Fibrillar biological molecules and biomolecular assemblies
of DNA, peptides, proteins and hybrid molecules have all been
employed as scaffolds for nanoparticle support.[55] A versatile system

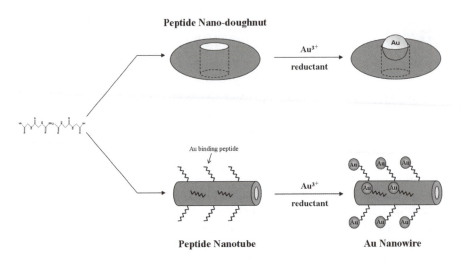

Peptide Nano-doughnut

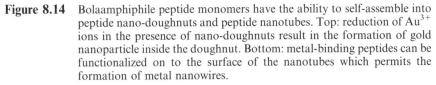

Figure 8.14 Bolaamphiphile peptide monomers have the ability to self-assemble into peptide nano-doughnuts and peptide nanotubes. Top: reduction of Au^{3+} ions in the presence of nano-doughnuts result in the formation of gold nanoparticle inside the doughnut. Bottom: metal-binding peptides can be functionalized on to the surface of the nanotubes which permits the formation of metal nanowires.

developed by Matsui and co-workers employs peptide–amphiphiles (Figure 8.14). These fibers are formed by the self-assembly of building blocks comprised of a hydrophilic peptide component connected to a hydrophobic aliphatic tail. Such nanotubular constructs have been used to for 1D arrays of gold nanoparticles.[56] This versatile system can also be used to form other structures, such as the nanodoughnut reactor capable of supporting a single nanoparticle whose size is defined by the diameter of the doughnut hole.[57]

Although these are just a small selection of the uses of biology in nanotechnology, it should be apparent that the principles of biological organization provide valuable insights and tools for the creation of functional hybrid biological–nanoscale structures. Biological systems represent genuine examples of nanoscale devices. Even the simplest living organism contains functional complex components such as motors, pumps and cables that function on the nanoscale. These objects are self-assembled through molecular recognition events between the building blocks to create a larger, functional hierarchy. Inspired by these examples, future research efforts will focus on the use of the biological tools for nanotechnological applications in the areas of electronics, fluidics and electromechanical systems.

8.6 CONCLUSIONS

Once the concepts of the behavior of a material being dependent on a specific length scale and that biology operates on that length scale are linked, that a field such as bionanotechnolgy should arise seems obvious. In the past decade, there have been tremendous advances in both our understanding of the fundamental nanoscaling laws of materials and in their practical application. There is little doubt that the horizon of bionanotechnology stretches far beyond the borders outlined in this introduction to some of the most exciting future challenges facing researchers today. The development of nanotechnologically enabled biosensors for multiplex analysis will find critical applications in clinical diagnostics, homeland security, environmental control and forensic applications. Metallic or semiconductor circuitry constructed on biomolecular templates is expected to provide the new logic elements for miniaturized computers. Biomolecular hybrid nanoparticles will become improved carriers for drugs, imaging agents for single cellular events and ordered biomolecular structures acting as nanoscale information storage and processing systems. While these prospects suggest a rich and exciting field for years to come, their achievement will only be accomplished through the continued interdisciplinary research of chemists, physicists and life scientists.

REFERENCES

1. A. P. Alivisatos, P. F. Barbara, A. W. Castleman, J. Chang, D. A. Dixon, M. L. Klein, G. L. McLendon, J. S. Miller, M. A. Ratner, P. J. Rossky, S. I. Stupp and M. E. Thompson, *Adv. Mater.*, 1998, **10**, 1297.
2. N. C. Seedman and A. M. Belcher, *Proc. Natl. Acad. Sci. USA*, 2002, **99**, 6451.
3. C. P. Poole Jr and F. J. Owens, *Introduction to Nanotechnology*, Wiley-Interscience, Hoboken, NJ, 2003.
4. M. R. Warnement, I. D. Tomlinson and S. J. Rosenthal, *Curr. Nanosci.*, 2007, **3**, 273–284.
5. T. Kippeny, L. Swafford and S. J. Rosenthal, *J. Chem. Educ.*, 2002, **79**, 1094.
6. F. Pinaud, 8. Michaler, L. A. Bentolila, J. Tsay, J. J. Doose, G. Iyer and S. Weiss, *Biomaterials*, 2006, **27**, 1679–1687.
7. E. L. Bentzen, F. House, T. J. Utley, J. E. Crowe Jr and D. W. Wright, *NanoLetters*, 2005, **5**, 591.
8. B. Dubertret, P. Skourides, D. J. Norris, V. Noireaux, A. H. Brivanlou and A. Libchaber, *Science*, 2002, **298**, 1759.

9. M. Dahan, S. Levi, C. Luccardini, P. Rostaing, B. Riveau and A. Triller, *Science*, 2003, **302**, 442.

10. S. Kim, Y. T. Lim, E. G. Soltesz, A. M. De Grand, J. Lee and A. Nakayama, *Nat. Biotechnol.*, 2004, **22**, 93.

11. A. Hoshino, K. Fujioka, T. Oku, S. Nakamura, M. Suga and Y. Tamaguchi, *Microbiol. Immunol.*, 2004, **48**, 985.

12. A. M. Drefus, W. C. W. Chan and S. N. Bhatia, *Adv. Mater.*, 2004, **16**, 961.

13. I. L. Medintz, A. R. Clapp, H. Mattoussi, E. R. Goldman, B. Fisher and J. M. Mauro, *Nat. Mater.*, 2003, **2**, 630.

14. T. D. Lacoste, 8. Michalet, F. Pinaud, D. S. Chemla, A. P. Alivisatos and S. Weiss, *Proc. Natl. Acad. Sci. USA*, 2000, **97**, 9461.

15. M. Kuno, D. P. Fromm, H. F. Hamann, A. Gallagher and D. J. Nesbitt, *J. Chem. Phys.*, 2000, **112**, 3117.

16. J. M. Slocik, M. R. Knecht and D. W. Wright, in *The Encyclopedia of Nanoscience and Nanotechnology*, ed. H. S. Nalwa, American Scientific Publishers, Stevenson Ranch, CA, 2003, p. 239.

17. Y. W. Jun, J. W. Seo and J. Cheon, *Acc. Chem. Res.*, 2008, **41**, 179.

18. D. Jales, *Introduction to Magnetism and Magnetic Materials*, CRC Press, Boca Raton, FL, 1998.

19. M. P. Morales, S. Veintemillas-Verdaguer, M. I. Montero and C. J. Serna, *Chem. Mater.*, 1999, **11**, 3058.

20. Y. Jun, Y.-M. Huh, J.-S. Choi, J.-H. Lee, H.-T. Song, S. J. Kim, S. Yoon, K.-S. Kim, J.-S. Shin, J.-S. Suh and J. Cheon, *J. Am. Chem. Soc.*, 2005, **127**, 5732.

21. S. Sun, C. B. Murray, D. Weller, L. Folks and A. Moser, *Science*, 2000, **287**, 1989.

22. J.-I. Park, M. G. Kim, Y. Jun, J. S. Lee, W.-R. Lee and J. Cheon, *J. Am. Chem. Soc.*, 2004, **126**, 9072–9078.

23. J.-H. Lee, Y.-M. Huh, Y. Jun, J.-W. Seo, J.-T. Jang, H.-T. Song, S. J. Kim, E.-J. Cho, H.-G. Yoon, J.-S. Suh and J. Cheon, *Nat. Med.*, 2007, **13**, 95–99.

24. E. L. Bentzen, D. W. Wright and J. E. Crowe, *Future Virology*, 2006, **1**, 769–781.

25. M. A. Hayes, N. A. Polson, A. N. Phayre and A. A. Garcia, *Anal. Chem.*, 2001, **73**, 5896.

26. Y. K. Hahn, Z. Jin, H. Kang, E. Oh, M.-K. Han, H.-S. Kim, J.-T. Jang, J.-H. Lee, J. Cheon, S.-H. Kim, H. S. Park and J.-K. Park, *Anal. Chem.*, 2007, **79**, 2214.

27. M. Quinten, *Appl. Phys. B*, 2001, **73**, 317.

28. T. Tani, *J. Dispers. Sci. Technol.*, 2004, **25**, 375.

29. C. J. Orendorf, T. K. Sau and C. J. Murphy, *Small*, 2006, **5**, 636.

30. S. Schultz, D. R. Smith, J. J. Mock and D. A. Schultz, *Proc. Natl. Acad. Sci. USA*, 2000, **97**, 996–1001.

31. M. A. El-Sayed, *Acc. Chem. Res.*, 2001, **34**, 257–264.

32. K. L. Kelly, T. R. Jensen, A. A. Lazarides and G. C. Schatz, in *Metal Nanoparticles: Synthesis Characterization and Applications*, D. Feldheim and C. Foss (ed.), Marcel Dekker, New York, 2002, p. 89.

33. J. J. Mock, M. Barbic, D. R. Smith, D. A. Schultz and S. Schultz, *J. Chem. Phys.*, 2002, **116**, 6755.

34. R. Jin, Y. Cao, C. A. Mirkin, K. L. Kelly, G. C. Schatz and J. G. Zheng, *Science*, 2001, **294**, 1901–1903.

35. A. Merkoci, *FEBS J.*, 2007, **274**, 310.

36. J. F. Hicks, D. T. Miles and R. W. Murray, *J. Am. Chem. Soc.*, 2002, **124**, 13322.

37. R. R. Peterson and D. C. Cliffel, *Langmuir*, 2006, **22**, 10307.

38. I. Willner, B. Basner and B. Willner, *FEBS J.*, 2007, **274**, 302.

39. Y. Xiao, F. Patolsky, E. Katz, J. F. Hainfeld and I. Willner, *Science*, 2003, **299**, 1877.

40. J. J. Storhoff, R. Elghanian, R. C. Mucic, C. A. Mirkin and R. L. Letsinger, *J. Am. Chem. Soc.*, 1998, **120**, 1959.

41. T. A. Taton, C. A. Mirkin and R. L. Letsinger, *Science*, 2000, **289**, 1757.

42. S.-J. Park, T. A. Taton and C. A. Mirkin, *Science*, 2002, **295**, 1503–1506.

43. F. E. Round, R. M. Crawford and D. G. Mann, *The Diatoms: Biology and Morphology of the Genera*, Cambridge University Press, Cambridge, 1990.

44. N. Kröger, R. Deutzmann and M. Sumper, *Science*, 1999, **286**, 1129.

45. N. Kröger, R. Deutzmann and M. Sumper, *J. Biol. Chem.*, 2001, **276**, 26066.

46. M. R. Knecht and D. W. Wright, *Chem Commun.*, 2003, 3038.

47. L. F. Deravi, J. D. Swartz and D. W. Wright, *The Biomimetic Synthesis of Metal Oxide Nanomaterials, Nanomaterials for the Life Sciences*; Vol 2. Ed. Challa Kumar: Wiley-VCH Verlag GmbH & Co. KgaA, In Press, Pages 1–59.

48. M. R. Knecht and D. W. Wright, *Langmuir*, 2004, **20**, 4728.

49. M. R. Knecht and D. W. Wright, *Chem. Mater.*, 2004, **16**, 4890.

50. J. M. Whitling, G. Spreitzer and D. W. Wright, *Adv. Mater.*, 2000, **12**, 1377.

51. C. T. Dameron, R. N. Reese, R. K. Mehra, A. R. Kortan, P. J. Carroll, M. L. Steigerwald, L. E. Brus and D. R. Winge, *Nature*, 1989, **338**, 596.

52. U. Kriplani and B. K. Kay, *Curr. Opin. Biotechnol.*, 2005, **16**, 470.
53. S.-W. Lee, C. Mao, C. E. Flynn and A. M. Belcher, *Science*, 2002, **296**, 892.
54. C. E. Flynn, C. Mao, A. Hayhurst, J. L. Williams, G. Georgiou, B. Iverson and A. M. Belcher, *J. Mater. Chem.*, 2003, **13**, 2414.
55. E. Gazit, *FEBS J.*, 2007, **274**, 317.
56. R. Djalali, Y. Chen and H. Matsui, *J. Am. Chem. Soc.*, 2002, **124**, 13660.
57. R. Djalali, J. Samson and H. Matsui, *J. Am. Chem. Soc.*, 2004, **126**, 7935.

CHAPTER 9

Molecular Engineering of Antibodies

JAMES D. MARKS

Department of Anesthesia and Pharmaceutical Chemistry, University of California, San Francisco, San Francisco General Hospital, 1001 Potrero Avenue, San Francisco CA 94110, USA

9.1 INTRODUCTION

Antibodies are one of the effector molecules of the vertebrate humoral immune system. They are generated *in vivo* in response to the presence of foreign pathogens or molecules (antigens), bind specifically to the antigen and result in its neutralization and elimination. One of the characteristics of antibodies is that they can bind with high affinity and specificity to only the target antigen and not to any of the tens of thousands of other proteins and potential antigens in the circulation. This specific and high affinity binding led to the appreciation that antibodies could be the so-called 'magic bullets' proposed by Ehrlich at the turn of the 20th century: molecules that could selectively target a disease-causing organism and deliver a toxic payload, killing only the organism targeted.

As will be described below, antibodies in the serum of immunized animals were some of the first therapeutics used for infectious diseases, at a time before antibiotics had been discovered. Yet it took more than 100 years from this initial use of serum therapy for antibodies to begin to be approved by the US Food and Drug Administration (FDA) for the treatment of human diseases. The first such antibodies entered clinical practice in the 1990s and today there are currently 21 therapeutic antibodies

Molecular Biology and Biotechnology, 5th Edition
Edited by John M Walker and Ralph Rapley
© Royal Society of Chemistry 2009
Published by the Royal Society of Chemistry, www.rsc.org

approved by the FDA which had sales of more than $15 billion in 2006. Most of these antibodies have been approved for clinical use within the last 10 years and it is estimated that there are at least 100 antibodies in the different phases of human clinical trials for a wide range of diseases, including cancer, inflammatory diseases and infectious diseases.[1–3]

The era of antibodies as therapeutics became possible with the advent of hybridoma technology in 1975, a technological breakthrough that resulted in the ability to clone single B-cells and the single antibody made by that B-cell (Figure 9.1).[4] Such monoclonal antibodies (mAbs), unlike the hundreds to thousands of antibodies in serum, recognized only a single antigen and could be made in virtually unlimited quantities. Unfortunately, the technology was developed to make mAbs from the B-cells of immunized mice and it has proven technically challenging to apply the technology to generate human mAbs.[5] As it turns out, when murine mAbs are administered to humans, they elicit an immune response, called the human anti-mouse antibody response (HAMA) that either results in unacceptable systemic reactions or results in rapid clearance of the mAb from the bloodstream.[6,7] While many mAbs from hybridomas entered clinical trials, few were approved by the FDA due to the limitations described above.

With the advent of molecular cloning and protein engineering technologies in the late 1980s, it has proven possible to engineer murine mAbs to have sequences more similar to human mAbs, so called chimeric[8–10] and humanized antibodies. Such antibodies have proven significantly less immunogenic than murine mAbs and many of these are now approved for clinical use. More recently, it has proven possible to make mAbs that are fully human in sequence using antibody gene diversity libraries and display technologies (see Refs 11 and 12 for reviews), and also mice that are transgenic for the human immunoglobulin loci.[13]

In the following sections, we will review the increasing importance of antibodies as a therapeutic class and review antibody structure, generation and function. We will use this information as background to describe the molecular engineering techniques of chimerization and humanization that have yielded the first widely successful therapeutic antibodies. We will then describe how the more recent techniques using diversity libraries and display technologies can be used to generate full human antibodies and to evolve antibody affinity to values not typically generated by the humoral immune system.

9.2 ANTIBODIES AS THERAPEUTICS

Antibody therapy began approximately a century ago with the discovery that serum from animals immunized with toxins, *e.g.* diphtheria toxin or

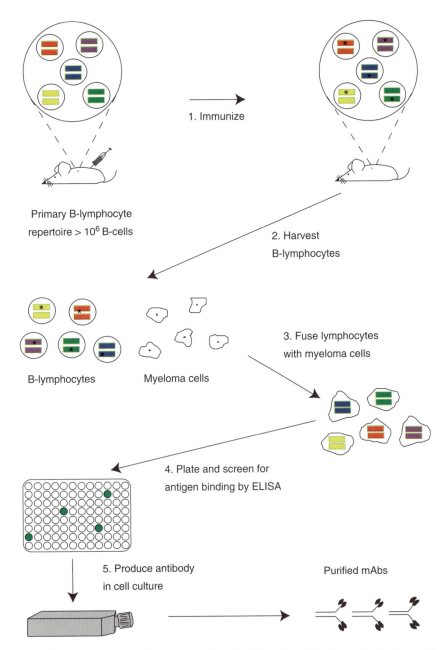

Figure 9.1 Generation of monoclonal antibodies using hybridoma technology. The
naïve mouse generates a primary repertoire of more than 10^6 rearranged
V_H and V_L genes (colored bars) in B-cells, coding for antibodies that are
displayed as membrane-bound molecules. Immunization (step 1) causes
antigen-driven proliferation and somatic hypermutation ('stars' within V
genes represent mutations introduced by the somatic hypermutation
machinery). To make hybridomas, B-cells are harvested from the spleen
(step 2) or marrow of the mouse and fused with immortal myeloma cells
(wrinkled edged cells, step 3) to generate immortalized, antibody secreting
hybridomas. Hybridomas are plated into microtiter plates and the
supernatants containing secreted antibody are screened by ELISA for
antigen binding (step 4). Hybridomas are expanded into tissue culture
flasks and the secreted monoclonal antibodies purified (step 5).

viruses, was an effective therapy for the disease caused by the same agent in humans. While capable of potent antigen neutralization, such sera contain hundreds to thousands of different antibodies with only approximately 1% of the antibodies binding to the immunizing antigen. In the 1880s, von Behring developed an antitoxin that neutralized the toxin that the Diphtheria bacillus released into the body and was awarded the first Nobel Prize in Medicine in 1901 for his role in the discovery and development of serum therapy for diphtheria.

Following the initial successes in the late 1800s, sera from humans or animals containing antibodies were widely used for prophylaxis and therapy of viral and bacterial diseases.[14,15] Serum therapy of most bacterial infections was abandoned in the 1940s, however, after antibiotics became widely available.[14] Polyclonal antibody preparations have continued to be used for some toxin-mediated infectious diseases and venomous bites.[16] Serum immunoglobulin has also continued to be used for viral diseases where there are few treatments available, although immunoglobulin is largely used for pre- or post-exposure prophylaxis.[17,18]

Although serum polyclonal antibody preparations have been clinically effective in many cases, they have problems related to toxicity, including a significant risk of allergic reactions including anaphylactic shock and serum sickness.[19] Other limitations of polyclonal antibodies include lot to lot variations in potency and side-effects and uncertainty in dosing due to these effects.[16] In addition, the active antigen-specific antibodies in a polyclonal preparation typically represent a relatively small proportion of the total antibodies (1%); the rest of the antibodies are not only ineffective but could even be toxic or immunogenic. However, until the 1970s it was not possible to produce large amounts of antibodies with the desired specificity from a single antibody-producing B-cell.

Development of hybridoma technology made it possible to clone single B-cells and the single antibody made by that B-cell (Figure 9.1).[4] Molecular cloning techniques then made it possible to replace genetically the mouse constant regions of the mouse antibody with human constant regions yielding chimeric antibodies, antibodies that are approximately 90% human in sequence (Figure 9.2) (refs. 8–10). Chimeric antibodies are far less immunogenic than murine antibodies. Therapeutic antibodies were ushered into a 'take-off' phase by the 1997 launch of the chimeric antibody Rituxan (rituximab) for non-Hodgkin's lymphoma (NHL). Rituxan represented the first mAb product to succeed commercially in a high-revenue/high-growth market (oncology) and to provide significant enhancements in the efficacy of treatment versus existing non-mAb therapies. As a result, Rituxan rapidly became

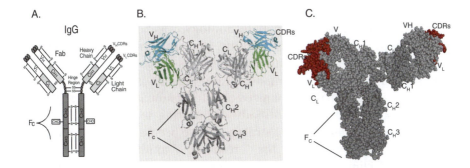

Figure 9.2 Structure of IgG antibody and chimeric and humanized antibodies. (A) IgG structure. IgG antibody consists of a pair of heavy and light chains. Each chain consists of the antigen binding variable domains (V_H and V_L) and one or more constant domains (C_H1, C_H2, C_H3 and C_L). Each V or C domain contains an intramolecular disulfide bond (S–S). A single glycosylation site (CHO) exists in the C_H2 domain of the heavy chain. The Fab consists of the light chain and the V_H–C_H1 domains; each IgG consists of two Fab arms. The Fv is the minimal antigen binding unit, consisting of the V_H and V_L domains. The V_H and V_L domains contact antigen via the amino acids in the complementarity determining regions (CDRs). The Fc region elicits antibody effector functions such as ADCC and CDC. (B) Alpha carbon backbone tracing of a chimeric antibody. The alpha carbon backbone tracing of a chimeric IgG antibody is shown. The chimeric antibody consists of murine V_H and V_L domains (colored cyan and green, respectively) and human constant domains (colored shades of gray). (C) Space filling model of a humanized antibody. Murine CDRs in each of the Fab arms are colored red. The framework regions of the V domains and all of the C-domains are colored gray.

established as the gold standard therapy for NHL and the first launched mAb product which went on to achieve blockbuster status (revenues above \$1 billion per year). Rituxan approval by the FDA was followed by approval of Remicade (infliximab), which binds tumor necrosis factor-alpha (TNF-α) and which is used to treat rheumatoid arthritis and other inflammatory diseases mediated by TNF, such as psoriasis and Crohn's disease.

Further advances in antibody engineering resulted in humanized antibodies, mAbs where the variable region framework regions and also the constant regions were replaced with human sequences (Figure 9.2). Humanized antibodies, which are greater than 90% human in sequence, are potentially less immunogenic than chimeric antibodies. A number of blockbuster humanized antibodies have been approved by the FDA, including Herceptin (trastuzumab) for treatment of breast cancer, Avastin (bavacizumab) for treatment of colon cancer and Synagis (palivizumab) for prevention of respiratory syncytial virus (RSV) infection. Recently Humira (adalimumab), the first fully human antibody produced via phage display technology, was licensed by the FDA. It, too, has achieved blockbuster status. Currently, more than 100 mAbs are in various stages of clinical trials to treat a range of human diseases, including cancers and inflammatory and infectious diseases.[2,3] These antibodies are primarily humanized and fully human antibodies generated and optimized using antibody engineering technologies.[3] Moreover, a number of engineering strategies have been deployed to enhance antibody potency and the resulting antibody-based drugs have entered clinical trials. These include antibody–drug conjugates,[20] antibody–toxin fusion proteins[21] and antibody where the Fc portion of the antibody has been engineered to elicit more effectively antibody-dependent cellular cytotoxicity (ADCC) or complement dependent cytotoxicity (CDC).[22] The engineering techniques used to generate therapeutic antibodies will be discussed in the following sections, after a review of antibody structure and function.

9.3   ANTIBODY STRUCTURE AND FUNCTION

Antibodies are Y-shaped dichain glycoproteins of molecular mass approximately 150 kDa (Figure 9.2). They consist of two heavy chains and two light chains, with each chain consisting of two or more domains that share the immunoglobulin fold. Each domain is a pleated beta sheet with an intramolecular disulfide bond. The light chain consists of two domains, a light chain variable domain (V_L) and a light chain constant domain (C_L). There are two types of light chains, kappa and lambda. In humans, approximately two-thirds of light chains are kappa and one-third are lambda. The heavy chain consists of four to five domains, depending on the antibody isotype, and includes a heavy chain variable domain (V_H) and three to four constant domains (C_H1, C_H2, C_H3 and C_H4). There are a number of different isotypes of antibodies, which include IgG, IgD, IgA, IgM and IgE. The major isotype present in the circulation is IgG, which is also the isotype most frequently used for therapeutic antibodies. The light chains are disulfide linked to the V_H–C_H1 domains to make the antigen binding fragment (Fab). The two Fabs are linked via the hinge region to the Fc domain, which consists of C_H2–C_H3 or C_H2–C_H4 domains. The Fc region is involved in the elicitation of antibody effector functions including the ability of antibody to mediate cellular killing by immune effector cells [antibody-dependent cellular cytotoxicity (ADCC)] or complement [antibody-dependent cellular cytotoxicity (CDC)].

The smallest antibody fragment capable of binding antigen is the fragment variable or Fv. The Fv consists of the V_H and V_L domains (Figure 9.2). Within each of the V domains are three complementarity-determining regions (CDRs), which form loops that comprise the antigen binding region of the V domains. CDRs were first identified by Kabat and Wu as regions with the greatest sequence diversity when different V domains were compared to each other and are also called the hypervariable regions.[23] The six CDRs contribute the majority of the amino acid side-chains that make direct contact with antigen.[24,25] Each CDR is flanked by regions of more conserved sequence that make up the variable region framework regions.[26,27] The frameworks fold into the β-strands that comprise the two β-sheets of each V domain and which support the CDRs.

9.4 CHIMERIC ANTIBODIES

As described in Section 9.2, the development of chimeric antibodies was a significant leap forward for the therapeutic antibody field.[8,28,29] To reduce the immunogenicity of murine antibodies, the murine variable regions were grafted on to human kappa light chain and gamma heavy chain constant regions (Figure 9.2B). This is relatively straightforward as the V regions and C regions are contiguous pieces of DNA and each domain folds independently.

For chimerization, the DNA encoding the antibody heavy (V_H) and light (V_L) chain variable region genes must be cloned from the hybridoma cell. Prior to the development of the polymerase chain reaction (PCR), this required cDNA cloning of the variable region genes. With the advent of PCR, 5′ PCR primers were initially designed based on N-terminal protein sequencing of the variable regions from the heavy and light chain.[30] Design of the 3′ primer was more straightforward, as it could be based in the constant regions. A major breakthrough was the development of partially degenerate universal V region primers based on sequence databases that could be used to amplify and directionally clone most murine variable regions without the need for protein sequencing[31,32] (Figure 9.3). Additional groups have designed sets of universal V gene primers containing internal or appended restriction sites suitable for amplification of murine,[33–38] human,[39–43] chicken[44] and rabbit[45] V genes. Once the V_H and V_L genes have been cloned, they can be directly forced into cloning vectors by incorporating restriction sites into the primers.[40,46] Alternatively, PCR can be used to combine V_H and V_L genes directly with linker DNA using splicing by overlap extension to generate antigen binding antibody fragment constructs such as those encoding

1. First strand cDNA synthesis

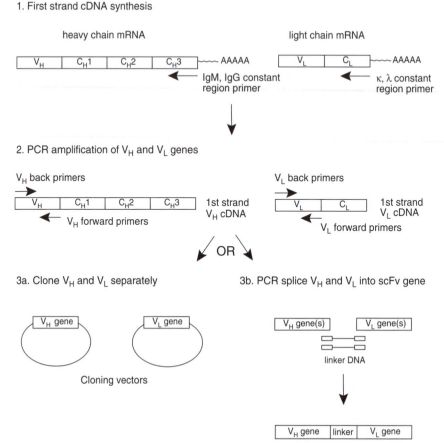

2. PCR amplification of V$_H$ and V$_L$ genes

3a. Clone V$_H$ and V$_L$ separately

3b. PCR splice V$_H$ and V$_L$ into scFv gene

Figure 9.3 PCR cloning of antibody V genes. (1) mRNA is isolated from hybridoma cells, peripheral blood lymphocytes, spleen or bone marrow and antibody genes are reverse transcribed (by using reverse transcriptase) using IgG, IgM, κ or λ constant-region specific primers, creating first strand cDNA. (2) The V$_H$ and V$_L$ variable region genes are amplified using PCR and universal primer mixtures specific for the 5' (back primers) and 3' (forward primers) ends of the heavy and light chain V genes. (3a) Amplified V$_H$ and V$_L$ gene DNA is purified and cloned into cloning vectors for DNA sequencing. By incorporating restriction sites into the primers, it is possible to directionally clone the V genes. (3b) Alternatively, the amplified V$_H$ and V$_L$ genes can be combined with a short 'linker DNA' which overlaps the 3' and 5' ends of the V$_H$ and V$_L$ genes, respectively, and PCR amplified to yield one continuous DNA fragment. The linker DNA encodes a flexible peptide that links the V$_H$ and V$_L$ gene to generate s single-chain Fv (scFv) gene. A final PCR reaction (not shown) adds flanking restriction sites to the assembled scFv gene for cloning into bacterial secretion vectors or display vectors.

Secreted Antibody Fragments

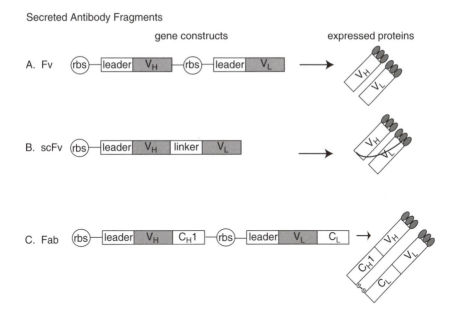

Figure 9.4 Antibody fragments that can be expressed in *E. coli*. Fv, single-chain Fv (scFv) and Fab antibody fragments can be expressed in *E. coli* by directing the V domains into the bacterial periplasm using a leader sequence. rbs = ribosome binding site; leader = bacterial secretion signal directing the expressed protein to the bacterial periplasm; V_H = heavy chain variable domain; V_L = light chain variable domain; linker = flexible scFv linker between V_H and V_L domains; C_H1 = heavy chain constant domain 1; C_L = light chain constant domain; S–S indicates the disulfide bond between the C_H1 and the C_L domains.

single chain Fv antibodies (scFv, Figure 9.3 3b).[40] By incorporating restriction sites into the primers, the scFv genes can be directionally cloned into bacterial secretion vectors (see below and Figure 9.4), which allow antibody fragment expression and screening for antigen binding.[40,46]

Although PCR greatly simplified the cloning of V genes, mutations introduced by the somatic hypermutation machinery into the regions where the primers anneal may make PCR amplification difficult or impossible, necessitating another amplification approach such as RACE or oligoligation PCR.[47–49] Cloning the correct V_H and V_L can also be complicated by the presence of several immunoglobulin transcripts, some of them arising from the fusion partner.[50]

In the original examples of chimerization, vectors containing murine V regions fused to human constant regions were constructed and used to transfect mouse myeloma cell lines to produce antibody to confirm antigen binding, affinity and specificity.[29] This process, which ensures that the correct V_H and V_L genes have been cloned from the hybridoma, takes several months if one constructs stable mammalian cell lines, for example in myeloma cells. An alternative would be to express the chimeric antibody transiently in mammalian cells such as COS-7 cells, cutting the time from cloning to screening for antigen binding to several weeks. Commercially, full-length IgG (chimeric, humanized or human) are typically expressed in Chinese hamster ovary (CHO) cells at yields greater than $1\,g\,l^{-1}$.

A major step forward in antibody engineering occurred with the discovery that antibody fragments could be expressed in bacteria, such as *E. coli* (Figure 9.4). Due to the rapid growth of bacteria, time from V gene cloning to screening of antibody fragments for binding can as short as several days. It is now common practice after cloning V genes from hybridomas to express them as antibody fragments in bacteria to verify antigen binding prior to construction of chimeric IgG or prior to humanization (see below). Initial attempts to express full-length antibody in the cytoplasm of *E. coli* resulted in very low yields of insoluble protein that required solubilization and refolding.[51,52] A significant breakthrough occurred when it was discovered that antigen binding fragments of antibodies could be produced in properly folded soluble form if expression of the antibody fragment was directed into the periplasmic space. Secretion into the periplasm was directed by attaching signal sequences such as pelB to the N-terminus of the antibody fragment genes (Figure 9.4). The oxidizing environment of the periplasm results in proper folding and intramolecular disulfide bond formation. Both the Fab and the Fv antigen binding fragments can be expressed in *E. coli* where they can be harvested from the bacterial periplasm.[53,54] Fvs, however, are not particularly stable; the V_H and V_L domains are not covalently linked and they tend to dissociate at typically expressed concentrations.[55,56] Fvs can be stabilized by physically linking the V_H and V_L domains together with a flexible peptide linker to create the single chain Fv antibody fragments (scFv).[57,58] Alternatively, cysteines can be engineered into both the V_H and V_L domains, resulting in disulfide bond formation when the V domains pair, a so-called disulfide linked Fv.[59]

As described in Section 9.2, a number of chimeric antibodies have been approved by the FDA and entered clinical practice, including rituximab and infliximab. Chimerization reduces the immunogenicity of murine mAbs and allows multiple and repeated dosing.[60,61] Replacing the murine Fc with a human Fc also results in more efficient effector functions such as ADCC and CDC as the human Fc interacts with human Fc receptors with higher affinity than the murine Fc. Similarly, chimeric antibodies have a longer half-life in humans than murine antibodies, due to more efficient interaction of the human Fc with FcRn. Despite their successes, however, chimeric antibodies can still be immunogenic. A human anti-chimeric antibody (HACA) response is frequently observed and in some cases can be severe and require discontinuation of the antibody or result in its ineffectiveness.[60,61] As a result, techniques to generate more fully human antibodies were developed.

9.5 ANTIBODY HUMANIZATION

As illustrated in Figure 9.2, the V_H and V_L CDRs comprise the antigen binding site and contain the majority of amino acids that make direct contact with antigen. Jones *et al.* hypothesized that the CDRs from a murine antibody could be 'grafted' on to human frameworks to make a so called 'humanized' antibody.[62] Grafting the CDRs from the V_H domain of a murine antibody to an hapten antigen on to human frameworks resulted in an antibody that bound the hapten with comparable affinity when combined with the murine light chain variable domain.[62] Such CDR grafting has been applied to an anti-lymphocyte mAb and to an antibody to respiratory syncytial virus (RSV), but is not always successful due to the critical role that framework residues may play in both supporting the proper conformation of the CDRs and contributing antigen contacting amino acids.[63,64]

With the determination of a large number of antibody X-ray crystallographic structures, many now take a more directed approach to humanization.[65] After the murine V_H and V_L genes have been sequenced, human frameworks are selected that are the most homologous to the murine V genes. This is straightforward given the large number of antibody sequence databases, including those of Kabat (http://www.bioinf.org.uk/abs/seqtest.html) and IMGT (http://imgt.cines.fr/). Modeling the human frameworks and mouse CDRs on to a homologous antibody structure allows the identification of murine residues that may either contact antigen or directly or indirectly influence the conformation of the CDRs. Finally, amino acids in the human frameworks that are 'rare' or 'unusual' are replaced with residues more typical of those positions. Approaches such as this tend to retain more murine residues in the framework regions than with CDR grafting and have been used to humanize antibodies to the interleukin receptor and interferon-gamma.[65,66]

A number of variations to the above approach have been developed. These include replacing murine CDR residues that are outside the antigen binding loops with human residues, as done by Presta and colleagues,[67,68] the use of human germline V genes as the acceptor frameworks and grafting the murine CDRs into human frameworks with the most homologous CDRs, as opposed to most homologous V genes or frameworks (superhumanization).[69] Alternatively, library approaches, as described below, can be used to humanize murine antibodies.[70,71] In these approaches, one of the murine V domains, for example the V_H domain, is paired with a library of human light chains and a chimeric antibody containing murine V_H and human V_L is selected. The murine V_H is then replaced with a library of human V_Hs paired to the new human V_L and a fully human antibody is selected. With any of these methodologies, it can be anticipated that the binding constants of the humanized antibodies may be less than those of the parental antibodies. If so, additional constructs can be generated and evaluated or library approaches can be used to return affinity to that of the parental antibody (see below).

More than half of the antibody therapeutics licensed by the FDA are humanized antibodies. These include such 'blockbuster' antibodies as Herceptin (trastuzmab) for the treatment of breast cancer, Avastin (bavacizumab) for treatment of colon cancer and Synagis (palivizumab) for prevention of respiratory syncytial virus (RSV) infection. Humanized antibodies are clearly less immunogenic than murine antibodies. They also appear to be less immunogenic in some instances than chimeric antibodies. Humanized antibodies are not completely free of immunogenicity and anti-humanized antibody responses (HAHA) can be detected in some patients. Such responses may be related to the number of non-human amino acids retained in the humanized antibodies, in addition to the dose, the immunocompetence of the individual and the specific target of the antibody.

9.6 ANTIBODIES FROM DIVERSITY LIBRARIES AND DISPLAY TECHNOLOGIES

The pharmaceutical industry discovers small molecule drugs by screening very large compound libraries to obtain lead 'hits' with affinities in the micromolar range. The tools of medicinal chemistry are then applied to these leads to improve their binding to the drug target, resulting in a new drug entity. In the late 1980s, advances in molecular biology made it possible to apply the same type of strategy to the discovery of fully human antibodies. Libraries of millions to billions of different antibodies could be generated and methods were developed that allowed the isolation of rare lead antibodies binding to a specific target antigen. As with the medicinal chemists, similar methods could be applied to diversify the structure of lead antibodies and allow the selection of antibodies with considerably higher affinities. The methods used to create such libraries are called display technologies and include phage display, yeast display, ribosome display and a number of other less frequently used technologies. Display technologies have three characteristics in common: (1) a method for generating antibody gene diversity; (2) a method for linking the antibody genotype with the expressed antibody phenotype; and (3) a method for isolating rare antigen binding antibodies (and their genes) from a majority of non-binding antibodies.

9.6.1 Antibody Phage Display

Phage display was the first antibody display technology to be developed and is the one most widely used today (see Figure 9.5 for an overview and Refs. 11–12, 72 and 73 for reviews). Antibody phage display resulted from concurrent progress in prokaryotic expression of antibody fragments, PCR cloning of antibody gene repertoires and display of peptides and proteins on filamentous bacteriophages. PCR cloning of antibody genes from hybridomas was covered in Section 9.4. By applying PCR and V region-specific primers to RNA prepared from human peripheral blood lymphocytes or mouse splenocytes, it is possible to amplify repertoires of V_H and V_L genes (see Figures 9.3 and 9.5).[40–41,46] The V gene repertoires can then be either separately cloned sequentially or spliced by overlap extension to create human or murine scFv or Fab gene repertoires (see Figures 9.3 and 9.5).[40,46] An alternative to the use of naturally occurring V gene repertoires is to synthesize V genes *in vitro* using cloned or synthetic V gene segments and random oligonucleotides encoding part of the antigen combining site.[72,74] Using this approach, it is possible to ensure that the potential V gene repertoire diversity is large enough that a library will only contain a single member of each sequence. In contrast, there may be considerable duplication of sequences using naturally occurring V genes.

The filamentous bacteriophages that infect *E. coli*, such as M13 and fd phage, are the phages used for antibody display. Such phage are single-stranded DNA viruses where the phage contains one copy of the viral DNA which is covered by a protein coat. The coat consists of approximately 3000 copies of the major coat protein pVIII, three to five copies of the minor coat protein pIII and the coat proteins pVII and pIX (Figure 9.6). scFv and Fab antibody fragments have been successfully displayed on pIII, pVII, pVIII and pIX[75–77] (Figure 9.6). Antibody phage display is accomplished by cloning antibody genes into a vector where they are in-frame with one of the coat proteins. They are then

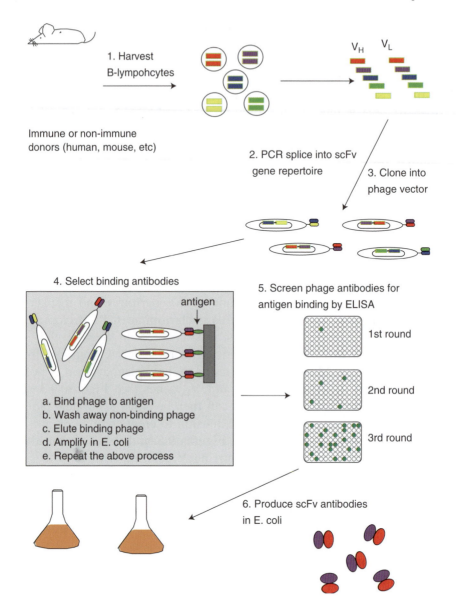

1. Harvest
   B-lympohcytes

Immune or non-immune
donors (human, mouse, etc)

V_H V_L

2. PCR splice into scFv
   gene repertoire

3. Clone into
   phage vector

4. Select binding antibodies

antigen

a. Bind phage to antigen
b. Wash away non-binding phage
c. Elute binding phage
d. Amplify in E. coli
e. Repeat the above process

5. Screen phage antibodies for
   antigen binding by ELISA

1st round

2nd round

3rd round

6. Produce scFv antibodies
   in E. coli

Figure 9.5 For phage display, B-cells are isolated from immunized mice (as in panel A) or naive or immunized humans. Heavy and light chain V genes (shaded bars) are amplified by PCR and assembled as single-chain Fv antibody genes (scFvs). Alternatively, rearranged V genes can be generated entirely *in vitro* from cloned V segments and synthetic oligonucleotides. The repertoire of scFv genes are cloned into a phage display vector, where the encoded scFv proteins (colored ovals) are displayed as fusion proteins to one of the phage coat proteins. The phage contain the appropriate scFv gene within. Multiple rounds of selection with immobilized antigen allow the isolation of even rare antigen binding phage antibodies, which are identified by ELISA. Native scFvs can be expressed from *E. coli* and purified for characterization and use in assays.

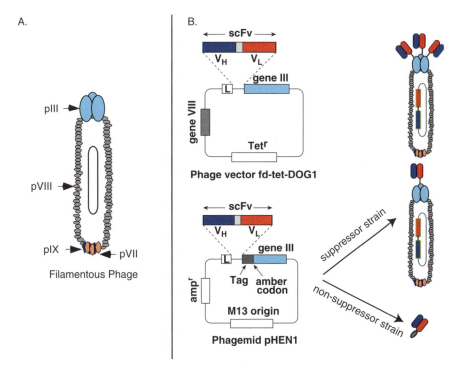

Figure 9.6 Features of phage display vectors. (A) Illustration of phage proteins pIII, pVII, pVIII and pIX, which can be used for antibody fragment display. (B) Representative phage and phagemid vectors. Phage vector fd-tet-DOG1 (top panel) and phagemid vector pHEN1 (bottom panel). Both vectors display scFv (V_H–linker–V_L) as fusions to the amino terminus of the pIII protein. Both vectors have leader sequences (pelB or gene III) to direct the expressed fusion protein to the bacterial periplasm. In phage fd-tet-DOG1, all copies of pIII are scFv fusions, leading to 3–5 copies displayed per phage particle. With phagemid pHEN1, expression in suppressor strains of *E. coli* allows the amber codon following the scFv-tag to be read as a glutamine, causing the scFv to be fused to the pIII protein. In phagemids, both wild-type pIII (from the helper phage) and fusion pIII (from the phagemid) compete for inclusion in the viral particle. In non-suppressers *E. coli* strains, the scFv is expressed as a soluble protein with a myc epitope tag for detection of binding in ELISA.

expressed as fusion proteins on the phage surface (Figure 9.6). The phage provides a physical link between the antibody fragment on its surface and the gene that encodes the antibody contained within the phage. The first example of antibody phage display was reported in 1990 by McCafferty, who demonstrated that an scFv fragment of the anti-lysozyme Mab D1.3 could be functionally displayed on pIII in the phage vector fd and that the scFv retained antigen binding[77] (Figure 9.6). Moreover, phage displaying the scFv could be enriched from a mixture with wild-type phage by affinity chromatography on a lysozyme column. Enrichment factors of 10^3 for one round of selection and 10^6 for two rounds of selection were achieved.

A large number of vector systems have subsequently been described for the display of antibody fragments (Figure 9.6). The vectors differ primarily with respect to the type of antibody fragment displayed, Fab[76,78,79] or scFv.[40,46,77] the fusion partner pIII[37,79] or pVIII[76,80] and whether the vector is a phage[46,77] or phagemid.[37,40,78,79] The most widely used vectors by far are those that display scFv or Fab antibody fragments as pIII fusions in phagemid vectors (Figure 9.6). pIII display appears to be more robust than display on the other phage proteins; phagemids result in much higher transformation efficiencies, leading to larger antibody library sizes. Use of a phagemid vector results in 'monovalent' antibody fragment display compared with display in true phage vectors such as fd (Figure 9.6). This is because phagemid vectors do not contain the genes for phage proteins other than pIII. The other genes, and regulatory DNA, are provided *in trans* by infecting *E. coli* with helper phage. Wild-type pIII from the helper phage competes with the antibody fragment–pIII fusion for incorporation into the phage coat.

scFv or Fab gene repertoires can be cloned into phage or phagemid display vectors to create what are called 'libraries' of phage, each with a different antibody fragment on its surface and the gene for the antibody fragment inside (see Figure 9.5). Given the high transformation of *E. coli*, it is possible to create libraries containing millions to billions of different antibodies (Figure 9.5). Rare antibodies binding specific antigens can be isolated from non-binding antibodies by a range of different types of affinity chromatography. For example, antigen can be adsorbed on a plastic plate and then phage incubated with immobilized antigen, non-binding phage removed by washing and bound phage eluted. A single round of selection will result in a 20–1000-fold enrichment for binding phage.[77] Eluted phage are used to infect *E. coli*, which produce more phage for the next round of selection. Repetition of the selection process makes it possible to isolate binding phage present at frequencies of less than one in a billion.

Adsorption of protein antigens on plastic may lead to partial or complete antigen denaturation. The problem of antigen denaturation can be overcome by selecting a soluble antigen in solution. The antigen can be chemically tagged, for example by biotinylation, and captured along with bound phage using avidin or streptavidin magnetic beads. Alternatively, the antigen can be genetically tagged with a hexahistidine tag with capture on Ni-NTA agarose or expressed as GST or maltose binding protein fusions with capture on either glutathione or maltose columns. Selections can be performed on more complex mixtures of antigen, including intact cells, provided that measures are taken to prevent enrichment of phage antibodies which bind to non-relevant cell surface antigens. Cell surface antigen specific antibodies have been isolated by selecting phage antibody libraries on adult erythrocytes,[81] fetal erythrocytes,[82] lymphocytes,[83] melanoma cells[84] and breast tumor cells.[85] It has also proven possible to select directly phage antibodies that trigger receptor-mediated endocytosis.[85] Such antibodies can be used for the targeted delivery of therapeutics to the cytosol.

9.6.1.1 Bypassing Hybridoma Technology Using Phage Display

Phage display can be used to bypass hybridoma technology to make scFv or Fab antibodies from immunized mice, rabbits, chickens or humans or from humans who have mounted an immune response as a result of infection or a disease process. Immune lymphocytes are harvested from peripheral blood lymphocytes, spleen or bone marrow and RNA prepared. scFv or Fab gene repertoires are prepared using PCR as described above and used to construct phage antibody libraries from which antigen specific antibodies are isolated using affinity chromatography.

In the first example, an scFv phage antibody library was created from the V genes of a mouse immunized with the hapten phenyloxazolone.[46] An scFv phage antibody library was created from murine V genes and binding phage antibodies were isolated by selection on an antigen column. After two rounds of selection, more than 20 unique scFvs were isolated. The K_d of the highest affinity phage antibodies (1.0×10^{-8} M) were comparable to the affinities of IgG from hybridomas constructed from mice immunized with the same hapten. Similar panels of scFv have been obtained using phage display and mice immunized with EGF receptor[86] and botulinum neurotoxin type A.[33] This approach has also been used to produce monoclonal chicken[44] and rabbit[45] antibody fragments using species specific primers.

Antigen specific antibodies can also be isolated from phage antibody libraries constructed from the V genes of immunized humans or from humans mounting an immune response to an infection or disease process. Since hybridoma technology has not worked well to generate human hybridomas, this approach is especially useful for the generation of human antibodies where an immune response can be induced or is mounted. Human monoclonal antibody fragments have been isolated from immunized volunteers or infected patients against tetanus toxin,[78] botulinum neurotoxin,[87] HIV-1,[88] hepatitis B,[89] hepatitis C,[90] respiratory syncytial virus[91] and hemophilus influenza.[92] Autoimmune antibodies have been isolated from patients with SLE[93] and myasthenia gravis,[94] and also other autoimmune diseases. There has been success in isolating antibodies to self-antigens, including tumor antigens, from libraries constructed from patients with disease[95] or after vaccination with tumor antigen.[84]

9.6.1.2 Bypassing Immunization using Phage Display

In many instances, it is not feasible or ethical to immunize humans with antigens to which one desires to produce human antibodies. In addition, many antigens are evolutionarily conserved and not immunogenic. Fortunately, phage display technology offers a route to produce human antibodies directly without immunization. Very large and diverse scFv or Fab phage antibody libraries are constructed from which it is theoretically possible to generate panels of antigens to virtually any antigen. In the first example, a human scFv gene repertoire was cloned into the phagemid pHEN-1 to create a non-immune phage antibody library of 3.0×10^7 members.[40] From this single non-immune library, scFvs were isolated against more than 20 different antigens, including a hapten, three different polysaccharides and 16 different proteins. The scFvs were highly specific for the antigen used for selection and had affinities typical of the primary immune response with dissociation equilibrium constants ranging from 1 µM to 15 nM. Larger or more diverse phage antibody libraries theoretically provide higher affinity antibodies against a greater number of epitopes on all antigens used for selection.[96] At least five published examples of such very large libraries exist, with sizes ranging between 10^9 and 10^{11}. For example, we have constructed a 6.7×10^9 member scFv phage antibody library from the V genes of healthy humans.[97] An average of nine scFvs with K_d as high as 3.7×10^{-10} M were isolated to 10 different protein antigens. Antibodies from non-immune libraries have been used for Western blotting, epitope mapping, cell agglutination assays, cell staining and FACS. A number have also entered human clinical trials for the treatment of a range of different diseases. The non-immune libraries described above represent the current state of the art in antibody engineering; the ability to generate high-affinity human monoclonal scFv or Fab antibody fragments to any antigen within weeks and without immunization.

9.6.2 Alternative Display Technologies

Although a number of alternative display technologies have been developed, the two most widely used are yeast display and ribosome display. For yeast display, scFv antibody gene repertoires are fused to the

Saccharomyces cell surface protein AgaII.[98] Antigen specific scFvs are isolated by subjecting the yeast antibody library to fluorescent activated cell sorting (FACS) using flow cytometry. Yeast display has been used to generate lead antibodies from both immune and non-immune antibody libraries.[99,100] Its major use, however, has been to increase the affinity of antibodies (antibody affinity maturation, see Section 9.7 below).[101,102] In the most impressive example of affinity maturation using yeast display, the affinity of an scFv was increased more than 1000-fold to 48 fM.[101] Potential advantages of yeast compared with phage display include the ability to select more efficiently higher affinity antibodies using flow cytometry and the ability to measure antibody fragment affinity with the antibody in the display format. This obviates the need for antibody fragment expression and purification. Potential disadvantages of yeast display include the lower transformation efficiencies of yeast compared with bacteria.

In ribosome display, scFv antibody gene repertoires encoded by mRNA are translated *in vitro* in cell-free systems such that the mRNA remaining attached to the ribosome along with the scFv.[103] The ribosome provides the physical link between genotype and phenotype, taking the place of the phage. Antigen binding scFvs are selected on antigen and the mRNA encoding the scFv genes amplified by RT-PCR. Transcription and translation of the scFv DNA provide the display library for the next round of selection. The major advantage of ribosome display is that no cloning is required, resulting in libraries much greater in size than those created by cloning. Ribosome display has been successfully applied to generate scFvs from immune[104] and non-immune[105] libraries and for affinity maturation of existing antibodies.[105]

9.7 ENGINEERING ANTIBODY AFFINITY

Display technologies have been widely used to increase antibody affinity to values not typically achievable using hybridoma technology.[106,107] Mutations are introduced into the DNA of an antigen specific antibody fragment either randomly using, for example, error prone PCR[108] or specifically using spiked oligonucleotides.[109] The mutated antibody fragment DNA is then used to create a display library either in phage or yeast or by using ribosome display. Higher affinity antibody fragments are selected from lower affinity ones by a variety of different affinity chromatography approaches, including flow cytometry for yeast displayed libraries. For example, we reported increasing the affinity of an HER2 scFv using phage display by more than 1000-fold from 16 nM to 13 pM.[109] Others have reported comparable levels of affinity maturation to values down to low picomolar to femtomolar affinities using either phage or yeast display.[101,110] These *in vitro* affinity maturation approaches have now become routine in the biotechnology field.

9.8 ENHANCING ANTIBODY POTENCY

Although antibodies are an effective class of therapeutics, there is certainly room for increasing their potency. This is especially true in the field of cancer therapy. The response rate of antibodies, either as single drugs or in combination with chemotherapy, ranges from 10 to 50%. Existing antibody oncologics work by interrupting essential signaling pathways, such as the epidermal growth factor receptor signaling family, and/or by eliciting ADCC, CDC or apoptosis. There are a number of approaches under investigation to enhance antibody potency by either conjugating antibodies to toxic compounds or by engineering the Fc portion of the antibody to increase ADCC or CDC.

The concept of enhancing antibody potency via delivery of a toxic payload dates to the time of Ehrlich, at the turn of the 20th century. Ehrlich conceptualized magic bullets, molecules that could selectively target a disease-causing organism and deliver a toxic payload. One of the first examples of such antibody delivery of toxic payloads are immunotoxins, antibodies or antibody fragments fused to toxins.[111] Toxins that have been utilized, such as *Pseudomonas* exotoxin, are highly toxic, with just a few molecules being capable of cell killing. Therapeutic development of immunotoxins has required overcoming a number of obstacles, including immunogenicity of bacterial and plant toxins when administered to humans, the high toxicity of the toxins used for immunotoxin construction and the development of methods to fuse toxins to antibodies or antibody fragments.[112] scFv antibody fragments are natural fusion partners for toxins allowing the immunotoxin to be encoded in a single gene and expressed as a single polypeptide chain.[113] Although immunogenicity still remains an issue, immunotoxins are now in clinical trials for cancer and may be especially efficacious against hematological malignancies.[114] An alternative approach is to fuse potent chemotherapeutic drugs directly to antibodies rather than toxins. Such antibody–drug conjugates also show dramatic therapeutic affects in preclinical models and are currently in human clinical trials as cancer therapies.[115]

An alternative way to increase the potency of antibodies is to enhance their inherent cell killing activities that occur via the Fc portion of the antibody and Fc receptors. This interaction is responsible for antibody-dependent cellular cytotoxicity (ADCC) and complement-dependent cellular cytotoxicity (CDC).[116,117] The Fc binds to both activating and inhibitory Fc receptors and, depending on the affinity of the specific interactions, cell killing occurs. Recently, it has proven possible to engineer the Fc to bind activating Fc receptors more efficiently compared with inhibitory Fc receptors.[22] Such engineered Fcs show enhanced ADCC in preclinical models of cancer and will be evaluated for enhanced efficacy in human clinical trials.

9.9 CONCLUSION

The field of antibody engineering effectively began with the engineering of the first chimeric antibodies back in 1984. In the ensuing 25 years, we have seen the development of humanized antibodies and also library approaches to generate directly fully human antibodies to virtually any antigen, including those that are evolutionarily conserved. Library techniques have also developed to the point where it is possible to tune the antibody binding site to virtually any affinity for antigen. At the same time, these bench-side scientific achievements have translated into a novel therapeutic pipeline, biological antibodies. There are now 21 therapeutic antibodies approved by the FDA and in clinical use, with at least 100 more in the various stages of clinical trials. While once the purview of biotechnology companies, essentially every major pharmaceutical company now has a significant effort in the field of therapeutic antibodies. Antibodies, as protein therapeutics, are capable of interrupting protein–protein interactions, something difficult to achieve with traditional small-molecule drugs. It is therefore reasonable to expect that antibodies will continue to be an important and evolving novel class of therapeutics. We are now seeing the next generation of engineered antibodies entering clinical trials. These antibodies are 'armed' to be more potent by conjugation to toxins or toxic drugs or by increased effector functions resulting from engineering of the antibody Fc. It is reasonable to expect that these advances, and others, will result in antibodies with even greater therapeutic efficacy.

REFERENCES

1. A. Casadevall, E. Dadachova and L. A. Pirofski, *Nat. Rev. Microbiol.*, 2004, **2**, 695.
2. J. M. Reichert, C. J. Rosensweig, L. B. Faden and M. C. Dewitz, *Nat. Biotechnol.*, 2005, **23**, 1073.
3. J. M. Reichert and V. E. Valge-Archer, *Nat. Rev. Drug Discov.*, 2007, **6**, 349.

4. G. Kohler and C. Milstein, *Nature*, 1975, **256**, 495.
5. K. James and G. T. Bell, *J. Immunol. Methods*, 1987, **100**, 5.
6. G. J. Jaffers, T. C. Fuller, A. B. Cosimi, P. S. Russel, H. J. Winn and R. B. Colvin, *Transplantation*, 1986, **41**, 572.
7. C. H. J. Lamers, J. W. Gratama, S. O. Warnaar, G. Stoter and R. L. H. Bolhuis, *Int. J. Cancer*, 1995, **60**, 450.
8. M. S. Neuberger, G. T. Williams, E. B. Mitchell, S. S. Jouhal, J. G. Flanagan and T. H. Rabbitts, *Nature*, 1985, **314**, 268.
9. J. Sharon, M. L. Gefter, T. Manser, S. L. Morrison, V. T. Oi and M. Ptashne, *Nature*, 1984, **309**, 364.
10. S. Takeda, T. Naito, K. Hama, T. Noma and T. Honjo, *Nature*, 1985, **314**, 452.
11. A. R. Bradbury and J. D. Marks, *J. Immunol. Methods*, 2004, **290**, 29.
12. J. D. Marks, H. R. Hoogenboom, A. D. Griffiths and G. Winter, *J. Biol. Chem.*, 1992, **267**, 16007.
13. M. J. Mendez, L. L. Green, J. R. Corvalan, X. C. Jia, C. E. Maynard-Currie, X. D. Yang, M. L. Gallo, D. M. Louie, D. V. Lee, K. L. Erickson, J. Luna, C. M. Roy, H. Abderrahim, F. Kirschenbaum, M. Noguchi, D. H. Smith, A. Fukushima, J. F. Hales, S. Klapholz, M. H. Finer, C. G. Davis, K. M. Zsebo and A. Jakobovits, *Nat. Genet.*, 1997, **15**, 146.
14. A. Casadevall and M. D. Scharff, *Antimicrob. Agents Chemother.*, 1994, **38**, 1695.
15. A. Casadevall and M. D. Scharff, *Clin. Infect. Dis.*, 1995, **21**, 150.
16. A. Casadevall, *Clin. Immunol.*, 1999, **93**, 5.
17. J. Bayry, S. Lacroix-Desmazes, M. D. Kazatchkine and S. V. Kaveri, *Trends Pharmacol. Sci.*, 2004, **25**, 306.
18. L. A. Sawyer, *Antiviral Res.*, 2000, **47**, 57.
19. R. E. Black and R. A. Gunn, *Am. J. Med.*, 1980, **69**, 567.
20. A. M. Wu and P. D. Senter, *Nat. Biotechnol.*, 2005, **23**, 1137.
21. I. Pastan, R. Hassan, D. J. Fitzgerald and R. J. Kreitman, *Nat. Rev. Cancer*, 2006, **6**, 559.
22. G. A. Lazar, W. Dang, S. Karki, O. Vafa, J. S. Peng, L. Hyun, C. Chan, H. S. Chung, A. Eivazi, S. C. Yoder, J. Vielmetter, D. F. Carmichael, R. J. Hayes and B. I. Dahiyat, *Proc. Natl. Acad. Sci. USA*, 2006, **103**, 4005.
23. E. A. Kabat and T. T. Wu, *Ann. N. Y. Acad. Sci.*, 1971, **190**, 382.
24. I. M. Tomlinson, J. P. Cox, E. Gherardi, A. M. Lesk and C. Chothia, *EMBO J.*, 1995, **14**, 4628.
25. I. M. Tomlinson, G. Walter, J. D. Marks, M. B. Llewelyn and G. Winter, *J. Mol. Biol.*, 1992, **227**, 776.

26. C. Chothia and A. M. Lesk, *J. Mol. Biol.*, 1987, **196**, 901.

27. C. Chothia, A. M. Lesk, E. Gherardi, I. M. Tomlinson, G. Walter, J. D. Marks, M. Llewelyn and G. Winter, *J. Mol. Biol.*, 1992, **227**, 799.

28. G. L. Boulianne, N. Hozumi and M. J. Shulman, *Nature*, 1984, **312**, 643.

29. S. L. Morrison, M. J. Johnson, L. A. Herzenberg and V. T. Oi, *Proc. Natl. Acad. Sci. USA*, 1984, **81**, 6851.

30. J. W. Larrick, Y. L. Chiang, R. Sheng-Dong, G. Senck and P. Casali, in *Generation of Specific Human Monoclonal Antibodies by In vitro Expansion of Human B Cells: a Novel Recombinant DNA Approach*, C. A. K. Borrebaeck ed., Elsevier, Amsterdam, 1988.

31. J. W. Larrick, L. Danielsson, C. A. Brenner, M. Abrahamson, K. E. Fry and C. A. Borrebaeck, *Biochem. Biophys. Res. Commun.*, 1989, **160**, 1250.

32. R. Orlandi, D. H. Gussow, P. T. Jones and G. Winter, *Proc. Natl. Acad. Sci. USA*, 1989, **86**, 3833.

33. P. Amersdorfer, C. Wong, S. Chen, T. Smith, S. Desphande, R. Sheridan, R. Finnern and J. D. Marks, *Infect. Immun.*, 1997, **65**, 3743.

34. S. A. Iverson, L. Sastry, W. D. Huse, J. A. Sorge, S. J. Benkovic and R. A. Lerner, *Cold Spring Harbor Symp. Quant. Biol.*, 1989, **1**, 273.

35. C. A. Kettleborough, J. Saldanha, K. H. Ansell and M. M. Bendig, *Eur. J. Immunol.*, 1993, **23**, 206.

36. R. D. LeBoeuf, F. S. Galin, S. K. Hollinger, S. C. Peiper and J. E. Blalock, *Gene*, 1989, **82**, 371.

37. H. Orum, P. S. Andersen, A. Oster, L. K. Johansen, E. Riise, M. Bjornvad, I. Svendsen and J. Enberg, *Nucleic Acids Res.*, 1993, **21**, 4491.

38. L. Sastry, M. M. Alting, W. D. Huse, J. M. Short, J. A. Sorge, B. N. Hay, K. D. Janda, S. J. Benkovic and R. A. Lerner, *Proc. Natl. Acad. Sci. USA*, 1989, **86**, 5728.

39. D. R. Burton, C. F. Barbas, M. A. A. Persson, S. Koenig, R. M. Chanock and R. A. Lerner, *Proc. Natl. Acad. Sci. USA*, 1991, **88**, 10134.

40. J. D. Marks, H. R. Hoogenboom, T. P. Bonnert, J. McCafferty, A. D. Griffiths and G. Winter, *J. Mol. Biol.*, 1991, **222**, 581.

41. J. D. Marks, M. Tristrem, A. Karpas and G. Winter, *Eur. J. Immunol.*, 1991, **21**, 985.

42. M. A. Persson, R. H. Caothien and D. R. Burton, *Proc. Natl. Acad. Sci. USA*, 1991, **88**, 2432.

43. D. Sblattero and A. Bradbury, *Nat. Biotechnol.*, 2000, **18**, 75.
44. E. Davies, J. Smith, C. Birkett, J. Manser, D. Anderson-Dear and J. Young, *J. Immunol. Methods*, 1995, **186**, 125.
45. I. Lang, C. R. Barbas and R. Schleef, *Gene*, 1996, **172**, 295.
46. T. Clackson, H. R. Hoogenboom, A. D. Griffiths and G. Winter, *Nature*, 1991, **352**, 624.
47. J. B. Edwards, J. Delort and J. Mallet, *Nucleic Acids Res.*, 1991, **19**, 5227.
48. A. Heinrichs, C. Milstein and E. Gherardi, *J. Immunol. Methods*, 1995, **178**, 241.
49. F. Ruberti, A. Cattaneo and A. Bradbury, *J. Immunol. Methods*, 1994, **173**, 33.
50. S. M. Kipriyanov, O. A. Kupriyanova and G. Moldenhauer, *J. Immunol. Methods*, 1996, **13**, 51.
51. M. A. Boss, J. H. Kenten, C. R. Wood and J. S. Emtage, *Nucleic Acids Res.*, 1984, **12**, 3791.
52. S. Cabilly, A. D. Riggs, H. Pande, J. E. Shively, W. E. Holmes, M. Rey, L. J. Perry, R. Wetzel and H. L. Heyneker, *Proc. Natl. Acad. Sci. USA*, 1984, **81**, 3273.
53. M. Better, C. P. Chang, R. R. Robinson and A. H. Horwitz, *Science*, 1988, **240**, 1041.
54. A. Skerra and A. Pluckthun, *Science*, 1988, **240**, 1038.
55. R. Glockshuber, M. Malia, I. Pfitzinger and A. Pluckthun, *Biochemistry*, 1990, **29**, 1362.
56. C. Horne, M. Klein, I. Polidoulis and K. J. Dorrington, *J. Immunol.*, 1982, **129**, 660.
57. R. E. Bird, K. D. Hardman, J. W. Jacobson, S. Johnson, B. M. Kaufman, S. M. Lee, T. Lee, S. H. Pope, G. S. Riordan and M. Whitlow, *Science*, 1988, **242**, 423.
58. J. S. Huston, D. Levinson, H. M. Mudgett, M. S. Tai, J. Novotny, M. N. Margolies, R. J. Ridge, R. E. Bruccoleri, E. Haber, R. Crea and H. Oppermann, *Proc. Natl. Acad. Sci. USA*, 1988, **85**, 5879.
59. Y. Reiter, U. Brinkmann, S. Jung, I. Pastan and B. Lee, *Protein Eng.*, 1995, **8**, 1323.
60. M. B. Khazaeli, R. M. Conry and A. F. LoBuglio, *J. Immunother. Emphasis Tumor Immunol.*, 1994, **15**, 42.
61. K. Kuus-Reichel, L. S. Grauer, L. M. Karavodin, C. Knott, M. Krusemeier and N. E. Kay, *Clin. Diagn. Lab. Immunol.*, 1994, **1**, 365.
62. P. T. Jones, P. H. Dear, J. Foote, M. S. Neuberger and G. Winter, *Nature*, 1986, **321**, 522.

63. L. Riechmann, M. Clark, H. Waldmann and G. Winter, *Nature*, 1988, **332**, 323.
64. P. R. Tempest, P. Bremner, M. Lambert, G. Taylor, J. M. Furze, F. J. Carr and W. J. Harris, *Biotechnology (N.Y.)*, 1991, **9**, 266.
65. C. Queen, W. P. Schneider, H. E. Selick, P. W. Payne, N. F. Landolfi, J. F. Duncan, N. M. Avdalovic, M. Levitt, R. P. Junghans and T. A. Waldmann, *Proc. Natl. Acad. Sci. USA*, 1989, **86**, 10029.
66. A. B. Thakur and N. F. Landolfi, *Mol. Immunol.*, 1999, **36**, 1107.
67. P. Carter, L. Presta, C. M. Gorman, J. B. Ridgway, D. Henner, W. L. Wong, A. M. Rowland, C. Kotts, M. E. Carver and H. M. Shepard, *Proc. Natl. Acad. Sci. USA*, 1992, **89**, 4285.
68. L. G. Presta, H. Chen, S. J. O'Connor, V. Chisholm, Y. G. Meng, L. Krummen, M. Winkler and N. Ferrara, *Cancer Res.*, 1997, **57**, 4593.
69. P. Tan, D. A. Mitchell, T. N. Buss, M. A. Holmes, C. Anasetti and J. Foote, *J. Immunol.*, 2002, **169**, 1119.
70. M. Figini, J. D. Marks, G. Winter and A. D. Griffiths, *J. Mol. Biol.*, 1994, **239**, 68.
71. L. S. Jespers, A. Roberts, S. M. Mahler, G. Winter and H. R. Hoogenboom, *Biotechnology (N.Y.)*, 1994, **12**, 899.
72. H. R. Hoogenboom, J. D. Marks, A. D. Griffiths and G. Winter, *Immunol. Rev.*, 1992, **130**, 41.
73. C. Marks and J. D. Marks, *N. Engl. J. Med.*, 1996, **335**, 730.
74. A. Nissim, H. R. Hoogenboom, I. M. Tomlinson, G. Flynn, C. Midgley, D. Lane and G. Winter, *EMBO J.*, 1994, **13**, 692.
75. C. Gao, S. Mao, C. H. Lo, P. Wirsching, R. A. Lerner and K. D. Janda, *Proc. Natl. Acad. Sci. USA*, 1999, **96**, 6025.
76. A. S. Kang, C. F. Barbas, K. D. Janda, S. J. Benkovic and R. A. Lerner, *Proc. Natl. Acad. Sci. USA*, 1991, **88**, 4363.
77. J. McCafferty, A. D. Griffiths, G. Winter and D. J. Chiswell, *Nature*, 1990, **348**, 552.
78. C. F. Barbas, A. S. Kang, R. A. Lerner and S. J. Benkovic, *Proc. Natl. Acad. Sci. USA*, 1991, **88**, 7978.
79. H. R. Hoogenboom, A. D. Griffiths, K. S. Johnson, D. J. Chiswell, P. Hudson and G. Winter, *Nucleic Acids Res.*, 1991, **19**, 4133.
80. W. Huse, T. Stinchcombe, S. Glaser, L. M. Starr, M. K. Hellstrom, I. Hellstrom and D. Yelton, *J. Immunol.*, 1992, **149**, 3914.
81. J. D. Marks, W. H. Ouwehand, J. M. Bye, R. Finnern, B. D. Gorick, D. Voak, S. Thorpe, N. C. Hughes-Jones and G. Winter, *Bio/Technology*, 1993, **11**, 1145.

82. M. A. Huie, M. C. Cheung, M. O. Muench, B. Becerril, Y. W. Kan and J. D. Marks, *Proc. Natl. Acad. Sci. USA*, 2001, **98**, 2682.

83. J. de Kruif, L. Terstappen, E. Boel and T. Logtenberg, *Proc. Natl. Acad. Sci. USA*, 1995, **92**, 3938.

84. X. Cai and A. Garen, *Proc. Natl. Acad. Sci. USA*, 1995, **92**, 6537.

85. M. A. Poul, B. Becerril, U. B. Nielsen, P. Morisson and J. D. Marks, *J. Mol. Biol.*, 2000, **301**, 1149.

86. C. Kettleborough, K. Ansell, R. Allen, E. Rosell-Vives, D. Gussow and M. Bendig, *Eur. J. Immunol.*, 1994, **24**, 952.

87. P. Amersdorfer, C. Wong, T. Smith, S. Chen, S. Deshpande, R. Sheridan and J. D. Marks, *Vaccine*, 2002, **20**, 1640.

88. C. F. Barbas, T. A. Collet, W. Amberg, P. Roben, J. M. Binley, D. Hoekstra, D. Cababa, T. M. Jones, A. Williamson, G. R. Pilkington, N. L. Haigwood, E. Cabezas, A. C. Satterthwait, I. Sanz and D. R. Burton, *J. Mol. Biol.*, 1993, **230**, 812.

89. S. L. Zebedee, C. F. Barbas, Y.-L. Hom, R. H. Cathoien, R. Graff, J. DeGraw, J. Pyatt, R. LaPolla, D. R. Burton and R. A. Lerner, *et al., Proc. Natl. Acad. Sci. USA*, 1992, **89**, 3175.

90. S. Chan, J. Bye, P. Jackson and J. Allain, *J. Gen. Virol.*, 1996, **10**, 2531.

91. C. Barbas, J. Crowe, D. Cababa, T. Jones, S. Zebedee, B. Murphy, R. Chanock and D. Burton, *Proc. Natl. Acad. Sci. USA*, 1992, **89**, 10164.

92. D. Reason, T. Wagner and A. Lucas, *Infect. Immun.*, 1997, **65**, 261.

93. S. Barbas, H. Ditzel, E. Salonen, W. Yang, G. Silverman and D. Burton, *Proc. Natl. Acad. Sci. USA*, 1995, **92**, 2529.

94. Y. Graus, M. de Baets, P. Parren, S. Berrih-Aknin, J. Wokke, V. Breda, P. Vriesman and D. Burton, *J. Immunol.*, 1997, **158**, 1919.

95. M. A. Clark, N. J. Hawkins, A. Papaioannou, R. J. Fiddes and R. L. Ward, *Clin. Exp. Immunol.*, 1997, **109**, 166.

96. A. S. Perelson and G. F. Oster, *J. Theor. Biol.*, 1979, **81**, 645.

97. M. D. Sheets, P. Amersdorfer, R. Finnern, P. Sargent, E. Lindquist, R. Schier, G. Hemingsen, C. Wong, J. C. Gerhart and J. D. Marks, *Proc. Natl. Acad. Sci. USA*, 1998, **95**, 6157.

98. E. T. Boder and K. D. Wittrup, *Nat. Biotechnol.*, 1997, **15**, 553.

99. D. R. Bowley, A. F. Labrijn, M. B. Zwick and D. R. Burton, *Protein Eng. Des. Sel.*, 2007, **20**, 81.

100. M. J. Feldhaus, R. W. Siegel, L. K. Opresko, J. R. Coleman, J. M. Feldhaus, Y. A. Yeung, J. R. Cochran, P. Heinzelman, D. Colby, J. Swers, C. Graff, H. S. Wiley and K. D. Wittrup, *Nat. Biotechnol.*, 2003, **21**, 163.

101. E. T. Boder, K. S. Midelfort and K. D. Wittrup, *Proc. Natl. Acad. Sci. USA*, 2000, **97**, 10701.
102. A. Razai, C. Garcia-Rodriguez, J. Lou, I. N. Geren, C. M. Forsyth, Y. Robles, R. Tsai, T. J. Smith, L. A. Smith, R. W. Siegel, M. Feldhaus and J. D. Marks, *J. Mol. Biol.*, 2005, **351**, 158.
103. C. Schaffitzel, J. Hanes, L. Jermutus and A. Pluckthun, *J. Immunol. Methods*, 1999, **231**, 119.
104. J. Hanes, L. Jermutus, S. Weber-Bornhauser, H. R. Bosshard and A. Pluckthun, *Proc. Natl. Acad. Sci. USA*, 1998, **95**, 14130.
105. J. Hanes, C. Schaffitzel, A. Knappik and A. Pluckthun, *Nat. Biotechnol.*, 2000, **18**, 1287.
106. J. Foote and H. N. Eisen, *Proc. Natl. Acad. Sci. USA*, 1995, **92**, 1254.
107. J. Foote and C. Milstein, *Nature*, 1991, **352**, 530.
108. R. E. Hawkins, S. J. Russell and G. Winter, *J. Mol. Biol.*, 1992, **226**, 889.
109. R. Schier, A. McCall, G. P. Adams, K. Marshall, M. Yim, H. Merritt, R. S. Crawford, W. L. M. C. Marks and J. D. Marks, *J. Mol. Biol.*, 1996, **263**, 551.
110. W.-P. Yang, K. Green, S. Pinz-Sweeney, A. T. Briones, D. R. Burton and C. F. Barbas, *J. Mol. Biol.*, 1995, **254**, 392.
111. I. Pastan, M. C. Willingham and D. J. FitzGerald, *Cell*, 1986, **47**, 641.
112. I. Pastan, R. Hassan, D. J. FitzGerald and R. J. Kreitman, *Annu. Rev. Med.*, 2007, **58**, 221.
113. V. K. Chaudhary, J. K. Batra, M. G. Gallo, M. C. Willingham, D. J. FitzGerald and I. Pastan, *Proc. Natl. Acad. Sci. USA*, 1990, **87**, 1066.
114. R. J. Kreitman and I. Pastan, *Curr. Drug Targets*, 2006, **7**, 1301.
115. D. Schrama, R. A. Reisfeld and J. C. Becker, *Nat. Rev. Drug Discov.*, 2006, **5**, 147.
116. F. Nimmerjahn and J. V. Ravetch, *Nat. Rev. Immunol.*, 2008, **8**, 34.
117. F. Nimmerjahn and J. V. Ravetch, *Curr. Opin. Immunol.*, 2007, **19**, 239.

CHAPTER 10

Plant Biotechnology

MICHAEL G. K. JONES

School of Biological Sciences and Biotechnology, Murdoch University, Perth, Western Australia, Australia

10.1 INTRODUCTION

The last decade has been a period of remarkable change which has taken plant biotechnology from study of the science itself to large-scale commercial applications. This is true for almost every aspect of plant biotechnology, both applying basic knowledge of molecular biology and gene organisation such as development of molecular markers to speed up plant breeding practices and using knowledge of genes and how to control expression of those genes to generate and commercialise transgenic crops. In general, the application of plant biotechnology can be divided into two categories: those directed towards the same goals as conventional plant breeding (*e.g.* improved yield, quality, resistance to pests and diseases, tolerance to abiotic stresses) and completely novel applications (such as the use of plants as bioreactors to generate pharmaceuticals, vaccines or biodegradable plastics). The emphasis of this chapter therefore reflects these changes and is focused more on the application of plant biotechnology rather than the detailed molecular biology which underlies those applications.

Molecular Biology and Biotechnology, 5th Edition
Edited by John M Walker and Ralph Rapley
© Royal Society of Chemistry 2009
Published by the Royal Society of Chemistry, www.rsc.org

10.2 APPLICATIONS OF MOLECULAR BIOLOGY TO SPEED UP THE PROCESSES OF CROP IMPROVEMENT

Plant breeding is based on the principles of Mendelian genetics. In the past, plant breeding was something of an art and selection of superior genotypes of a particular crop depended to a great extent on subjective decisions made by the breeder. With increasing knowledge of the genes underlying useful traits, plant breeding has become a more directed and scientific activity. This is in part a result of the generation of molecular maps of crop genomes, extensive sequencing of expressed sequences [expressed sequenced tags (ESTs)] and of genomic sequences and of study of genome organisation, repetitive and non-coding sequences and the ability to identify polymorphisms at particular loci which can be exploited as molecular markers if they are closely linked to a useful trait.

10.2.1 Molecular Maps of Crop Plants

Arabidopsis thaliana had been used as a model plant for mutagenic and genetic studies in the 1970s. The advantages of this species as a model for molecular studies became apparent in the 1980s because of its small nuclear genome size, low repetitive DNA content, short life-cycle, large seed production and later its amenability to transformation. As a result, through international collaborations, a genomic sequencing programme was established and the complete DNA sequence of the *Arabidopsis* genome was completed in 2000.[1]

The rice genome is only about four times the size of the *Arabidopsis* genome and, as a model cereal and important food crop, sequencing of this genome is also well advanced. Massive sequencing efforts of ESTs of wheat, barley, soybean, rice, *Medicago truncatula and* other crops are also in progress, mainly driven by major life sciences companies.

Much interesting information on genome organisation has resulted from this work, including knowledge of location of genes, gene clustering and repetitive and non-coding sequences.[2] For cereals, gene arrangements show 'synteny' in that major blocks of genes are arranged in similar sequences in rice, maize, barley, wheat, *etc.* The major differences in genome size is the result of different amounts of repetitive/non-coding sequences and, for wheat, the fact that it is hexaploid and contains three sets of progenitor genomes.

For *Arabidopsis*, it emerges that there are about 22 000 genes required to contain all the information for this organism. For other plants, we might expect the number of genes present to be between this figure and about 50 000 genes.

Research on mapping genome organisation, sequences and synteny has led to many practical applications, some of which are discussed below.

10.2.2 Molecular Markers

A genetic marker is any character that can be measured in an organism which provides information on the genotype (*i.e.* genetic make-up) of that organism. A genetic marker may be a recognizable phenotypic trait (*e.g.* height, colour, response to pathogens), a biochemical trait (*e.g.* an isozyme) or a molecular trait (*i.e.* DNA based). Whereas phenotypic markers depend on expression of genes and are limited to those genes expressed at a particular time or under particular developmental or environmental conditions, DNA-based markers provide an almost unlimited supply of markers that identify specific sequences across the genome.[3] Their advantages are:

(i) Single base changes in DNA can be identified, providing many potential marker sites across a genome.
(ii) They are independent of developmental stage, environment or expression.
(iii) Markers can be found in non-coding or repetitive sequences.
(iv) Most DNA marker sequences are selectively neutral.

Thus, for example, because about 80% of the wheat genome is non-coding DNA, only molecular markers can be used to identify polymorphisms and to map 'loci' in these regions of the genome.

10.2.3 Types of Molecular Markers

There are many potential approaches to identify molecular markers. Most are based on using the polymerase chain reaction (PCR) to amplify specific DNA sequences.[3-5] They include:

(i) RFLPs (restriction fragment length polymorphisms);
(ii) RAPD-PCR (random amplified polymorphic DNA);
(iii) microsatellites or simple sequence repeats (SSRs);
(iv) AFLP (amplified fragment length polymorphisms).

RFLPs rely on the combination of a probe and restriction enzymes to identify polymorphic DNA sequences using Southern blotting. This approach requires either radioactive or non-radioactive detection methods to identify polymorphic DNA bands and is therefore more time consuming than PCR-based methods.

RAPD-PCR does not require sequence information and involves amplifying random pieces of DNA in which PCR is primed by a single 10 base primer at low stringency, such that random sequences of DNA are amplified based on homologous sequences to the primer being present in the target DNA. It is a useful initial approach to identify polymorphisms, but is not regarded as reproducible enough between laboratories.

Microsatellites or SSRs are groups of repetitive DNA sequences that are present in a significant proportion of plant genomes. They consist of tandemly repeated mono-, di-, tri-, tetra- or pentanucleotide units. The number of repeats varies in different individuals and so the different repeats can be regarded as 'polymorphic' alleles at that 'locus'. To reveal polymorphic microsatellite sequences, it is necessary to sequence the conserved flanking DNA and to design PCR primers that will amplify the repeat sequences. (Because of the repetitive nature of the amplified sequences, typically the main amplified PCR band and additional 'stutter' bands are generated.) For example, at microsatellite locus *Hspl76* of soybean, there is an AT repeat with 13 different numbers of bases in the repeated units in different soybean accessions.[4,5] Microsatellites provide reliable, reproducible molecular markers.

AFLP is also a PCR-based technique, in which selective pre-amplification and amplification steps are carried out to amplify a subset of fragments of the genome, depending on the linkers added and primers used. Many potentially polymorphic fragments are generated by this approach. Polymorphic bands between parents can be identified and linked to useful traits.

Both microsatellite and AFLP markers can be analysed using autoradiography or a DNA sequencer, using fluorescent tags. The latter allows multiplexing such that three different coloured tags plus a size marker can be run in each lane. A single multiplexed AFLP gel can generate 100 polymorphic bands.

10.2.4 Marker-assisted Selection

Using one of the above approaches to identify molecular markers, in combination with an appropriate mapping population of plants plus or minus the trait of interest, many markers have been identified which are closely linked to genes for agronomic traits of interest.[3-7] These include markers for genes for:

(i) pest and disease resistance (against viruses, fungi, bacteria, nematodes, insects);

(ii) quality traits (*e.g.* malting quality barley, bread or noodle wheats, alkaloid levels, *etc.*);

(iii) abiotic stresses (*e.g.* tolerance to salinity or toxic elements such as boron or aluminium);

(iv) developmental traits (*e.g.* flowering time, vegetative period).

If the molecular marker is in the target gene itself, it has been called a 'perfect' marker. Clearly, the closer the molecular marker is linked to the target gene, the better. The overall process of developing a marker thus involves setting up appropriate mapping populations, looking for polymorphic DNA sequences closely linked to the trait of interest, conversion of the polymorphism to a routine marker (usually PCR based), validation and implementation.

Quantitative trait loci (QTLs) are the genes which control quantitative traits such as yield for which the final character is controlled by several genes.[6] To identify and map QTLs, a defined mapping population is required which is screened for polymorphisms by RFLPs, AFLPs and SSRs which can be mapped. Statistical approaches are then used to identify associations between the traits of interest and specific markers. Although the location of QTLs is usually not known exactly, the association of a genotype at a marker/locus and a contribution to the trait indicates that there is a QTL near that marker.

The promise of molecular marker-assisted selection for crop improvement is in the following: increased speed and accuracy of selection; stacking genes, including minor genes; following genes in backcross populations; and reduced costs of field-based selection. Thus, rather than growing breeding lines in the field and challenging or testing for important traits over the growing season, it is possible to extract DNA from 50 mg of a seedling leaflet and test for the presence or absence of a range of traits in that DNA sample in one day. Plants lacking the required traits can then be removed early in the breeding programme.[7] With the availability of more validated molecular markers, marker-assisted selection therefore becomes a highly cost-effective and efficient process.

10.2.5 Examples of Marker-assisted Selection

There are now many examples of the use of molecular markers for selection in plant breeding.[3] Examples include (i) the microsatellite locus *HSP176* of soybean (Section 10.2.3), which is closely linked to a gene (*Rsv*) conferring resistance to soybean mosaic virus, (ii) a perfect marker for noodle quality starch in wheat[7] and (iii) a marker for early flowering in lupins.

Western Australia exports specialty wheat to the Asian market to make white alkaline salted noodles. This segment of the export trade is worth $250 million per annum. White noodles require specific swelling properties of starch. Noodle quality wheats all have two rather than three copies of granule bound starch synthase (GBSS), the enzyme which synthesizes amylose, the linear polymer of starch. This reduces the ratio of amylose to amylopectin by 1.5–2%, increasing the flour swelling volume. A 'perfect' PCR molecular marker was developed which identified presence or absence of the GBSS gene on chromosome 4A, *i.e.* 'bad' or 'good' noodle starch. This molecular marker test is now used as a primary screen for all noodle wheat breeding lines in Western Australia and has resulted in the accelerated production of a series of new noodle quality wheat varieties.[7]

Narrow-leafed lupin is the major grain legume grown in Australia. Early flowering is required for the crop to complete its life-cycle before the rain limits growth in areas of Mediterranean climate where it is grown. Using fluorescent AFLPs, a marker linked to early flowering of lupins was identified (using a DNA sequencer). The AFLP was then run as a radioactive version, the polymorphic band isolated, cloned and sequenced and a co-dominant PCR-based marker developed for routine implementation.

10.2.6 Molecular Diagnostics

The same principles as used in developing molecular markers can be applied for a range of molecular diagnostic purposes in plants, including:

  (i) identification of plant pathogens (viruses, fungi, nematodes, bacteria, insects);
 (ii) studying population structure/variations in pathogens;
(iii) identifying the presence and quantifying the presence of transgenes in transgenic foods;
(iv) following possible pollen transfer of transgenes.

All that is needed is to identify specific nucleic acid (RNA or DNA) sequences unique to the target organism and then to develop a reliable extraction/PCR analysis system such that a DNA fragment is only amplified if the target organism or target sequence is present in a sample. The methods and scale by which such analyses (and also marker-assisted selection) can be carried out are advancing rapidly. Analysis can be by:

  (i) PCR and gel electrophoresis;
 (ii) real-time fluorescent PCR (e.g. using an ABI TaqMan 7700) to
       quantify the original amount of target sequence without gel
       electrophoresis;
(iii) matrix-assisted laser desorption/ionisation time-of-flight mass
       spectrometry (MALDI-TOF MS);
(iv)   use of DNA chips and hybridisation of labelled samples to
       bound DNA sequences.

The last two methods (MALDI-TOF MS and chip technology),
including microarrays (Section 10.2.8), promise to speed up all DNA
fragment analysis applications by an order of magnitude over current
gel-based DNA separation technologies.

Typical examples of applications of molecular diagnostics are the
following. (1) Routine analysis of farmers' seed samples (such as lupin)
for the presence of seed-borne cucumber mosaic virus (CMV).[8] This can
done by RT-PCR (to detect viral RNA, sensitivity <1 infected seed per
1000 seeds) or by real-time fluorescent PCR. Farmers must buy clean
seed if infection levels of CMV are above 0.5%. (2) Routine analysis of
farmers' lupin seed for the fungal disease anthracnose, caused by *Col-
letotrichum acutatum*. A PCR test, based on repeated ribosomal DNA
sequences specific to the fungal pathogen, allows infection levels of one
seed in 10 000 to be detected. In this case, only clean anthracnose-free
seed can be sown.

10.2.7 DNA Fingerprinting, Variety Identification

The same processes of DNA fragment production and analysis can
be applied to DNA fingerprint plants. The main use is in variety identification and quality control, but the techniques can equally be applied
to study plant populations, taxonomy, conservation biology and rehabilitation of mine sites or cleared forests. For rehabilitation studies,
DNA fingerprinting data can be used to ensure that an appropriate
range of genotypes of species removed is used to rehabilitate cleared
land.[9] With the advent of end-point levies on delivered bulk grains
rather than on royalties from seed sales (in some countries), there
is a need for rapid and accurate identification of crop varieties at
the receiving depots. This can be achieved by rapid analysis of DNA
fingerprints of specific crop varieties.

10.2.8 DNA Microarrays

DNA microarrays can be set up robotically by depositing specific fragments of DNA at indexed locations on microscope slides.[10] With current technology, cDNAs, EST clones or open reading frames (ORFs) from sequenced genomes can be set up in microarrays of 10 000 spots per 3.24 cm^2, thus the whole genome of *Arabidopsis* could be displayed on one microscope slide. Fluorescently labelled mRNA probes are hybridised on to the array and specific hybridising sequences are identified by their fluorescent signals. Microarray technology can be used to study gene expression patterns, expression fingerprints, DNA polymorphisms and, in theory, as a breeding tool to evaluate new genetic materials, for a specific trait (*e.g.* drought tolerance), together with phenotypic tests.[11]

10.2.9 Bioinformatics

The generation of massive amounts of molecular information on plant genomes and their products from large-scale sequencing programmes, DNA fingerprinting and fragment analyses, mapping, molecular diagnostics, marker-assisted selection and DNA microarrays requires a concomitant increase in the ability to handle and analyse such data. The handling and interpretation of molecular data are generally referred to as bioinformatics and marries requirements of computing power, data handling and appropriate software. Depending on the level of analysis, the area can be divided into 'genomics' (DNA level), 'proteomics' (protein expression level) and 'metabolomics' (metabolic level). In many cases, it is useful to 'mine' DNA or other databases to look for new or useful genes or sequences, using specific software, programs and algorithms.

10.3 TRANSGENIC TECHNOLOGIES

It is now possible to generate transgenic plants from every major crop plant species. Some are easier to transform than others, but if there are sufficient economic reasons to fund research and development, then, for almost any crop, transgenic plants can be produced. Detailed procedures to generate transgenic plants have been described in the literature, with many incremental improvements in efficiencies of transformation or applications to new genotypes. Two approaches have survived the test of time: transformation using *Agrobacterium tumefaciens* as a vector and particle bombardment.

10.3.1 *Agrobacterium*-mediated Transformation

The use of *A. tumefaciens* (or *A. rhizogenes*) as a gene vector is the method of choice for plant transformation. The development of disarmed binary vectors, in which the virulence (*vir*) genes are separated from the genes to be transferred ('T-DNA'), with the latter on a small easily manipulated plasmid, made the processes of recombinant DNA manipulations more routine. The process of transfer of gene(s) of interest and selectable marker genes have been extensively reviewed,[12] and will not be covered further here. However, one topic that deserves further discussion is that of the selectable marker genes used to identify and extract transgenic cells from the non-transgenic cells and new reporter genes used to monitor gene expression.

10.3.2 Selectable Marker and Reporter Genes

The neomycin phosphotransferase (*nptll*) gene is still the most widely used selectable marker gene and has been used extensively to generate many transgenic dicotyledonous plants and also some monocotyledonous plants. It confers tolerance to aminoglycoside antibiotics (kanamycin, geneticin, puromycin). For various reasons (including intellectual property considerations, fear of transfer of kanamycin tolerance to gut bacteria, *etc.*), it would be useful to have alternative selectable marker genes. Some alternatives include *hpt* (hygromycin tolerance), *bar* or *pat* (phosphinothricin/Basta tolerance),[13] *aroA* (EPSP synthase/glyphosate tolerance),[14] modified *als* (chlorsulfuron tolerance),[15] bromoxynil tolerance and carbon utilising genes mannose phosphate isomerase (mannose utilisation) and xylose isomerase (xylose utilisation).[16]

To study specificity of expression, widely used reporter genes include those encoding β-glucuronidase (*gus*), luciferase (*luc*) and more recently green fluorescent protein (*gfp*) from the jellyfish *Aequoria victoriae*. There are now engineered variants of the latter for improved plant expression (cryptic intron removed) or altered colours of fluorescence (*e.g.* blue, red, yellow). An advantage of *gfp* is that no substrate is required and so *in vitro* expression can be followed in the same cells or tissues over long periods (weeks/months) and high-resolution studies in live cells can be undertaken using confocal scanning laser microscopy.[17]

10.3.3 Particle Bombardment

Many different strategies have been tried to introduce naked DNA into plant cells (including microinjection, DNA pollination, silicon whiskers, electroporation of cells or tissues, electroporation or chemically induced introduction of DNA into protoplasts), but the major alternative to *Agrobacterium* transformation is particle bombardment. This approach now normally involves coating 1 μm diameter particles of tungsten or gold with DNA, putting them on a support, accelerating them to high speed using a pulse of high pressure helium into an evacuated chamber containing the target tissues. Particles can penetrate up to about six cell layers and the DNA released from the particles in surviving cells may be expressed transiently and, in a small proportion of cases, it becomes integrated into the nuclear genome of that cell. With appropriate tissue culture and selection, transgenic plants can be regenerated.

Particle bombardment is usually limited to the generation of transgenic cereals and its use has decreased since *Agrobacterium* transformation now works for cereals such as rice and barley[18] It is still used widely for maize and wheat transformation, although there is one report of wheat transformation using *Agrobacterium* as a vector.[19]

10.4 APPLICATIONS OF TRANSGENIC TECHNOLOGIES

The application of transgenic technologies in plant biotechnology has the potential to exceed the technical advances that have taken place in previous 'revolutions' in production agriculture. It has only taken 4 years of commercial growth of transgenic crops in North America for 52% of the soybean, 30% of the maize and 9% of both cotton and canola grown in 1999 to be transgenic, with increasing production of a wide range of other transgenic crops such as rice, wheat, barley, sorghum, sugar cane, sugar beet, tomato, potato, sunflower, peanut, papaya, tree species and horticultural crops such as carnations. Major companies involved in commercialisation of transgenic crops and their alliances are given in Table 10.1.[20]

Of the 69.5 million acres of transgenic crops grown (27.8 million hectares) in 1998, 74% were in the USA, 15% in Argentina, 10% in Canada and 1% in Australia. The transgenic traits grown commercially were dominated by herbicide tolerance (71%) and insect resistance (28%), with only 1% for the other traits.[20]

However, the transgenic plants currently being commercialised are the first generation of transgenic crops and three generations of transgenic

Table 10.1 Major companies involved in commercialisation of transgenic crops and their alliances (modified from ref. 20).

Company	Partner	Technology basis of collaboration
AgroEvo	Gene Logic	Genomics for crops/crop protection
	Kimeragen	Gene modification technology
	Lynx Therapeutics	Genetics for crops/crop protection
American	Acacia Biosciences	Compounds for agrochemicals
Cyanamid	AgriPro Seeds	Herbicide-tolerant wheat
	Zeneca Seeds	Transgenic canola
BASF	Metanomics	Functional plant genomics
	SunGene	Testing genes in crops
Bayer	Exelixis Pharmaceuticals	Screening targets for agrochemicals
	Lion Bioscience	Genomics for crop protection products
	Oxford Asymmetry	Compounds for agrochemicals
	Paradigm Genetics	Screening targets for herbicides
Dow Chemical	Biosource Technologies	Functional genomics for crop traits
	Demegen	Technology to increase protein content
	Oxford Asymmetry	Compounds for agrochemicals
	Performance Plants	Gene technology to increase yield/content
	Proteome Systems	Protein production in plants
	Ribozyme Pharmaceuticals	Technology to modify oil/starch content
	SemBiosys Genetics	Commercialise proteins produced in plants
DuPont	CuraGen	Genomics for crop protection products
	3-D Pharmaceuticals	Compounds for agrochemical target
	Lynx Therapeutics	Genetics for crops/crop protection
	Pioneer Hybrid	Crop breeding/varieties
FMC	Xenova	Compounds for agrochemicals
Monsanto	ArQule	Compound libraries for agrochemicals
	GeneTrace	Genomics for crops
	Incyte Pharmaceuticals	Plant, bacterial, fungal genomics
	Mendel Biotechnology	Functional genomics in plants
	Cereon Genomics	Plant genomics
Novartis	Chiron	Compounds for agrochemicals
	CombiChem	Compounds for agrochemicals
	Diversa	Plant genetics for transgenic crops
Rhône-Poulenc	Agritope	Genomics joint venture for plant traits
	Celera AgGen	Corn genomics
	Mycogen/Dow	Genetic traits in crops, marketing
	RhoBio	Genetics for disease resistance
Zeneca	Alanex	Compound libraries for agrochemicals
	Incyte Pharmaceuticals	Plant genomics
	Rosetta Inpharmatics	Compounds for crop protection products

crops can be envisaged:

- First generation – production traits (*e.g.* herbicide tolerance, insect/disease resistance).
- Second generation – stacked genes for multiple traits (*e.g.* combinations of disease resistance genes plus quality traits).
- Third generation – varieties tailored for specific end uses (*e.g.* food, fibre, fuel, lubricants, plastics, pharmaceuticals and raw materials for industrial processes).

Already, the production of second-generation transgenic crops is in progress and some specific examples of applications are given in the following sections. However, to achieve the potential benefits of transgenic plant biotechnology, there are many additional factors to consider, which include regulation of biotechnology, intellectual property, food safety, public acceptance, allergenicity, labelling, choice, the environment, segregation of transgenic products and international trade.

The emphasis here is on application and potential benefits of transgenic technologies. As in any developing technologies, the aim is to maximize the benefits and minimize the risks.

10.5 ENGINEERING CROP RESISTANCE TO HERBICIDES

Herbicides are the method of choice to control weeds in most broadscale agricultural systems. They play a major role in maximising crop yields by reducing competition from weeds for space, light, water and nutrients and help control erosion by enabling weed control where crops are drilled directly without ploughing. Weeds can also act as a reservoir for crop pathogens.

Because herbicide resistance genes are also effective selectable marker genes in culture, herbicide-tolerant crop varieties were the first major transgenic trait to be produced and commercialised and herbicide tolerant varieties are still the most widely grown transgenic crops.

Based either on expression of a herbicide insensitive gene, degradation of the herbicide or overexpression of the herbicide target gene product, engineered resistance is now available to a range of herbicides,[20,21] including: glyphosate ('Roundup Ready', 'Touchdown'), glufosinate ('Liberty Link'), imidazolonones (IMI), protoporphyrinogen oxidase inhibitors ('Acuron'), bromoxynil, triazines, 2,4-D, chlorsulfuron/sulfonylureas and isoxazoles.

There is now good evidence that far from increasing the application of herbicides, the control that transgenic herbicide resistant crops provide to the farmer has resulted in a reduction in glyphosate usage of 33% on Roundup Ready soybeans and a reduction in glufosinate usage of about 20% for Liberty Link canola.[20]

10.6   ENGINEERING RESISTANCE TO PESTS AND DISEASES

10.6.1   Insect Resistance

Chemical control of insect pests is both expensive and environmentally unfriendly. Worldwide expenditure on insecticides and the value of crop losses from insect predations[21,22] have been estimated in Table 10.2.

Transgenic cotton, maize and potato crops are being grown commercially which express *Bacillus thuringiensis* (*Bt*) toxins to confer resistance to chewing insects. On sporulation, *B. thuringiensis* synthesizes *S*-endotoxin crystalline proteins encoded by *Cry* genes. On ingestion by an insect, prototoxins are cleaved in the alkaline midgut to the active toxin. This binds to specific receptors in the gut epithelial cells, which results in the formation of pores and eventually to the death of the insect. Some advantages of *Bt* toxins include:

(i)   Specificity – each Cry protein is active against only one or a few insect species.
(ii) Diversity – many different Cry proteins have been identified.
(iii) Reduced or no detrimental effects identified on non-target insects or natural enemies of insects.
(iv) Very low mammalian toxicity.
(v)   Easily degradable.

The production of transgenic *Bt* expressing insect resistant crops has been a high commercial priority, but effective production of *Bt* toxins required re-engineering of *Cry* genes for plant codon usage and removal of cryptic signals (*e.g.* splice sites, polyadenylation signals). These changes permit efficient expression of *Bt* toxins in plants, with both full-length and truncated versions of *Cry* genes used successfully for insect

Table 10.2   Worldwide costs of insecticides and losses caused by insects.

Cost of insecticides (US$ millions)		Losses caused by insects (US$ millions)	
Cotton	1870	Fruit	20000
Fruit and vegetables	2465	Vegetables	25000
Rice	1190	Rice	45000
Maize	620	Maize	8000
Other crops	1965		

resistance. Now, more than 40 different genes containing insect resistance have been incorporated into transgenic crops with several commercialised in different countries such as the USA and Australia.[23]

Given the usefulness of *Bt* toxins for insect control, various management strategies must be adopted to delay development of insect resistance to *Bt*. These include:

(i) setting aside areas of non-*Bt* cotton as refuges to reduce the selection pressure towards insect resistance;
(ii) deploying different insect resistance genes (e.g. protease inhibitors);
(iii) using multiple *Bt* toxins which target different receptors;
(iv) use of spray inducible promoters to control expression of *Bt* genes;
(v) use of tissue-specific promoters such that insects can feed unharmed on economically less important parts of the plant.

It is mandatory to set aside non-transgenic refuges when growing *Bt* cotton, maize and potatoes. Even with additional costs associated with this and other agronomic management practices, the growth of *Bt* cotton gives higher returns to farmers, environmental benefits (50–80% less chemicals used, spraying reduced from 10–12 per year to 4–5 per year) and less occupational exposure of farm workers to sprays.[24]

There are other approaches under development for transgenic insect resistance,[23] including those based on: protease inhibitors, a amylase inhibitors, lectins, chitinases, cholesterol oxidase, cloned insect viruses, tryptophan decarboxylase, anti-chymotrypsin, anti-elastase, bovine pancreatic trypsin inhibitor and spleen inhibitor.

10.6.2 Engineered Resistance to Plant Viruses

Viruses cause significant losses in most major food and fibre crops worldwide. A range of strategies can be used to control virus infection, including chemical treatments to kill virus vectors, identification and introduction of natural resistance genes from related species and use of diagnostics and indexing to ensure propagation of virus-free starting material (seeds, tubers, *etc.*). However, the major development has been the exploitation of pathogen-derived resistance, *i.e.* the use of virus-derived sequences expressed in transgenic plants to confer resistance to plant viruses.[25] This approach is based on earlier observations that inoculation or infection of a plant initially with a mild strain of a virus confers protection against subsequent inoculation with a virulent strain of the same or a closely related virus.[25] Pathogen-derived resistance thus involves transformation of plants with virus-derived sequences; host resistance appears to result from two different mechanisms: (i) protection thought to be mediated by expression of native or modified viral proteins (*e.g.* coat protein, replicase, defective replicase) and (ii) protection mediated at the transcriptional level ('RNA-mediated resistance') which requires transcription of RNA either from full or partial sequences derived from the target virus (including genes for coat protein, replicase, defective replicase, protease, movement protein, *etc.*).

The molecular events which underlie pathogen-derived resistance are the subject of intensive research.[26] The bases of RNA-mediated virus resistance and post-transcriptional gene silencing are probably similar and reflect fundamental activities in plant cells to detect, inactivate and eliminate foreign DNA or RNA.[27] For example, endogenous plant genes inserted into viruses such as PVX can silence expression of the endogenous plant gene.[28] It is probable that low molecular weight double-stranded RNA sequences homologous to the gene message to be silenced or degraded travel systemically through the plant from the site of induction, to ensure that viruses with homologous sequences are degraded when they arrive elsewhere in the plant.[29] Understanding the basis of resistance is needed to ensure practical applications of transgenic virus resistance are stable and have the least environmental risks when deployed on a wide scale.

Most of the major crops have been transformed with genes from major viral pathogens based on the concept of pathogen-derived resistance. For example, by expression of viral replicase derived sequences, host resistance has to be obtained to 13 genera of viruses representing 11 plant virus taxa.[30]

Major crops transformed for virus resistance include potato, tomato, canola, soybean, sugar beet, rice, barley, sugar cane, papaya, melons/ cucurbits, peanut, horticultural and tree species. Effective resistance against a wide range of viruses has been achieved,[21] including PVX, PVY, PLRV, CMV, BYMV, PRSV, ACMV, CPMV, TYLCV, PPV, PMMV, TMV, PEBV, CymRSV, BYDV, RTV and BBTV.

A good practical application of this technology is that of effective protection of transgenic papaya (*Carica papaya*) against papaya ring-spot virus (PRSV).[31] In Hawaii, PRSV has devastated papaya production. Resistance to PRSV (mediated *via* viral coat protein constructs) has held up under field conditions in Hawaii and these results suggest that long-term protection of perennial crops, such as papaya, will be possible using pathogen-derived resistance.

10.6.3 Resistance to Fungal Pathogens

Plants react to attack by fungal and other pathogens by activating a series of defence mechanisms, both locally and throughout the plant.[32] The responses may be non-specific induction of defence reactions to pathogens or specific responses based on the race of the pathogen and the genotype of the host plant. Local resistance may appear as a hypersensitive response in which a local necrotic lesion restricts the growth and spread of a pathogen. Systemic resistance, which may take several hours or days to develop, provides resistance to pathogens in parts of the plant remote form the initial site of infection and longer term resistance to secondary challenge by the initial pathogen and also to unrelated pathogens.

The hypersensitive response is characterised by rapid reactions to invasion by a potential pathogen, through recognition of pathogen or cell wall derived elicitors.[33] It involves:

  (i) opening of specific ion channels;
 (ii) membrane potential changes;
(iii) oxidative burst (generation of reactive oxygen species);
 (iv) synthesis of peroxidase;
  (v) production of secondary metabolites (phenylpropanoids and phytoalexins);
 (vi) synthesis of pathogenesis-related (PR) proteins (*e.g.* β-1,3-glucanases, chitinases);
(vii) cell wall changes (*e.g.* suberin, lignin).

It results in death of host cells, formation of a necrotic lesion and restriction or death of the pathogen.

Transduction of a signal following pathogen recognition can be both local and systemic and involves a number of different pathways.[34,35] The synthesis and accumulation of salicylic acid appears to be necessary for both local and systemic induction of defence responses and salicylic acid (or methyl salicylate) is a major signalling molecule. However, other compounds can activate plant defence genes[32] (*e.g.* 2,6-dichloro-isonicotinic acid, benzothiadiazole, ethylene, abscisic acid, jasmonic acid and systemin). Systemic signals may lead to induction of systemic acquired resistance.

For salicylic acid signalling, salicylic acid moves in the phloem and its presence may be required to establish and maintain systemic acquired resistance. Its arrival in tissues leads to expression of plant defence-related genes in sites distant from the initial challenge, such as PR proteins and production of hydrogen peroxide and reactive oxygen species, cross-linking of cell wall proteins and lignin synthesis.

There are a series of other defence systems that plants use to combat pathogens and these include natural resistance genes and antifungal proteins.

10.6.4 Natural Resistance Genes

The hypersensitive response often results from a specific interaction between a biotrophic pathogen and its host plant. This is known as a 'gene-for-gene' interaction between pathogen and host, in which an avirulence gene product from the pathogen is recognised by a resistance gene in the host plant. This recognition leads to induction of hypersensitive defence responses.[36]

Avirulence (*Avr*) genes encode a variety of polypeptides, some of which may be required for pathogenicity but have then become avirulence factors once they have been detected by the plant. The best characterised *Avr* genes are *Avr4* and *Avr9* of the fungal pathogen *Cladosporium fulvum*.[37] These encode pre-proteins which are processed to mature, extracellular cysteine-rich peptides of 86 and 28 amino acids, respectively. These peptides induce a hypersensitive response in plants which contain the matching resistance genes *Cf-4* and *Cf-9*, respectively.

A series of plant resistance (R) genes active against a range of pathogens have been cloned and characterised. The R genes *Cf-4* and *Cf-9* encode transmembrane glycoproteins, in which the extracellular portion has characteristic leucine-rich repeats (LRR), a transmembrane domain and a C-terminal cytoplasmic domain. The *Avr* gene products are recognised by the LRR receptor regions, which results in signal transduction, gene activation and the hypersensitive response.

With the characterisation of R-genes against fungal, bacterial, viral and nematode pathogens, common patterns have emerged.[36] Depending on the site of recognition of the elicitor, the R gene products may either span the plasma membrane and detect the elicitor extracellularly or be located in the cytoplasm for intracellular elicitor recognition. Intracellular recognition would be expected for virus infection, but pathogens growing extracellularly appear to be recognised by the presence of extracellular elicitors or signal molecules from the pathogens that cross the host plasma membrane.

Five classes of R-genes have been proposed (Table 10.3 and Figure 10.1), in which common features occur – leucine-rich repeats (LRR), nucleotide binding site (NBS), leucine zippers (LZ), toll and interleukin-

Table 10.3 Natural resistance genes characterised.

Class	R-gene	Feature(s)	Location	Pathogen
I	*Pto*	Kinase site	Intracellular	*P. syringae* pv *tomato*
II	*RPS2*	LRR, NBS, LZ	Intracellular	*P. syringae* pv *tomato*
	RPMl	LRR, NBS, LZ	Intracellular	*P. syringae* pv *maculicula*
	12	LRR, NBS, LZ	Intracellular	*F. loxsporium* f.sp
	Mi	LRR, NBS, LZ	Intracellular	*Lycopersici Meloidogyne* spp. (root-knot nematodes)
III	*N*	LRR, NBS, TIR	Intracellular	TMV
	L6	LRR, NBS, TIR	Intracellular	*Melamspora lini*
	RPP5	LRR, NBS, TIR	Intracellular	*Peronospora parasitica*
IV	*Cf2, Cf4, Cf5, Cf9, HS1^pro*	LRR	Transmembrane	*C. fulvum, Heterodera schachtii*
V	*Xa21*	LRR, kinase site	Transmembrane	*Xanthomonas campestris pv vesicatoria*

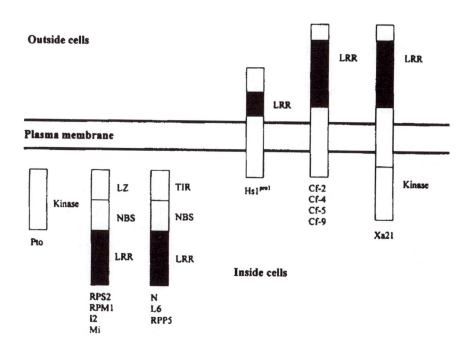

Figure 10.1 Diagnostic representation of different classes of resistance (R)-genes.

like receptors (TIR) and kinase sites.[36] Some of these are shown in Table 10.3.

Natural R-genes of this type can be transferred to other plants and, depending on the presence of elicitors/races of pathogen, may confer resistance in other plants. Modification of R-genes to alter the specificity of recognition of elicitors could also be achieved and various schemes have been developed to convert the specific recognition (*e.g.* of the *avr9/Cf9* system) to a general defence response switched on by damage caused by non-specific pathogens.[38]

As indicated above, most of the R-genes identified so far have common sequence and structural motifs. It is possible to align published sequences and synthesize PCR primers complementary to the conserved sequences and to amplify classes of 'resistance gene analogues' (RGAs).[39] RGAs can be mapped and exist in local clusters of related sequences in the genome. This approach can aid identification of natural resistance genes and mapping of RGAs can also provide molecular markers closely linked to known resistance genes.

However, not all R-genes operate via a gene-for-gene mechanism. The *HM1* R-gene encodes a reductase which inactivates toxins produced by the fungal pathogen *Cochliobulus carbonum* and the *mlo* gene for powdery mildew resistance encodes a negative regulator of cell death.[40]

10.6.5   Engineering Resistance to Fungal Pathogens

The strategy used to engineer resistance to fungal pathogens often depends on the nature of the host–pathogen interaction. As indicated in the previous section, for biotrophic fungal pathogens, a specific R-gene approach can be used since there is often a gene-for-gene interaction between pathogen and host and natural or modified R-genes may be transferred to other genotypes of the same species or to other species, which may confer resistance to the race of pathogen which they recognised in the host plant. However, for necrotrophic fungal pathogens, which kill tissues in advance of hyphal invasion, other approaches are required. These include induction of systemic acquired resistance, production of a range of antifungal proteins,[41] or introduction of genes which can degrade fungal toxins. Examples include:

(i)   genes for toxin inactivation (*e.g. HM1*);
(ii) genes encoding anti-fungal proteins (*e.g.* plant defensins such as radish anti-fungal protein, thionins such as macadamia nut antifungal protein);
(iii) genes encoding PR proteins (*e.g.* chitinases, β,3-glucanases);
(iv) genes that will activate the systemic acquired resistance response;
(v)   artificially induced hypersensitive reaction.

In general, the approaches which involve transformation of plants with genes for anti-fungal proteins do not give complete resistance to fungal pathogens. As a result, it is envisaged that stacking of such resistance genes will be required to provide effective fungal resistance.[41] This may be achieved by multiple transformations or by joining the coding sequences of different anti-fungal protein genes with linkers for peptides recognised by proteases, such that the anti-fungal proteins are translated as one polyprotein and subsequently cleaved to their separate active constituents by protease digestion.

10.6.6 Resistance to Bacterial Pathogens

Pathogenic bacteria have developed pathogenicity or virulence factors which allow the pathogen to multiply in infected tissues. As indicated above for fungal pathogens, in which R-genes resistant to bacterial pathogens are also given, host R-genes recognize some of these factors, encoded by *avr* genes, which trigger a similar range of defence responses.[42,43] Approaches to engineer resistance to bacterial pathogens include:

(i) production of anti-bacterial proteins of non-plant origin (*e.g.* antibacterial proteins *Shival*, *MB39*, Attacin E from the giant silk moth, lysozymes of various origins, *lactoferrin* and *tachyplesin* from the horseshoe crab);

(ii) inhibition of bacterial pathogenicity or virulence factors (*e.g.* tab toxin-resistance protein, phaseolotoxin-insensitive ornithine carbamoyltransferase from *P. syringae*);

(iii) enhanced plant defences *e.g.* pectate lyase, R-genes *Xa21* and *Pto*, glucose oxidase, thionins);

(iv) artificial induction of the hypersensitive response.

The insect lytic peptides (*e.g.* cecropins from the giant silk moth and synthetic analogues) form pores in bacterial membranes. Transgenic potato, tobacco and apples with genes for these lytic peptides have been generated, which confer some protection against bacterial wilt and fireblight (*Erwinia amylovora*). Lysozymes digest bacterial cell wall peptidoglycans and expression of different lysozyme genes can give partial resistance to *E. caratova atroseptica* and *P. syringae* pv *tabaci* in potato and tobacco, respectively.

The R-gene *Xa21* confers resistance to bacterial leaf blight (*Xanthomonas oryzae* pv *oryzae*) and has been transferred into susceptible rice genotypes to confer resistance to *X. oryzae* and overexpression of the resistance conferred by the R-gene *Pto* to *Pseudomonas syringae* pv in tomato confers resistance to other pathogens.[44]

10.6.7   Resistance to Nematode Pathogens

The major nematode pathogens of crop plants are the sedentary endo-parasites root-knot (*Meloidogyne* spp.) and cyst-nematodes (*Hetero-dera/Globodera* spp.). Root-knot nematodes are more prevalent in subtropical and tropical regions, whereas cyst-nematodes are more of a problem in temperate regions. There are also semi-endoparasites and ectoparasites.

The endoparasites develop an intimate association with hosts, with the formation of specific feeding cells (giant cells for *Meloidogyne* spp. and syncytia for *Heterodera* and *Globodera* spp.), which take the form of multinucleate transfer cells.[45] The nematodes are dependent on these cells for nutrients and will not complete their life-cycles if the feeding cells are damaged. The feeding cells therefore provide a major target for engineered resistance.[46] This can take the form of genes that prevent feeding cell induction, genes that prevent or inhibit feeding cell function or toxic compounds delivered to the nematode *via* the feeding cells. For endo-, semi-endo- and ectoparasitic nematodes, it is also possible to attack the body wall of the nematode itself, by expression of specific cuticle-degrading enzymes

Some resistance to nematode attack has been achieved by expression of protease inhibitors in feeding cells,[47] and by damaging feeding cells by feeding cell specific expression of a ribonuclease (*barnase*). Because of the clear requirement for feeding cell formation and function and the differences between the nematode and its host plant, it is probable that effective broadly applicable synthetic resistance genes to nematodes will soon be available.

It is also of interest that cloned nematode R-genes (to sugar beet and cereal cyst-nematodes and to root-knot nematodes) have been characterised. The *Mi* R-gene[47] also confers resistance to aphids,[48] and this is the first example of an R-gene which gives resistance to two different classes of pests (*i.e.* nematodes and aphids).

10.7   MANIPULATING MALE STERILITY

For many crops, it was either impossible or very difficult to generate commercial hybrid seed by conventional means. Hybrid seed is attractive to seed companies because farmers must purchase new seed from them each year, since hybrid varieties do not breed true. It is now possible to engineer male sterility by expression of a ribonuclease gene (*barnase*) specifically during development of the tapetal layer that nourishes developing pollen grains. Developmental regulation of the ribonuclease (by the TA29 tapetum specific promoter) kills the tapetal cells leading to male sterility. Male sterile plants can be used as the female parent to produce hybrid seed. Fertility can be restored by expression of the *barstar* gene, which inactivates *barnase*. This technology can be used to produce hybrids of crops such as maize or sugar beet or canola/rapeseed. It is not possible to produce hybrid canola conventionally and hybrid canola can exhibit hybrid vigour and increased yields. The same basic technology can also be used to induce sterility in transgenic trees, to prevent gene flow *via* cross-pollination.

10.8 TOLERANCE TO ABIOTIC STRESSES

The full potential yield of crops is rarely met because of environmental stresses. These include drought, cold and salt stresses and mineral toxicities. In most cases, advances to generate stress-tolerant plants by traditional breeding have been slow. This is partly because tolerance to these stresses usually involves many genes and physiological processes.

For cold and drought tolerance, recent research has shown that a series of functionally different cold (COR) and drought response genes show common promoter regulatory sequences (CRT/DRE).[49,50] Transcription factors which bind to these regulatory sequences have been identified which switch on the stress response genes in concert. Over-expression of the transcription factor (CBFI or DREB) in *Arabidopsis* by two separate groups conferred tolerance to freezing.[51,52] In one case, the plants also withstood high salt stress.[53] These results are significant because they show that introduction of a single regulatory gene can confer tolerance to stresses.

It may be that freezing injury in plants is mainly caused by dehydration as ice crystals form, so that drought and freezing tolerance may share common protective mechanisms.

Other approaches to stress tolerance include expression of compatible solutes[54] (*e.g.* trehalose or glycine betaine or the amino acid proline[55]) or antifreeze proteins, which order the water in cell cytoplasm and so reduce the freezing point or size of ice crystals which may form. They may also help in maintaining membrane integrity.

The application of microarray technology has major applications in helping to identify genes that are up-regulated when plants encounter environmental stresses, including genes involved in stress perception, signalling and tolerance. For example, this approach is being used to identify genes involved in response to salt tolerance.

10.9 MANIPULATING QUALITY

Quality may be defined simply as the nutritional or technological properties of a product. However, the required quality of a harvested product depends on its intended use, whether it be fresh produce for human or animal consumption, for processing as food or as a raw material for an industrial or other commercial process. It is therefore a complex subject. Improved taste, storage life and quality for current uses, novel uses and aspects such as partitioning of metabolites when products are altered are considered here.

10.9.1 Prolonging Shelf Life

Much produce is lost between harvesting and the point of sale to consumers and before final consumption. A delay in ripening of climacteric fruit (*e.g.* tomato, melon, in which ripening is accompanied by a burst of ethylene production and respiration) and longer storage life is therefore useful, because it allows produce to be left on the plant longer to ripen (providing better flavour) and reduces losses in transport and storage.

Approaches in commercial use to prolong shelf-life include: switching off polygalacturonase genes (*e.g.* 'Flavr-Savr' tomato); switching off genes in the ethylene biosynthesis pathway or degrading intermediates in that pathway; and expression of cytokinin genes.

Initial commercialisation was done with antisense polygalacturonase tomatoes, with both fresh fruit ('Flavr-Savr' sold under the 'Macgregor' label) in the USA and engineered canned tomato purée in the UK. The latter was carefully marketed in two major supermarkets in 1996 by Zeneca and all production was sold.[57] This was the first food product from a transgenic plant sold in the European Union.

In Australia, transgenic long vase-life carnations have been commercialised, with ethylene production inhibited by down-regulation of the ACC synthase gene. Similar technology is now being applied to a wide range of other crop and horticulture plants to enhance storage life.

10.9.2 Nutritional and Technological Properties

10.9.2.1 Proteins
Animals, including humans, cannot synthesize 10 of the 20 essential amino acids, which must therefore be obtained from the diet.[58] It is a paradox that in the past the nutritional balance of foods for human consumption was of less concern to breeders than nutritional properties of products for animal feed. Thus, limiting factors for animal feed for

grains such as barley, maize and wheat were the levels of the amino acids lysine, threonine and tryptophan, whereas sulfur-containing amino acids (methionine and cysteine) were limiting in grain legumes and pulses. It is now possible to overexpress genes encoding proteins with high proportions of limiting amino acids, either from natural sources or as synthetic proteins.

One example is the synthesis of enhanced levels of the sulfur-rich amino acid methionine in grain legumes (*e.g.* lupin), by transfer of a sulfur-rich sunflower seed albumin into lupins. Feeding trials of sulfur-enriched lupins gave increased live weight gain, showing that effective improvements in nutritive value of grain crops are achievable by genetic engineering.[59]

The use of plants to produce speciality proteins is considered in Section 10.10.

The breadmaking quality of wheat results from the viscoelastic properties of the storage proteins. Wheat gluten is a complex mixture of more than 50 proteins, mainly prolamins. The most important of these are high molecular weight subunits of glutenin, because these determine much of the elastic properties of gluten for wheat doughs to make bread, pasta and other foods. High molecular weight glutenin genes have now been transferred to a number of wheat varieties, with modification to dough elasticity.[60,61] This work shows that the technological properties of wheat storage proteins can be modified usefully by genetic engineering.

10.9.2.2 Oils

Plant oils are normally stored as triacylglycerols, with fatty acids and glycerol separated in downstream processing. Oil crops are second in importance to cereals as food sources for humans and provide many industrial products. The major oil products are derived from: soybean (18 Mt), oilpalm (15 Mt), canola/rapeseed (10 Mt), sunflower (8 Mt) and other sources (20 Mt), with a total value of US$45 billion per annum.[62] This value is expected to increase to US$70 billion by about 2010.

Rapeseed (or canola) has been a model oil crop, in great part because it is closely related to *Arabidopsis* and information from the study of *Arabidopsis* can be applied directly to rapeseed. Identification and isolation of genes involved in pathways of oil synthesis have led to a range of transgenic rapeseed varieties with modified oils. The term 'designer oils' has been coined to indicate that chain length, degree of saturation and position of double bonds can be manipulated. Field trials and commercial release of many new oil quality lines are in progress and are shown in Table 10.4.[62]

Table 10.4 Transgenic rapeseed varieties under development

Seed product	*Industrial product*
40% stearic (18:0)	Margarine, cocoa butter
40% lauric (12:0)	Detergents
60% lauric (12:0)	Detergents
80% oleic (18:1 A_9)	Food, lubricants, inks
Petroselinic (18:1 A_6)	Polymers, detergents
Jojoba wax (C20, C22)	Cosmetics, lubricants
40% myristate (14:0)	Detergents, soaps, personal care
90% erucic (22:1)	Polymers, cosmetics, inks, pharmaceuticals
Ricinoleic (18:1–OH)	Lubricants, plasticisers, cosmetics, pharmaceuticals
Polyhydroxybutyrate	Biodegradable plastics
Phytase	Animal feeds
Industrial enzymes	Fermentation, paper manufacture, food processing
Novel peptides	Pharmaceuticals

The percentage content of a specific oil can probably be raised up to at least 90% of the oil product. With depletion of world hydrocarbon reserves, in the future it is probable that plant oils (*e.g.* biodiesel) will compete in terms of price and quality with oil, coal and gas and will become the major large-scale source of renewable industrial hydrocarbons.

Oilseed meal, which is high in protein, is also a valuable commodity used to supplement animal and fish feeds and its protein can be used for human consumption in products such as soy milk and textured vege-table meat substitutes.

Although most advances in biotechnology of oils have come from rapeseed/canola improvement, the same principles can be applied to other oil crops, although this is technically more difficult to achieve, for example, with oilpalms.

10.9.2.3 Manipulation of Starch

Starch constitutes 50–80% of the dry weight of starch-storing organs (*e.g.* potato tubers, cereal endosperm). It occurs in granules as amylose, a linear α1–4 glucan polymer (15–35%) and amylopectin, a linear α1–4 glucan chains connected with α1–6 branches (65–85%). Transgenic expression of enzymes in sucrose metabolism have shown that it is possible, for example in potato, to modify tuber size, number, yield and starch content.[63]

There has been increased commercial interest in amylose-free (waxy) or high-amylopectin starches, because they make processing easier and because of their gel stability and clarity.[64] Amylose-free 'waxy' mutants have been found for many species (*e.g.* maize, rice, barley, wheat, potato, pea) and this correlates with loss of granule-bound starch synthase (GBSS) enzyme activity. Waxy starches can also be generated using antisense technology by switching off GBSS gene expression and its relative accumulation and form can be manipulated (*e.g.* in potato tubers). Applications include: generating noodle quality wheat starch (lacking GBSS on chromosome 4A); reduced amylose or waxy wheat starch (*e.g.* two or all three GBSS loci – 4A, 7A, 7D – inactive); and generating starches with novel structures.

Modified starches have novel properties that can be used to develop new applications in the food processing industry.

10.9.2.4   Fructans

Sugar beet normally stores sucrose, but by expression of a 1-sucrose–sucrose fructosyl transferase (1-ISST) gene in sugar beet transgenic plants now store fructans. The significance of this work is that humans cannot digest fructans, so that they can be used as low-calorie food ingredients. Low molecular weight fructans (up to five monosaccharide units) taste like sugar and can be used as low-calorie sweeteners, whereas long-chain fructans form emulsions which can be used to replace fats in foods such as creams and spreads.[65,66]

10.9.3   Manipulation of Metabolic Partitioning

The flow of metabolites through metabolic and biosynthetic pathways is regulated to respond to developmental and environmental conditions. The distribution of metabolites (partitioning) controls the flux of carbon compounds towards the synthesis of sugars, starch and oils in storage organs and may change to meet developmental states, environmental constraints or in the activation of the diverse array of defence responses when a plant is attacked by pathogens. Metabolic fluxes can be manipulated and an understanding of the control of partitioning between metabolic pathways is required for metabolic engineering – this knowledge can be used to increase the production or yield of commercially important metabolites and products.[67,68] Strategies to modify partitioning include:

(i)   changing levels of signalling metabolites and hormones;
(ii) removal of end products to change reaction equilibria;
(iii) expression of heterologous enzymes to bypass endogenous regulatory mechanisms;
(iv) expression of transcriptional regulators that control pathways;
(v)   switching off gene expression by sense/antisense suppression.

Most commercial applications of metabolic partitioning involve altered starch and oil biosynthesis, reducing lignin content or synthesis of biodegradable plastic (polyhydroxybutyrate, PHB), but this aspect is also relevant to all uses of plants as bioreactors to produce more of current or novel compounds efficiently.

10.10 PRODUCTION OF PLANT POLYMERS AND BIODEGRADABLE PLASTICS

Plastics are important materials in everyday use. They are polymers that can be moulded using heat and pressure. However, they present serious problems of disposal, persistence and environmental pollution and it would be preferable if biodegradable plastics could be generated from agricultural products, which could be degraded biologically to carbon dioxide and water.[69] Potential sources of biological polymers include starch, cellulose, pectins, proteins and PHB. Various native starches from wheat, rice, maize and potato have been used to make plastics and a range of shaped products have been manufactured.[70] However, production of PHB or related polyhydroxyalkanoates in plants provides the most attractive sources of biodegradable plastics at low cost. These crops will enhance the value of production and benefit land use and waste disposal.

A combination of microbiology, polymer chemistry and plant biotechnology is likely to provide new crops providing valuable non-food products. For example, there is the prospect of growing 'easy-care' polyester cotton, in which PHB is deposited in the hollow central lumen of the cotton fibre. Less than 1% of PHBs can improve the insulating property of the cotton fibre by 8–9%.[71]

10.11 PLANTS AS BIOREACTORS: BIOPHARMING AND NEUTRACEUTICALS

Plant products have long been used in medicine, and now plant biotechnology is entering the medical field in a spectacular way, with the creation of plants which produce proteins of pharmaceutical value or subunit vaccines.

10.11.1 Edible Vaccines

Traditionally, attenuated strains of pathogenic organisms have been injected or delivered orally to invoke an immune response. Now it is possible to clone genes encoding immunogenic subunits of pathogen proteins and to express these either in transgenic plants or in plant viruses. Transgenic plants have a permanent capacity to express the vaccine, whereas engineered viruses permit transient production of large quantities of immunogenic protein.[72] The aim of this work is that by eating fresh fruit (*e.g.* banana) containing an antigen, individuals can develop immunity to the pathogenic organism *via* the gut immune system. In some cases human trials are in progress to test the efficiency of this approach for the following diseases and applications: hepatitis B; cholera; *E. coli* heat-labile enterotoxin; Norwalk virus; rhinovirus; HIV; rabies; antimalarial parasites; immunocontraception (*ZP3* zona pellucida protein); and inhibition of late-onset diabetes.[73]

The same approach is being used to engineer plant vaccines for domestic animals and livestock (*e.g.* parvoviruses such as feline pan-leukopenia virus, canine virus, mink enteritis virus);[74] swine fever, foot and mouth disease and rabies.[75]

10.11.2 Production of Antibodies in Plants

The 'biopharming' approach indicated above can also be used to produce either large quantities of antibodies in plants as bioreactors or to deliver pharmacologically active antibodies in food. A good example of the latter is the synthesis in plants of antibodies against the bacterium (*Streptococcus mutatis*) that causes dental caries in teeth. Extracted antibodies confer protection in human trials.[76] Expressed in an apple, this transgenic apple a day would also keep the dentist away!

10.11.3 Plant Neutraceuticals

The concept of using plants as bio-factories to produce pharmaceuticals also extends to the potential production of a wide range of other products. These may be in the health care area (*e.g.* dermatology, cardiology, human metabolism, endocrinology, respiration, transplantation and oncology),[77] or production of compounds that improve human diet or health (the formation of a new 'wellness complex' by food). For example, Prodi Gene and Stauffer seeds are growing maize that produces avidin for use in health diagnostic kits,[78] and Brazzein, a low-calorie natural sweetener.[77] Other neutraceuticals include over-production of vitamin A in rice or phytosterols to reduce cholesterol in humans. Clearly, identity preservation and separate storage/transport and handling will be required for such products, as for most high-value transgenic crops.

10.12   PLANT BIOTECHNOLOGY IN FORESTRY

Forests are both economically and environmentally important. The demand for wood products, such as timber for construction, paper, pulp and energy, is increasing and expanded plantation forestry will be required to meet the global demand. The time lines for tree breeding have made progress in improving forest productivity slow. Now, the full range of biotechnological technologies is being applied to tree improvement. These include clonal micropropagation of superior trees, hybridisation, molecular markers and marker-assisted selection and transformation.

A range of tree species has been transformed using both *Agrobacterium* and particle bombardment approaches:[79] poplar, European larch, hybrid larch, Norway spruce, Scots pine, white spruce, black spruce, eastern larch, radiata pine, Tasmanian bluegum and a range of Eucalyptus species. The targets for tree improvement involve improved growth rate, wood characteristics, pulp quality, pest and disease resistance and tolerance to abiotic stresses.

A major target is to engineer reduced lignin content, since lignin represents about 25% of the wood biomass and lignins reduce the efficiency of pulp and paper production. High energy usage and production of chemical pollutants also result from delignification processes. This has been achieved by down-regulation of key enzymes in the lignin biosynthetic pathway [*e.g.* cinnamyl alcohol dehydrogenase (CAD); *O*-methyl transferase (OMT)].

Transgenic trees have also been generated with modified form, quality and performance using auxin biosynthetic genes, *rol* c genes and peroxidase,[79] and also with herbicide resistance and Bt insect resistance. Environmental considerations are particularly important for growth of transgenic trees. In order to reduce gene flow *via* pollen to native species, engineered sterility of transgenics is desirable.

10.13   INTELLECTUAL PROPERTY

Because of patenting of novel genes, promoters, generic technology and applications, the commercialisation of plant biotechnology can be a complex issue. The costs associated with establishing, contesting, defending and monitoring patents are a major reason why large life sciences companies rather than public organisations (universities and government-funded research institutes) are exploiting the new technologies. This trend will undoubtedly continue and the major companies listed in Table 10.1, such as Monsanto, DuPont-Pioneer and Novartis, will increase their hold on commercialisation. Plant breeding is already essentially in the private domain in Europe and is rapidly moving that way in North America and Australia.

Although intellectual property and patents must now be considered from the start of any scientific work in molecular biology that is intended to be commercialised and freedom to operate from gene to the sale of seeds established, there is another level to consider, that is, the ownership of germplasm. Breeding materials from many crops often originate in centres of origin which are frequently in developing countries, in which germplasm may be held by international research centres. The countries where the germplasm originated may well require some equity or payment for use of their germplasm and accusations of 'biopiracy' have been made against breeders in developed countries. Greater sensitivity will be needed in this area if access to new or wild germplasm is not to be blocked.

10.14   PUBLIC ACCEPTANCE

The speed at which genetically modified ('GM') crops and food have arrived has been unexpectedly rapid. In North America, where labelling of GM food or products is not undertaken unless there is a substantial difference from non-GM food, there are more than 300 million people routinely eating GM produce. In Europe, different forces prevail on public opinion, many of which are more connected with other events such as the 'mad cow disease' (BSE) outbreak in the UK, trust of scientists, politics and trade. These will slow acceptance and development of GM foods and biotechnology and put Europe at a competitive disadvantage compared with North America. Australia lies between Europe and North America in terms of public acceptance of GM crops and food, with transgenic cotton and carnations currently commercialised. Major questions in acceptance of GM technology arise:

- Is GM food safe?
- Are there any long-term effects from eating GM food?
- Should all GM food be labelled?
- Should consumers have the right to choose between GM and non-GM food?
- Will they contain new allergens?
- What are the environmental consequences?

There is, of course, no logical reason why any organisation would wish to produce unsafe food. At present, GM food is subjected to much more stringent testing than conventional foods and is probably safer as a result. The labelling issue is not as simple as it may seem. It is straightforward to label a transgenic tomato as such but, for example, soy products are used in differing amounts in thousands of processed foods and it is difficult to label these in a meaningful and informative way. Similarly, cotton oil is relatively pure and free of DNA or protein derived from transgenes conferring herbicide tolerance. Does it need to be labelled as GM? Is it substantially equivalent to the non-GM product or not – where should the line be drawn? There are also concerns about 'ethnically' sensitive genes (*e.g.* a gene derived from a pig inserted in a crop plant) and access of developing countries to GM technologies.

The large-scale consumption of GM food in North America by 300 million people since its introduction, without any reported problems, supports the general view based on the science that GM food is safe. Environmental issues should perhaps be of more concern. Pollination of organic crops by transgenic maize is an issue,[80] as is possible injury to Monarch butterfly caterpillars eating food plants dusted with Bt maize pollen. In contrast, there are reports of increased numbers of birds of prey after 3 years' growth of Bt insect-resistant crops, as fewer non-target insects are killed, and this effect is transmitted through the food chain.

At present, public debate often has little to do with the science and actual risks and more to do with sensationalism and disinformation. Banner headlines occur regularly in the press, on 'Demon Seeds', 'Mutant Foods from the Gene Giants', 'Frankenstein foods', *etc.* However, this phase is expected to pass fairly rapidly, as understanding of the benefits of the technologies in efficient food production, improved food quality, protection of the environment and of biodiversity are realised.

10.15   FUTURE PROSPECTS

Although transgenic technologies will not provide all the answers to generating sustainable food production in the coming decades, the 'gene revolution' will undoubtedly have more impact than the 'green revolution' on all aspects of agriculture and biotechnology. The potential is that 80% of the food eaten in developed countries will have transgenic content by the year 2010 and that up to 10% of the US maize crop will be devoted to bioreactors, biopharming and production of neutraceuticals. Whether or not these percentages will be achieved in practice will depend very much on public attitudes and acceptance of transgenic technologies over the next few years.

REFERENCES

1. E. M. Meyerowitz and C. R. Somerville (eds.), *Arabidopsis*, Cold Spring Harbor Laboratory Press, Cold Spring Harbor, NY, 1994.
2. C. Dean and R. Schmidt, *Annu. Rev. Plant Physiol. Plant Mol. Biol.*, 1995, **46**, 395.
3. M. Mohan, S. Nair, A. Bhagwat, T. G. Krishna, M. Yano, C. R. Bhatia and T. Sasaki, *Mol. Breeding*, 1997, **3**, 87.
4. W. Powell, G. C. Machray and J. Provan, *Trends Plant Sci.*, 1996a, **1**, 215.
5. W. Powell, M. Morgante, C. Andre, M. Hanafey, J. Vogel, S. Tingey and A. Rafalski, *Mol. Breeding*, 1996b, **2**, 225.
6. J.-M. Ribaut and D. Hoisington, *Trends Plant Sci.*, 1998, **3**, 236.
7. A. Briney, R. Wilson, R. H. Potter, I. Barclay, R. Appels and M. G. K. Jones, *Mol. Breeding*, 1998, **4**, 427.
8. S. J. Wylie, C. R. Wilson, R. A. C. Jones and M. G. K. Jones, *Aust. J. Agric. Res.*, 1993, **44**, 41.
9. A. Karp, K. J. Edwards, M. Bruford, S. Funk, B. Vosman, M. Morgante, O. Seberg, A. Kremer, P. Boursot, P. Arctander, D. Tantz and G. M. Hewitt, *Nat. Biotechnol.*, 1997, **15**, 625.
10. Desprez, J. Amselem, M. Caboche and H. Hofte, *Plant J.*, 1998, **14**, 643.
11. D. M. Kehoe, P. Villand and S. Somerville, *Trends Plant Sci.*, 1999, **4**, 38.
12. R. G. Birch, *Annu. Rev. Plant Physiol. Plant Mol. Biol.*, 1997 **48**, 297.
13. C. J. Thompson, N. R. Movva, R. Tichard, R. Crameris, J. E. Davies and M. Lauwereys, *EMBO J.*, 1987, **6**, 2519.
14. D. M. Shah, R. B. Horsch, H. J. Klee, G. M. Kishore, J. A. Winter, N. E. Turner, C. M. Hironaka, P. R. Sanders, C. S. Gasser, S. Aykent, N. R. Siegel, S. R. Rogers and R. T. Fraley, *Science*, 1986, **233**, 478.
15. G. W. Haughn, J. Smith, B. Mazur and C. Somerville, *Mol. Gen. Genet.*, 1998, **211**, 266.
16. A. Haldrup, S. G. Petersen and F. T. Okkels, *Plant Cell Rep.*, 1998, **18**, 76.
17. J. Haseloff, K. R. Siemering, D. C. Prasher and S. Hodge, *Proc. Natl. Acad. Sci. USA*, 1997, **94**, 2122.
18. S. Tingay, D. McElroy, R. Kalla, S. Fieg, M. Wang, S. Thornton and R. Brettel, *Plant J.*, 1997.
19. S. Tingay, D. McElroy, R. Kalla, S. Fieg, M. Wang, S. Thornton and R. Brettel, *Plant J.*, 1997.

20. A. M. Thayer, *Chem. Eng. News*, 1999, **77**, 21.
21. P. J. Dale, *TIBTech*, 1995, **13**, 398.
22. R. A. de Maagd, D. Bosch and W. Stiekma, *Trends Plant Sci.*, 1999, **4**, 9.
23. T. H. Schuler, G. M. Poppy, B. R. Kerry and I. Denholm, *TIBTech*, 1998, **16**, 168.
24. N. W. Forrester, in *Commercialisation of Transgenic Crops: Risk, Benefit and Trade Consideration*, ed. G. D. McLane, P. M. Waterhouse, G. Evans and M. J. Gibbs, Bureau of Resource Science, Canberra, 1997, p. 239.
25. J. C. Sanford and S. A. Johnstone, *J. Theor. Biol.*, 1985, **113**, 395.
26. A. van Kammen, *Trends Plant Sci.*, 1997, **2**, 409.
27. S. P. Kumpatla, M. B. Chandrasekharan, L. M. Iyer, G. Li and T. C. Hall, *Trends Plant Sci.*, 1998, **3**, 97.
28. J. J. English and D. C. Baulcombe, *Plant J.*, 1997, **12**, 1311.
29. H. Vaucheret, C. Beclin, T. Elmayan, F. Feuerbach, C. Godon, J.-B. Morel, P. Mourrain, J.-C. Palauqui and S. Vernheltes, *Plant J.*, 1998, **16**, 651.
30. R. Yang, PhD Thesis, Murdoch University, Perth, 1999.
31. S. Luis, R. M. Manshardt, M. M. M. Fitch, J. L. Slightom, J. C. Sanford and D. Gonsalves, *Mol. Breeding*, 1997, **3**, 161.
32. J. Durner, J. Shah and D. F. Klessig, *Trends Plant Sci.*, 1997 **2**, 266.
33. N. Benhamou, *Trends Plant Sci.*, 1996, **1**, 233.
34. C. Wasternack and B. Parthier, *Trends Plant Sci.*, 1997, **2**, 302.
35. E. Blumwald, G. S. Aharon and B. C.-H. Lam, *Trends Plant Sci.*, 1998, **3**, 342.
36. P. J. G. M. de Witt, *Trends Plant 5c/.*, 1997, **2**, 452.
37. M. Kooman-Gersmann, R. Vogelslang, E. C. M. Hoogendijk and P. J. G. M. de Witt, *Mol. Plant Microbe Interact.*, 1997, **10**, 821.
38. K. E. Hammond-Kosack, S. Tang, K. Harrison and J. G. Jones, *Plant Cell*, 1998, **10**, 1251.
39. V. Kanazin, L. F. Marek and R. C. Shoemaker, *Proc. Natl. Acad. Sci. USA*, 1996, **93**, 11746.
40. R. Buschges, K. Hollricher, R. Panstruga, G. Simons, M. Wolter, A. Frijters, R. van Daelen, T. van der Lee, P. Diergaarde, J. Groenendijk, S. Topsch, P. Vos, F. Salamini and P. Schulze-Lefert, *Cell*, 1997, **88**, 695.
41. W. Broekaert, B. P. A. Cammue, M. F. C. De Bolle, K. Thevissen, G. W. De Samblanx and R. W. Osborn, *Crit. Rev. Plant Sci.*, 1997, **16**, 297.

42. U. Bonas and G. Van den Ackerveken, *Plant J.*, 1997, **12**, 1.
43. F. Mourgues, M. Brisset and E. Chevreau, *TIBTech*, 1998, **16**, 203.
44. X. Tang, M. Xie, Y. J. Kim, J. Zhou, D. F. Klessig and G. B. Martin, *Plant Cell*, 1999, **11**, 15.
45. M. G. K. Jones, *Ann. Appl. Biol.*, 1981, **97**, 353.
46. V. M. Williamson and R. S. Hussey, *Plant Cell*, 1996, **8**, 1735.
47. P. E. Urwin, C.-J. Lilley, M. J. McPherson and H. J. Atkinson, *Plant J.*, 1997, **12**, 455.
48. M. Rossi, F. L. Goggin, S. B. Milligan, I. Kaloshian, D. E. Ullman and V. M. Williamson, *Proc. Natl. Acad. Sci., USA*, 1998, **95**, 9750.
49. E. J. Stockinger, S. J. Gilmour and M. F. Thomashow, *Proc. Natl. Sci. USA*, 1997, **94**, 1035.
50. S. J. Gilmour, D. G. Zarka, E. F. Stockinger, M. P. Salazar, J. M. Houghton and M. F. Thomashow, *Plant J.*, 1998, **16**, 433.
51. K. R. Jaglo-Ottosen, S. J. Gilmour, D. G. Zarka, O. Schabenberger and M. F. Thomashow, *Science*, 1998, **280**, 104.
52. Q. Liu, M. Kasuga, Y. Sakuma, H. Abe, S. Miura, K. Yamaguch-Shinozaki and K. Shinozaki, *Plant Cell*, 1998, **10**, 1391.
53. M. Kasuga, Q. Liu, K. Yamaguchi-Shinozaki and K. Shinozaki, *Nat. Biotechnol.*, 1999, **17**, 287.
54. H. J. Bohnert and R. G. Jensen, *Trends Biotechnol.*, 1996, **14**, 89.
55. P. B. K. Kishor, Z. Hong, G.-H. Miao, C.-A. A. Hu and D. P. S. Verma, *Plant Physiol.*, 108, 1387.
56. D. Worrall, L. Elias, D. Ashford, M. Smallwood, C. Sidebottom, P. Lillford, J. Telford, C. Holt and D. Bowles, *Science*, 1998, **282**, 115.
57. N. J. Poole, in *Commercialisation of Transgenic Crops: Risk, Benefit and Trade Considerations*, ed. G. D. McLane, P. M. Waterhouse, G. Evans and M. J. Gibbs, Bureau of Resource Sciences, Canberra, 1997, p. 17.
58. L. Tabe and T. J. V. Higgins, *Trends Plant Sci.*, 1998, **3**, 282.
59. L. Molvig, L. M. Tabe, B. O. Eggum, A. E. Moore, S. Craig, D. Spencer and T. J. V. Higgins, *Proc. Natl. Acad. Sci. USA*, 1997, **94**, 8393.
60. F. Barro, L. Rooke, F. Bekes, R. Gras, A. S. Tatham, R. Fido, P. A. Lazzeri, P. R. Shewry and P. Barcelo, *Nat. Biotechnol.*, 1997 **15**, 1295.
61. K. Vasil and O. D. Anderson, *Trends Plant Sci.*, 1997, **2**, 292.
62. D. J. Murphy, *TIBTech*, 1996, **14**, 206.
63. U. Sonnewald, M.-R. Hajirezaei, J. Kossman, A. Heyer, R. M. Trethewey and L. Willmitzer, *Nat. Biotechnol.*, 1997, **15**, 794.
64. S. G. Ball, M. H. B. J. van de Wal and R. G. F. Visser, *Trends Plant Sci.*, 1998, **3**, 462.

65. S. Smeekens, *Nat. Biotechnol.*, 1998, **16**, 822.
66. R. Sevenier, R. D. Hall, I. M. van der Meer, H. J. C. Hakkert, A. J. van Tunen and A. J. Koops, *Nat. Biotechnol.*, 1998, **16**, 843.
67. M. Stitt and U. Sonnewald, *Annu. Rev. Plant Physiol. Plant Mol. Biol.*, 1995, **46**, 341.
68. K. Herbers and U. Sonnewald, *TIBTech*, 1996, **14**, 198.
69. C. Nawrath, Y. Poirier and C. Somerville, *Mol. Breeding*, 1995 **1**, 105.
70. J. J. G. Van Soest and J. F. G. Vliegenthart, *TIBTech*, 1997, **15**, 208.
71. C. Bryne, *Chem. Ind.*, 1999, May, 343.
72. C. J. Arntzen, *Nat. Biotechnol.*, 1997, **15**, 221.
73. S. W. Ma, D.-L. Zhao, Z.-Q. Yin, R. Mukherjee, B. Singh, H.-Y. Qin, C. R. Stiller and A. M. Jevnikar, *Nat. Med.*, 1997, **3**, 793.
74. K. Dalsgaard, A. Uttenthal, T. D. Jones, F. Xu, A. Merryweather, W. D. O. Hamilton, J. P. M. Langeveld, R. S. Boshuizen, S. Kamstrup, G. P. Lomonossoff, C. Porta, C. Vela, J. I. Casal, R. H. Meloen and P. B. Rodgers, *Nat. Biotechnol.*, 1997, **15**, 248.
75. Modelska, B. Dietzschold, N. Sleysh, Z. F. Fu, K. Steplewski, D. C. Hooper, H. Koprowski and V. Yusibov, *Proc. Natl. Acad. Sci. USA*, 1998, **95**, 2481.
76. J. K.-C. Ma, B. Y. Hikmal, K. Wycoff, N.-D. Vine, D. Chargeleque, L. Yu, M. B. Hein and T. Lehmer, *Nature*, 1998, **4**, 601.
77. J. Olson, *Farm Ind. News*, 1999, **32**, 1.
78. E. E. Hood, D. Witcher, S. Maddock, T. Meyer, C. Baszczynski, M. Bailey, P. Flynn, J. Register, L. Marshall, D. Bond, E. Kulisek, A. Kusnadi, R. Evangelista, Z. Nikolov, C. Wooge, R. J. Mehigh, R. Hernan, W. K. Kappel, D. Ritland, C. P. Li and J. A. Howard, *Mol. Breeding*, 1997, **3**, 291.
79. T. Tzfira, A. Zuker and A. Altman, *TIBTech*, 1998, **16**, 439.
80. R. B. Jorgensen, T. Hanser, T. R. Mikkelsen and H. Ostergaard, *Trends Plant Sci.*, 1996, **1**, 356.

CHAPTER 11

Biotechnology-based Drug Discovery

K. K. JAIN

Jain PharmaBiotech, Blaesiring 7, 4057 Basel, Switzerland

11.1 INTRODUCTION TO DRUG DISCOVERY

11.1.1 Basics of Drug Discovery in the Biopharmaceutical Industry

Historically, pharmaceutical products have been developed primarily through the random testing of thousands of synthetic compounds and natural products. The traditional approach to drug discovery is based on the generation of a hypothesis based on biochemistry and a pharmacological approach to a disease. Targets are defined on the basis of this hypothesis and lead discovery is a matter of chance. The classical drug discovery effort also involves isolating and characterising natural products with some biological activity. These compounds are then 'refined' by redesigning their molecular structure to yield new entities with higher biological activity and lower toxicity/side-effects. The main limitation of such a process is that the discovery of natural products with defined biological activity is essentially a hit-or-miss approach and therefore lacks a rational basis.

The focus of this chapter is on the description of drug discovery in the biotechnology era. Genomics, proteomics, metabolomics, monoclonal antibodies, antisense, RNA interference (RNAi), molecular diagnostics, biomarkers and nanobiotechnology have a significant impact on the drug discovery process.

Molecular Biology and Biotechnology, 5th Edition
Edited by John M Walker and Ralph Rapley
© Royal Society of Chemistry 2009
Published by the Royal Society of Chemistry, www.rsc.org

11.1.2   Historical Landmarks in Drug Discovery and Development

Drug discovery is as old as human history. Ancient human cultures had medicines for illness that were discovered by trial and error from plant, animal and mineral sources. The effective constituents of some of the herbal medicines have been isolated and form the basis of some modern medicines. Modern medicine is considered to have started in the nineteenth century, although several important discoveries, notably smallpox vaccine, were made close to the end of the eighteenth century. Modern pharmaceuticals and drug discovery started to develop in the twentieth century. Five stages in the evolution of drug development are shown in Table 11.1.

In the earlier part of the twentieth century, drug discovery was based on chemistry and serendipity played a role. For the first generation of pharmaceutical products, studies were done *in vivo*; there were no clinical trials. Most of the drugs were used for palliation or symptomatic relief in a medical environment that was more art than science. Pioneer pharmaceutical companies were still growing.

The second generation of pharmaceutical products, which were based on biology, started to be introduced around the middle of the twentieth century. There was still a considerable amount of empiricism and *in vitro* methods started to develop for the study of drugs. The medicines were more effective and could modify the disease process. Laboratory methods started to be used increasingly to support clinical diagnosis in

Table 11.1   Five stages in the evolution of drug development.

Period	*Advances*
1820–1880	Discovery of 'active principles' in natural products, fermentations and simple coal-tar derivatives: analgesics, antipyretics, anaesthetics, hypnotics, sedatives
1880–1930	Experimental therapeutics and chemotherapy. Use of synthetic organic dyes to identify pathogenic microorganisms and to manufacture antiprotozoal medicines, serums, toxins and vaccines
1930–1960	Introduction of sulfa drugs, antibiotics, antihistamines, vitamins, corticosteroids and sex hormones
1960–1980	Drugs to treat cardiovascular diseases such as hypertension; drugs acting on the nervous system such as anti-anxiety drugs and antidepressants; oral contraceptives; semisynthetic penicillins, cephalosporins; and non-steroidal anti-inflammatory drugs
1980–present	Protein therapeutics, antineoplastics and antivirals. Cell therapy and gene therapy, novel drug delivery systems and diagnostic tests based on recombinant DNA and MAbs. Introduction of genomics, proteomics, nanobiotechnology, RNAi and biomarkers for drug discovery in the last decade of 20th century and the first years of 21st century

an environment of experience-based medicine, which was still a mixture of art and science. The last quarter of the century saw the development of biotechnology.

Towards the end of the twentieth century and in the twenty-first century, genomics and genetics started to play an important role in drug development, with a marked increase in the number of biotechnology-based drugs. Drug discovery activities have accelerated with the use of high-throughput methods, biochips and robotics. The role of chemistry is still considered to be important and proteomics is gradually assuming an important role in drug discovery in the post-genomic era. Bioinformatics and *in silico* methods are being used for drug discovery. The clinical environment is evidence-based medicine with the start of evolution of personalised medicines. With all the scientific advances in medicine, there is still some resurgence of alternative or complementary medicine and an evaluation of drugs from natural sources within the pharmaceutical industry.

11.1.3 Current Status of Drug Discovery

The drug discovery process is shown in Figure 11.1. The usual duration of this process is 5–6 years, which is half of the total development time (10–12 years) taken from target identification to marketing of a drug.

Traditional screening methods are slow and labour intensive and have limited the number and chemical diversity of the compounds and targets

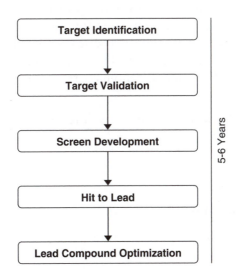

Figure 11.1 Drug discovery process.

that can be tested in a given assay. Even though many thousands of distinct chemical structures exist, it is not unusual for screening utilising this approach to be terminated at the end of several years with no lead compounds identified, having examined only a small fraction of available compounds. This limitation of speed and scale often restricts both the quality and quantity of lead compounds available for further testing and development, thereby hindering drug discovery.

Several 'hits' are produced as a result of high-throughput screening (HTS). The hit-to-lead stage has been added to the drug discovery. Multiple parameters are optimised in parallel with produce leads with a balanced profile of biological and physicochemical properties. New technologies are playing an increasing role in this process. It is desirable to have multiple series in hit optimisation so that more than one series is available for lead optimisation.

11.2 NEW BIOTECHNOLOGIES FOR DRUG DISCOVERY

Various biotechnologies that can be applied for improving drug discovery are listed in Table 11.2 and will be described briefly below. Currently available 'omics' technologies for drug discovery include genomics, proteomics and metabolomics.

11.3 GENOMIC TECHNOLOGIES FOR DRUG DISCOVERY

From the current knowledge of genomics, at least 500 targets are available for drug therapy. Opportunities provided by genomics include the potential for developing treatments for 100 of the most important multifactorial diseases with 500–1000 disease-related genes and 3000–10 000 new drug targets. Genomic technologies are built on basic tools of biotechnology. Functional genomics represents a new phase of

Table 11.2 Biotechnologies for improving drug discovery.

Omics technologies:
 Genomics
 Proteomics
 Metabolomics and metabonomics
Microarrays and biochips
Monoclonal antibodies (MAbs)
Antisense
RNA interference (RNAi)
Nanobiotechnology
Biomarkers
Bioinformatics

genome analysis, *i.e.* function of genes. It is characterised by high-throughput or large-scale experimental methods combined with statistical and computational analysis of the results. The fundamental strategy in a functional genomics approach is to expand the scope of biological investigation from studying single genes or proteins to studying all genes or proteins at once in a systematic fashion.

11.3.1 SNPs in Drug Discovery

The study of single nucleotide polymorphisms (SNPs) is crucial for characterising molecular targets and can also validate the role of these targets in disease. SNPs are important in the development of new pharmaceuticals with impact in the following areas:

- *Target identification.* Positional cloning looks for disease-susceptibility genes near markers that have an inheritance pattern similar to that of the disease. SNPs are used as simple genetic markers.
- *Target characterisation.* The degree of genetic variation within a target is important because it can alter gene function and influence drug interaction. Because modern methods of drug discovery involve high-throughput screening of large chemical libraries produced by combinatorial chemistry, it is important that the target of the screen is representative of the majority of the target population. This is even more important when SNPs affect the amino acid structure and function of the protein. Although drug targets are screened at nucleic acid level to determine their degree of genetic variation, identification of the variants may be inadequate and bioinformatic support is necessary.
- *Target validation.* There is an abundance of drug targets available but many have no clue as to their role in disease. Genetic epidemiology can be used to show the functional involvement of a particular drug target in the disease of interest. This approach can also be used to identify new therapeutic targets for existing drugs.
- *Pharmacogenetics.* This is the study of how genetic variations affect drug response and metabolism. Polymorphisms in enzymes that metabolise drugs are also responsible for unexpected adverse effects of normal doses of drugs. This information is now used for the development of personalised medicines.
- *Pharmacogenomics.* This term implies the use of genetic sequence and genomics information in patient management to enable therapy decisions to be made. The genetic sequence and genomics information can be that of the host (normal or diseased) or of the

pathogen. Pharmacogenomics will have an impact on all phases of drug development – from drug discovery to clinical trials. Pharmacogenomics is an important basis for the development of personalised medicines.

11.3.2 Gene Expression Profiling

Analysis of gene-expression patterns derived from large expressed sequence tag (EST) databases has become a valuable tool in the discovery of therapeutic targets and diagnostic markers. Sequence data derived from a wide variety of cDNA libraries offer a wealth of information for identifying genes for pharmaceutical product development. Collecting, storing, organising, analysing and presenting cDNA expression data require advanced bioinformatics methods and high-performance computational equipment. Comparison of expression patterns from normal and diseased tissues enables inferences about gene function to be made and medically relevant genes as candidates for therapeutics research and drug discovery.

11.3.3 Limitations of Genomics for Drug Discovery and Need for Other Omics

Although useful, DNA sequence analysis alone does not lead efficiently to new target identification, because one cannot easily infer the functions of gene products or proteins and protein pathways from a DNA sequence. It has become obvious that analysing genome sequences alone will not lead to new therapies for human diseases. Rather, an understanding of protein function within the context of complex cellular networks will be required to facilitate the discovery of novel drug targets and, subsequently, new therapies directed against them.

Functional genomics and proteomics have provided a huge amount of new drug targets. High-throughput screening and compound libraries produced by combinatorial chemistry have increased the number of new lead compounds. The challenge is to increase the efficiency of testing lead efficacy and toxicity. The traditional methods of toxicity testing in laboratory animals using haematological, clinical chemistry and histological parameters are inadequate to cope with this challenge. Gene and protein expression studies following treatment with drugs have shown that it is possible to identify changes in biochemical pathways that are related to a drug's efficacy and toxicity and precede tissue changes. The patterns of these changes can be used as efficacy or toxicity markers in high-throughput screening assay. Several other omics technologies play an important role in drug discovery. Notable among these are proteomics and metabolomics.

11.4 ROLE OF PROTEOMICS IN DRUG DISCOVERY

The term 'proteomics' indicates PROTEins expressed by a genOME and is the systematic analysis of protein profiles of tissues. The term 'proteome' refers to all proteins produced by a species, much as the genome is the entire set of genes. Proteomics parallels the related field of genomics. Proteomics is important for drug discovery because many of the pharmaceutically important regulation systems operate through proteins (*i.e.*, post-translationally). Major drugs act by binding to proteins. For example a 'protease inhibitor' drug is designed to disable the protease enzyme (which is a protein) that allows a particular virus to reproduce. A drug with the right shape can latch on to the surface of the protease protein and keep it from doing its job. If the protease is disabled, the virus cannot reproduce itself, so the damage it can inflict is limited. The role of proteomics in drug development can be termed 'pharmacoproteomics' and is an important component of personalised medicine.

11.4.1 Proteins as Drug Targets

The majority of drug targets are proteins that are encoded by genes expressed within tissues affected by a disease. It is estimated that there are approximately 10 000 different enzymes, more than 2000 different G-protein-coupled receptors, 200 different ion channels and 100 different nuclear hormone receptors encoded in the human genome. These proteins are key components of the pathways involved in disease and, therefore, are likely to be a rich source of new drug targets.

Proven drug targets share certain other characteristics, which can only be identified by understanding their expression levels in cells and cannot be determined by their gene sequences alone. Drug targets are (1) often expressed primarily in specific tissues, allowing for selectivity of pharmacological action and reducing the potential for adverse side-effects and (2) generally expressed at low abundance in the cells of the relevant organ. An effective target discovery system would therefore allow the detection of genes that encode for proteins expressed in specific tissues at low abundance, thereby permitting the rapid identification of proteins, which are likely to be targets for therapeutic and diagnostic development. Some proteomic technologies that are useful for drug discovery are described briefly.

11.4.2 Protein Expression Mapping by 2D Gel Electrophoresis

2D gel electrophoresis (2DGE) and image analysis may be used for the quantitative study of global changes in protein expression in tissues, cells or body fluids. This method has the advantages of direct determination of protein abundance and detection of post-translational modifications such as glycosylation or phosphorylation, which result in a shift in mobility; mass spectrometry may be used for the subsequent characterisation of proteins of interest. Because thousands of proteins are imaged in one experiment, a picture of the protein profile of the sample at a given point in time is obtained, allowing comparative proteome analysis. Protein expression changes may give clues to the role of certain proteins in disease and some of the identified proteins map to known genetic loci of a disease.

11.4.3 Liquid Chromatography-based Drug Discovery

Liquid chromatography (LC) shifts the high-resolution separation from front-end 2DGE to back end LC–mass spectrometry (MS). Proteins are kept in solution so that a higher percentage of the sample is analysed using affinity chromatography and having access to the entirely accurately assembled genome of the organism under study. Protein mixture may be digested to analyse at the peptide level. This high-efficiency proteomics technology is applied to small-molecule drug discovery.

Instead of separating individual proteins using 2DGE, complex mixture analysis can start with proteins pooled after partial fractionation by multi-dimensional LC, which involves serial protein separations over a variety of chromatographic matrices. The complex mixtures are treated with trypsin and the resulting peptides are separated by LC and measured by MS. This approach may be specifically applied to integral membrane proteins to obtain detailed biochemical information on this unwieldy class of proteins.

11.4.4 Matrix-assisted Laser Desorption/Ionisation Mass Spectrometry

Among several proteomic technologies used in drug discovery, matrix-assisted laser desorption/ionisation (MALDI)-MS and its variants, and related techniques, play an important role. MALDI time-of-flight (TOF) MS targets, when uniformly precoated with a thin film of matrix/nitrocellulose, make the sample preparation straightforward and permit the enrichment and analysis of proteins at low levels in proteomics samples. In general, the sensitivity for proteins and peptides can be enhanced 10–50 times compared with traditional MALDI sample preparation techniques. Tissue imaging mass spectrometry (IMS) by MALDI and ion trap MS with higher order MS scanning functions have been used for localisation of dosed drug or metabolite in tissues. Laser capture microscopy (LCM) is used to obtain related samples from tissue for analyses by standard MALDI-MS and HPLC-MS. IMS by MALDI ion trap MS has proved sensitive, specific and highly amenable to the image analysis of traditional small molecule drug candidates directly in tissue.[1]

A novel method for on-tissue identification of proteins in spatially discrete regions is described using tryptic digestion followed by MALDI-IMS with MS/MS analysis.[2] IMS is first used to reveal the protein and peptide spatial distribution in a tissue section and then a serial section is robotically spotted with small volumes of trypsin solution to carry out *in situ* protease digestion. After hydrolysis, 2,5-dihydroxybenzoic acid matrix solution is applied to the digested spots, with subsequent analysis by IMS to reveal the spatial distribution of the various tryptic fragments. Sequence determination of the tryptic fragments is performed using on-tissue MALDI-MS/MS analysis directly from the individual digest spots. This protocol enables protein identification directly from tissue while preserving the spatial integrity of the tissue sample.

11.4.5 Protein–Protein Interactions

Protein interactions can be monitored *in vivo* over the course of the cell cycle, drug treatments or other environmental stimuli. The development of green fluorescent protein derivatives has provided the opportunity to study protein–protein interactions in living cells. The structural organisation of macromolecular protein complexes, which may contain scores of protein interactions and may be difficult to study *in vitro*, can be analysed. Techniques to study protein–protein interactions in living subjects will allow the study of cellular networks, including signal transduction pathways, and also the development and optimisation of pharmaceuticals for modulating protein–protein interactions.

Protein–protein interaction networks are also called 'interactome' networks. The interactome is a map of all interactions that take place in an organism between all proteins, in all cells, all tissues, at all ages and in response to all possible environmental conditions. The ability to find links between sets of proteins involved in different genetic disorders offers a novel approach for more rapidly identifying new candidate genes involved in human diseases. Pharmaceutical investigators use interaction data to prioritise potential drug targets as these networks help the companies to weed out proteins that have several interaction, some of which are irrelevant to the target.

11.4.6 Use of Proteomic Technologies for Important Drug Targets

Protein kinases are encoded by more than 2000 genes and thus constitute the largest single enzyme family in the human genome. Most cellular processes are regulated by the reversible phosphorylation of proteins on serine, threonine and tyrosine residues. At least 30% of all proteins contain covalently bound phosphate. A novel method to determine if drugs and drug targets are effective in combating disease is by identifying the key regulatory protein 'switches" (phosphorylation sites) inside human cells. This allows the identification and characterisation of changes in the chemical modification of proteins that may arise in response to drug treatment. It can be used to identify novel targets in disease, to compare the effects of different drug candidates and to develop assays that can be used throughout preclinical and clinical development. Kinomics, omic for kinome (the kinase complement of the human genome), is a useful tool for identifying protein kinases that play an important role in disease. It also assists in the drug optimisation process. Kinomics can be used to understand both the mechanism of action and the specificity of potential drugs. This knowledge forms a crucial base from which to develop potent and selective compounds with minimal side-effects. Protein kinases are important drug targets in human cancers, inflammation and metabolic diseases.

Proteomic technologies have been used for the study of G-protein coupled receptors (GPCRs), which are an important class of drug targets that exist as proteins on the surface membranes of all cells. GPCRs are a superfamily of proteins accounting for approximately 1% of the human genome and are associated with a wide range of therapeutic categories, including asthma, inflammation, obesity, cancer, cardiovascular, metabolic, gastrointestinal and neurological diseases. Purified multiple GPCRs in a functional form can be used for the identification of tight binding ligands. There are estimated to be ∼2000 GPCRs within the human body with potential availability as drug discovery targets. GPCRs have historically been valuable drug targets, but to date there are only approximately 100 well-characterised GPCRs with known ligands, several of which are currently targets of commercial drugs. Approximately 60% of all currently available prescription drugs interact with these receptors.

11.5 METABOLOMIC AND METABONOMIC TECHNOLOGIES FOR DRUG DISCOVERY

Metabolomics is the study of small molecules and their interactions within an organism, which is critical to the drug-discovery process. The importance of metabolomic studies is indicated by the finding that a large proportion of the ~ 6000 genes present in the genome of *Saccharomyces cerevisiae* and of those sequenced in other organisms, encode proteins of unknown function. Many of these genes are 'silent', *i.e.* they show no overt phenotype, in terms of growth rate or other fluxes, when they are deleted from the genome. How the intracellular concentrations of metabolites can reveal phenotypes for proteins active in metabolic regulation has been demonstrated. Quantification of the change of several metabolite concentrations relative to the concentration change of one selected metabolite can reveal the site of action, in the metabolic network, of a silent gene. In the same way, comprehensive analyses of metabolite concentrations in mutants, providing 'metabolic snapshots', can reveal functions when snapshots from strains deleted for unstudied genes are compared to those deleted for known genes.

An approach combining proteomics with metabolomics creates a new platform for identifying and validating metabolite biomarkers important to the development of safe and effective drugs. The unique combination of gene expression profiling (determining the level of activity of genes in an organism at a specific time), metabolic profiling (determining the identity and quantities of chemicals in an organism at a specific time) and phenotypic profiling (measuring the physical and chemical characteristics of an organism at a specific time), with data from all systems and analysed by bioinformatics, creates a new paradigm for industrialising functional genomics.

Metabonomics is a systems approach to investigate the metabolic consequences of drug exposure, disease processes and genetic modification whereas metabolomics is the measurement of metabolite concentrations in cell systems. It is important to direct the timing of proteomic and genomic studies to maximise the probability of observing useful biological transitions that are indicative of function. Although a number of spectroscopic methods have been used, NMR spectroscopy is considered to be one of the most powerful methods for generating multivariate metabolic data. An NMR-based systems approach is used for drug toxicity screening to aid lead compound selection. Metabolic phenotyping (metabotyping) is also used for investigating the metabolic effects of genetic modification and modelling of human disease processes. One deliberate gene knockout can produce several metabolic disturbances. Metabonomics can thus be used as a functional genomics tool with applications in various stages of drug discovery and development.

11.6 ROLE OF NANOBIOTECHNOLOGY IN DRUG DISCOVERY

Nanotechnologies – nanoparticles and various nanodevices such as nanobiosensors and nanobiochips – have the potential to improve drug discovery. Microfluidics has already proven useful for drug discovery and, through further miniaturisation, nanotechnology will improve the ability to fabricate massive arrays in small spaces using nanofluidics. Nanoparticles such as gold nanoparticles and quantum dots have attracted considerable attention recently with their unique properties for potential use in drug discovery.[3]

11.6.1 Nanobiotechnology for Target Validation

Multivalent attachment of small molecules to nanoparticles can increase specific binding affinity and reveal new biological properties of such nanomaterial. Multivalent drug design has yielded antiviral and anti-inflammatory agents several orders of magnitude more potent than monovalent agents. Parallel synthesis of a library has been described, which is comprised of nanoparticles decorated with different synthetic small molecules.[4] Screening of this library against different cell lines led to the discovery of a series of nanoparticles with high specificity for endothelial cells, activated human macrophages or pancreatic cancer cells. This multivalent approach could facilitate the development of functional nanomaterials for applications such as differentiating cell lines, detecting distinct cellular states and targeting specific cell types. It has potential applications in high-throughput drug discovery, target validation, diagnostics and human therapeutics.

11.6.2 Nanotechnology-based Drug Design at Cell Level

To create drugs capable of targeting human diseases, one must first decode exactly how a cell or a group of cells communicates with other cells and reacts to a broad spectrum of complex biomolecules surrounding it. But even the most sophisticated tools currently used for studying cell communications suffer from significant deficiencies and typically can only detect a narrowly selected group of small molecules or, for a more sophisticated analysis, the cells must be destroyed for sample preparation. A nanoscale probe, the scanning mass spectrometry (SMS) probe, can capture both the biochemical makeup and topography of complex biological objects. The SMS probe can help map all those complex and intricate cellular communication pathways by probing cell activities in the natural cellular environment, which might lead to better disease diagnosis and drug design on the cellular level.

11.6.3 Nanomaterials as Drug Candidates

In addition to the use of nanobiotechnology for drug discovery, some drugs are being developed from nanomaterials. Well-known examples of these are dendrimers, fullerenes and nanobodies. Specialised chemistry techniques allow precise control over the physical and chemical properties of the dendrimers. Polyvalent dendrimers interact simultaneously with multiple drug targets. They can be developed into novel targeted cancer therapeutics. Polymer–protein and polymer–drug conjugates can be developed as anticancer drugs. Dendrimer conjugation with low molecular weight drugs has been of increasing interest recently for improving pharmacokinetics, targeting drugs to specific sites and facilitating cellular uptake. Opportunities for increasing the performance of relatively large therapeutic proteins such as streptokinase (SK) using dendrimers have been explored in one study.[5] Using the active ester method, a series of streptokinase–polyamidoamine (PAMAM) G3.5 conjugates were synthesised with varying amounts of dendrimer-to-protein molar ratios. All of the SK conjugates displayed significantly improved stability in phosphate buffer solution, compared with free SK. The high coupling reaction efficiencies and the resulting high enzymatic activity retention achieved in this study could lead to a desirable approach for modifying many bioactive macromolecules with dendrimers.

A key attribute of the fullerene molecules such as C_{60} is their numerous points of attachment, allowing for precise grafting of active chemical groups in 3D orientations. This attribute, the hallmark of rational drug design, allows for positional control in matching fullerene compounds to biological targets. In concert with other attributes, namely the size of the fullerene molecules, their redox potential and its relative inertness in biological systems, it is possible to tailor requisite pharmacokinetic characteristics to fullerene-based compounds and optimise their therapeutic effect. A number of water-soluble C_{60} derivatives have been suggested for various medical applications. These applications include neuroprotective agents, HIV-1 protease inhibitors, bone-disorder drugs, transfection vectors, X-ray contrast agents, photodynamic therapy agents and a C_{60}–paclitaxel chemotherapeutic.

Nanobodies, derived from naturally occurring single-chain antibodies, are the smallest fragments of naturally occurring heavy-chain antibodies that have evolved to be fully functional in the absence of a light chain. Like conventional antibodies, nanobodies show high target specificity and low inherent toxicity; however, like small-molecule drugs, they can inhibit enzymes and can access receptor clefts. Nanobodies can address therapeutic targets not easily recognised by conventional antibodies such as active sites of enzymes.

11.7 ROLE OF BIOMARKERS IN DRUG DISCOVERY

A biomarker is a characteristic that can be objectively measured and evaluated as an indicator of a physiological and also a pathological process or pharmacological response to a therapeutic intervention. Any specific molecular alteration of a cell on DNA, RNA, metabolite or protein level can be referred to as a molecular biomarker.

Among the current applications of biomarkers, those for drug discovery and development are one of the most important types. Biomarkers can be used to predict and confirm target binding, to determine mechanism of action of a drug, pharmacokinetics and toxicity. Their role continues into the drug development stage to determine treatment efficacy in clinical trials. It is worthwhile to develop biomarkers for exploring the pharmacology of new molecules and to develop potential biomarkers of efficacy. A molecule that does not have the intended pharmacological effect is unlikely to have the desired efficacy and its development should be terminated. Discovery efforts would then be directed at understanding the reasons for the lack of pharmacological effect and finding improved molecules. For those molecules that have the intended pharmacological effect but then fail to show efficacy, it is possible to say with confidence that the molecular target is ineffective and that discovery efforts should be directed to other targets. Such an approach will increase the overall success rates of the candidate molecules delivered into clinical development.[6]

11.8 SCREENING IN DRUG DISCOVERY

Screening in drug discovery is conducted in a streamline assay format aimed at hit identification. Screening may be low throughput (10 000–50 000 assay points), medium throughput (50 000–100 000 data points) or high throughput (100 000–500 000 + data points). Ideal screening procedure should be cost-effective and robust under varying conditions such as chemical types. Two well-established strategies, diversity-based and focused screening, have been compared and reviewed. Focused screening is more widely used and the most popular of these methods involves the use of 3D information on targets where crystal structure is available. Diversity screening means testing all the molecules that could be considered drug candidates and the number could go up to one million samples. This approach is based on a commitment to increase in capacity by automation, miniaturisation and throughput with reduction of cost.

11.8.1   Cell-based Screening System

Unlike traditional methods in the drug discovery process, the cell-based method provides deep biological information (time, space and activity) about a drug candidate's physiological impact on specific cellular targets within living cells. This approach provides significant insight into the potential efficacy and toxicity of a drug candidate on cells before initiating animal testing and human clinical trials, thereby saving substantial time and expense.

Cell-based assays have the advantage that they can be miniaturised to increase screening throughput and reduce costs. Functional cell-based assays would be useful for screening of modulators of ion channels. Because the binding sites may be unknown, formatting of these assays is difficult with the use of traditional biochemical or ligand-displacement methods. New optical assays employ fluorescent or optical readouts and permit cell-based assays for most targets, including ion channels. Further, genetically encoded probes offer the possibility of biosensors for intracellular biochemistry, specifically localised targets and protein–protein interactions.

11.8.2   Receptor Targets: Human *versus* Animal Tissues

In the traditional approach to drug discovery, compounds are screened against animal tissues containing many different types of receptors. Companies such as Lundbeck Research USA follow approaches involving use of human tissues. The advantages of using human tissues are:

1. By having an isolated receptor as a target, chemists are better able to design compounds that interact with only the target of interest and not with other receptors that may be responsible for side-effects.

2. Using human receptors as drug design targets will substantially reduce the number of problems that often arise during the drug development process as a result of differences in a compound's activity in humans compared with its activity in animal models.

3. Use of human tissues may be more cost-effective than traditional drug discovery because one can eliminate or redesign compounds that react poorly with human receptor targets prior to initiating the costly activities related to preclinical testing and clinical trials.

4. A compound targeted against specific receptor in human tissues is better validated for entry into clinical development.

11.8.3 Tissue Screening

Testing directly in human tissue rather than in animal models or yeast allows drug development teams to obtain vital information about what their gene of interest is doing in a human system. Because animal models are often not predictive of how drugs will behave in humans, obtaining information directly from human tissues is a critical step in choosing one target from many candidates. Even after the drug development process has begun, determining where a particular gene is expressed in other, perhaps unexpected, sites within the body may assist researchers in the design and interpretation of preclinical or clinical studies.

11.9 TARGET VALIDATION TECHNOLOGIES

Once a gene has been identified as a potential therapeutic target, its relevance to a disease process and its suitability as a target should be validated before starting the costly procedure of drug discovery. Target validation involves manipulating the target and confirming that the resulting effect is consistent with purported role. In practice, this is not easy because an infinite number of genes, proteins and other molecules interact with each other in signalling pathways to direct cell function. Finding a drug target that safely regulates disease without affecting normal function has proved very challenging. Various methods to achieve this are antisense and RNAi (vector-mediated gene inactivation and transient gene inactivation), proteomics, gene expression arrays and combinatorial biology approaches.

11.9.1 Animal Models for Genomics-based Target Validation Methods

Animal models of human disease are important for understanding the disease mechanism and for the development and evaluation of new therapies. Some animal models are available where a genetic defect occurs spontaneously. Animal models are invaluable for functional genomic studies. Spontaneously occurring genetic defects in animals are inadequate for serving as models of human disease and thus the need for induced mutations. Mutations can be established in the animal genome by one of two approaches: non-homologous recombinations (transgenic) or homologous recombination (knockout, null mutations).

11.9.2 Role of Knockout Mice in Drug Discovery

Most drugs act as inhibitors of their targets. Inactivating a gene in a knockout mouse can mimic the effect of the target's inhibitor. Mouse functional genomics is similar to that in humans. Thus the knockout mouse defines a drug target and its underlying physiology, permitting an insight into the disease, its diagnosis and treatment. For example, p53 gene knockouts have been used extensively to investigate tumorigenesis. The knockout mouse is becoming an invaluable addition to functional genomics-driven drug discovery. There are, however, some reservations about the value of mouse genetics in functional genomics. There are some mutagenesis experiments that result in no phenotype, which may be due to redundancies within the mouse genome or perhaps due to takeover of the function of missing members in certain tissues. This, however, occurs infrequently. If we believe the oft-quoted statement that 'the first company to show biological relevance in an animal model wins' in genomic drug development, there is little doubt that the mouse will play an important part in this venture.

11.10 ANTISENSE FOR DRUG DISCOVERY

Antisense molecules are synthetic segments of DNA or RNA, designed to mirror specific mRNA sequences and block protein production. The use of antisense drugs to block abnormal disease-related proteins is referred to as antisense therapeutics. Synthetic short segments of DNA or RNA are referred to as oligonucleotides. The literal meaning of this word is a polymer made of a few nucleotides. Naturally occurring RNA or DNA oligonucleotides may or may not have antisense properties. Antisense oligonucleotides are synthetic pieces of DNA (at least 15 nucleotides in length) that can hybridise to sequences in the RNA target by Watson–Crick or Hoogstein base pairing. An alternative antisense approach is the use of ribozymes that catalyse RNA cleavage and inhibit the translation of RNA into protein. Peptide nucleic acids (PNAs), which are DNA-like molecules, are potential antisense and antigene agents. Aptamers are synthetic chains of nucleotides that bind directly to target proteins, inhibiting their activity and are considered to be antisense compounds. A high-affinity DNA analogue, locked nucleic acid (LNA), confers several desired properties to antisense agents. LNA/DNA copolymers exhibit potent antisense activity on assay systems.

11.10.1   Antisense Oligonucleotides for Drug Target Validation

Antisense technology uses genetic sequence information to design rapidly inhibitors of any gene target. Because of their exquisite specificity, antisense oligonucleotides can inhibit the selected gene only, without an impact on other closely related genes. As a result, antisense inhibitors allow the identification of function of that single gene target more precisely than any other method. Several companies have integrated antisense technologies in functional genomics.

DNA microarrays have been used to evaluate thousands of genes simultaneously. Target mRNA and protein expression can be inhibited using antisense oligonucleotides to facilitate this process and determine genetic pathways. Antisense technology is considered to be a viable option for high-throughput determination of gene function and drug target validation.

Peptide nucleic acid (PNA) inhibits gene expression. If two different bacterial RNAs are targeted with a complementary PNA and a randomised control sequence, the complementary PNA inhibits its intended target, whereas the other does not. This provides an opportunity for novel antibiotic discovery in addition to target validation using antisense PNA. For validation of bacterial targets, the main advantages of PNA is it can be used for any bacterial species and the level of gene inhibition can be regulated. Genetic knockout is not applicable to all bacterial species and gene expression is an all-or-none phenomenon which cannot be regulated. Antisense PNA has the limitation that it cannot knockout 100% of the target whereas genetic knockout can do so. However, reduced levels of target activity mimic the therapeutic situation more realistically. Both techniques are synergistic and useful. In conclusion, antisense PNA technology makes good sense for application in antibiotic target validation.

11.10.2   Aptamers

Aptamers (derived from the Latin word *aptus* = fitting) are single-stranded DNA or RNA oligomers, which can bind to a given ligand with high affinity and specificity due to their particular 3D structure and thereby antagonise the biological function of the ligand. Aptamers are considered to be antisense compounds. The technology builds on the ability of aptamers to bind tenaciously to proteins and have been used to identify protein signatures. Streptavidin aptamers (streptavidin-binding RNA ligands) are also potentially powerful tools for the study of RNAs or ribonucleoproteins as a means for rapid detection, immobilisation and purification. Recent developments demonstrate that aptamers are valuable tools for diagnostics, purification processes, target validation, drug discovery and therapeutic.

Aptamer expression libraries do not depend on information from genome sequence. These can be used to find peptides with a desired biological activity. This can be caused by the expressed peptide activating or inhibiting a cellular factor. For stable expression of peptides in mammalian cells, libraries are constructed that express aptamers in the context of protease-resistant scaffold structures. A limitation of this approach is that highly complex mixtures of aptamers are required to identify active aptamers in any pathway. In addition, it can be cumbersome to identify the cellular factor that is affected by the biologically active peptide.

11.10.3 RNA as a Drug Target

RNA has a structural complexity rivalling that of proteins and thus provides an opportunity as a target for small-molecule drugs. Several steps are required to find and exploit RNAs as drug targets. Because all proteins are synthesised using RNA template, they can be inhibited by preventing the translation of mRNA. Advantages of targeting RNA are:

- Drugs that bind RNA can produce more selective action than those that bind proteins. For example, a drug can bind to a region of RNA which is relevant to the target tissue without affecting RNA in other tissues.
- Proteins are difficult to isolate or purify whereas RNA is easier to synthesise and use in assays.
- RNA can be easily synthesised in large quantities and is not extensively modified *in vivo*, whereas large-scale production of proteins is still limited.

11.10.4 Ribozymes

Ribozymes are enzymes comprised of RNA which can act both as a catalyst and as a genetic molecule. In Nature, ribozymes catalyse RNA cleavage and RNA splicing reactions and are sometimes called 'catalytic RNAs'. Ribozymes are being increasingly used for the sequence-specific inhibition of gene expression by the cleavage of mRNAs encoding proteins of interest. The possibility of designing ribozymes to cleave any specific target RNA has rendered them valuable tools in both basic research and therapeutic applications. In the thera-peutics area, they have been exploited to target viral RNAs in infectious diseases, dominant oncogenes in cancers and specific somatic mutations in genetic disorders. More recently, ribozymes have been used for transgenic animal research, gene target validation and pathway elucidation. For therapeutic purposes, a ribozyme can be considered to be a chimeric RNA molecule consisting of two stretches of antisense RNA flanking a nucleolytic motif. The antisense RNA component, referred to as the complementary flanking regions, provides target selectivity.

Unlike traditional pharmaceuticals, ribozymes disrupt the flow of genetic information rather than inhibit protein function. The therapeutic potential of ribozymes by targeting distinct mRNAs is tremendous and is a novel approach to curing disease. Diseases which result from undesirable expression of RNA, such as neoplastic disorders and viral infections, should be particularly amenable to this therapeutic approach. Use in genetic disorders is also being explored.

11.11   RNAI FOR DRUG DISCOVERY

RNA interference (RNAi) is a cellular mechanism to regulate the expression of genes and the replication of viruses. RNAi or gene silen-cing involves the use of a double-stranded RNA (dsRNA). Once in the cell, the dsRNAs are processed into short, 21–23 nucleotide dsRNAs termed small interfering RNAs (siRNAs) that are used in a sequence-specific manner to recognize and destroy complementary RNAs. There are several classes of naturally occurring small RNA species, including siRNAs, microRNAs (miRNAs) and repeat-associated siRNAs (rasiRNAs). The RNAi pathway has been exploited in simpler organ-isms to evaluate gene function by introducing dsRNAs that are specific to the targeted gene.

RNAi technology is being evaluated not only as an extremely pow-erful instrument for functional genomic analyses, but also as a poten-tially useful method to develop highly specific dsRNA-based gene-silencing therapeutics. RNAi is an important method for analysing gene function and identifying new drug targets that uses double-stranded

RNA to knock down or silence specific genes. The challenge has been to select reliably an siRNA segment that can efficiently silence the gene without triggering unwanted effects. One solution is to use algorithms to select highly functional siRNA sequences and then pool the best sequences for guaranteed gene knockdown. RNAi technology could considerably reduce the time needed for target validation and overall drug development, accelerating the drug discovery process. RNAi screening can identify high-value drugs targets such as kinases involved in cell proliferation.

11.11.1 Use of siRNA Libraries to Identify Genes as Therapeutic Targets

The ability of RNAi to provide relatively easy ablation of gene expression has opened up the possibility of using collections of siRNAs to analyse the significance of hundreds or thousands of different genes whose expression is known to be upregulated in a disease, given an appropriate tissue culture model of that disease. Perhaps more important still is the possibility of using genome-wide collections of siRNAs, whether synthetic or in viral vectors, as screening tools. The libraries of RNAi reagents can be used in one of two ways.

One is in a high-throughput manner, in which each gene in the genome is knocked down one at a time and the cells or organism scored for a desired outcome, *e.g.* death of a cultured cancer cell but not a normal cell. Owing to the very large numbers of assays needed to look at the involvement of all genes in the human genome, this approach is very labour intensive.

The other approach is to use large pools of RNAi viral vectors and apply a selective pressure that only cells with the desired change in behaviour can survive. The genes knocked down in the surviving cells can then be identified by sequencing the RNAi vectors that they carry. This method is being used to investigate genes involved in neurodegenerative diseases, diabetes and cancer. Both approaches show considerable promise in identifying novel genes that may make important therapeutic targets for inhibition either by conventional drug discovery methods or, more controversially, by RNAi itself.

Sets of siRNAs focused on a specific gene class (siRNA libraries) have the capacity to increase greatly the pace of pathway analysis and functional genomics. RNAi-based functional chemogenomics has been integrated into drug discovery programs.

11.11.2   RNAi as a Tool for Assay Development

RNAi can be a useful tool for assay development, hit selection and
specificity testing. RNAi can be used as a positive control to calibrate the
assay readout based on the effects of known levels of mRNA or protein
knockdown. Effects of compounds can then be tested against the known
inhibitory effects of RNAi reagents. Selectivity of certain RNAi reagents
can be used to advantage in assay development. By using RNAi reagents
with known specificity for individual mRNA isoforms, it should be
possible to predict the effects for compounds with similar specificity.

11.11.3   Challenges of Drug Discovery with RNAi

The advantages of cell-based RNAi screens over small-molecule
screening for target identification include the fact that most cell types are
amenable to RNAi and it is relatively easy to knock down any gene of
interest. So far, every gene tested has been susceptible to RNAi. However, one of the big issues is how to make siRNAs 'druggable'. Some of
the challenges are:

- To ensure that the candidate siRNA is appropriately stabilised in a
  'druggable' formulation or by chemical modifications. Stability of
  the RNA towards exo- and endonucleases can be resolved by
  appropriate chemical modifications.
- Safely and successfully delivering siRNA in an acceptable and
  effective manner.
- Scaling up siRNA synthesis in the near term and, ultimately,
  manufacturing reliably and effectively.
- Cell-based RNAi assays are particularly prone to edge effects because
  the cells in the outer wells of the plates grow at a different rate than
  the cells in the inner wells. One should ignore the outer wells.
- There are problems with the 'penetrance' of some RNAi screens, in
  which the level of GFP in the cells is heterogeneous, making it
  difficult to interpret. Actually, the expression levels of several proteins vary significantly within cells grown in culture. Therefore, the
  problem is not heterogeneity of the siRNA knockdown, but heterogeneity of protein expression, and is an artefact of the cell culture. Analysis of the data can be improved by looking at single cells
  rather than entire wells.

Once these issues have been resolved, there is potential for rapid early-
stage drug development as RNAi-based therapy development relies

predominantly on documented gene sequence data and leverages a natural process.

11.11.4 Role of MicroRNA in Drug Discovery

MicroRNAs (miRNAs), small and mostly non-coding RNA gene products, are molecules derived from larger segments of 'precursor' RNA that are found in all diverse multicellular organisms. miRNAs are 21–25 nucleotide transcripts that repress gene function through interactions with target mRNAs. miRNAs appear to regulate at least one-third of all gene expression and are also likely play significant roles in the manifestation of many disease states, including cancer and many metabolic and infectious diseases. Thus they represent a new class of drug targets for the pharmaceutical industry.

Investigators are seeking miRNA targets and functions with tools ranging from traditional genetics to computer-based genome scanning. Application to the *Drosophila melanogaster* and *Anopheles gambiae* genomes identifies several hundred target genes potentially regulated by one or more known miRNAs. These potential targets are rich in genes that are expressed at specific developmental stages and that are involved in cell fate specification, morphogenesis and the coordination of developmental processes, in addition to genes that are active in the mature nervous system. miRNAs can be used for rapid target gene identification and target validation. miRNAs and the genes they regulate are candidates for the development of new therapies. Several methods have been developed for computational prediction of miRNA targets. Online resources provide researchers with useful tools and data for assessing the impact of miRNAs on the gene or biological process of interest.[7]

11.12 BIOCHIPS AND MICROARRAYS FOR DRUG DISCOVERY

Biochip is a broad term indicating the use of microchip technology in molecular biology and can be defined as arrays of selected biomolecules immobilised on a surface. DNA microarray is a rapid method of sequencing and analysing genes. An array is an orderly arrangement of samples. The sample spot sizes in microarray are usually less than 200 μm in diameter. It is comprised of DNA probes formatted on a microscale (biochips) plus the instruments needed to handle samples (automated robotics), read the reporter molecules (scanners) and analyse the data (bioinformatic tools).

11.12.1 Finding Lead Compounds

With an emphasis on functional genomics rather than sequencing, drug discovery programmes are using custom chips to find lead compounds. It has already been shown that it is possible to treat cells with compounds and compare the resulting patterns of gene expression with patterns previously obtained when treating cells in known ways, thereby identifying which proteins or targets the compound is altering. Such *in vitro* target identification should greatly improve the inefficient conventional methods of developing drugs. Because animal testing of compounds is expensive, time consuming and has other negative aspects, DNA microarrays are likely to improve the efficiency of drug discovery by supplementing the information obtained by traditional animal testing.

11.12.2 High-throughput cDNA Microarrays

High-throughput gene expression analysis is playing an important part in the drug discovery process in a genomic-oriented atmosphere. This requires an ability to survey and compare rapidly gene expression levels between reference and test samples. In this setting, microarray technology is exploiting collections of known sequences to pinpoint drug targets. Assay miniaturisation and microfluidics have shown promise in high-throughput screening. Microfluidic lab-on-a-chip technology has been widely used to provide small volumes and fluid connections and could eventually outperform conventionally used robotic fluid handling.[8]

11.12.3 Use of Gene Expression Data to Find New Drug Targets

Comprehensive gene expression analysis data and powerful computational methods coupled with appropriate genetically modified organisms can be used to decipher the function of previously uncharacterised genes. Comprehensive gene expression profiles of cells have been used to generate databases with a wide variety of phenotypes and following different chemical treatments through the accurate and systematic analysis of gene expression *en masse*. Such compendia of gene expression profiles were used as a pattern matching tool to identify novel gene functions and to understand the biochemical basis of drug action. This approach can be applied for drug discovery and development. Gene expression data highlight meaningful differences between normal and disease-related genes and document the effects of drugs on gene function.

Gene expression analysis is the first new technology to be applied for many steps in the drug development process. Microarrays are being used

Table 11.3 Role of microarrays in drug discovery.

Selective tissue expression of a drug target is helpful in selecting the appropriate drug for effect on a specific organ and which avoids unwanted side-effects on other organs

Differential gene expression can be monitored by microarrays. A comparison of thousands of genes between 'disease' and 'normal' tissues and cells allows the identification of multiple potential targets

Microarray analysis permits a high-throughput analysis of gene expression in model organisms. A comparison of the gene expression patterns in normal mice can be compared with those in transgenic mice overexpressing selected genes

Microarray, by identifying genes that are turned on, can be used to study the response of the host to challenge with a pathogen

Microarrays are potentially powerful tools for investigating the mechanism of drug action

for genome-wide expression monitoring, large-scale polymorphism screening and mapping. These technologies permit the measurement of gene expression components of disease and the identification of promising new drug targets. Drug target validation and identification of secondary drug target effects can be facilitated by using DNA microarrays. Gene chip technology also provides a method of predicting side-effects of drugs and choosing those for development that have minimal or no adverse effects. Several ways in which microarray analysis is likely to affect drug discovery are listed in Table 11.3.

Gene expression analysis has an important application in analysis of signalling pathways of relevance to cancer and inflammation for drug target evaluation. Since activation of signalling pathways leads to mRNA expression, parallel measurement of mRNA expression is the most practical method of determining if a gene is expressed or not.

11.12.4 Investigation of the Mechanism of Drug Action

Analysis of genes can contribute to determination of the mechanism of action of a drug. Several events are triggered by the initial action of a drug. The ability to screen thousands of genes simultaneously may help in the identification of potential drug effectors. This allows the formulation of sound hypotheses of mechanism of action of drugs to be formed and tested in subsequent investigations.

11.13 APPLICATIONS OF BIOINFORMATICS IN DRUG DISCOVERY

The challenge for bioinformatics is to create short lists of targets most likely to be 'druggable' from long lists of genes. Bioinformatics can help with three aspects of target selection:

1. classification of protein families;
2. an understanding of the biochemical and cellular behaviour of the targets;
3. the development of targets involving predictions on detoxification, stratification of patient populations and other gene-based variations.

The term '*in silico* biology' is used for conducting biology before setting foot in the laboratory for traditional 'wet' biology. Bioinformatics has made this possible but it should not be misunderstood as a replacement for 'wet' biology. Using simple point-and-click commands, researchers can quickly perform a variety of biological analyses *in silico*. The detection of similarities between amino acid sequences is fundamental to pharmaceutical research *in silico*. Conventional techniques can only detect relationships between sequences when at least 25% of their residues match.

11.13.1 Combination of *In Silico* and *In vitro* Studies

Information obtained *in silico* combined with information provided by bacteria growing on Petri dishes may lead to novel drug targets and new insights into the nature of conserved and essential genes in organisms. It is recommended that, apart from prediction of the gene function, it is worthwhile to get as much information out of the computer whenever drug design is potentially involved. This narrows the range of necessary experiments and saves time and money.

Metabolic instability is a common limiting factor in obtaining acceptable ADME early in the drug discovery process. Metabolite identification is helpful but early access to information is limited. A high metabolic clearance may raise the question of where the drug is being metabolised. Early identification of the metabolite and structure–metabolism relationship may facilitate early discovery decisions. Bioinformatic support in this situation is valuable. The increasing volume and complexity of biological and chemical data require the use of bioinformatic systems to extract knowledge relevant to decision making for the drug discovery process. The largest and most important category of information is genomic and proteomic databases with further studies in gene/protein and disease associations and population genetics. Databases can now communicate with each other over the Internet. The daunting task of navigating through the massive amount of data requires simple and integrated tools to generate innovative but meaningful information for drug discovery. The bioinformatic tools should be able to cope with more than 100 drug discovery approaches that are now available.

11.14 ROLE OF MODEL ORGANISMS IN DRUG DISCOVERY

Simple and easily accessible organisms can be used for probing gene function because of the conservation of gene function and sequence between widely divergent species. Commonly used organisms include the mouse, chicken, frog, zebra fish, nematode worm (*C. elegans*) and unicellular yeast (*S. cerevisiae*). Particularly useful are *C. elegans* and *S. cerevisiae* because their genomes have been completely sequenced. The mouse has the advantages that it is a mammal, some genetics is known, transgenic knockouts can be made and some genomic data are already available. The disadvantages are that it is costly and the generation time is long. Evidence for the usefulness of model organisms in functional genomics comes from search of known human disease genes in model organisms and the finding that majority of known disease genes have counterparts in lower species. This common origin is reflected not only in the high degree of conservation of genes between organisms but also in the role of genes in signalling networks. In many cases, the same proteins interacting in the same manner are involved in analogous processes in different species. Comparative genomics enables tests to be performed quickly in organisms with simple genomes such as the fruit fly or algae to predict and guide the analysis of gene function in organisms with complex genomes such as humans.

The ease with which genes can be underexpressed or overexpressed in these model organisms enables studies of gene function to be performed in a short time. Because the early developmental stages of model organisms are easily accessible, it is possible to examine gene expression during embryogenesis to obtain clues to function. Functional homology can be tested in species such as *Drosophila* that are amenable to genetics. It is important to identify all components of a genetic pathway where a particular gene functions. This knowledge is important if the disease happens to be due to a mutation in a gene that cannot be corrected and it becomes necessary to activate or inactivate the genetic pathway downstream of the mutant gene as a strategy for treatment.

Multicellular organisms, such as *C. elegans*, offer numerous biological advantages in programmes such as drug discovery, toxicology and basic research because of its remarkable similarity to the human genome. Over 70% of ~ 20 000 genes in *C. elegans*, are also found in humans and 70% of the 300 most important human disease genes have homologues with that of *C. elegans*. More is known about the biochemistry and genetics of *C. elegans* than any other animal. Since much of the genetic makeup of *C. elegans* also occurs in humans, laboratories in pharmaceutical companies and universities worldwide are modelling human diseases in these organisms. Their goals are to find new drug targets and to screen drug compounds *in vivo* using these biological systems. This information has been used to manipulate the nematode genes to create a 'human disease model'. Disease-model *C. elegans* develops abnormally and has an abnormal size for its stage of development. These model organisms can be exposed to drugs and changes can be monitored.

11.15 CHEMOGENOMIC APPROACH TO DRUG DISCOVERY

Chemogenomics is the interdisciplinary approach that attempts to derive predictive links between the chemical structures of bioactive molecules and the receptors with which these molecules interact.[9] Chemogenomics thus unites medicinal chemistry with molecular biology and uses low-resolution sequence homology to identify genomic targets of interest. Insights from chemogenomics are used for the rational compilation of screening sets and for the rational design and synthesis of directed chemical libraries to accelerate drug discovery. Ligands identified by this method can be used in secondary biological assays to check for biological activity. In the case of targets of unknown function, low-throughput secondary assays can be used to determine the therapeutic relevance of a novel target. Chemogenomics can be applied to the discovery and description of all possible drug compounds (all of the chemical possibilities) directed at all possible drug targets. This approach has the potential to cover a broad range of therapeutic areas, because while gene families code for structurally similar proteins, each protein in a gene family can have a very different biological function. Different targets within a gene family may be implicated in widely different diseases. In practice, the chemogenomics approach is accelerating the drug discovery and will increase the flow of new drug candidates into development.

11.16 VIRTUAL DRUG DEVELOPMENT

In addition to drug design, computers are used for modelling human disease and response to therapy. Research and drug development can be carried out in a virtual laboratory using computer-based models of human disease, target validation and patient selection for clinical trials. Drug testers turn to 'virtual patients' as guinea pigs and this technology can reveal unseen biological interactions that explain conflicting clinical trial results.

Virtual screening is a strategy for introducing a more focused approach to HTS. Computational analysis of very large real or virtual chemical databases can readily identify compounds with appropriate properties for binding to the target receptor. Full receptor–ligand docking represents the most detailed – and the most time-consuming – approach to virtual screening. One of the aims of introducing computational methods in the drug discovery process is the early elimination of compounds which are chemically unsuitable for further development. This approach is needed because HTS is identifying a large number of hit compounds, many of which do not possess drug-like properties. Several methods for predicting drug likeness have been described but the focus is on the prediction of intestinal absorption and blood–brain barrier penetration. The routine use of experimental absorption systems in the pre-screening of compounds has provided data that can be used for construction of improved models of drug absorption. Use of such systems have the potential to reduce lead optimisation time and attrition rates in preclinical and clinical phases of drug development.

11.17 ROLE OF BIOTECHNOLOGY IN LEAD GENERATION AND VALIDATION

Structural characterisation of biomacromolecules, computer sciences and molecular biology have made rational drug design feasible. Molecular modelling has helped in the discovery process of new drugs, particularly lead generation and optimisation. With a host of methods available, lead generation is no longer a problem in drug discovery. However, there is still a need to improve the methods of lead validation for potential therapeutic candidates. None of the screening methods currently available has a very high degree of certainty but a success rate of over 80% is acceptable while efforts are in progress to improve it. The aim is to develop high-throughput screens that also provide clinically relevant information. There is limitation of cell models in that the human disease process is not reproduced reliably. An improved understanding of cell signalling pathways within the cell will provide a better basis for the development of specific assays for various diseases.

11.18 CONCLUSION

Biotechnology-based methods are being used to improve drug discovery and molecular targeting. Most of the new technologies are being used for this purpose. There is a considerable parallel activity in the development of drug delivery and personalised medicine based on pharmacogenetics, pharmacogenomics and pharmacoproteomics. Various omics, particularly combined genomic and proteomic approaches, have contributed considerably to this process. Bioinformatics plays an important role in integrating and analysing massive amounts of data produced from application of various technologies. It is hoped that biotechnology-based approaches will help in discovery of new drugs to meet the current shortage.

REFERENCES

1. D. M. Drexler, T. J. Garrett, J. L. Cantone, R. W. Diters, J. G. Mitroka, M. C. Prieto Conaway, S. P. Adams, R. A. Yost and M. Sanders, *J Pharmacol. Toxicol. Methods*, 2007, **55**, 279.
2. M. R. Groseclose, M. Andersson, W. M. Hardesty and R. M. Caprioli, *J. Mass Spectrom.*, 2007, **42**, 254.
3. K. K. Jain, in *Handbook of Drug Screening*, ed. R. Seethala, Informa Healthcare, New York, 2008.
4. R. Weissleder, K. Kelly, E. Y. Sun, T. Shtatland and L. Josephson, *Nat. Biotechnol.*, 2005, **23**, 1418.
5. X. Wang, R. Inapagolla, S. Kannan, M. Lieh-Lai and R. M. Kannan, *Bioconjug. Chem.*, 2007, **18**, 791.
6. R. W. Peck, *Drug Discov. Today*, 2007, **12**, 289.
7. P. Mazierea and A. J. Enright, *Drug Discov. Today*, 2007, **12**, 452.
8. P. S. Dittrich and A. Manz, *Nat. Rev. Drug. Discov.*, 2006, **5**, 210.
9. T. Klabunde, *Br. J. Pharmacol.*, 2007, **152**, 5.

CHAPTER 12

Vaccines

NIALL MCMULLAN

School of Life Sciences, University of Hertfordshire, Hatfield, Hertfordshire
AL10 9AB, UK

12.1 AN OVERVIEW OF VACCINES AND VACCINATION

Some two centuries ago, Edward Jenner introduced the concept of modern
vaccination when he successfully used cowpox virus to induce protective
immunity to smallpox. The pioneering work of Jenner and his con-
temporary Louis Pasteur paved the way for the concept of vaccination.
The term vaccination (from the Latin *vacca* – cow) was coined by Pasteur
as a lasting tribute to Jenner's work. The 1970s saw the global eradication
of smallpox, the last naturally occurring case being in Somalia in 1977.

Through the efforts of the WHO, other diseases have been targeted for
eradication, most notably measles and polio. Like smallpox, these diseases
are caused by viruses that only infect humans, thus increasing the odds of
eradication. Despite global eradication programmes, however, there are
still pockets of infection around the world. The need for effective vaccines
and vaccination programmes is brought into sharp focus when considering the fact that over half the human population of the world are
infected by one or more pathogens. The scale of malaria infection coupled
with the re-emergence of tuberculosis and the increasing incidence of HIV/
AIDS is an ever-present reminder of the burden of infectious disease.

Time and space do not permit the inclusion of animal diseases, an
issue in its own right, or the attempts to develop vaccines against cancer.

Molecular Biology and Biotechnology, 5th Edition
Edited by John M Walker and Ralph Rapley
© Royal Society of Chemistry 2009
Published by the Royal Society of Chemistry, www.rsc.org

Table 12.1 The main types of bacterial and viral vaccines for use in humans

Vaccine type	Bacterial	Viral
Live attenuated	BCG	Sabin polio (OPV)
Inactivated/killed	*Salmonella typhi* (Ty21a)	Measles[b]
Toxoid	*Bordetella pertussis* (acellular)[a]	Mumps[b]
Subunit	*Vibrio cholerae*	Rubella[b]
Conjugate	*Yersinia pestis*	Adenovirus
	Coxiella burnetii	Rotavirus
	Corynebacterium diphtheriae[a]	Varicella
	Clostridium tetani[a]	Yellow fever
	Streptococcus pneumoniae (23 subunit)	Salk polio (IPV)[a]
	Neisseria meningitides A, C, W135, Y	Influenza
	Haemophilus influenzae type b (HIb)[a]	Hepatitis A
	Neisseria meningitides C	Rabies
	Streptococcus pneumoniae (7 subunit)	Japanese encephalitis
		Hepatitis B
		Influenza

[a]Or combined DTaP/IPV.
[b]Or combined MMR vaccine

This chapter will focus on infectious diseases affecting humans; however, some aspects of new approaches to vaccine development are pertinent to other areas of health.

12.2 TYPES OF VACCINES IN CURRENT USE

Current vaccines approved for use in humans fall into three main types: live, attenuated or inactivated whole organism vaccines; subunit vaccines, either isolated macromolecules or recombinant proteins; and toxoid vaccines. The major vaccines in current use are listed in Table 12.1.

12.2.1 Live, Attenuated Vaccines

These vaccines use attenuated strains of the pathogen. The attenuated strain is able to replicate inside the host but has lost one or more features which contribute to its pathogenicity. Attenuation can be achieved by growing the pathogen under abnormal culture conditions. The mutants generated through this process are then tested for loss of virulence while retaining immunogenicity. Attenuation by this approach can be a long and time-consuming process. The current vaccine for tuberculosis, bacillus Calmette–Guérin (BCG), an attenuated strain of *Mycobacterium bovis*, was isolated by Calmette and Guérin in 1921 after 13 years of growing *M. bovis* in medium containing increasing concentrations of bile. *M. bovis* confers immunity to *M. tuberculosis*, the primary cause of tuberculosis in humans due to similarities between the two organisms. Although primarily associated with tuberculosis in cattle, *M. bovis* was also a significant cause of tuberculosis in humans prior to the introduction of pasteurised milk. The BCG vaccine is the most commonly administered vaccine, with over 5 billion doses having been given since its introduction in the 1940s. However, the efficacy of BCG is variable due in part to the fact that BCG, *per se*, has never been cloned and consequently there are variations between BCG preparations.[1]

Attenuated viral vaccines are more common, but the basic strategy has been the same. Typically, attenuated strains of the virus are isolated by passage through cells which are not the normal host cell. The Sabin polio vaccine contains three different attenuated strains of poliovirus developed by growth in monkey kidney epithelial cells. Attenuated strains of the viruses responsible for measles, mumps and rubella were produced by similar approaches. These attenuated strains are used in the combined MMR vaccine. The attenuated measles virus may also used on its own; consequently, the WHO data refer to measles-containing vaccine (MCV).

The molecular attenuation or deletion of virulence genes should permit more rapid production of attenuated strains for use of vaccines. One current example is a typhoid vaccine, which uses the attenuated *Salmonella typhi* strain Ty21a. Recently, the production of a biologically contained Ebola virus has been reported.[2] The gene encoding an essential viral transcription factor (VP30) required for viral replication was replaced by a reporter gene. Vero cells expressing the critical VP30 protein permit completion of the viral life cycle but the virus is contained within the cell type. These VP30-negative viruses are genetically stable and morphologically indistinguishable from the wild-type virus. The production of this biologically contained form of Ebola virus will permit studies of the virus outside of biosafety level-4 containment and should accelerate vaccine development for this highly lethal and feared virus.

12.2.2 Inactivated Vaccines

Several inactivated vaccines are in current use in humans (Table 12.1). The approach is chemical inactivation of the pathogen using formaldehyde or β-propiolactone. This approach usually ensures that the immunogenic features of the pathogen are retained. Production of inactivated vaccines requires large quantities of pathogenic material, with associated risks for the workers involved. Furthermore, it is essential that the inactivation process is complete, otherwise the pathogen will be introduced into the population. This occurred with early batches of the

Salk polio vaccine, resulting in paralytic polio in a number of recipients. The advantage over live, attenuated vaccines is that there is no risk of reversion to virulence. It is for this reason that the Salk vaccine is preferred over the Sabin polio vaccine in Scandinavian countries.

Inactivated vaccines generally produce significant antibody responses but weaker cellular immunity, in particular cytotoxic T cell responses, than live vaccines. This is due mainly to the absence of microbial protein synthesis inside host cells. In turn, this leads to reduced or absent cytosolic processing required for MHC class I presentation of peptides and poor activation of cytotoxic CD8+ T cells. Furthermore, the inability to replicate inside the host reduces immunostimulation and a need for more booster immunisations.

12.2.3 Subunit Vaccines

Subunit vaccines are used widely to protect against a number of important pathogens. Subunit vaccines typically use isolated antigenic macromolecules from the external surface of the pathogen. Vaccines against *Haemophilus influenzae* type b (HIb), *Neisseria meningitidis* and *Streptococcus pneumoniae* contain capsular polysaccharides derived from the bacteria. Alternatively, the vaccine may be a recombinant protein such as the hepatitis B surface antigen (HBsAg) used in the hepatitis B vaccine. Different factors influence or even dictate the use of subunit vaccines as opposed to whole organism vaccines. Gram-negative bacterial pathogens such as *Haemophilus influenzae* type b and *Neisseria meningitidis* represent a significant health threat even when dead due to the presence of lipopolysaccharide moieties (endotoxin) in their cell wall structures. Endotoxin is a powerful inducer of proinflammatory cytokines in macrophages leading to septic shock. In the case of hepatitis B virus, the inability to culture the viruses has so far made attenuation impossible.

Despite the limitations described above, subunit vaccines are among the most efficacious vaccines due to their ability to elicit highly appropriate immune responses.

12.2.3.1 Capsular Polysaccharide Vaccines and Protein Carriers (Conjugate Vaccines)

Haemophilus influenzae type *b* (HIb) and *Neisseria meningitidis* are the major cause of bacterial meningitis in infants, toddlers and young adults. Phagocytosis is the primary defence mechanism against extracellular pathogens. A major pathogenic feature of *Haemophilus influenzae* type b (HIb), *Neisseria meningitidis* and *Streptococcus pneumoniae* is the presence of hydrophilic polysaccharides which possess anti-phagocytic properties. This feature allows the pathogens to establish infection within the host. The production of opsonising antibodies of the IgG isotype that promote phagocytosis is a key process in eliminating these pathogens. This is the basis of the current vaccines. However, polysaccharides are generally weak inducers of antibody responses due to poor T-helper cell responses to polysaccharides. This leads to poor immunological memory and diminished antibody class switching to IgG. Poor immune responses were a feature of the first polysaccharide-based vaccines. This problem has been overcome by the use of carrier proteins conjugated to the polysaccharides. Such vaccines are referred to as conjugate vaccines and are now routinely used in vaccination against *Haemophilus influenzae* type b (HIb) and *Neisseria meningitides*. The use of carrier proteins has greatly enhanced the immunogenicity of these vaccines. The basis for this is that the carrier protein is processed by host cells and activates T-helper cells resulting in the cytokines required to produce memory and antibody isotype switching. The carrier proteins used are inactivated bacterial exotoxins (toxoids), in particular tetanus toxoid. This exploits the fact that individual will have been immunised against tetanus (see Section 12.2.3.3) and will have memory T cells to the carrier protein, in turn increasing the effectiveness of the T cell contribution.

S. pneumoniae displays considerable variability in the polysaccharide capsule. The vaccine contains up to 23 different polysaccharides from the major serovars. Alternatively, a conjugate vaccine containing 7 major polysaccharides is available. A polysaccharide-based typhoid vaccine is also available.

12.2.3.2 Recombinant Protein Vaccines

The hepatitis B virus (HBV) vaccine is the major example of a recombinant protein vaccine in current use and the first vaccine produced by recombinant DNA technology. The vaccine contains a highly purified form of recombinant HBsAg, encoded by a single gene and expressed in yeast.[3] As with most subunit vaccines, it requires multiple boosters but has proven to be highly efficacious. Humans are the only host for HBV. Prior to the recombinant vaccine becoming available, the only available source of HBV antigen was blood from infected individuals, a risky procedure for those involved. This is a highly infectious virus and the number of cases of infection prevented and indeed hepatic cancer cases prevented is incalculable. WHO estimates suggest that up to 2 billion of the global population are infected with HBV, with 360 million chronically infected. Due to the large scale of HBV infection, the vaccine has been introduced into infant immunisation programmes in over 160 countries. In the UK, the vaccine is retained for use in high-risk groups, in particular healthcare workers and appropriate laboratory staff.

Recently, two recombinant protein vaccines have been developed for human papilloma virus (HPV), the primary cause of cervical cancer and genital warts. These vaccines combine HPV proteins in self-associating virus-like particles. The vaccines are Gardisal (Merck), which contains proteins from HPV 16, 18, 6 and 11 produced by recombinant yeast technology, and Cervarix (GlaxoSmithKline), which contains proteins from HPV 16 and 18 produced by recombinant baculovirus technology.[4]

12.2.3.3 Toxoid Vaccines

Toxoids are inactivated toxins. In some instances, bacterial exotoxins are the sole or primary virulence factor and the strategy here is to immunise against the exotoxin not the actual bacterium. Tetanus toxin produced by *Clostridium tetani* is one such example. The bacterium is an obligate anaerobe widely distributed in the environment and is non-invasive. Tetanus exotoxin is a very potent toxin which interferes with neuronal transmission and in extreme cases can prove fatal. Neonatal tetanus kills thousands of children each year. The problem is particularly acute in technologically underdeveloped countries, where home births in poor sanitary conditions are often the norm. The vaccine is highly effective in preventing the onset of tetanus. Diphtheria caused by infection with *Corynebacterium diphtheriae*, another common pathogen, is also prevented by use of a toxoid vaccine. Toxoid vaccines are expensive due to the large-scale cultures of organisms required to obtain exotoxin.

12.3 THE NEED FOR NEW VACCINES

There is a constant demand for new and improved vaccines. Despite the success of current vaccines in controlling widespread epidemics, none of them demonstrates complete efficacy. In the case of cholera, the vaccine efficacy is around 50% and, as stated earlier, efficacy for BCG varies considerably. WHO figures for 2007 put the number of deaths from tuberculosis at 1.5 million. Although herd immunity may ultimately lead to eradication of some pathogens due to a reduced reservoir of infection, this is probably only attainable for pathogens where the only natural host is the human population. This was the case with smallpox and raises a similar prospect for polio and measles. Furthermore, there are no vaccines available for major diseases such as HIV/AIDS and malaria. In 2007, there were 2.5 million new cases of HIV infection and 2.1 million deaths recorded. One child under 5 years of age dies every 30 s from malaria and there were 500 million cases of illnesses reported in 2007. The extent of these diseases inevitably results in co-infection. There were 200 000 deaths associated with HIV/tuberculosis in 2007, and the emergence of an extensively drug-resistant form of tuberculosis, XDR-TB, compounds the existing problem.

12.4 NEW APPROACHES TO VACCINE DEVELOPMENT

The rapid development of molecular biological techniques coupled with major advances in genomics and proteomics has formed the bedrock of new vaccine development strategies. The techniques used in identifying gene sequences and in whole genome sequencing are discussed in other chapters. These techniques have been applied to the identification of potential antigenic targets for vaccine development and facilitated new methods for the expression and delivery of candidate genes. The common strategies arising from this include the development of recombinant live vector vaccines expressing heterologous genes, as delivery systems, and DNA vaccines (eukaryotic antigen expression vectors).

12.4.1 Recombinant Live Vectors

Recombinant live vectors expressing heterologous proteins have the potential benefits of live vaccines without the need to isolate and attenuate the actual pathogen. Several attenuated bacterial and viral vectors have been constructed with varying levels of complexity to deliver proteins from a range of pathogens. In principle, the introduction of these living vaccines should stimulate strong cellular responses as the transgene products will be expressed and processed within the host leading to MHC presentation. The stimulation of cytotoxic T cells via MHC class I presentation is of particular importance in immunity to viruses and other intracellular pathogens.

12.4.2 Recombinant BCG Vectors

BCG is the most commonly used vaccine in the world. Several recombinant BCG (rBCG) candidate vaccines containing genes from several different pathogens have been developed and tested,[5] and in general these proved effective at stimulating both B and T cell responses. Protective immunity in animal models has been demonstrated for rBCG vaccines expressing the outer surface protein (OspA) of *Borrelia* burg-dorferi,[6] the merozoite surface protein 1 (MSP-1)[7] and the *Leishmania* protective antigen (LRC1).[8] Other studies have shown immunogenicity for rBCG vaccines but no protection was demonstrated. Likewise, varying levels of heterologous protein have been reported.[9–11] The variability in protective responses and protein production may be due to differences in the shuttle vectors used, stability of the rBCG vaccines used and the route of administration.[12] Despite the variability in responses, the immunogenicity of rBCG vector vaccines shows promise.

12.4.3 Recombinant *Salmonella* Vectors

Salmonella spp. invade the gut and as such are good candidates for oral delivery of vaccines. In common with other Gram-negative bacteria, *Salmonella* possess type 3 secretion secretions (T3SS), which promote their invasiveness. One system is used to infect non-phagocytic cells while the second is required for intracellular replication and spread within the host. *Salmonella* Pathogenicity Island 2 (SPI2-T3SS) creates a needle-like complex through which bacterial effector proteins are delivered into the cell. It has been shown that the SPI2-T3SS can be used to deliver heterologous antigenic proteins.[13] The introduction of pro-teins through this route should result in intracellular processing of proteins for presentation to T cells. Significantly, the SPI2-TSS has been shown to be active in dendritic cells, the major professional antigen presenting cells of the immune system.[14] In one study,[15] using *S. typhimurium* strain, translational fusions between *Salmonella* effector proteins (SseF or SseJ) and ovalbumin and *L. monocytogenes* proteins listeriolysin O (Llo) and p60. The last two are protective antigens in various models of listerial infection.[16] Mice orally immunised with the recombinant *S. typhimurium* carrying the fusion proteins showed reduced *Listeria* burden. As found in other studies,[17] this protection was significantly greater than that observed with constitutively expressed proteins. In clinical studies,[18] mucosal delivery of an attenuated *S. typhimurium* carrying an SopE-HIV Gag fusion protein from a lethal plasmid showed some level of Il-2 response to the Gag protein as determined by ELISPOT. However, the response was low compared with that to *Salmonella* antigen alone. Only one shot of vaccine was used, so it may be possible to enhance the responsiveness. Importantly, the vaccine was well tolerated in the trial subjects, suggesting good safety prospects.

A recombinant *S. typhimurium* strain which secretes the *M. tubercu-losis* virulence protein ESAT-6 was tested in mice and demonstrated significant protection to subsequent challenge with *M. tuberculosis*[19] as measured by a reduction in the number of tubercle bacilli in the lungs. Co-immunisation with a DNA vaccine using ESAT-6 ligated to mammalian expression vectors showed no enhancement, although priming with the *S. typhimurium* vaccine followed by mixed immunisation showed a synergistic effect with some enhanced protection in the spleens of the *M. tuberculosis*-infected mice.

12.4.4 Recombinant Adenovirus Vectors

Adenoviruses, the primary cause of the common cold, show some promise as delivery systems for heterologous antigenic proteins. Attenuated human adenoviruses (HAd7) have been used for many years to immunise US military personnel and their families with no significant side-effects.[20] More recently, trials of attenuated human adenovirus 5 (HAd5) in non-human primates have shown protective immunity against Ebola virus.[21] Protective immunity has also been shown against HIV-1 in non-human primates using a HAd5 vector expressing the HIV-1 *gag* gene.[22] Adenovirus vectors are stable and generally safe vectors with good efficacy.[23] Furthermore, adenovirus vectors have been shown to be reliable and induce strong expression of transgenes. As they are replication-defective, horizontal transmission is very unlikely.[24] The major disadvantages of using HAd vectors in humans is the presence of pre-existing immunity to adenoviruses themselves. This arises from routine exposure to the wild-type viruses. HAd vectors are highly immunogenic in their own right and memory responses coupled with prior exposure to the wild types is likely to be the cause of reduced responses to heterologous proteins encoded in the vectors by neutralising antibodies to the vector itself. There may also be destruction of transduced cells expressing vector antigens. This type of memory response is the probable reason for the recent withdrawal from clinical trials of an adenovirus vector vaccine against HIV-1.[25] Different approaches have been undertaken to bypass this problem, such as prior boosting with the recombinant DNA,[26] microencapsulation of the vector[27] and the use of rarer types of adenoviruses.[28] A further strategy to overcome the preimmune interference is the use of non-human adenoviruses as vectors in humans.[29] Pre-existing antibodies to bovine adenovirus 3 (BAd3) are not prevalent in the human population.[30] Replication-defective BAd3 vectors have been shown to express heterologous proteins in various human cell types *in vitro*.[31] The use of non-human adenoviruses shows promise in the development of new vector vaccine delivery systems in humans.

12.4.5   Recombinant Vaccinia Vectors

Vaccinia has been used in humans since the pioneering work of Edward Jenner at the end of the 18th century. As the prototype vaccine, it is perhaps unsurprising that recombinant vaccinia vectors expressing a multitude of genes from across the microbial world have been described.[32] However, vaccinia is not recommended for use in individuals with compromised immune systems. In view of the ever-increasing population of HIV-positive individuals and ageing populations coupled with the immunogenicity of the vaccine and its ability to replicate in humans, the use of vaccinia as a vector vaccine in humans is limited without significant attenuation as seen in the modified vaccinia virus Ankara (MVA). MVA has been demonstrated to be safe in immune-suppressed non-human primates, suggesting a possible future role in vaccine development.[33] One alternative is to use other members of the pox family of viruses such as the avipoxviruses. These are replication-deficient in mammalian cells and offer some possibility for use in humans.

12.4.6   DNA Vaccines

The basis of this concept is the introduction of plasmid DNA encoding antigenic transgenes into the host cells in which the transgenes are expressed. In practice, although they do elicit specific immune responses, such vaccines have shown low immunogenicity, probably through reduced ability of the proteins to enter antigen processing pathways or loss of the plasmid prior to sufficient concentrations of protein being expressed. Immunogenicity can be augmented by including cytokine genes in the plasmid. Another approach is to target the vaccine to the antigen processing pathways using a gene encoding a targeting protein for the processing pathways. This has been demonstrated using the gene encoding the lysosome-associated membrane protein (LAMP). The use of an LAMP/gag chimera has been shown to induce long-term immunological memory in mice.[34] Furthermore, a single dose of the plasmid was sufficient to induce a secondary CD8+ T cell response without CD4+ T-helper cell assistance. The inclusion of genes encoding immunomodulatory molecules, in particular cytokines, would not just enhance the immunogencity of the vaccine but also allow manipulation of the immune response by altering the T-helper cell polarisation into the subset, *i.e.* TH_1 or TH_2 appropriate to the pathogen. Additionally, the inclusion of other molecules that influence the maturation of dendritic cells or molecules, such as chemokines that promote recruitment of other antigen-presenting cells to the immunisation site, open up possibilities not available in more conventional vaccine preparations. Such considerations may also apply to recombinant vector vaccines.

12.5 ADJUVANTS

The term adjuvant covers a multitude of potential substances with diverse chemical properties. An adjuvant is a substance that increases the effectiveness of the vaccine (from the Latin *ad juvare* – to help). Adjuvants may create depots of vaccine, thus prolonging its availability at a particular site, increase uptake of the vaccine by antigen presenting cells or they may be immunostimulatory in their own right. There are many different adjuvants available for experimental vaccines but the only ones universally approved for use in humans are the aluminium salt-based adjuvants including aluminium hydroxide, aluminium phosphate and potassium aluminium sulfate (alum). The probable adjuvant activity of alum is the creation of a vaccine depot at or near the site of immunisation. Other adjuvants used in experimental vaccines are described below.

12.5.1 Immune-stimulating Complexes (ISCOMs) and Liposomes

ISCOMs are a mixture of cholesterol, phospholipid and Quil A that form a matrix structure described rather well as being similar in shape to a practice golfball.[35] The proteins are added to these lipid carriers and orientate with their hydrophilic residues exposed. Liposomes containing protein antigens are prepared by mixing the proteins with a suspension of phospholipids and under appropriate conditions form a complex with the proteins protruding out from the centre. Liposomes and ISCOMs probably fuse with the plasma membrane of cells, permitting the antigen to enter the cell where they enter the cytosolic pathway. They are then degraded and presented with MHC class I molecules to cytotoxic T cells. A variation on the use of liposomes is the use of virosomes. These are functionally reconstituted viral membranes which retain the cell-binding and fusion properties of the virus. Virosome-encapsulated OVA elicited a strong CTL response in mice with as little as 0.75 µg of OVA when administered by an intramuscular, intraperitoneal or subcutaneous route.[36]

12.5.2 Freund-type Adjuvants

Freund's complete adjuvant (FCA) is the stalwart of experimental vaccines. The adjuvant is a mixture of mineral oils containing heat-killed *M. tuberculosis* cells together with an emulsifier. This is a very potent adjuvant, stimulating both antibody and cellular immune responses in its own right. This is mainly due to the immunogenic properties of the mycobacterial component. Due to potential formation of granuloma, a mixture of epithelioid macrophages and T cells and the chronic inflammation induced, it is not used in vaccines intended for application in humans. Although it undoubtedly has a major role in immunisation for the purposes of inducing immune responses in experimental animals and in the production of antibodies as reagents, its use in vaccine development may be questionable as the immune responses induced may not be representative of the likely outcomes using preparations approved for use in humans.

12.5.3   CpG Oligonucleotides (CpG ODNs)

CpG ODNs are unmethylated nucleotide sequences typically less than 20 nucleotides in length. There are three types of CpG ODNs distinguished by the backbone structure. Type A have a mixed phosphodiester–phosphorothioate backbone, type B have a phosphorothioate backbone and type C also have a phosphorothioate backbone with a TCG dimer at the 5' end.[37] These unmethylated sequences bind to one of the pattern recognition receptors, toll-like receptor 9 (TLR 9) expressed in intracellular endosomal compartments of dendritic cells, macrophages and to a lesser extent on other cell types. TLR9 typically binds such sequences derived form bacteria following phagocytosis of bacteria. TLR9 induces cellular activation of dendritic cells inducing cytokine production and promoting T_H1 responses. The promotion of this subset is primarily due to the release of Il-12. The ability to activate dendritic cells and other antigen-presenting cells has generated much interest in CpG ODNs as potential adjuvants for both DNA vaccines to boost the low immunogenicity often observed with such vaccines and also for more classical vaccine preparations. CpG ODNs have been shown both to accelerate and to magnify immune responses, particularly at lower vaccine doses. Differences in the immune responses between type A and B CpG ODNs have been reported.[38] Type B induced Ag-specific CTL numbers in human PBMCs whereas type A did not. Soluble factors induced by type A but not type B increased the cytotoxicity of established CD8 + clones.

Combining CpG ODNs with biphasic vesicles carrying the outer membrane lipoprotein A (OmlA) from *Actinobacillus pleuropneumoniae* delivered by nasal immunisation induced both systemic and localised antibody responses in pigs, whereas the protein alone or with the CpG ODN did not induce mucosal responses.[39] The use of these

Vaccines 349

immunostimulatory molecules coupled with new DNA-based vaccines offers the prospect of a new era in vaccine design.

REFERENCES

1. P. E. M. Fine, I. A. M. Carneiro, J. B. Milstien and C. J. Clements, *WHO Discussion Document*, 1999.
2. P. Halfman, J. H. Kim, H. Ebihara, T. Noda, G. Neumann, H. Feldmann and Y. Kawaoka, *Proc. Natl. Acad. Sci. USA*, 2008, **105**(4), 1129–1133.
3. P. Valenzuala, A. Medina, W. J. Rutter, G. Ammerer and B. D. Hall, *Nature*, 1982, **298**, 347–350.
4. Network for Education and Support in *Immunisation, Immunisation Curriculum Review and Update Workshop*, 2007.
5. M. Dennehy and A. Williamson, *Vaccine*, 2005, **23**(10), 1209–24.
6. C. K. Stover, G. P. Bansal and M. S. Hanson, *J. Exp. Med.*, 1993, **178**(1), 197–209.
7. S. Matsumoto, H. Yukitake, H. Kanbara and T. Yamada, *J. Exp. Med.*, 1998, **188**, 845–854.
8. J. A. Streit, T. J. Recker, J. E. Donelson and M. E. Wilson, *Exp. Parasitol.*, 2000, **94**(1), 33–41.
9. S. Langermann, S. R. Palaszynski and J. E. Burlein, *et al., J. Exp. Med.*1994, **180**, 2277–2286.
10. A. Aldovini and R. A. Young, *Nature*, 1991, **351**(6326), 479–482.
11. B. Abomoelak, K. Huygen, L. Kremer, M. Turneer and C. Locht, *Infect. Immunol.*, 1999, **67**(10), 5100–5105.
12. N. Ohara and T. Yamada, *Vaccine*, 2001, **1**(19), 4089–4098.
13. H. Rûssmann, H. Shams, F. Poblete, Y. Fu, J. E. Galan and R. O. Donis, *Science*, 1998, **281**, 565–568.
14. J. Jantsch, C. Cheminay, D. Chakravortty, T. Lindig, J. Hein and M. Hensel, *Cell. Microbiol.*, 2003, **5**, 933–945.
15. M. I. Husseiny, F. Wartha and M. Hensel, *Vaccine*, 2007, **25**(1), 185–193.
16. M. Lara-Tejero and E. G. Pamer, *Curr. Opin. Microbiol.*, 2004, **7**(1), 45–50.
17. M. I. Husseiny and M. Hensel, *Vaccine*, 2004, **23**(1), 185–193.
18. C. N. Kotton, A. J. Lankowski, N. Scott, D. Sisul, L. M. Chen, K. Raschke, G. Borders, M. Boaz, A. Spentzou, J. E. Galán and E. L. Hohmann, *Vaccine*, 2006, **24**, 37–39.
19. H. Mollenkopf, D. Groine-Triebkorn, P. Andersen, J. Hess and S. H. E. Kaufmann, *Vaccine*, 2001, **19**(28–29), 4028–4035.

20. F. H. Top Jr, R. A. Grossman, P. J. Bartelloni, H. E. Segal, B. A. Dudding, P. K. Russell and E. L. Buescher, *J. Infect. Dis.*, 1971, **124**(2), 148–154.

21. N. J. Sullivan, T. W. Geisberg, J. B. Geisberg, L. Xu, Z. Y. Zang and M. Roederer, *et al., Nature*, 2003, **424**, 681–684.

22. D. R. Casimiro, A. Tang, L. Chen, T. M. Fu, R. K. Evans and M. E. Davies, *et al., J. Virol.*, 2003, **77**(13), 7663–8.

23. N. Tatsis and H. C. Ertl, *Mol. Ther.*, 2004, **10**(4), 616–29.

24. D. S. Bangeri and S. K. Mittal, *Vaccine*, 2006(24) 849–862.

25. National Institutes of Allergy a Infectious Disease Press Release, 2007.

26. Z. Y. Yang, L. S. Wyatt, W. P. Kong, Z. Moodie, B. Moss and G. J. Nabel, *J. Virol.*, 2003, **77**(1), 799–803.

27. G. Sailaja, H. HogenEsch, A. North, J. Hays and S. K. Mittal, *Gene Ther.*, 2002, **9**(24), 1722–1729.

28. R. Vogels, D. Zuijdgeest, E. Hartkoorn, I. Damen and M. P. de Bethune, *et al., J. Virol.*, 2003, **77**(15), 8263–8271.

29. D. S. Bangari and S. K. Mittal, *Vaccine*, 2006, **24**, 849–862.

30. S. Moffatt, J. Hays, H. HogenEsch and S. K. Mittal, *Virology*, 2000, **272**(1), 159–167.

31. U. B. Rasmussen, M. Benchaibi, V. Meyer, Y. Schlesinger and K. Schughart, *Hum. Gene Ther.*, 1999, **10**(16), 2587–99.

32. B. Moss, *Proc. Natl. Acad. Sci. USA*, 1996, **93**, 11341–11348.

33. K. J. Stittelaar, T. Kuiken, R. L. de Swart, G. Amerongen, H. W. Vos, H. Niesters, et al., *Vaccine*, **19**(27), 3700–9.

34. L. Barros de Arruda, P. R. Chikhlikar, J. T. August and E. T. A. Marques, *Immunology*, 2004, 126–135.

35. D. E. S. Stewart-Tull, Vaccine protocols, in *Methods in Molecular Medicine*, 87, 175–94.

36. L. Bungener, A. Huckriede, A. de Mare, J. de Vries-idema, J. Wilschut and T. Daemen, *Vaccine*, 2005, **23**(10), 1232–41.

37. D. M. Klinman, *Int. Rev. Immunol.*, 2006, **25**, 135–154.

38. S. Rothenfusser, V. Horning, M. Ayyoub, S. Britsch, A. Towarowski, A. Krug, A. Sarris, N. Lubenow, D. Speiser, S. Endres and G. Hartmann, *Blood*, 2004, **103**(6), 2162–9.

39. V. Alcon, M. Baca-Estrada, M. Vega-Lopez, P. Willson, L. A. Babiuk, P. Kumar, R. Hecker and M. Foldvari, *AAPS J.*, 2005, **7**(3), 566–71.

CHAPTER 13

Tissue Engineering

NILS LINK AND MARTIN FUSSENEGGER

Institute for Chemical and Bioengineering (ICB), ETH Zurich, HCI F115, Wolfgang-Pauli-Strasse 10, CH-8093, Zurich, Switzerland

13.1 INTRODUCTION

13.1.1 Economic Impact of Healthcare

For thousands of years, medical practitioners have attempted to reduce pain, cure disease, extend lifespan and generally improve quality of life. The immense technological progress of the past 30 years has enabled great advances to be made in the treatment of disease. This is reflected statistically in life expectancies, which have increased significantly over the past decades. However, longer life expectancy creates new medical issues, particularly with regard to the longevity of organs and tissues that must now function for longer periods of time. At present, organ transplantation is the only possible therapy for some patients. Thus, as life expectancy increases, so too does the demand for donor organs (Table 13.1). In the USA alone, 90 000 patients were waiting for organs in 2006.[1] Furthermore, the treatment and care of patients undergoing such treatment are costly [more than US$400 billion per year in the USA (Tables 13.1 and 13.2)]. Although the use of synthetic materials aids in keeping people alive, many implants function only to a limited extent like living tissues. Unlike real tissues, which are in a constant state of renewal, synthetic implants begin to show signs of decomposition within

Molecular Biology and Biotechnology, 5th Edition
Edited by John M Walker and Ralph Rapley
© Royal Society of Chemistry 2009
Published by the Royal Society of Chemistry, www.rsc.org

Table 13.1 Number of transplants performed in 2005 and impact for the general public. Source: Milliman Research Report 2005.

Transplant	No. of transplants	Total charge per year (US$ millions)
Bone marrow	16890	5629
Cornea	32840	627
Heart only	1960	939
Heart–lung	40	26
Intestine only	60	49
Kidney only	16150	3392
Kidney–heart	50	29
Kidney–pancreas	890	259
Liver only	5980	2349
Liver–intestine	50	42
Liver–kidney	290	149
Liver–pancreas–intestine	50	44
Lung only	1205	472
Pancreas only	570	154
Pancreas–intestine	6	5
Other multi–organ	20	15
Total	77001	14180

Table 13.2 Cost of treating the most expensive diseases in 2005. Source: Forbes 2005.

Indication	Total charge per year (US$ billions)
Heart conditions	68
Trauma	56
Cancer	48
Mental illnesses	48
Respiratory ailments	45
Hypertension	32.5
Arthritis and joint disorders	32
Diabetes	28
Back problems	23
Total	380.5

10–15 years and must eventually be replaced. Therefore, a key goal of regenerative medicine is to develop artificially generated tissues, which can take over the function of the original tissue without being rejected by the patient's immune system.

13.1.2 Tissue Engineering

As early as 1987, the US National Science Foundation (NSF) approved the proposal submitted by Y. C. Fung of the Granlibakken Workshop to initiate research into tissue engineering to overcome the problem of donor organ shortage. This project is regarded as the birth of tissue engineering, the goal of which, according to the NSF, is 'the application of principles and methods of engineering and life sciences toward fundamental understanding of structure-function relationships in normal and pathological mammalian tissues and the development of biological substitutes to restore, maintain or improve tissue function'. This statement makes it clear that tissue engineering is an interdisciplinary and heterogeneous research field, where biologists, chemists, material scientists and medical practitioners must work together to achieve success.[2] After 50 years of progress in cell culture, immunology, the cell cycle, proliferation, (trans-)differentiation and extracellular matrices (ECM),[3,4] we are currently in a position to extract, culture and expand, *in vitro*, all the existing cell types of an organism.

The groups of Hay, Heath and Ikehara have performed outstanding investigations of the different growth factors, their effects and the chronology of the markers expressed by cells during differentiation[5–8] (Table 13.3). Nevertheless, the cell model developed thus is far from complete. It is not difficult to keep cells alive in culture, but it is challenging to multiply them without altering their differentiation and phenotype.[9–12] Therefore, research must continue to identify the exact biochemical markers and growth factors that will finally enable the tissue engineer to determine unambiguously whether the cells behave in exactly the same way as cells in living organs. Only then can artificial tissues be engineered which express the same phenotype as the desired organ.[9]

Within the field of tissue engineering, there are four main areas of research: (i) cell therapy, which is based mainly on self-organization of free cells that are injected into damaged tissue,[13,14] (ii) bio-artificial devices, in which cells of a certain type are encapsulated to take on certain functions of the organ, *e.g.* liver cells, pancreatic islet cells and kidney cells,[15–18] (iii) scaffold-assisted tissue engineering, in which cells are grown on a matrix until they reach a certain level of stability,[19,20] and (iv) scaffold-free tissue engineering, in which cells form *ex vivo* after aggregation of their natural extracellular matrix has occurred.[21–23] A discussion of all the topics is beyond the scope of this chapter. Therefore, we will focus on the last two approaches, which involve the growth of real tissues *in vitro*.

13.1.3 Treating Disease Through Tissue Engineering

In the short time since the initiation of tissue engineering programs, immense progress has been made. In principle, all diseases associated with the failure of tissue functions are currently being addressed. Many of the tissues, generated *in vitro* for the replacement of cartilage,

Table 13.3   Known markers and methods for identification of different cell
types.

Cell type	Marker/method
Adipocyte	Adipocyte lipid-binding protein (ALBP)
	Fatty acid transporter (FAT)
Astrocyte	Glial fibrillary acidic protein (GFAP)
Bone marrow fibroblast	Fibroblast colony-forming unit (CFU-F)
	Muc-18 (CD146)
Cardiomyocyte	Myosin heavy chain
Chondrocyte	Collagen types II and IV
	Sulfated proteoglycan
Ectoderm	Neuronal cell-adhesion molecule (N-CAM)
	Pax6
Ectoderm, neural and pancreatic progenitor	Nestin
	Vimentin
Embryoid body (EB)	Neurosphere
Embryonic stem (ES), embryonal carcinoma (EC)	Alkaline phosphatase
	Cripto (TDGF-1) [cardiomyocyte]
	Cluster designation 30 (CD30)
	GCTM-2
	Germ cell nuclear factor
	Oct-4
	Stage-specific embryonic antigen-3 (SSEA-3)
	Stage-specific embryonic antigen-4 (SSEA-4)
	Telomerase
	TRA-1-60
	TRA-1-81
	Stem cell factor (SCF or c-Kit ligand)
Endoderm	Alpha-fetoprotein (AFP)
	GATA-4
	Hepatocyte nuclear factor-4 (HNF-4)
	Fetal liver kinase-1 (Flk1)
Hematopoietic stem cell (HSC)	CD34
	Lineage surface antigen (Lin)
	c-Kit
	Stem cell antigen (Sca-1)
	Thy-1
Hepatocyte	Albumin
	B-1 integrin
Keratinocyte	Keratin
Mesenchymal	CD44
Mesenchymal stem cell (MSC)	Bone morphogenic protein receptor (BMPR)
	CD34
	Sca-1
	Lin
Mesoderm	Bone morphogenic protein-4
	Brachyury

Table 13.3 (*Continued*).

Cell type	Marker/method
Myoblast	MyoD
	Pax7
Neural stem cell	Nestin
	CD133
Neuron	Microtubule-associated protein–2 (MAP–2)
Oligodendrocyte	Myelin basic protein (MPB)
	O4
	O1
Osteoblast	Bone-specific alkaline phosphatase (BAP)
	Hydroxyapatite
	Osteocalcin (OC)
Pancreatic islet	Glucagon
	Insulin
	Insulin-promoting factor-1 (PDX-1)
	Pancreatic polypeptide
	Somatostatin
Pancreatic progenitor	Nestin
Skeletal myocyte	Myogenin
	MR4
	Myosin light chain
Skeletal muscle	Smooth muscle cell-specific myosin heavy chain
	Vascular endothelial cell cadherin
White blood cell (WBC)	Stro-1 antigen
	CD4
	CD8

articulation, heart, liver and pancreatic islets, have already undergone clinical tests. Cartilage has already reached the stage of clinical application[24] due to the robustness and phenotypic stability of the cells and also the fact that cartilage and bone tissues can be grown *in vitro* with relative ease. These tissues do not require strong vascularization – still a major challenge in tissue engineering – since there is a sufficient diffusion of nutrients and oxygen through the scaffolds used for this type of cell. Several tests have been conducted with three-dimensional hyaline cartilage to remedy articular damage or abrasions when other treatments fail (*e.g.* chondroitin sulfate (Chondrosulf)).[25] Other tissues, such as heart valves and heart-muscle tissues, have also been produced *in vitro*.[26,27] However, they have shown poor integration after transplantation into the organ and have not reliably maintained their function. Researchers and medics are working on a cure for liver malfunction (liver cirrhosis and fibrosis), induced by either alcohol abuse or viral infections such as hepatitis. Although the liver can regenerate itself, it can also become so weakened that transplantation is the only solution. To overcome the perpetual shortage of donor organs, an attempt has been made to propagate hepatocytes *in vitro*. However, only the culturing of these cells as scaffold-free microtissues or on three-dimensional scaffolds that mimic the natural extracellular matrix of the liver has been successful; growing hepatocytes as monolayers changes their phenotype and results in the loss of their natural function.[23] In addition, research into the generation of pancreatic islet tissues to combat diabetes, the main cause of hormone dysfunction in developed countries, is ongoing.[28,29]

It is not possible here to list all the organs that are currently being reconstructed in laboratories. We have, therefore, concentrated on the most important diseases, for which a clinical therapy is standard or soon will be. Although researchers have made great advances in the engineering of tissues with simple structures (*e.g.* cartilage and liver), many tissues do not consist of a single cell type but rather a specific architecture of many cell types that enables them to perform their tasks, as is the case, for example, with the kidney, heart and skin. Hence it is very important to understand the structure and cellular composition of native tissue in order to reconstruct them for therapeutic applications.

13.2   CELL TYPES

In recent years, in-depth research has been performed to identify the cells responsible for tissue-/organ-specific functions. Four classes of (stem)-cell variants can be defined, namely (i) embryonic stem cells (ESC), which are sub-divided into totipotent early embryonic stem cells (eESC) and pluripotent blastocyst embryonic stem cells (bESC), (ii) multipotent umbilical cord stem cells (uCSC), (iii) multipotent adult stem cells (ASC) and (iv) differentiated cells (DC). Each of these variants possess advantages and disadvantages.

13.2.1   Embryonic Stem Cells

For over 20 years, researchers have been able routinely to extract and cultivate mouse embryonic stem cells.[30] This cultivation expertise has also permitted the isolation and cultivation of human embryonic stem cell lineages.[31] Embryonic stem cells are non-specialized cells which are found in the inner cell mass of 7–30-day-old blastocysts or embryos. At this stage of development, the three main cell lineages of ectoderm, mesoderm and endoderm have not yet formed. Thus, embryonic stem cells are pluripotent and are able to differentiate into every cell type of a tissue or organism. This is one of the most important advantages for tissue engineering when considering suitable cell types. Compared with adult stem cells, embryonic stem cells are capable of long-term renewal without differentiation, thus allowing the production of the cell mass required for tissue engineering applications.

As a general rule, a stem cell can do one of the following: divide and generate a new stem cell or differentiate into a particular tissue lineage (Figure 13.1). Symmetric division gives rise to two identical daughter cells, both endowed with stem cell properties, whereas asymmetric division produces only one stem cell and a progenitor cell with limited potential for self-renewal. Progenitors can go through several rounds of cell division before terminally differentiating into a mature cell. The switch between division and differentiation is usually triggered easily by external factors such as the cell environment, hormones or signals from neighboring cells. As a result, it is difficult to expand embryonic stem cells in culture without them differentiating spontaneously in one direction. Therefore, embryonic stem cell cultures must be monitored carefully and continuously. Due to lack of knowledge, the research community has not yet established a protocol for the absolute and complete characterization of stem cells, although the following methods

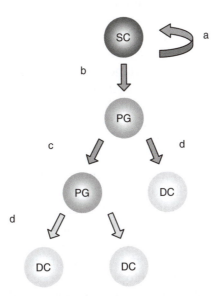

Figure 13.1 Pathways of stem cells (SCs). SCs often divide asymmetrically, thereby generating a daughter SC (a) and a committed progenitor cell (PG) (b). The PG either divides to produce a PG (c) or fully matures to a differentiated cell (DC) (d).

are commonly used simultaneously to ensure that the cells display the required stem cell traits:[9,32]

- visual characterization by microscopy;
- formation of embryoid bodies;
- analysis of surface markers, such as the stage-specific embryonic antigen (SSEA)-3, SSEA-4, TRA-1-60, TRA-1-81 and alkaline phosphatase;
- determination of whether Oct-4 – a transcription factor typical of undifferentiated cells is expressed and whether cells display high telomerase activity;
- long-term cultivation to ensure long-term self-renewal, a special characteristic of embryonic stem cells;
- microscopic analysis of chromosome damage;
- determination of whether sub-culturing is possible after freezing/ thawing;
- differentiation assays leading to a specific cell type;
- proteomic analysis.[33]

Using these methods, researchers are in a position to vary and improve culture techniques in order to study the fundamental properties of embryonic stem cells, including the precise determination of why embryonic stem cells are not specialized, how they renew themselves over many years and which factors cause stem cells to undergo differentiation. The determination of these signals and the chronology of their occurrence are of prime importance for the production of the desired phenotype for tissue reconstruction. To trigger differentiation, the surface of the culture dish, the scaffold, the chemical composition of the culture medium (Figure 13.2) or the genetic expression pattern of the cells can all be manipulated.

Embryonic stem cells are the only naturally occurring cells that are able to divide frequently enough without manipulation to generate a sufficient number of cells for therapeutic application. Were it not for ethical concerns, restrictions or bans and also the risk of developing teratomas or rejection by the patient, these cells would be the optimal choice for tissue engineering. Furthermore, we still possess only a rudimentary understanding of most of the differentiation pathways and the culture conditions required to achieve precise lineage control of stem cells. As a result, scientists often obtain differentiated cells, which have a phenotypic resemblance to the cells of a certain tissue but which express not only the usual markers, but also tissue-atypical proteins and do not fulfill all the functions of the native cells in a tissue.

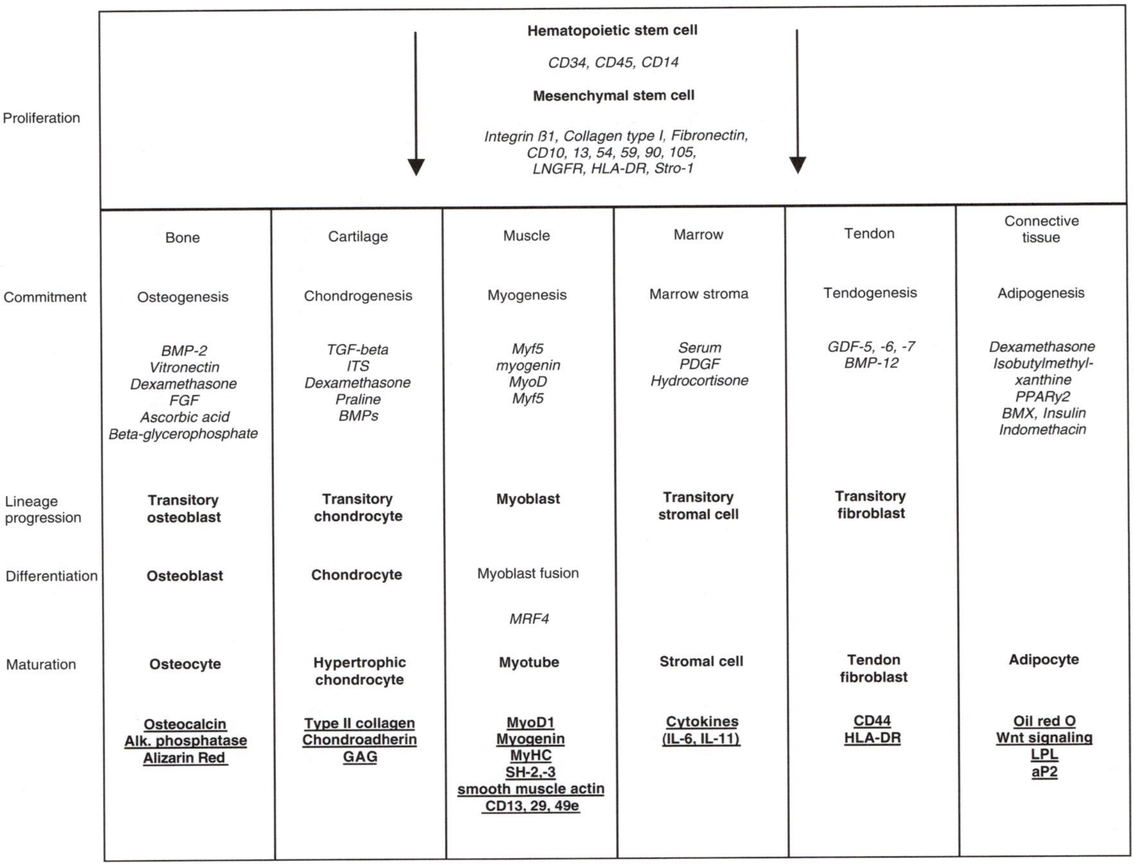

Hematopoietic stem cell

CD34, CD45, CD14

Mesenchymal stem cell

Integrin β1, Collagen type I, Fibronectin, CD10, 13, 54, 59, 90, 105, LNGFR, HLA-DR, Stro-1

	Bone	Cartilage	Muscle	Marrow	Tendon	Connective tissue
Proliferation						
Commitment	Osteogenesis	Chondrogenesis	Myogenesis	Marrow stroma	Tendogenesis	Adipogenesis
	BMP-2 Vitronectin Dexamethasone FGF Ascorbic acid Beta-glycerophosphate	*TGF-beta ITS Dexamethasone Praline BMPs*	*Myf5 myogenin MyoD Myf5*	*Serum PDGF Hydrocortisone*	*GDF-5, -6, -7 BMP-12*	*Dexamethasone Isobutylmethyl-xanthine PPARγ2 BMX, Insulin Indomethacin*
Lineage progression	**Transitory osteoblast**	**Transitory chondrocyte**	**Myoblast**	**Transitory stromal cell**	**Transitory fibroblast**	
Differentiation	**Osteoblast**	**Chondrocyte**	Myoblast fusion *MRF4*			
Maturation	**Osteocyte**	**Hypertrophic chondrocyte**	**Myotube**	**Stromal cell**	**Tendon fibroblast**	**Adipocyte**
	<u>Osteocalcin</u> <u>Alk. phosphatase</u> <u>Alizarin Red</u>	<u>Type II collagen</u> <u>Chondroadherin</u> <u>GAG</u>	<u>MyoD1</u> <u>Myogenin</u> <u>MvHC</u> <u>SH-2-3</u> <u>smooth muscle actin</u> <u>CD13, 29, 49e</u>	<u>Cytokines (IL-6, IL-11)</u>	<u>CD44</u> <u>HLA-DR</u>	<u>Oil red O</u> <u>Wnt signalling</u> <u>LPL</u> <u>aP2</u>

Figure 13.2   Mesengenic process, differentiation factors and markers of mesenchymal stem cells (MSCs). The value of MSCs for tissue engineering is apparent after analysis of the different cell types that can be obtained. The diagram displays the stepwise cellular transition upon induction with differentiation factors (*italic*) from MSCs to highly differentiated phenotypes expressing characteristic cell markers (underlined).

13.2.2 Adult Stem Cells

Adult stem cells, also referred to as somatic stem cells, represent an ethically acceptable alternative to embryonic stem cells in that they do not require harvesting of embryos. These cells constitute a very small population of undifferentiated cells that can be found amongst the differentiated cells of an organ. They renew themselves and differentiate into the major specialized cells of the organ in which they reside. Their primary tasks are to maintain the proper functioning and repair of the tissue. According to the literature, adult stem cells have been found in the brain, skin, bone marrow, peripheral blood, blood vessels, skeletal muscle and liver. Some of the most important and best characterized adult stem cells include:

- *Brain:* neural stem cells can develop into four major cell types: (i) neurons, (ii) astrocytes, (iii) oligodendrocytes and (iv) glial cells.[34,35]
- *Bone marrow:* containing two types of stem cells – first, the hematopoietic stem cells that give rise to all types of blood cells, *i.e.* (i) red blood cells, (ii) B-lymphocytes, (iii) T-lymphocytes, (iv) natural killer cells, (v) neutrophils, (vi) basophils, (vii) eosinophils, (viii) monocytes, (ix) macrophages and (x) platelets; and second, the mesenchymal stem cells (bone marrow stromal cells) which give rise to different cell types, such as (i) osteocytes (bone cells), (ii) chondrocytes (cartilage cells), (iii) adipocytes and (iv) some types of connective tissue cells (Figure 13.2).
- *Digestive tract:* epithelial stem cells differentiate into several cell types such as (i) absorptive cells, (ii) goblet cells, (iii) paneth cells and (iv) entero-endocrine cells.[36,37]
- *Skin:* epidermal stem cells can be found in the basal layer of the epidermis. These stem cells give rise to keratinocytes. Follicular stem cells give rise to both the hair follicle and to the epidermis at the base of hair follicles.[38,39]

Unfortunately, not all essential human organs contain adult stem cells. As in the case of the heart, the stem cells may not divide or differentiate rapidly enough if the organ has been damaged.[40–42] Most adult stem cells are present only in small numbers and are, therefore, difficult to extract and separate from other cells of the organs. In contrast to embryonic stem cells, adult stem cells have lost their ability to differentiate naturally into all the tissues of an organism. However, recent publications indicate that adult stem cells exhibit plasticity or undergo transdifferentiation under certain conditions.[43–45] The discovery of plasticity among adult stem cells is very important for tissue engineering, as it permits the generation of new types of cells through transdifferentiation.[4] Another advantage of adult stem cells is that it may be possible to use the patient's own cells, to expand them, to generate the desired tissues *in vitro* and to transplant the tissues back into the patient, thereby avoiding the risk of immunological rejection of the transplant. Therefore, adult stem cells, treated to differentiate into specific cell types, could be a renewable source of cells and tissues for the efficient treatment of diseases such as cystic fibrosis, renal failure, stroke, burns, heart disease, diabetes and arthritis.

13.2.3 Mature Cells

In the human body, there are approximately 300 different types of differentiated cells, each with its own specific function. During differentiation, these cells lose the ability to divide and, by adapting to a specific task, can no longer proliferate. Since cell therapy and tissue engineering rely on the implantation of considerable quantities of well-characterized autologous cells that display defined properties, differentiated cells must be manipulated so that they multiply *in vitro*. Taking primary cells from the healthy tissues of choice is mandatory for therapeutic applications. It is unimaginable that immortalized cells could be reintroduced into humans, as they could develop into a cancer. There are different methods for re-initiating the cell division of differentiated cells. One way is to modify the cells genetically and conditionally immortalize them[46] to proliferate for a limited period of time without inducing epigenetic changes or changes in the cell properties.[46,47] Ideal candidate transgenes to control conditional expansion of therapeutic cell populations could be Notch[48] or antisense $p27^{Kip1}$.[49] However, since these strategies enable proliferation induction in a limited number of cell lines viral genes, such as the simian virus large T antigen, the Herpes virus-16 E6/7 or the ubiquitous telomerase reverse transcriptase, are sometimes used generically to trigger proliferation.[50,51] In any case, proliferation-inducing transgenes must be tightly and timely controlled using either state-of-the-art transcription or translation control modalities[52–54] or site-specific recombination technology to excise transgene expression loci prior to implantation.[55–58]

Another way to trigger the proliferation of differentiated cells is to reprogram them under controlled conditions. The phenotypic conversion of a cell or a tissue type into another cell or tissue is referred to as 'metaplasia', with many examples prevalent in humans.[59] Metaplasia frequently begins with the de-differentiation of the cell, whereby endogenous proliferation genes are turned on and cell mass can be generated for tissue engineering. This can be achieved either by adding extracts of pluripotent cells to the media[60–62] or by implanting the nucleus of the cell in an ovum, from which the nucleus has been removed.[63,64] However, a drawback of this approach is that the de-differentiated cells must differentiate correctly after the proliferation step.

An alternative to obtaining the desired cell type is trans-differentiation, which is a specific form of metaplasia in which the phenotype of a mature cell is converted into another fully differentiated cell type. Mammalian cells have been trans-differentiated by co-culture, by modification of cell culture conditions or by genetic modification of the cells. Hu *et al.* reported the transdifferentiation of myoblasts into mature adipocytes by ectopic expression of the adipogenic transcription factors PPAR gamma and C/EBPα under conditions conductive to adipogenesis.[65] Schiller *et al.* showed that inhibition of gap-junctional communication between osteoblasts results in adipocyte formation as well.[66] Furthermore, it has been shown that pancreatic cells can be converted into hepatocytes by the induction of hepatic transcription factors using dexamethasone[67] and that myoblast transdifferentiation can be controlled precisely by introducing differentiation determinants such as MyoD and msx1.[4] The microenvironment of the cells, including the extracellular matrix, the surrounding cells, the local milieu and growth or differentiation factors, plays a crucial role in redirecting cell fate. Injection of endothelial cells into damaged heart tissue has led to their trans-differentiation into beating cardiomyocytes.[68] By the same means, pancreatic epithelial cells have also trans-differentiated successfully into hepatocytes.[69] Trans-differentiation plays an essential role in tissue engineering since large amounts of certain cell types, such as adipocytes, can be grafted from patients with reintroduction of the tissue into the body not resulting in immunogenic rejection.

13.3 EXTRACELLULAR MATRIX

13.3.1 Biological Extracellular Matrices

The second important component in tissue engineering is the extracellular matrix (ECM). The ECM represents the secreted product of the resident cells of each tissue and organ. Therefore, the ECM of each tissue and organ has a unique structure and composition, providing structural stability for the tissue. It includes information about the position and alignment of the different cell types and ensures that the relevant growth factors are provided at the right level, time and place to coordinate organ morphogenesis and repair. Although all the cells in an organism are embedded in ECM, sufficient amounts of ECM for use in tissue engineering can only be extracted from a few tissues such as skin, the pericardium, small intestine, urinary bladder, liver and Achilles tendon.[70,71] ECM is by no means static and uniform but is rather a structure that adapts continuously to the requirements of the tissue. The composition and structure of the ECM are directly coupled to its location within the organ, the function of the tissue and the age of the individual.[72–75] For example, kidney has very little ECM compared with its cellular component, whereas tissue that is primarily structural, such as tendons and ligaments, displays large amounts and a differential composition of ECM.[76,77] ECM is composed mainly of collagen, of which more than 20 types have been identified thus far.[78] The most common is type 1 collagen that has been highly conserved during the course of evolution. Thus, allogenic and xenogenic sources of type 1 collagen are both relevant for tissue engineering, making collagen the most widely used biologic scaffold in therapeutic interventions. The 12 subtypes of collagen are responsible for the distinctive biological activity of the ECM. In combination with laminin they form a three-dimensional mesh-like structure that is adapted to the specific function of a tissue and provides optimal strength, rigidity or plasticity.[79] Laminin, the second most abundant protein in ECM, is a complex trimeric, cross-linked adhesion protein with separate binding domains for collagen IV, heparin, heparin sulfate and direct cell binding.[80] Laminin exists in different isoforms, depending on the particular mixture of the peptide chains[81,82] and plays an important role in the vascularization and maintenance of vascular structures.[83,84] Since vascularization of scaffolds for tissue repair is the most rate-limiting step, laminin is considered to be an important component of cell-friendly scaffold material.[85]

A very important peptide motif found in most proteins which form the ECM (*e.g.* the glycoproteins fibronectin or vitronectin) is the arginine, glycine and asparaginic acid sequence (RGD).[86,87] This motif binds to cellular adhesion molecules (CAM), known as integrins, thereby anchoring the cells mechanically in the ECM. The different RGD sequences adopt different conformations in the different matrix proteins, so that these sequences are recognized by different integrin subtypes expressed by specific cell types which favors a tissue-specific cellular organization. Furthermore, the generation of focal adhesions generates cell responses such as the polarization of the cells, the production of survival signals and factors for the remodulation of the ECM. Therefore, synthetic scaffolds produced for tissue engineering are often modified with special RGD–peptide sequences to render them biocompatible, to provide better integration and to control the setup of the tissue.[88–90]

Important non-protein components of ECM are the glycosaminoglycans (GAGs), which do not have a structural function. However, they substantially modulate the gel properties of the ECM by retaining water and binding growth factors and cytokines. Also, their ability to mediate ECM–cell interactions makes heparin-rich GAGs a valuable component for tissue engineering-compatible scaffolds.[91] In addition to structural proteins, the ECM also contains trace amounts of a variety of bioactive proteins which functionalize the bio-scaffold. Non-limiting examples include vascular endothelial growth factor, basic fibroblast growth factor, epidermal growth factor, hepatocyte growth factor, keratinocyte growth factor, transforming growth factor beta and platelet-derived growth factor. Due to the presence of such growth factor cocktails, natural ECM grafts are often used to functionalize synthetic scaffolds,[92] which are otherwise biologically inert and fail to degrade and promote infiltration of cells from neighboring tissues. In contrast, the use of unmodified, de-cellularized ECM promotes rapid cell infiltration, scaffold degradation, deposition of neo-matrix and tissue organization with a minimum of scar tissue.[93] Today, over one million patients have been treated successfully with xenogeneic ECM scaffolds[94] to heal skin lesions,[95] promote vascular reconstruction[96,97] and re-establish the urinary tract,[98–100] the intestine,[101] diaphragm,[102] rotator cuff[103] and muscle structures.[104]

The complex three-dimensional structure and composition of ECM have yet to be fully elucidated, exemplifying how difficult it is to design ECM-containing scaffold mimetics. Bottom-up approaches to functionalize synthetic scaffolds with well-characterized components (*e.g.* laminin, fibronectin, hyaluronic acid, vascular endothelial growth factors)[105] have resulted in some success for very specialized applications, but there is still a long way to go until fully functional ECM mimetics can be used as synthetic scaffolds.

13.3.2 Artificial Extracellular Matrices

It will be extremely challenging to reproduce exactly a biological extracellular matrix through artificial means in the coming decades. To obtain a scaffold resembling biological ECM, many researchers currently use Matrigel, a cocktail of substances extracted from natural ECM.[106] The advantage of Matrigel is that it is a liquid at temperatures below 4 °C, where it can be mixed with desired cell populations to result in a pre-seeded ready-to-implant scaffold upon gelation at physiological temperatures. Despite the superior attributes of ECM, it is sometimes advised to use artificial scaffolds for tissue engineering to achieve reproducibility and control of the individual parameters.

A protocol for the manufacture of a so-called ideal scaffold has yet to be developed. Therefore, special attention must be paid to the essential

Table 13.4   Ideal parameters for tissue scaffolds.

Characteristics	Comments
Architecture	3D architecture should assist in the guidance and arrangement of the cells
Angiogenic	Scaffold should support fast vascularization of the growing tissue
Biocompatible	Scaffolds should be flexible, rigid, strong and should not induce rejection or immune responses
Biodegradable	Degradation should occur at the same rate as tissue regeneration; degradation products must be non-toxic
Co-casting with cells	It should be possible to produce an optimal scaffold in the presence of the cells
Economically producible	The scaffold should be generally affordable
High surface-to-volume ratio	A critical factor is to generate scaffolds with a high cell load
Homogeneic	Scaffold should be homogeneous and show homogeneous cell distribution
Mechanical strength	Should be adapted to the biological forces in the body
Non-corrosive	Scaffold should withstand body fluids and body temperature
Non-immunogenic	Scaffold should not induce immunogenic responses to the tissue
Non-toxic	Scaffold should not induce toxicity or inhibit tissue development
Porous	Pores should be of correct size to accommodate the infiltrating cells and should be interconnected to allow diffusion of nutrients and oxygen
Specificity	Scaffold should allow growth of different cells types at the same time
Withstands sterilization	Contamination with pathogens should be avoided
Storable	Scaffold should not degrade during storage
Surface chemistry	Surface chemistry should enable easy modification or improved cell adhesion

parameters of ECM, which are to be mimicked. Irrespective of the application, the scaffold should (i) be bio-compatible, (ii) provide a 3D template for guided growth, (iii) have a highly porous structure for maximum cell load, nutrition and oxygen diffusion, (iv) degradation dynamics should match *de novo* ECM synthesis, (v) be mechanically resistant to biological forces, (vi) withstand sterilization and (vii) production should be economically feasible (Table 13.4). The shape of the starting material, including the fibers,[107] microspheres[108,109] and sheets and films,[110,111] must be determined and the generation of the scaffold microstructure studied.[112] For this purpose, different materials, including ceramics, polymers and composites, are under investigation. Ceramics such as hydroxyapatite,[113,114] tricalcium phosphate,[115,116] glass ceramics and glass[117,118] are generally used for the reconstruction of hard tissue, whereas polymers are implemented for the formation of soft tissues.[119,120] Polymers are usually the material of choice as their properties can be modulated with ease. Gelatin,[121] elastin,[19,122] collagen,[123,124] fibrin glue[125,126] and hyaluronic acid,[80-82] all of biological origin, are the most important polymers for production of artificial scaffolds. Synthetic polymers are also available including poly(lactic acid), poly(glycolic acid), polycaprolactone poly(lactic-co-glycolic acid), polyethylene, poly(ethylene glycol), poly(ethylene oxide)-block-poly-caprolactone, poly(ethylene terephthalate), polytetrafluoroethylene and polyurethane.[127] Both degradable and non-degradable synthetic polymers exist. Although degradable polymers are usually selected, such as the most widely used poly(lactic acid), the degradation products must be either biologically inert or easily metabolized, but above all non-toxic.

As a result of refining the production process for scaffold material, the spectrum and quality of scaffold types have increased considerably. Latest-generation scaffolds have average pore sizes ranging from a few nanometers[128] to several micrometers.[129] In many cases the pores are interconnected to allow the transport of nutrients and metabolic waste[124] and to provide a high surface-to-mass ratio for the promotion of cell differentiation and proliferation. These scaffolds are produced by various techniques, including (i) electrospinning, (ii) phase separation, (iii) foaming processes, (iv) microsphere sintering, (v) solid free-form fabrication, (vi) shape deposition manufacturing, (vii) fused deposition modeling, (viii) non-fused liquid deposition modeling, (ix) 3D printing, (x) selective laser sintering, (xi) stereo-lithographic processes, (xii) leaching processes and (xiii) self-assembly.[130]

Electrospinning is one of the most commonly used techniques to develop scaffolds for tissue engineering, because it is possible to modulate the fiber size from the nanometer to the micrometer scale and to guide the arrangement of fibers in the non-woven 3D fiber network.[131,132] The process consists of forcing a polymer melt through an electrically charged nozzle (10–20 kV), thereby forming a thin polymer string. The solvent quickly evaporates *en route* to the collector, which can be a simple grounded plate for random orientation of the fibers or a drum for their parallel alignment.[132] Different processing parameters such as the polymer or solvent chosen, viscosity, surface tension, the charge applied, the polymer mass flux through the nozzle, the processing temperature and the humidity, all control the diameter and the ultrastructure of the generated fibers.[133] Another advantage of electrospinning is that fibers of different polymers can be co-deposited into a multilayered structure.[134] Such scaffolds display a range of favorable properties for cell attachment (*e.g.* a wide range of pore sizes, high porosity and parallel fiber orientation).[135,136] Yang *et al.* showed that cells tend to grow in the direction of the fiber and that spatial orientation of the fibers therefore plays a key role in guiding cell growth and the subsequent organization of the tissues.[137]

Phase separation is another valuable and cost-effective process to form scaffolds for tissue engineering. The polymer is usually dissolved in a solvent that is brought into contact with a non-solvent. During the casting process, the solvent diffuses out and the non-solvent diffuses into the polymer solution until the polymer becomes unstable and precipitates, thereby forming the final scaffold after a only a few hundred milliseconds.[138,139] The pore size of such scaffolds can be varied from nanometers to several micrometers[140] and can be extended by other processes such as salt leaching.[139] These scaffolds are ideal for the engineering of special tissues such as skin and bone, the pore size of which must be in the range 0.04–0.4 mm.[141]

Self-assembly systems are gaining in importance in the development of artificial scaffolds.[125,142] They utilize a bottom-up approach, whereby molecules undergo self-association to form a higher order structure without external manipulation. Special attention should be paid to the monomer, which must be well designed to induce chemical reactivity and structural compatibility. Given these prerequisites, various building blocks have been used, such as nanofiber scaffolds,[143] peptides[144,145] and dipolar molecules.[146] Although the design of scaffolds by molecular self-assembly is extremely challenging, it is a promising process that can be performed under physiological conditions which allow pre-seeding of the scaffolds with desired cell populations.

The choice of a specific scaffold for tissue engineering is usually based on the type of tissue to be engineered. Several researchers have provided specific information concerning each type of scaffold and their application.[132,147–151] However, this information should be treated with caution because the described scaffolds were, in part, developed by material scientists and chemists and have not been tested in depth with cell cultures. For example, foaming processes may result in inclusion of air bubbles which compromise cell colonization,[141] salt-leaching processes do not completely remove entrapped salt crystals[128] and sterilization radically changes the microstructure of polymer scaffolds. In conclusion, only successful testing in *in vivo* models will finally prove whether a scaffold is suitable for tissue engineering or not. Despite all the obstacles that scientists have encountered in producing natural or synthetic scaffolds for tissue engineering, many scaffolds have already been approved by the FDA and are being used in a clinical setting (Table 13.5).

Table 13.5 A selection of commercially available scaffolds in 2005.

Product	Company	Material	Form
AlloDerm	Lifecell	Human skin	Cross-linked Dry sheet
Axis dermis	Mentor	Human dermis	Natural Dry sheet
CuffPatch	Arthrotek	Porcine small intestinal submucosa (SIS)	Cross-linked Hydrated sheet
DurADAPT	Pegasus Biologicals	Horse pericardium	Cross-linked
Dura-Guard	Synovis Surgical	Bovine pericardium	Hydrated sheet
Durasis	Cook SIS	Porcine SIS	Natural Dry sheet
Durepair	TEI Biosciences	Fetal bovine skin	Natural Dry sheet
Fortaderm	Organogenesis	Highly purified collagen	Dry sheet
FortaFlex	Organogenesis	Highly purified collagen	Cross-linked Dry sheet
FortaPerm	Organogenesis	Highly purified collagen	High-level cross-linked Resistant to degradation
Gracilis tendon	Regeneration Technologies	Patellar tendon	Frozen tendon
Graft Jacket	Wright Medical Tech	Human skin	Cross-linked Dry sheet
Grafton	Osteotech	Demineralized bone	Natural Dry sheet
Hydrix XM	Caldera Medical	Bovine pericardium	Natural Hydrated sheet
Oasis	Healthpoint	Porcine SIS	Natural Dry sheet
OrthADAPT	Pegasus Biologicals	Horse pericardium	Cross-linked
Peri-Guard	Synovis Surgical	Bovine pericardium	
Permacol	Tissue Science Laboratories	Porcine skin	Cross-linked Hydrated sheet
PriMatix	TEI Biosciences	Fetal bovine skin	Natural Dry sheet
Repliform	Boston Scientific	Human dermal allograft	Natural
Restore	DePuy	Porcine SIS	Natural Sheet
Stratasis	Cook SIS	Porcine SIS	Natural Dry sheet
Straumann Bone Ceramic	Straumann	Hydroxyapatite and tricalcium phosphate	Powder
SurgiMend	TEI Biosciences	Fetal bovine skin	Natural Dry sheet
Surgisis	Cook SIS	Porcine SIS	Natural Dry sheet

Table 13.5 (*Continued*).

Product	Company	Material	Form
Suspend	Mentor	Human fascia lata	Natural Dry sheet
TissueMend	TEI Biosciences	Fetal bovine skin	Natural Dry sheet
VeriCart	Histogenics	Collagen matrix	
Veritas	Synovis Surgical	Bovine pericardium	Hydrated sheet
Xelma	Molnlycke	ECM protein, PGA, water	Gel
Xenoform	TEI Biosciences	Fetal bovine skin	Natural Dry sheet

13.4 TISSUE ENGINEERING CONCEPTS

In addition to an optimal scaffold, a suitable bioreactor and bioprocess are required to produce the cell mass required to assemble artificial tissues. A variety of disposable plates and flasks are commercially available for standard cell culture but only a few systems are available for cultivating artificial tissues.

13.4.1 Cultivation of Artificial Tissues

Although many publications have reported on novel bioreactor configurations,[152,153] only very few have covered the production of life-sized tissues such as those intended for the reconstruction the liver, kidney and heart. *In vitro* fabrication of mammalian tissues for human therapeutic use has become standard practice for small-sized prototype tissues. However, considerable improvements in culture techniques will be required to produce artificial tissues beyond the cubic millimeter size, since larger sized tissue suffers from limited oxygen diffusion, which induces hypoxia and compromises cell viability in the center of the tissue.[154] Most tissue engineers have reached the conclusion that oxygen supply is the most critical factor in limiting tissue growth.[155,156] Therefore, the bioreactor should be optimized to modulate mass transfer into the tissue, which is essential both for the nutrient supply and the elimination of metabolites if optimal tissue viability is to be maintained. Once these limitations have been overcome, it should become possible to produce life-sized tissues for clinical use.

Growing mammalian tissues under *in vitro* conditions is particularly challenging because of their nutrient requirements, their sensitivity to metabolic waste and their susceptibility to shear stresses.[157] The required nutrients and growth parameters and the cells' susceptibility to stress vary considerably, depending on the type of tissue.[158,159] These differences must be accounted for when designing a bioreactor and a bioprocess for a particular type of tissue.

Here we present several cultivation techniques that are currently being applied in tissue engineering. When scaffold design is the primary research interest, it is best to cultivate the artificial tissues in a Petri dish. The scaffold is placed in the dish, covered with culture medium and the cells are seeded on to the scaffold where they quickly migrate and attach. However, the static milieu in the dish may rapidly lead to localized oxygen limitation and insufficient removal of metabolic waste products from the tissue. The thickest bone tissue obtained by this method has been 0.5 mm.[160] Static culture conditions have always resulted in a shell of cells around the scaffold with poor migration into the interior of the scaffold.[161,162] Diffusion can be improved by cultivating tissue samples in medium-containing spinner flasks. Spinner flasks allow constant mixing of the culture and therefore provide better supply of nutrients and oxygen to the tissue. Nevertheless, the culture medium still becomes depleted over time and 50% of it must be exchanged every 3 days.[163] For this kind of bioreactor, the typical mixing rate is 50–80 rpm, a compromise between optimal mass transfer and minimal shear stress.[163] Using spinner flasks, cartilage tissue has been grown to a thickness of 0.5 mm, which is almost five times thicker than isogenic tissues grown in Petri dishes.[158] However, typical cartilage implants used today are 2–5 mm thick.[158]

Although hollow-fiber bioreactors are not suitable for the production of implantable tissues, they have been extensively studied for the design of extracorporeal devices which could provide liver and kidney function in a dialysis-like therapy.[164,165] The hollow-fiber bioreactor consists of a closed container filled with a cell-containing matrix, into which a bundle of semi-permeable hollow fibers is inserted. A constant flow of culture medium through the fiber provides nutrients and eliminates metabolic waste products. With this method, the interface-to-tissue mass ratio is very high and provides a more homogeneous nutrient supply throughout the tissue. This type of bioreactor more closely resembles the situation in a vertebrate body, where cells are usually never more than 200 μm away from the next blood vessel. Studies with hepatocytes cultured in a hollow-fiber bioreactor revealed that, if the distance between the fibers exceeds 250 μm, then a perpendicular flow to the fibers is necessary to achieve a sufficient stream of nutrients to the cells.[166]

An important recent development in the construction of hollow-fiber bioreactors is the use of degradable fiber materials. After degradation of the hollow fibers, the tissue is in principle ready for implantation.[167] Poly(D,L-lactide-co-glycolide) fibers maintain their structural integrity for 4 weeks and degrade homogenously until they disappear completely by 8 weeks, thereby maintaining the structure of the tissue.[167] Although artificial tissues produced by hollow-fiber bioreactors have not yet been used in clinical trials, the bioreactor itself is already in clinical use as cartridges for dialysis/plasma separation. These cartridges contain 50–200 g of primary hepatocytes and are connected for 6–8 h per day in a typical dialysis setting.[17]

One of the most commonly applied reactors for the generation of tissues is the rotating-wall bioreactor. The tissue floats freely in the chamber and the rotation speed, usually 15–30 rpm, is adjusted so that the tissue remains in a state of zero gravity. Cartilage tissues up to 5 mm thick [168] and liver tissues with a thickness of up to 3 mm have been produced.[169] Rotating-wall bioreactors have been used to produce many other tissues, of which the most important has been myocardial tissues.[170,171]

Arguably, the best bioreactor system to cultivate artificial tissues is the perfusion bioreactor, which typically contains a small chamber through which a flow of fresh and defined medium is pumped at a constant rate. The scaffold is usually fixed to a porous support in the middle of the chamber. The flow of medium through the scaffold enhances cell growth inside the scaffold and provides mechanical stimulation in the form of shear force as the media is forced through the scaffold.[172] A disadvantage of this bioreactor is that the orientation of the cells follows the direction of the liquid flow. Although alignment of the cells is desired in tissue engineering, engineers would prefer the cells to align perpendicular rather than parallel to the liquid flow.[173] Perfusion bioreactors are of particular interest for growing tissues for skin replacement and consist of two chambers, through one of which is pumped medium optimized for the growth of epithelial cells and through the other is pumped medium optimized for the growth of connective tissue. This configuration permits the production of two-layer skin tissues.[174] The perfusion system can even be employed to simulate biological forces. For instance, Watanabe *et al.* applied intermittent hydrostatic pressure (0–5 MPa) when cultivating cartilage,[175] whereas Seidel *et al.* applied mechanical compression forces.[176] Other perfusion bioreactors with a modified configuration have also been used for the tissue engineering of skeletal muscle and oesteochondral composites.

13.4.2   Design of Scaffold-free Tissues

Although state-of-the-art artificial scaffolds allow an adequate flow of nutrients to the cells residing at the center, and growth of specific cell types, the physical, chemical and biological properties of the scaffold material are far from being optimal for every tissue.[177] Here are summarized the latest trends in designing scaffold-free artificial tissues, which, in our opinion, represent a valuable extension of current scaffold-based approaches.

The principles of generating scaffold-free microtissue spheroids are straightforward and have been applied for decades to test anticancer drugs in a more realistic tissue-like model[178–180] and also for the analysis of cell differentiation.[181,182] Monodispersed cells aggregate as spheroids whenever intercellular adhesion forces, most often mediated by homotypic interactions between surface proteins of the cadherin family, exceed those of cell–surface interactions.[183] Therefore, microtissue spheroids are produced by (i) cultivating the cells in culture dishes, spinner flasks and roller bottles with non-adhesive surfaces,[184–186] (ii) centrifugation-based pelleting of the cells,[187] (iii) growth of the cells in small containers[188] and (iv) gravity-enforced re-aggregation of the cells in hanging drops.[23] The hanging drop technology is by far the gentlest aggregation strategy and permits positioning of individual cells within a microtissue without coming in contact with any synthetic material. Using gravity-enforced self-assembly of cells in hanging drops, Kelm and co-workers have successfully designed heart, liver, neuronal and cartilage tissues with unmatched *in vivo* characteristics [21,23,189–191]

For example, hepatic microtissues generated by the hanging drop technology have shown increased levels of detoxifying enzymes and a more perfect hepatic ultrastructure, including formation of correct polarity and bile canaliculi, compared with hepatocyte monolayer or other 3D cultures.[23] Although it was possible to generate microtissues from a variety of cell types, not all of them will produce the correct extracellular matrix and the desmosome-based intercellular communication network required for correct positioning of individual cells within a microtissue and for the formation of fully functional microtissues.[23,183]

Although cell movement during development has been studied in great detail, the precise positioning of individual cells during the formation of artificial tissues remains largely elusive. Kelm's group has successfully used microtissue spheroids assembled from different cell types to study the relative positioning of different cell populations inside a forming microtissue. For example, gravity-enforced self-assembly of a cell mixture mimicking the natural composition of the heart suggested the presence of molecular forces which position cardiomyocytes preferentially at the periphery of beating 'microhearts'.[21] Microhearts stimulated by addition of phenylephrine or by ectopic expression of bone morphogenetic protein 2 (BMP2) reproduced electrogenic profiles reminiscent of fully functional hearts.[189]

Repositioning of individual cells and assembly of tissue substructures could also be observed in microtissues produced by co-cultivation of human hepatocytes (HepG2) and umbilical vein endothelial cells (HUVECs). In these, spheroids migrate from the surface to the center thereby forming tubular structures reminiscent of vascular structures.[192] After implantation into a chicken embryo, these vascular structures successfully connected to the chicken vasculature and chicken hemoglobin managed oxygen supply for the implant, which was seamlessly integrated into the embryo tissue without showing any scar structures.[192] Using vascularized microtissues as minimal building blocks, Kelm and co-workers also succeeded in producing fully functional larger sized tissues in the cubic millimeter range. These macrotissues could be assembled into custom shapes by cultivating and fusing prevascularized microtissues in agarose moulds.[22] The design of custom-shaped, scaffold-free, fully vascularized tissues of implantable size will significantly advance tissue engineering in the not-so-distant future.

13.5 CONCLUSIONS

Tissue engineering has grown rapidly as a research discipline in recent years and the number of engineered tissues used in a clinical setting is constantly growing. However, the challenges remain significant. Non-limiting examples include (i) precise reprogramming of lineage control in different cell types, (ii) increased knowledge of differentiation circuits and, more importantly, de-differentiation networks, (iii) control of position and molecular cross-talk of different cell types within a tissue, (iv) engineering of a fully functional vasculature into larger-sized tissues and (v) industrial-scale production of precursor cells for the assembly of artificial tissues. Given the dramatic progress made in recent years, tissue engineering may well revolutionize medical treatment in the new millennium.

REFERENCES

1. M. P. Aulisio, M. Devita and D. Luebke, Taking values seriously: ethical challenges in organ donation and transplantation for

critical care professionals, *Crit. Care Med.*, 2007, **35**(2 Suppl), S95–101.

2. R. Langer and J. P. Vacanti, *Tissue Eng. Sci.*, 1993, **260**(5110), 920–6.

3. M. Ehrbar, A. Metters, P. Zammaretti, J. A. Hubbell and A. H. Zisch, Endothelial cell proliferation and progenitor maturation by fibrin-bound VEGF variants with differential susceptibilities to local cellular activity, *J. Control. Release*, 2005, **101**(1–3), 93–109.

4. C. Fux, D. Langer and M. Fussenegger, Dual-regulated myoD- and msx1-based interventions in C2C12-derived cells enable precise myogenic/osteogenic/adipogenic lineage control, *J. Gene Med.*, 2004, **6**(10), 1159–69.

5. D. C. Hay, D. Zhao, A. Ross, R. Mandalam, J. Lebkowski and W. Cui, Direct differentiation of human embryonic stem cells to hepatocyte-like cells exhibiting functional activities, *Cloning Stem Cells*, 2007, **9**(1), 51–62.

6. J. K. Heath and A. G. Smith, Regulatory factors of embryonic stem cells, *J. Cell Sci. Suppl.*, 1988, **10**, 257–66.

7. J. K. Heath, A. G. Smith, L. W. Hsu and P. D. Rathjen, Growth and differentiation factors of pluripotential stem cells, *J. Cell Sci. Suppl.*, 1990, **13**, 75–85.

8. S. Ikehara, Pluripotent hemopoietic stem cells in mice and humans, *Proc. Soc. Exp. Biol. Med.*, 2000, **223**(2), 149–55.

9. A. Alhadlaq and J. J. Mao, Mesenchymal stem cells: isolation and therapeutics, *Stem Cells Dev.*, 2004, **13**(4), 436–48.

10. I. Jasmund, S. Schwientek, A. Acikgoz, A. Langsch, H. G. Machens and A. Bader, The influence of medium composition and matrix on long-term cultivation of primary porcine and human hepatocytes, *Biomol. Eng.*, 2007, **24**(1), 59–69.

11. S. H. Khoo and M. Al-Rubeai, Metabolomics as a complementary tool in cell culture, *Biotechnol. Appl. Biochem.*, 2007, **47**(Pt 2), 71–84.

12. L. Yao, C. S. Bestwick, L. A. Bestwick, N. Maffulli and R. M. Aspden, Phenotypic drift in human tenocyte culture, *Tissue Eng.*, 2006, **12**(7), 1843–9.

13. A. I. Caplan, Review: mesenchymal stem cells: cell-based reconstructive therapy in orthopedics., *Tissue Eng.*, 2005, **11**(7–8), 1198–211.

14. N. Kimelman, G. Pelled, G. A. Helm, J. Huard, E. M. Schwarz and D. Gazit, Review: gene- and stem cell-based therapeutics for bone regeneration and repair, *Tissue Eng.*, 2007, **13**(6), 1135–50.

15. W. H. Fissell, Developments towards an artificial kidney, *Expert Rev. Med. Devices*, 2006, **3**(2), 155–65.
16. M. Miyamoto, Current progress and perspectives in cell therapy for diabetes mellitus, *Hum. Cell*, 2001, **14**(4), 293–300.
17. J. K. Park and D. H. Lee, Bioartificial liver systems: current status and future perspective, *J. Biosci. Bioeng.*, 2005, **99**(4), 311–9.
18. A. Saito, Development of bioartificial kidneys, *Nephrology (Carlton)*, 2003, **8**(Suppl), S10–5.
19. W. H. Zimmermann and T. Eschenhagen, Cardiac tissue engineering for replacement therapy, *Heart Fail. Rev.*, 2003, **8**(3), 259–69.
20. V. L. Tsang and S. N. Bhatia, Fabrication of three-dimensional tissues, *Adv. Biochem. Eng. Biotechnol.*, 2007, **103**, 189–205.
21. J. M. Kelm, E. Ehler, L. K. Nielsen, S. Schlatter, J. C. Perriard and M. Fussenegger, Design of artificial myocardial microtissues, *Tissue Eng.*, 2004, **10**(1–2), 201–14.
22. J. M. Kelm, V. Djonov, L. M. Ittner, D. Fluri, W. Born, S. P. Hoestrup and M. Fussenegger, Design of custom-shaped vascularized tissues using microtissue spheroids as minimal building units, *Tissue Eng.*, 2006, **12**(8), 2151–60.
23. J. M. Kelm and M. Fussenegger, Microscale tissue engineering using gravity-enforced cell assembly, *Trends Biotechnol.*, 2004, **22**(4), 195–202.
24. S. Marlovits, P. Zeller, P. Singer, C. Resinger and V. Vecsei, Cartilage repair: generations of autologous chondrocyte transplantation, *Eur. J. Radiol.*, 2006, **57**(1), 24–31.
25. D. W. Hayes Jr, R. L. Brower and K. J. John, Articular cartilage. Anatomy, injury and repair, *Clin. Podiatr. Med. Surg.*, 2001, **18**(1), 35–53.
26. P. V. Kochupura, E. U. Azeloglu, D. J. Kelly, S. V. Doronin, S. F. Badylak, I. B. Krukenkamp, I. S. Cohen and G. R. Gaudette, Tissue-engineered myocardial patch derived from extracellular matrix provides regional mechanical function, *Circulation*, 2005, **112**(9 Suppl), I144–9.
27. T. C. Flanagan, C. Cornelissen, S. Koch, B. Tschoeke, J. S. Sachweh, T. Schmitz-Rode and S. Jockenhoevel, The *in vitro* development of autologous fibrin-based tissue-engineered heart valves through optimised dynamic conditioning, *Biomaterials*, 2007, **28**(23), 3388–97.
28. A. Cohen and E. S. Horton, Progress in the treatment of type 2 diabetes: new pharmacologic approaches to improve glycemic control, *Curr. Med. Res. Opin.*, 2007, **23**(4), 905–17.

29. M. Nair, Diabetes mellitus, Part 1: physiology and complications, *Br. J. Nurs.*, 2007, **16**(3), 184–8.

30. M. J. Evans and M. H. Kaufman, Establishment in culture of pluripotential cells from mouse embryos, *Nature*, 1981, **292**(5819), 154–6.

31. J. A. Thomson, J. Itskovitz-Eldor, S. S. Shapiro, M.A. Waknitz, J. J. Swiergiel, V. S. Marshall and J. M. Jones, Embryonic stem cell lines derived from human blastocysts, *Science*, 1998, **282**(5391), 1145–7.

32. I. Singec, R. Jandial, A. Crain, G. Nikkhah and E. Y. Snyder, The leading edge of stem cell therapeutics, *Annu. Rev. Med.*, 2007, **58**, 313–28.

33. D. Van Hoof, C. L. Mummery, A. J. Heck and J. Krijgsveld, Embryonic stem cell proteomics, *Expert Rev. Proteomics*, 2006, **3**(4), 427–37.

34. D. Mondal, L. Pradhan and V. F. LaRussa, Signal transduction pathways involved in the lineage-differentiation of NSCs: can the knowledge gained from blood be used in the brain?, *Cancer Invest.*, 2004, **22**(6), 925–43.

35. K. Nakashima and T. Taga, Mechanisms underlying cytokine-mediated cell-fate regulation in the nervous system, *Mol. Neurobiol.*, 2002, **25**(3), 233–44.

36. S. J. Leedham, M. Brittan, S. A. McDonald and N. A. Wright, Intestinal stem cells, *J. Cell Mol. Med.*, 2005, **9**(1), 11–24.

37. J. R. Walters, Cell and molecular biology of the small intestine: new insights into differentiation, growth and repair, *Curr. Opin. Gastroenterol.*, 2004, **20**(2), 70–6.

38. L. Alonso and E. Fuchs, Stem cells of the skin epithelium, *Proc. Natl. Acad. Sci. USA*, 2003, **100** Suppl 1, 11830–5.

39. G. Cotsarelis, Epithelial stem cells: a folliculocentric view, *J. Invest. Dermatol.*, 2006, **126**(7), 1459–68.

40. P. Anversa, A. Leri, M. Rota, T. Hosoda, C. Bearzi, K. Urbanek, J. Kajstura and R. Bolli, Concise review: stem cells, myocardial regeneration and methodological artifacts, *Stem Cells*, 2007, **25**(3), 589–601.

41. A. Leri, J. Kajstura and P. Anversa, Cardiac stem cells and mechanisms of myocardial regeneration, *Physiol. Rev.*, 2005, **85**(4), 1373–416.

42. K. Urbanek, D. Cesselli, M. Rota, A. Nascimbene, A. De Angelis, T. Hosoda, C. Bearzi, A. Boni, R. Bolli, J. Kajstura, P. Anversa and A. Leri, Stem cell niches in the adult mouse heart, *Proc. Natl. Acad. Sci. USA*, 2006, **103**(24), 9226–31.

43. K. A. Jackson, S. M. Majka, G. G. Wulf and M. A. Goodell, Stem cells: a minireview, *J. Cell Biochem. Suppl.*, 2002, **38**, 1–6.

44. A. Vescovi, A. Gritti, G. Cossu and R. Galli, Neural stem cells: plasticity and their transdifferentiation potential, *Cells Tissues Organs*, 2002, **171**(1), 64–76.

45. H. E. Young, C. Duplaa, M. Romero-Ramos, M. F. Chesselet, P. Vourc'h, M. J. Yost, K. Ericson, L. Terracio, T. Asahara, H. Masuda, S. Tamura-Ninomiya, K. Detmer, R. A. Bray, T. A. Steele, D. Hixson, M. el-Kalay, B. W. Tobin, R. D. Russ, M. N. Horst, J. A. Floyd, N. L. Henson, K. C. Hawkins, J. Groom, A. Parikh, L. Blake, L. J. Bland, A. J. Thompson, A. Kirincich, C. Moreau, J. Hudson, F. P. Bowyer, 3rd, T. J. Lin and A. C. Black, Jr., Adult reserve stem cells and their potential for tissue engineering, *Cell Biochem. Biophys.*, 2004, **40**(1), 1–80.

46. M. Heyde, K. A. Partridge, R. O. Oreffo, S. M. Howdle, K. M. Shakesheff and M. C. Garnett, Gene therapy used for tissue engineering applications, *J. Pharm. Pharmacol.*, 2007, **59**(3), 329–50.

47. T. May, P. P. Mueller, H. Weich, N. Froese, U. Deutsch, D. Wirth, A. Kroger and H. Hauser, Establishment of murine cell lines by constitutive and conditional immortalization, *J. Biotechnol.*, 2005, **120**(1), 99–110.

48. B. Varnum-Finney, L. Xu, C. Brashem-Stein, C. Nourigat, D. Flowers, S. Bakkour, W. S. Pear and I. D. Bernstein, Pluripotent, cytokine-dependent, hematopoietic stem cells are immortalized by constitutive Notch1 signaling, *Nat. Med.*, 2000, **6**(11), 1278–81.

49. C. Fux, S. Moser, S. Schlatter, M. Rimann, J. E. Bailey and M. Fussenegger, Streptogramin- and tetracycline-responsive dual regulated expression of p27 (Kip1) sense and antisense enables positive and negative growth control of Chinese hamster ovary cells, *Nucleic Acids Res.*, 2001, **29**(4), E19.

50. M. B. Goldring, Immortalization of human articular chondrocytes for generation of stable, differentiated cell lines, *Methods Mol. Med.*, 2004, **100**, 23–36.

51. C. Priesner, F. Hesse, D. Windgassen, R. Klocke, D. Paul and R. Wagner, Liver-specific physiology of immortal, functionally differentiated hepatocytes and of deficient hepatocyte-like variants, *In vitro Cell Dev. Biol. Anim.*, 2004, **40**(10), 318–30.

52. M. Fussenegger, R. P. Morris, C. Fux, M. Rimann and B. von Stockar C. J., Streptogramin-based gene regulation systems for mammalian cells, *Nat. Biotechnol.*, 2000, **18**(11), 1203–8.

53. M. Gossen and H. Bujard, Tight control of gene expression in mammalian cells by tetracycline-responsive promoters, *Proc. Natl. Acad. Sci. USA*, 1992, **89**(12), 5547–51.

54. W. Weber, B. P. Kramer, C. Fux, B. Keller and M. Fussenegger, Novel promoter/transactivator configurations for macrolide- and streptogramin-responsive transgene expression in mammalian cells, *J. Gene Med.*, 2002, **4**(6), 676–86.

55. M. Fussenegger, J. E. Bailey, H. Hauser and P. P. Mueller, Genetic optimization of recombinant glycoprotein production by mammalian cells, *Trends Biotechnol.*, 1999, **17**(1), 35–42.

56. Z. Ivics, A. Katzer, E. E. Stuwe, D. Fiedler, S. Knespel and Z. Izsvak, Targeted Sleeping Beauty transposition in human cells, *Mol. Ther.*, 2007, **15**(6), 1137–44.

57. C. Miskey, Z. Izsvak, R. H. Plasterk and Z. Ivics, The Frog Prince: a reconstructed transposon from Rana pipiens with high transpositional activity in vertebrate cells, *Nucleic Acids Res.*, 2003, **31**(23), 6873–81.

58. M. Narushima, N. Kobayashi, T. Okitsu, Y. Tanaka, S. A. Li, Y. Chen, A. Miki, K. Tanaka, S. Nakaji, K. Takei, A. S. Gutierrez, J. D. Rivas-Carrillo, N. Navarro-Alvarez, H. S. Jun, K. A. Westerman, H. Noguchi, J. R. Lakey, P. Leboulch, N. Tanaka and J. W. Yoon, A human beta-cell line for transplantation therapy to control type 1 diabetes, *Nat. Biotechnol.*, 2005, **23**(10), 1274–82.

59. D. Tosh and J. M. Slack, How cells change their phenotype, *Nat. Rev. Mol. Cell Biol.*, 2002, **3**(3), 187–94.

60. P. Collas, C. K. Taranger, A. C. Boquest, A. Noer and J. A. Dahl, On the way to reprogramming cells to pluripotency using cell-free extracts, *Reprod. Biomed. Online*, 2006, **12**(6), 762–70.

61. P. Mali and L. Cheng, Reprogramming somatic cells without fusion or ethical confusion, *Regen. Med.*, 2006, **1**(6), 837–840.

62. K. Takahashi and S. Yamanaka, Induction of pluripotent stem cells from mouse embryonic and adult fibroblast cultures by defined factors, *Cell*, 2006, **126**(4), 663–76.

63. K. Hochedlinger and R. Jaenisch, Nuclear reprogramming and pluripotency, *Nature*, 2006, **441**(7097), 1061–7.

64. R. Jaenisch, K. Hochedlinger and K. Eggan, Nuclear cloning, epigenetic reprogramming and cellular differentiation, *Novartis Found. Symp.*, 2005, **265**, 107–18; discussion 118–28.

65. E. Hu, P. Tontonoz and B. M. Spiegelman, Transdifferentiation of myoblasts by the adipogenic transcription factors PPAR gamma and C/EBP alpha, *Proc. Natl. Acad. Sci. USA*, 1995, **92**(21), 9856–60.

66. P. C. Schiller, G. D'Ippolito, R. Brambilla, B. A. Roos and G. A. Howard, Inhibition of gap-junctional communication induces the trans-differentiation of osteoblasts to an adipocytic phenotype *in vitro*, *J. Biol. Chem.*, 2001, **276**(17), 14133–8.

67. C. N. Shen, J. M. Slack and D. Tosh, Molecular basis of trans-differentiation of pancreas to liver, *Nat. Cell Biol.*, 2000, **2**(12), 879–87.

68. G. Condorelli, U. Borello, L. De Angelis, M. Latronico, D. Sirabella, M. Coletta, R. Galli, G. Balconi, A. Follenzi, G. Frati, M. G. Cusella De Angelis, L. Gioglio, S. Amuchastegui, L. Adorini, L. Naldini, A. Vescovi, E. Dejana and G. Cossu, Cardiomyocytes induce endothelial cells to trans-differentiate into cardiac muscle: implications for myocardium regeneration, *Proc. Natl. Acad. Sci. USA*, 2001, **98**(19), 10733–8.

69. M. D. Dabeva, S. G. Hwang, S. R. Vasa, E. Hurston, P. M. Novikoff, D. C. Hixson, S. Gupta and D. A. Shafritz, Differentiation of pancreatic epithelial progenitor cells into hepatocytes following transplantation into rat liver, *Proc. Natl. Acad. Sci. USA*, 1997, **94**(14), 7356–61.

70. T. W. Gilbert, T. L. Sellaro and S. F. Badylak, Decellularization of tissues and organs, *Biomaterials*, 2006, **27**(19), 3675–83.

71. J. C. Myers, P. S. Amenta, A. S. Dion, J. P. Sciancalepore, C. Nagaswami, J. W. Weisel and P. D. Yurchenco, The molecular structure of human tissue type XV presents a unique conformation among the collagens, *Biochem. J.*, 2007, **404**(3), 535–44.

72. A. M. Pizzo, K. Kokini, L. C. Vaughn, B. Z. Waisner and S. L. Voytik-Harbin, Extracellular matrix (ECM) microstructural composition regulates local cell-ECM biomechanics and fundamental fibroblast behavior: a multidimensional perspective, *J. Appl. Physiol.*, 2005, **98**(5), 1909–21.

73. P. Schedin, T. Mitrenga, S. McDaniel and M. Kaeck, Mammary ECM composition and function are altered by reproductive state, *Mol. Carcinog.*, 2004, **41**(4), 207–20.

74. M. Maatta, A. Liakka, S. Salo, K. Tasanen, L. Bruckner-Tuderman and H. Autio-Harmainen, Differential expression of basement membrane components in lymphatic tissues, *J. Histochem. Cytochem.*, 2004, **52**(8), 1073–81.

75. A. C. Bellail, S. B. Hunter, D. J. Brat, C. Tan and E. G. Van Meir, Microregional extracellular matrix heterogeneity in brain modulates glioma cell invasion, *Int. J. Biochem. Cell Biol.*, 2004, **36**(6), 1046–69.

76. P. Lin, W. C. Chan, S. F. Badylak and S. N. Bhatia, Assessing porcine liver-derived biomatrix for hepatic tissue engineering, *Tissue Eng.*, 2004, **10**(7–8), 1046–53.

77. J. H. Miner, Renal basement membrane components, *Kidney Int.*, 1999, **56**(6), 2016–24.

78. K. Gelse, E. Poschl and T. Aigner, Collagens – structure, function and biosynthesis, *Adv. Drug Deliv. Rev.*, 2003, **55**(12), 1531–46.

79. J. Huxley-Jones, D. L. Robertson and R. P. Boot-Handford, On the origins of the extracellular matrix in vertebrates, *Matrix Biol.*, 2007, **26**(1), 2–11.

80. J. Schwarzbauer, Basement membranes: Putting up the barriers, *Curr. Biol.*, 1999, **9**(7), R242–4.

81. R. Timpl, Macromolecular organization of basement membranes. Curr. Opin. Cell Biol., 1996, **8**(5), 618–24.

82. R. Timpl and J. C. Brown, Supramolecular assembly of basement membranes, *Bioessays*, 1996, **18**(2), 123–32.

83. R. Folberg and A. J. Maniotis, Vasculogenic mimicry, *Apmis*, 2004, **112**(7–8), 508–25.

84. Z. Werb, T. H. Vu, J. L. Rinkenberger and L. M. Coussens, Matrix-degrading proteases and angiogenesis during development and tumor formation, *Apmis*, 1999, **107**(1), 11–8.

85. T. Neumann, S. D. Hauschka and J. E. Sanders, Tissue engineering of skeletal muscle using polymer fiber arrays, *Tissue Eng.*, 2003, **9**(5), 995–1003.

86. M. D. Pierschbacher and E. Ruoslahti, Cell attachment activity of fibronectin can be duplicated by small synthetic fragments of the molecule, *Nature*, 1984, **309**(5963), 30–3.

87. K. M. Yamada and D. W. Kennedy, Dualistic nature of adhesive protein function: fibronectin and its biologically active peptide fragments can autoinhibit fibronectin function, *J. Cell Biol.*, 1984, **99**(1 Pt 1), 29–36.

88. S. Miyamoto, B. Z. Katz, R. M. Lafrenie and K. M. Yamada, Fibronectin and integrins in cell adhesion, signaling and morphogenesis, *Ann. N. Y. Acad. Sci.*, 1998, **857**, 119–29.

89. M. Mochizuki, N. Yamagata, D. Philp, K. Hozumi, T. Watanabe, Y. Kikkawa, Y. Kadoya, H. K. Kleinman and M. Nomizu, Integrin-dependent cell behavior on ECM peptide-conjugated chitosan membranes, *Biopolymers*, 2007, **88**(2), 122–30.

90. J. E. Schwarzbauer, Fibronectin: from gene to protein, *Curr. Opin. Cell Biol.*, 1991, **3**(5), 786–91.

91. J. P. Hodde, S. F. Badylak, A. O. Brightman and S. L. Voytik-Harbin, Glycosaminoglycan content of small intestinal submucosa:

a bioscaffold for tissue replacement, *Tissue Eng.*, 1996, **2**(3), 209–217.

92. W. M. Elbjeirami, E. O. Yonter, B. C. Starcher and J. L. West, Enhancing mechanical properties of tissue-engineered constructs via lysyl oxidase crosslinking activity, *J. Biomed. Mater. Res. A*, 2003, **66**(3), 513–21.

93. M. A. Cobb, S.F. Badylak, W. Janas, and F.A. Boop, Histology after dural grafting with small intestinal submucosa. *Surg. Neurol.*, 1996, 46 (4), 389–93; discussion 393–4.

94. S. F. Badylak, The extracellular matrix as a scaffold for tissue reconstruction, *Semin. Cell Dev. Biol.*, 2002, **13**(5), 377–83.

95. S. Badylak, S. Meurling, M. Chen, A. Spievack and A. Simmons-Byrd, Resorbable bioscaffold for esophageal repair in a dog model, *J. Pediatr. Surg.*, 2000, **35**(7), 1097–103.

96. S. Nemcova, A. A. Noel, C. J. Jost, P. Gloviczki, V. M. Miller and K. G. Brockbank, Evaluation of a xenogeneic acellular collagen matrix as a small-diameter vascular graft in dogs – preliminary observations, *J. Invest. Surg.*, 2001, **14**(6), 321–30.

97. K. J. Zehr, M. Yagubyan, H. M. Connolly, S. M. Nelson and H. V. Schaff, Aortic root replacement with a novel decellularized cryo-preserved aortic homograft: postoperative immunoreactivity and early results, *J. Thorac. Cardiovasc. Surg.*, 2005, **130**(4), 1010–5.

98. C. Danielsson, S. Ruault, A. Basset-Dardare and P. Frey, Modified collagen fleece, a scaffold for transplantation of human bladder smooth muscle cells, *Biomaterials*, 2006, **27**(7), 1054–60.

99. B. P. Kropp, S. Badylak and K. B. Thor, Regenerative bladder augmentation: a review of the initial preclinical studies with porcine small intestinal submucosa, *Adv. Exp. Med. Biol.*, 1995, **385**, 229–35.

100. P. A. Merguerian, P. P. Reddy, D. J. Barrieras, G. J. Wilson, K. Woodhouse, D. J. Bagli, G. A. McLorie and A. E. Khoury, Acellular bladder matrix allografts in the regeneration of functional bladders: evaluation of large-segment (>24 cm) substitution in a porcine model, *BJU Int.*, 2000, **85**(7), 894–8.

101. S. Q. Liu, *Gastrointestinal Regenerative Engineering. Bioregenerative Engineering: Principles and Applications*, Wiley, Hoboken, NJ, 2007.

102. J. R. Fuchs, A. Kaviani, J.T. Oh, D. LaVan, T. Udagawa, R.W. Jennings, J.M. Wilson, and D.O. Fauza, Diaphragmatic reconstruction with autologous tendon engineered from mesenchymal amniocytes, *J. Pediatr. Surg.*, 2004, 39 (6), 834–8; discussion 834–8.

103. C. G. Zalavras, R. Gardocki, E. Huang, M. Stevanovic, T. Hedman and J. Tibone, Reconstruction of large rotator cuff

tendon defects with porcine small intestinal submucosa in an animal model, *J. Shoulder Elbow Surg.*, 2006, **15**(2), 224–31.

104. P. G. De Deyne and S. M. Kladakis, Bioscaffolds in tissue engineering: a rationale for use in the reconstruction of musculoskeletal soft tissues, *Clin. Podiatr. Med. Surg.*, 2005, **22**(4), 521–32, v.

105. C. C. Weber, H. Cai, M. Ehrbar, H. Kubota, G. Martiny-Baron, W. Weber, V. Djonov, E. Weber, A.S. Mallik, M. Fussenegger, K. Frei, J. A. Hubbell and A. H. Zisch, Effects of protein and gene transfer of the angiopoietin-1 fibrinogen-like receptor-binding domain on endothelial and vessel organization, *J. Biol. Chem.*, 2005, **280**(23), 22445–53.

106. W. L. Rust, A. Sadasivam and N. R. Dunn, Three-dimensional extracellular matrix stimulates gastrulation-like events in human embryoid bodies, *Stem Cells Dev.*, 2006, **15**(6), 889–904.

107. V. Ella, M. E. Gomes, R. L. Reis, P. Tormala and M. Kellomaki, Studies of P (L/D)LA 96/4 non-woven scaffolds and fibres; properties, wettability and cell spreading before and after intrusive treatment methods, *J. Mater. Sci. Mater. Med.*, 2007, **18**(6), 1253–61.

108. J. P. Rubin, J. M. Bennett, J. S. Doctor, B. M. Tebbets and K. G. Marra, Collagenous microbeads as a scaffold for tissue engineering with adipose-derived stem cells, *Plast. Reconstr. Surg.*, 2007, **120**(2), 414–24.

109. T. Suciati, D. Howard, J. Barry, N. M. Everitt, K. M. Shakesheff and F. R. Rose, Zonal release of proteins within tissue engineering scaffolds, *J. Mater. Sci. Mater. Med.*, 2006, **17**(11), 1049–56.

110. A. S. Rowlands, S. A. Lim, D. Martin and J. J. Cooper-White, Polyurethane/poly(lactic-co-glycolic) acid composite scaffolds fabricated by thermally induced phase separation, *Biomaterials*, 2007, **28**(12), 2109–21.

111. T. Takezawa, K. Ozaki and C. Takabayashi, Reconstruction of a hard connective tissue utilizing a pressed silk sheet and type-I collagen as the scaffold for fibroblasts, *Tissue Eng.*, 2007, **13**(6), 1357–66.

112. R. Izquierdo, N. Garcia-Giralt, M.T. Rodriguez, E. Caceres, S.J. Garcia, J.L. Gomez Ribelles, M. Monleon, J.C. Monllau, and J. Suay, Biodegradable PCL scaffolds with an interconnected spherical pore network for tissue engineering, *J. Biomed. Mater. Res. A*, 2007, **85**(1), 25–35.

113. J. G. Dellinger, J. Cesarano III and R. D. Jamison, Robotic deposition of model hydroxyapatite scaffolds with multiple architectures and multiscale porosity for bone tissue engineering, *J. Biomed. Mater. Res. A*, 2007, **82**(2), 383–94.

114. I. K. Jun, Y. H. Koh, S. H. Lee and H. E. Kim, Novel hydroxyapatite (HA) dual-scaffold with ultra-high porosity, high surface area and compressive strength, *J. Mater. Sci. Mater. Med.*, 2007, **18**(6), 1071–7.

115. J. O. Eniwumide, H. Yuan, S. H. Cartmell, G. J. Meijer, and J. D. de Bruijn, Ectopic bone formation in bone marrow stem cell seeded calcium phosphate scaffolds as compared to autograft and (cell seeded) allograft, Eur. Cell Mater., 2007, 14, 30–8; discussion 39.

116. Li, Y. and W. Weng, In vitro synthesis and characterization of amorphous calcium phosphates with various Ca/P atomic ratios, *J. Mater. Sci. Mater. Med.*, 2007, **18**(12), 2303–8.

117. C. V. Brovarone, E. Verne and P. Appendino, Macroporous bioactive glass-ceramic scaffolds for tissue engineering, *J. Mater. Sci. Mater. Med.*, 2006, **17**(11), 1069–78.

118. J. R. Jones, O. Tsigkou, E. E. Coates, M. M. Stevens, J. M. Polak and L. L. Hench, Extracellular matrix formation and mineralization on a phosphate-free porous bioactive glass scaffold using primary human osteoblast (HOB) cells, *Biomaterials*, 2007, **28**(9), 1653–63.

119. A. R. Boccaccini and J. J. Blaker, Bioactive composite materials for tissue engineering scaffolds, *Expert Rev. Med. Devices*, 2005, **2**(3), 303–17.

120. G. C. Engelmayr Jr and M. S. Sacks, A structural model for the flexural mechanics of nonwoven tissue engineering scaffolds, *J. Biomech. Eng.*, 2006, **128**(4), 610–22.

121. C. W. Yung, L. Q. Wu, J. A. Tullman, G. F. Payne, W. E. Bentley, and T. A. Barbari, Transglutaminase crosslinked gelatin as a tissue engineering scaffold. *J. Biomed. Mater. Res. A*, 2007, **83**(4), 1039–46.

122. J. B. Leach, J. B. Wolinsky, P. J. Stone and J. Y. Wong, Cross-linked alpha-elastin biomaterials: towards a processable elastin mimetic scaffold, *Acta Biomater.*, 2005, **1**(2), 155–64.

123. K. A. Faraj, T. H. Van Kuppevelt and W. F. Daamen, Construction of collagen scaffolds that mimic the three-dimensional architecture of specific tissues. *Tissue Eng.*, 2007, 13(0), 2387–94.

124. F. J. O'Brien, B. A. Harley, M. A. Waller, I. V. Yannas, L. J. Gibson and P. J. Prendergast, The effect of pore size on permeability and cell attachment in collagen scaffolds for tissue engineering, *Technol. Health Care*, 2007, **15**(1), 3–17.

125. E. Alsberg, E. Feinstein, M. P. Joy, M. Prentiss and D. E. Ingber, *et al.*, Magnetically-guided self-assembly of fibrin matrices with ordered nano-scale structure for tissue engineering, *Tissue Eng.*, 2006, **12**(11), 3247–56.

...

126. D. Eyrich, H. Wiese, G. Maier, D. Skodacek, B. Appel, H. Sarhan, J. Tessmar, R. Staudenmaier, M. M. Wenzel, A. Goepferich, and T. Blunk, *In vitro* and *in vivo* cartilage engineering using a combination of chondrocyte-seeded long-term stable fibrin gels and polycaprolactone-based polyurethane scaffolds, *Tissue Eng.*, 2007, **13**(9), 2207–18.

127. J. Velema and D. Kaplan, Biopolymer-based biomaterials as scaffolds for tissue engineering, *Adv. Biochem. Eng. Biotechnol.*, 2006, **102**, 187–238.

128. S. Lin-Gibson, J. A. Cooper, F. A. Landis and M. T. Cicerone, Systematic investigation of porogen size and content on scaffold morphometric parameters and properties, *Biomacromolecules*, 2007, **8**(5), 1511–8.

129. D. A. Wahl, E. Sachlos, C. Liu and J. T. Czernuszka, Controlling the processing of collagen–hydroxyapatite scaffolds for bone tissue engineering, *J. Mater. Sci. Mater. Med.*, 2007, **18**(2), 201–9.

130. T. Weigel, G. Schinkel and A. Lendlein, Design and preparation of polymeric scaffolds for tissue engineering, *Expert Rev. Med. Devices*, 2006, **3**(6), 835–51.

131. U. Boudriot, R. Dersch, A. Greiner and J. H. Wendorff, Electrospinning approaches toward scaffold engineering – a brief overview., *Artif Organs*, 2006, **30**(10), 785–92.

132. R. Murugan and S. Ramakrishna, Design strategies of tissue engineering scaffolds with controlled fiber orientation. *Tissue Eng.*, 2007, **13**(8), 1845–66.

133. L. S. Nair, S. Bhattacharyya and C. T. Laurencin, Development of novel tissue engineering scaffolds via electrospinning, *Expert Opin. Biol. Ther.*, 2004, **4**(5), 659–68.

134. C. M. Vaz, S. van Tuijl, C. V. Bouten and F. P. Baaijens, Design of scaffolds for blood vessel tissue engineering using a multilayering electrospinning technique, *Acta Biomater.*, 2005, **1**(5), 575–82.

135. Z. Ma, M. Kotaki, R. Inai and S. Ramakrishna, Potential of nanofiber matrix as tissue-engineering scaffolds, *Tissue Eng.*, 2005, **11**(1–2), 101–9.

136. R. Murugan and S. Ramakrishna, Nano-featured scaffolds for tissue engineering: a review of spinning methodologies, *Tissue Eng.*, 2006, **12**(3), 435–47.

137. F. Yang, R. Murugan, S. Wang and S. Ramakrishna, Electrospinning of nano/micro scale poly(L-lactic acid) aligned fibers and their potential in neural tissue engineering, *Biomaterials*, 2005, **26**(15), 2603–10.

138. J. Guan, K. L. Fujimoto, M. S. Sacks and W. R. Wagner, Preparation and characterization of highly porous, biodegradable polyurethane scaffolds for soft tissue applications, *Biomaterials*, 2005, **26**(18), 3961–71.

139. P. Roychowdhury and V. Kumar, Fabrication and evaluation of porous 2,3-dialdehydecellulose membrane as a potential biodegradable tissue-engineering scaffold, *J. Biomed. Mater. Res. A*, 2006, **76**(2), 300–9.

140. J. Nakamatsu, F. G. Torres, O. P. Troncoso, Y. Min-Lin and A. R. Boccaccini, Processing and characterization of porous structures from chitosan and starch for tissue engineering scaffolds, *Biomacromolecules*, 2006, **7**(12), 3345–55.

141. P. Sarazin, X. Roy and B. D. Favis, Controlled preparation and properties of porous poly(L-lactide) obtained from a co-continuous blend of two biodegradable polymers, *Biomaterials*, 2004, **25**(28), 5965–78.

142. S. Zhang, Emerging biological materials through molecular self-assembly, *Biotechnol. Adv.*, 2002, **20**(5–6), 321–39.

143. F. Gelain, A. Lomander, A. L. Vescovi and S. Zhang, Systematic studies of a self-assembling peptide nanofiber scaffold with other scaffolds, *J. Nanosci. Nanotechnol.*, 2007, **7**(2), 424–34.

144. E. Garreta, D. Gasset, C. Semino and S. Borros, Fabrication of a three-dimensional nanostructured biomaterial for tissue engineering of bone, *Biomol. Eng.*, 2007, **24**(1), 75–80.

145. T. C. Holmes, S. de Lacalle, X. Su, G. Liu, A. Rich and S. Zhang, Extensive neurite outgrowth and active synapse formation on self-assembling peptide scaffolds, *Proc. Natl. Acad. Sci. USA*, 2000, **97**(12), 6728–33.

146. M. Altman, P. Lee, A. Rich and S. Zhang, Conformational behavior of ionic self-complementary peptides, *Protein Sci.*, 2000, **9**(6), 1095–105.

147. S. J. Hollister, Porous scaffold design for tissue engineering, *Nat. Mater.*, 2005, **4**(7), 518–24.

148. H. H. Lu and J. Jiang, Interface tissue engineering and the formulation of multiple-tissue systems, *Adv. Biochem. Eng. Biotechnol.*, 2006, **102**, 91–111.

149. A. G. Mikos, S. W. Herring, P. Ochareon, J. Elisseeff, H. H. Lu, R. Kandel, F. J. Schoen, M. Toner, D. Mooney, A. Atala, M. E. Van Dyke, D. Kaplan and G. Vunjak-Novakovic, Engineering complex tissues, *Tissue Eng.*, 2006, **12**(12), 3307–39.

150. A. S. Mistry and A. G. Mikos, Tissue engineering strategies for bone regeneration, *Adv. Biochem. Eng. Biotechnol.*, 2005, **94**, 1–22.

151. K. Rezwan, Z. Chen, J. J. Blaker and A. R. Boccaccini, Biode-gradable and bioactive porous polymer/inorganic composite scaffolds for bone tissue engineering, *Biomaterials*, 2006, **27**(18), 3413–31.

152. N. Karim, K. Golz and A. Bader, The cardiovascular tissue-reactor: a novel device for the engineering of heart valves, *Artif. Organs*, 2006, **30**(10), 809–14.

153. Y. Martin and P. Vermette, Bioreactors for tissue mass culture: design, characterization and recent advances, *Biomaterials*, 2005, **26**(35), 7481–503.

154. R. M. Sutherland, B. Sordat, J. Bamat, H. Gabbert, B. Bourrat and W. Mueller-Klieser, Oxygenation and differentiation in mul-ticellular spheroids of human colon carcinoma, *Cancer Res.*, 1986, **46**(10), 5320–9.

155. J. M. Piret and C. L. Cooney, Immobilized mammalian cell cultiva-tion in hollow fiber bioreactors, *Biotechnol. Adv.*, 1990, **8**(4), 763–83.

156. F. Zhao, P. Pathi, W. Grayson, Q. Xing, B. R. Locke and T. Ma, Effects of oxygen transport on 3-d human mesenchymal stem cell metabolic activity in perfusion and static cultures: experiments and mathematical model, *Biotechnol. Prog.*, 2005, **21**(4), 1269–80.

157. N. L. Parenteau and J. Hardin-Young, The use of cells in reparative medicine, *Ann. N. Y. Acad. Sci.*, 2002, **961**, 27–39.

158. R. P. Lanza, R. Langer and J. P. Vacanti, *Principles of Tissue Engineering*, Academic Press, San Diego, CA, 1997.

159. U. Meyer, U. Joos and H. P. Wiesmann, Biological and biophy-sical principles in extracorporal bone tissue engineering, *Part I. Int. J. Oral Maxillofac. Surg.*, 2004, **33**(4), 325–32.

160. I. Martin, R. F. Padera, G. Vunjak-Novakovic and L. E. Freed, *In vitro* differentiation of chick embryo bone marrow stromal cells into cartilaginous and bone-like tissues, *J. Orthop. Res.*, 1998, **16**(2), 181–9.

161. S. L. Ishaug, G. M. Crane, M. J. Miller, A. W. Yasko, M. J. Yaszemski and A. G. Mikos, Bone formation by three-dimensional stromal osteoblast culture in biodegradable polymer scaffolds, *J. Biomed. Mater. Res.*, 1997, **36**(1), 17–28.

162. I. Martin, B. Obradovic, L. E. Freed and G. Vunjak-Novakovic, Method for quantitative analysis of glycosaminoglycan distribu-tion in cultured natural and engineered cartilage, *Ann. Biomed. Eng.*, 1999, **27**(5), 656–62.

163. E. M. Bueno, B. Bilgen and G. A. Barabino, Wavy-walled bio-reactor supports increased cell proliferation and matrix deposition in engineered cartilage constructs, *Tissue Eng.*, 2005, **11**(11–12), 1699–709.

164. H. D. Humes, S. M. MacKay, A. J. Funke and D. A. Buffington, Tissue engineering of a bioartificial renal tubule assist device: *in vitro* transport and metabolic characteristics, *Kidney Int.*, 1999, **55**(6), 2502–14.

165. H. F. Lu, W. S. Lim, P. C. Zhang, S. M. Chia, H. Yu, H. Q. Mao and K. W. Leong, Galactosylated poly(vinylidene difluoride) hollow fiber bioreactor for hepatocyte culture, *Tissue Eng.*, 2005, **11**(11–12), 1667–77.

166. J. M. Macdonald, M. Grillo, O. Schmidlin, D. T. Tajiri and T. L. James, NMR spectroscopy and MRI investigation of a potential bioartificial liver, *NMR Biomed.*, 1998, **11**(2), 55–66.

167. X. Wen and P. A. Tresco, Fabrication and characterization of permeable degradable poly(DL-lactide-co-glycolide) (PLGA) hollow fiber phase inversion membranes for use as nerve tract guidance channels, *Biomaterials*, 2006, **27**(20), 3800–9.

168. L. E. Freed, R. Langer, I. Martin, N. R. Pellis and G. Vunjak-Novakovic, Tissue engineering of cartilage in space, *Proc. Natl. Acad. Sci. USA*, 1997, **94**(25), 13885–90.

169. V. I. Khaoustov, G. J. Darlington, H. E. Soriano, B. Krishnan, D. Risin, N. R. Pellis and B. Yoffe, Induction of three-dimensional assembly of human liver cells by simulated microgravity, *In vitro Cell Dev. Biol. Anim.*, 1999, **35**(9), 501–9.

170. G. Vunjak-Novakovic and L. E. Freed, Culture of organized cell communities, *Adv. Drug Deliv. Rev.*, 1998, **33**(1–2), 15–30.

171. L. E. Freed and G. Vunjak-Novakovic, Microgravity tissue engineering, *In vitro Cell Dev. Biol. Anim.*, 1997, **33**(5), 381–5.

172. H. L. Holtorf, J. A. Jansen and A. G. Mikos, Flow perfusion culture induces the osteoblastic differentiation of marrow stroma cell-scaffold constructs in the absence of dexamethasone, *J. Biomed. Mater. Res. A*, 2005, **72**(3), 326–34.

173. E. M. Darling and K. A. Athanasiou, Articular cartilage bioreactors and bioprocesses, *Tissue Eng.*, 2003, **9**(1), 9–26.

174. A. Ratcliffe and L. E. Niklason, Bioreactors and bioprocessing for tissue engineering, *Ann. N. Y. Acad. Sci.*, 2002, **961**, 210–5.

175. S. Watanabe, S. Inagaki, I. Kinouchi, H. Takai, Y. Masuda and S. Mizuno, Hydrostatic pressure/perfusion culture system designed and validated for engineering tissue, *J. Biosci. Bioeng.*, 2005, **100**(1), 105–11.

176. J. O. Seidel, M. Pei, M. L. Gray, R. Langer, L. E. Freed and G. Vunjak-Novakovic, Long-term culture of tissue engineered cartilage in a perfused chamber with mechanical stimulation, *Biorheology*, 2004, **41**(3–4), 445–58.

177. E.R. Ochoa and J.P. Vacanti, An overview of the pathology and approaches to tissue engineering. *Ann. N. Y. Acad. Sci.*, 2002, **979**, 10–26; discussion 35–8.

178. J. P. Burgues, L. Gomez, J. L. Pontones, C. D. Vera, J. F. Jimenez-Cruz, and M. Ozonas, A chemosensitivity test for superficial bladder cancer based on three-dimensional culture of tumour spheroids. *Eur. Urol.*, 2007, **51** (4), 962–9; discussion 969–70.

179. L. A. Kunz-Schughart, J. P. Freyer, F. Hofstaedter and R. Ebner, The use of 3-D cultures for high-throughput screening: the multicellular spheroid model, *J. Biomol. Screen.*, 2004, **9**(4), 273–85.

180. P. L. Olive and R. E. Durand, Drug and radiation resistance in spheroids: cell contact and kinetics, *Cancer Metastasis Rev.*, 1994, **13**(2), 121–38.

181. B. J. Conley, J. C. Young, A. O. Trounson and R. Mollard, Derivation, propagation and differentiation of human embryonic stem cells, *Int. J. Biochem. Cell Biol.*, 2004, **36**(4), 555–67.

182. T. Korff, T. Krauss and H. G. Augustin, Three-dimensional spheroidal culture of cytotrophoblast cells mimics the phenotype and differentiation of cytotrophoblasts from normal and pre-eclamptic pregnancies, *Exp. Cell Res.*, 2004, **297**(2), 415–23.

183. R. C. Bates, N. S. Edwards and J. D. Yates, Spheroids and cell survival, *Crit. Rev. Oncol. Hematol.*, 2000, **36**(2–3), 61–74.

184. H. Okubo, M. Matsushita, H. Kamachi, T. Kawai, M. Takahashi, T. Fujimoto, K. Nishikawa and S. Todo, A novel method for faster formation of rat liver cell spheroids, *Artif. Organs*, 2002, **26**(6), 497–505.

185. A. Rothermel, T. Biedermann, W. Weigel, R. Kurz, M. Ruffer, P. G. Layer and A. A. Robitzki, Artificial design of three-dimensional retina-like tissue from dissociated cells of the mammalian retina by rotation-mediated cell aggregation, *Tissue Eng.*, 2005, **11**(11–12), 1749–56.

186. X. Wang, G. Wei, W. Yu, Y. Zhao, X. Yu and X. Ma, Scalable producing embryoid bodies by rotary cell culture system and constructing engineered cardiac tissue with ES-derived cardiomyocytes *in vitro*, *Biotechnol. Prog.*, 2006, **22**(3), 811–8.

187. A. Ivascu and M. Kubbies, Rapid generation of single-tumor spheroids for high-throughput cell function and toxicity analysis, *J. Biomol. Screen.*, 2006, **11**(8), 922–32.

188. Y. Hasebe, N. Okumura, T. Koh, H. Kazama, G. Watanabe, T. Seki and T. Ariga, Formation of rat hepatocyte spheroids on agarose, *Hepatol. Res.*, 2005, **32**(2), 89–95.

189. J. M. Kelm, V. Djonov, S. P. Hoerstrup, C. I. Guenter, L. M. Ittner, F. Greve, A. Hierlemann, C. D. Sanchez-Bustamante, J. C. Perriard, E. Ehler and M. Fussenegger, Tissue-transplant fusion and vascularization of myocardial microtissues and macrotissues implanted into chicken embryos and rats, *Tissue Eng.*, 2006, **12**(9), 2541–53.
190. J. M. Kelm, L. M. Ittner, W. Born, V. Djonov and M. Fussenegger, Self-assembly of sensory neurons into ganglia-like microtissues, *J. Biotechnol.*, 2006, **121**(1), 86–101.
191. J. M. Kelm, N. E. Timmins, C. J. Brown, M. Fussenegger and L. K. Nielsen, Method for generation of homogeneous multi-cellular tumor spheroids applicable to a wide variety of cell types, *Biotechnol. Bioeng.*, 2003, **83**(2), 173–80.
192. J. M. Kelm, C. Diaz Sanchez-Bustamante, E. Ehler, S. P. Hoerstrup, V. Djonov, L. Ittner and M. Fussenegger, VEGF profiling and angiogenesis in human microtissues, *J. Biotechnol.*, 2005, **118**(2), 213–29.

CHAPTER 14

Transgenesis

ELIZABETH J. CARTWRIGHT AND XIN WANG

Faculty of Medical and Human Sciences and Faculty of Life Sciences,
University of Manchester, Oxford Road, Manchester M13 9PT, UK

14.1 INTRODUCTION

14.1.1 From Gene to Function

The sequencing of the human genome has had a major impact on our
understanding of our genetic make-up. Since its launch in 1990, the
Human Genome Project has worked towards the goal of sequencing
every gene in the human genome and mapping them to their precise
chromosomal locations. In 2001, draft sequences of the full human
genome were published and in 2004 were refined to reveal that there are
between 20 000 and 25 000 protein-encoding human genes.[1,2] It is now
vital that we fully address the next big challenge – to identify the
function of these genes in both health and disease. A number of technologies are currently used to analyse mammalian gene function,
including the techniques that will be the focus of this chapter – transgenesis technologies.

This chapter will outline the transgenic techniques that are currently
used to modify the genome in order to extend our understanding of the
in vivo function of these genes in normal development and physiology,
and also in pathogenesis.

Molecular Biology and Biotechnology, 5th Edition
Edited by John M Walker and Ralph Rapley
© Royal Society of Chemistry 2009
Published by the Royal Society of Chemistry, www.rsc.org

Transgenesis is a general term which covers several ways of modifying the genome of intact organisms, for example, it is possible to eliminate a gene of interest or introduce extra pieces of DNA (overexpress) in a whole organism or a specific tissue. Transgenesis can be carried out in numerous organisms, including plants, the fruit fly (*Drosophila*), worm (*Caenorabditis elegans*), frog (*Xenopus*), zebrafish, large mammals including sheep and small mammals such as the rat and mouse. Each of these model organisms is valuable in its own right; the fruit fly and the worm have been used extensively to provide information regarding the basic functional processes of organisms such as cell proliferation, metabolic pathways and developmental processes. The zebrafish is particularly suited to use in studies to elucidate vertebrate gene function and large mammals have even been used as 'bio incubators' to produce recombinant proteins. Although historically the rat has been used in many physiological experiments, the ability to modify its genome is much more limited than in the mouse. Subsequently, it is the mouse that has become by far the most popular organism in which to study mammalian gene function and in particular human disease.

This chapter will outline the most commonly used transgenic technologies applied to the mouse genome, including overexpression transgenesis, gene targeting and conditional gene targeting. In addition, it will outline how we determine *in vivo* function in the mice that are generated by these transgenic techniques and, finally, we address some of the ethical issues which must be considered when using animals as a model system for research.

14.2 TRANSGENESIS BY DNA PRONUCLEAR INJECTION

Transgenesis by DNA pronuclear injection is the oldest and still the most widely used method of introducing foreign genes into the mammalian germline, subsequently to generate mammalian transgenic animals (including mice, rats and livestock). The basis of this technique, which was developed in the early 1980s by, amongst others, Gordon, Ruddle, Brinster, Palmiter and colleagues,[3–5] is the microinjection of DNA of interest (the transgene) into the pronucleus of fertilised oocytes (single-cell embryos) leading to the ectopic expression (overexpression) of the transgene in the genome.

14.2.1 Generation of a Transgenic Mouse

This is a multi-step process (Figure 14.1) which begins with the construction of the transgene, involves collection of fertilised oocytes,

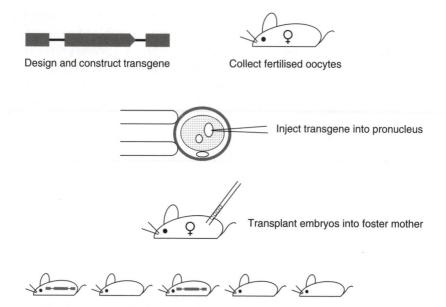

Design and construct transgene   Collect fertilised oocytes

Inject transgene into pronucleus

Transplant embryos into foster mother

Identify transgenic 'founder' pups by PCR and Southern blotting

Figure 14.1   Generation of transgenic mice by DNA pronuclear injection. Fertilised oocytes are collected from superovulated females. Transgene DNA is injected into one of the pronuclei. Embryos are re-implanted into pseudo-pregnant foster females and the resulting pups are screened for integration of the transgene. Founder (transgene positive) mice are then mated with wild-type mice to test for germline transmission and transgene activity.

microinjection of the transgene DNA into the pronucleus of the oocytes, reimplantation of those eggs into a foster mother, identification of pups carrying the transgene and establishment of the transgenic line.

14.2.1.1   Step 1. Construction of the Transgene

The transgene can be any DNA sequence/gene of interest to be added to the mouse genome, *i.e.* it will be present in addition to the normal complement of mouse genes. It is possible to use overexpression of a transgene in several ways: (i) to determine the function of a gene or (ii) to utilise its properties. For example, transgenes are often generated by linking a gene of interest with the regulatory regions of other known genes in order to express the gene of interest and its product at a higher level of expression or in a specific tissue or stage of development. Such experiments can be used to elucidate the normal function of the gene. For example, Figure 14.2 shows the transgene constructs used to

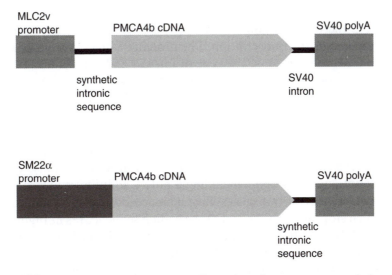

Figure 14.2 Construction of transgene. Examples of two transgenes designed to overexpress isoform 4 of the plasma membrane calcium pump (PMCA4b) in different tissues. Both constructs contain the same essential elements, *i.e.* a promoter sequence, cDNA of the gene of interest, a region of intronic sequence and a polyA tail. The top panel will result in the overexpression of PMCA4b in the heart, whereas the transgene in the bottom panel was designed to overexpress PMCA4b in vascular smooth muscle cells.

overexpress the plasma membrane calcium pump isoform 4 (PMCA4) in the cardiomyocytes (under the regulation of the MLC2v promoter) and in the vascular smooth muscle (driven by the SM22α promoter) – analysis of these transgenic lines has revealed that PMCA4 is involved in contractility in the heart and in the maintenance of peripheral blood pressure and vascular tone.[6,7] It is possible to use DNA pronuclear injection experiments to determine the function of regulatory elements of genes by fusing them to a reporter gene such as LacZ or GFP, whose expression can be readily detected.[8,9] It is also possible to use a transgene to express the wild-type form of a gene in a mouse that carries an endogenous mutant form *i.e.* it can be used to rescue a genetic defect.[10]

Transgene design has a major influence on the successful expression of the microinjected DNA; there are a number of elements that must be incorporated into the transgene construct, in addition to the protein coding sequence of the gene of interest, that are critical for gene expression. Essentially a transgene should also contain a promoter, an intron and a transcriptional stop sequence. The promoter is the regulatory sequence which determines in which cells and at what time the transgene is active and is usually derived from sequences upstream of, and including, the transcriptional start site which contain the necessary regulatory elements for transcription. The protein coding sequence is usually derived from the cDNA of the gene of interest and will contain both the ATG start codon and a translational stop codon. It is important to include an intron in the transgene construct as this increases transgene activity;[11] however, this does not need to be an intron from the gene of interest. It is not clear why the inclusion of an intron has a positive effect on transgene expression, but it is hypothesised to be due to a functional link between transcription and splicing. The transgene must also contain a transcriptional stop signal including a polyA addition sequence (AAA-UAA). A number of transgenes have included both the intron and transcriptional stop signal at the end of the coding sequence (as in Figure 14.2: SM22α–PMCA4b transgene), although they can be included separately (see Figure 14.2: MLC2v–PMCA4b transgene).

It is clear that the construction of a transgene is a multi-step cloning process which is most conveniently carried out in a bacterial plasmid. It has been shown that it is important to remove all of the prokaryotic plasmid sequences from the transgene prior to microinjection as its inclusion may inhibit its expression.[12] It is also clear that DNA that has been linearised prior to microinjection, rather than being left in its circular or supercoiled state, will integrate with greater efficiency.[13] Interestingly, although a large transgene ($>30\,$kb) may present technical difficulties when cloning and isolating the DNA, their large size does not appear to effect the frequency of integration into the genome. The final aspect to consider is the purity of the DNA to be microinjected as it must be free of any contaminants which might be harmful to the oocytes to be injected.

14.2.1.2 Step 2. Collection of Fertilised Oocytes

To improve the efficiency of the generation of the transgenic mice, it is essential to have a large number of viable fertilised eggs (oocytes) available for microinjection. Normally female mice will generate 6–10 eggs when naturally ovulating; however, it is possible to recover 20–30 eggs from a single female by inducing superovulation. The consequence is that less female mice are required for the production of a suitable number of quality eggs. Young female mice are injected with a combination of hormones, which mimic natural mouse hormones, and these ensure that large numbers of mature eggs are released from the ovaries simultaneously; treatment with follicle stimulating hormone is followed 44–48 h later by treatment with leutinising hormone. Fertilised oocytes are collected from the oviducts of the females following overnight mating with stud males. The oocytes will be collected at 0.5 days *post coitum* (dpc) as this allows time for the sperm to complete fertilisation following mating. In addition, at this stage the pronuclei from both gametes will be visible for several hours after embryo collection, which is necessary to allow successful injection of the transgene DNA.

14.2.1.3 Step 3. Pronuclear Injection of DNA Transgene

Microinjection of the transgene requires specialist, expensive equipment and highly trained personnel. The whole microinjection process is visualised under an inverted microscope which is attached to two micromanipulators, one to manipulate the glass pipette used to hold and secure the oocyte during the injection process (known as the holding pipette) and the other to control the fine glass pipette used to inject the DNA transgene into the pronucleus (Figure 14.3). For an excellent and detailed technical description of the microinjection process, see Hogan *et al.*[14]

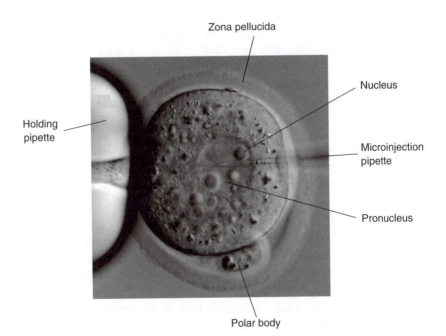

Figure 14.3 Microinjection of DNA into the pronucleus. The oocyte is secured by gentle suction on to the holding pipette. The microinjection pipette is advanced through the zona pellucida, it pierces the plasma membrane and is then pushed into the pronucleus. The oocyte has two pronuclei, either of which can be injected; the male pronucleus tends to be the largest and therefore easiest to visualise and inject. The DNA is then expelled from the injection pipette causing the pronucleus to swell.

It is usual that between 60 and 80% of the microinjected embryos will survive the process (viability can be tested by culturing the embryos to the two-cell stage) and be suitable for re-implantation into a surrogate female. Surrogate/recipient females are rendered pseudo-pregnant by mating with either a vasectomised or genetically sterile male. Of these re-implanted embryos, between 10 and 25% will survive to term. It is this high attrition rate of embryos at each stage which necessitates the use of such large numbers of fertilised eggs at the beginning of the whole procedure.

The transgene will integrate randomly into the genome, into any chromosome, including the sex chromosomes and often in multiple copies in a head-to-tail array.[15] Interestingly, it appears that integration most often occurs at just one site in the genome, even if multiple copies integrate.

Experience has shown that when microinjecting linearised DNA, approximately 25% of the resulting pups carry the transgene, these are known as founder transgenics. Due to the precise timing of injection and therefore integration of the transgene into the pronucleus, prior to replication, a high proportion (approximately 70%) of the transgenic mice will carry the transgene in every cell of the body. It is essential that the transgene integrates into the germ cells to be able to transmit the gene modification to the next generation. In those mice in which integration of the transgene occurred after DNA replication, the transgene will integrate into a proportion of the somatic and germ cells only; these mice are known as mosaics.

14.2.1.4 Step 4. Generation of a Transgenic Line

Founder transgenic mice are identified by DNA analysis including PCR and Southern blotting. It is common for DNA to be isolated from small pieces of tissue taken from the ears of the mice or from a small section of the end of the tail. The presence or absence of the transgene can be identified by PCR analysis using transgene-specific primers which will not amplify the endogenous gene locus. Southern blot analysis is then used to verify the integration of the transgene and to identify the number of copies that have inserted into the genome.

Once founders have been identified, they must be bred to wild-type mice to determine if the transgene can be passed to the next generation; in other words, they will be tested for germline transmission. The resultant pups must then be screened by PCR to identify heterozygous ($Tg/+$) transgenic mice from their wild-type ($+/+$) littermates – these would be expected in a 50:50 ratio.

One of the main drawbacks of generating transgenic mice by pronuclear injection of DNA is the so called position effect, *i.e.* the transgene can integrate at any chromosomal location and the site of integration itself could effect its expression. For example, a transgene could insert into a region of the chromosome that suppresses gene activity or could even insert into a functioning endogenous gene. Both of these scenarios would lead to a phenotype that is not solely due to the effect of the transgene. It is therefore essential that the phenotype of at least three independent founder lines carrying the same transgene are analysed to determine the effect of overexpression of the gene of interest.

14.2.2 Summary of Advantages and Disadvantages of Generating Transgenic Mice by Pronuclear Injection of DNA

Although the generation of transgenic mice by DNA pronuclear injection has several disadvantages and the list of potential technical problems is lengthy (for example, see Ref. 14), it remains a very rewarding technique which provides essential information regarding the biological function of a gene. Table 14.1 summarises the major advantages and disadvantages of this process.

14.3 GENE TARGETING BY HOMOLOGOUS RECOMBINATION IN EMBRYONIC STEM CELLS

Gene targeting was developed to ablate (knock out) the function of a specific gene. Although gene targeting is still most commonly used to generate this type of mutation, the method is now the basis of many techniques which can produce a whole variety of DNA modifications to

Table 14.1 Advantages and disadvantages of generating transgenic mice by pronuclear injection of DNA.

Advantages
Conceptually straightforward
Cloned DNA from any species can be microinjected
Integration of transgene is relatively efficient
Process is relatively quick – taking as little as 3 months to generate a construct, microinject DNA, analyse genotype of founder pups and test for germline transmission of the transgene

Disadvantages
Microinjection requires specialist, expensive equipment and highly trained personnel
Integration site is random which may affect the level and pattern of expression. Therefore, need to analyse the phenotype (effect of the gene deletion) of mice from several founder lines
The endogenous gene remains intact in the genome which may complicate molecular and phenotypic analysis

the mouse genome, including the deletion/rearrangement of large regions of chromosomes containing multiple genes,[16,17] the introduction of subtle mutations into the genome,[18,19] the knock-in of a gene of interest[20–22] and conditional gene targeting (see Section 14.4).

The technique of gene targeting in the mouse has enabled major advances to be made in our understanding of mammalian gene function, in normal developmental, biochemical and physiological processes, and also in disease processes. The technique was established by bringing together two important findings (i) that homologous recombination can be used to modify specifically a target mammalian gene and (ii) that embryonic stem cells isolated from pre-implantation stage mouse embryos can be maintained in culture and then used to contribute to the germline of another mouse. The importance of this technique to biomedical research was acknowledged when Mario R. Capecchi, Martin J. Evans and Oliver Smithies jointly received the Nobel Prize in Physiology or Medicine 2007 for their discoveries of the '*principles for introducing specific gene modifications in mice by the use of embryonic stem cells*'.

14.3.1 Basic Principles

14.3.1.1 Homologous Recombination
Homologous recombination is used to introduce modifications into a gene by exchanging endogenous DNA sequences in the chromosome with cloned DNA sequences (targeting vector). Gene targeting by homologous recombination in mammalian cells was first reported in 1985,[23,24] allowing for the stable insertion of DNA into the genome at a predictable site.

14.3.1.2 Embryonic Stem Cells
Embryonic stem (ES) cells are isolated from the inner cell mass of pre-implantation stage embryos at 3.5 days *post coitum*; at this stage embryos are known as blastocysts. The crucially important factor about the progenitor cells of these early embryos is that they are *pluripotent* – they have the potential to differentiate into any cell type, including the germ cells, of the subsequent embryo. It was in 1986 that Evans and colleagues demonstrated that these ES cells could be maintained in culture, where they could be genetically modified and then used to repopulate the germline of another mouse, thereby generating chimeric offspring containing genetic material from both the recipient mouse and the ES cells.[25] The chimeric mice, when mated, could pass on the genetic mutation to the next generation; with the mutation being inherited according to Mendel's laws of genetics.

It is interesting to note that the ability of ES cells to populate the germline has so far only been identified in the mouse and not even in all mouse strains.[26]

Smithies, Capecchi and their colleagues brought together the homologous recombination and ES cell technologies to develop the process known as *gene targeting*.[27,28]

It is essential that ES cells to be used in gene targeting experiments maintain their pluripotency in order to contribute to the germline of chimeric mice. It has been found that the presence of the cytokine leukaemia inhibitory factor (LIF) is essential to ensure that ES cells do not differentiate *in vitro*. For this reason, ES cells are generally grown on a feeder layer of fibroblasts which secrete LIF into the culture medium. When ES cells are grown in optimal culture conditions, they grow as discrete colonies of densely packed cells, with no obvious sign of differentiation (Figure 14.4). Other culture conditions have been shown to be essential in maintaining the ES cells in their undifferentiated state, including growing at a high cell density and ensuring that the colonies are thoroughly disaggregated when passaging (splitting) the cells, since clumps of cells are likely to differentiate.

Most ES cells lines currently in use have been derived from the 129 strain of mouse which has an agouti coat colour genotype (for a summary of mouse coat colour genetics, see Ref. 14); this is useful when identifying chimeric mice (see step 5 below). Another interesting feature of many of the ES cell lines currently in use is that they have an XY (male) genotype, which leads to the production of mainly male chimeras (even when XY ES cells are injected into a female blastocyst the resulting chimeric pup will tend to be male).

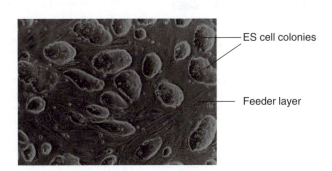

ES cell colonies

Feeder layer

Figure 14.4 ES cell colonies growing on a layer of fibroblast feeder cells. Healthy, undifferentiated ES cells form densely packed colonies.

14.3.2   Generation of a Knockout Mouse

As with transgenesis by pronuclear injection, gene targeting by homologous recombination in embryonic stem cells is a multi-step process. It
begins with the generation of the targeting vector, which is transferred by
electroporation into the ES cells. The ES cells are cultured and analysed
for the presence of the homologously recombined DNA sequence; the
targeted ES cells are then injected into blastocyst stage embryos. The
resultant pups are screened for the level of chimerism (percentage contribution from the targeted ES cells) and the chimeric mice are then tested
for their ability to pass the targeted mutation through the germline and to
generate mice heterozygous for the mutation. Heterozygous mice can then
be mated to generate the homozygous knockout mouse (Figure 14.5).

14.3.2.1   *Step 1. Generation of a Targeting Vector*
When designing and constructing a targeting vector, a number of factors
must be considered which will influence the type of mutation to be

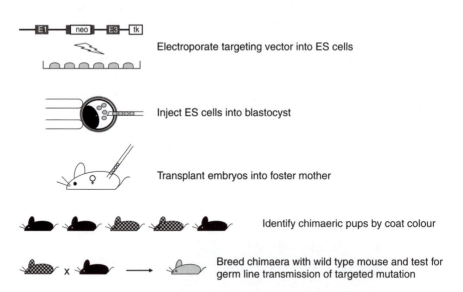

Figure 14.5   Generation of gene knockout mice by gene targeting in ES cells. The
targeting vector is electroporated into the ES cells. ES cells that have
undergone homologous recombination are injected into blastocyst stage
embryos and these embryos are then transplanted to pseudo-pregnant
foster mothers. Chimeric offspring can be identified by their coat colour;
these pups will carry the targeted mutation carried by the injected ES
cells. Chimeric offspring can then be mated to wild type mice to determine whether they transmit the targeted mutation through the germline
to give pups heterozygous for the mutation. The heterozygous offspring
can then be intercrossed to mice homozygous for the mutation.

introduced, the efficiency of targeting and the ease with which successful targeting can be detected.

DNA homologous with the chromosomal/gene site of interest. For successful and efficient targeting, the vector must contain at least 5–10 kb of isogenic DNA homologous with the sequence to be targeted. This homologous sequence is divided between the short arm of homology (1–1.5 kb) and a long arm of homology (4–8 kb); this permits easy screening of the ES clones. It is ideal to identify gene targeted colonies by PCR designed to span the short arm of homology. It is known that the efficiency of homologous recombination is decreased when there are base pair differences between the donor and recipient DNA.[29] For this reason, it is now common practice for the DNA used to construct the targeting vector to originate from the same mouse strain as the ES cells (*i.e.* isogenic DNA).

Positive and negative selection cassettes. Since gene targeting by homologous recombination occurs at low frequencies (typically 10^{-5}–10^{-6} of ES cells treated with construct DNA) and the targeting construct is much more likely to insert randomly into the genome, it is essential to be able to screen ES cell colonies quickly and efficiently for successful targeting. For this reason, most targeting vectors will be designed to insert a positive selection cassette into the gene of interest. For example, the neomycin phosphotransferase gene (neo) is often used as a positive marker, which when expressed in the ES cell genome will render the cells resistant to treatment with the antibiotic neomycin sulfate (G418). A negative selection marker can also be used to enrich for gene targeted colonies. The selection marker is cloned outside of the homologous sequence in the targeting vector and will therefore not insert into the genome when homologous recombination occurs, but will insert into the genome if random integration of the targeting vector occurs. For example, the herpes simplex virus thymidine kinase gene (HSVtk) when expressed in ES cells will produce a toxic product in the presence of gancyclovir (a thymidine analogue), killing ES cells expressing this gene (Figure 14.6).

14.3.2.2 Step 2. ES Cell Transfection

The most efficient method for introducing the targeting vector into the ES cells is by electroporation. The linearised vector DNA is electroporated into a large number of ES cells in a single cell suspension; the cells are then plated on to fresh feeder cells. Then, 24 h after electroporation, the selection process can begin, which will kill cells which have not incorporated the targeting vector by homologous recombination. The ES cells are cultured in media containing the drugs used for

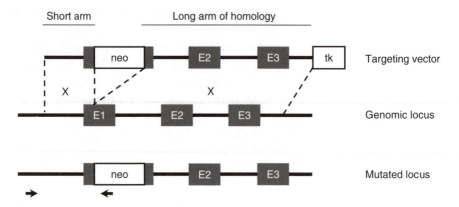

Figure 14.6   Gene targeting strategy. The top panel depicts a conventional targeting vector in which the positive selection cassette (neo) has been cloned into the first exon (E1) of the target gene and the negative selection cassette has been cloned at the end of the targeting construct after exons 2 and 3 (E2 and E3). The middle panel represents the genomic locus and homologous recombination will take place between the genomic locus and the two arms of homologous sequence in the targeting vector. The crossover points are depicted by the X. The bottom panel depicts the mutated locus when homologous recombination has occurred. The two small arrows represent the PCR primers which will be used to amplify the sequence across the short arm of homology to identify ES cells which have undergone homologous recombination.

selection for 7–10 days; this will enrich the population with cells that have undergone homologous recombination; however, it must be noted that this process is not 100% efficient.

14.3.2.3   Step 3. Identification of ES Cells Targeted
           by Homologous Recombination
To identify the ES cells that have undergone gene targeting by homologous recombination, discrete colonies are identified and picked. The colonies are dissociated into single cells by treatment with trypsin, divided between two wells on duplicate microtitre plates and cultured. The purpose of dividing the cells between duplicate plates is to allow one plate of cells to be used to prepare DNA to identify targeted ES cells and the cells from the second plate can be used to inject into blastocysts. Genomic DNA is prepared from each ES cell clone, which is then screened by PCR to identify clones in which homologous recombination has occurred. It is likely that several hundred ES cell clones will require screening, so it is important that a robust and easy PCR strategy is used – it is for this reason that it is usual to amplify the region across the short

arm of homology (Figure 14.6). Positive clones must then be further analysed, usually by Southern blotting and DNA sequencing, to verify that all regions of the targeting vector have undergone the desired recombination event.

14.3.2.4   Step 4. Injection of ES cells into Blastocysts

Blastocysts, which are 3.5-day-old embryos, are collected from the uterus of the donor female. It is usual when using ES cells from the 129 strain of mouse to collect blastocysts from a C57Bl/6 mother; this mouse line has a black coat colour (which is helpful when identifying chimeric mice – see step 5 below). ES cells carrying the desired mutation are treated to give a single cell suspension. The ES cells are drawn up into the injection pipette by gentle suction and the blastocyst to be injected is held by suction on the holding pipette. The injection pipette is advanced into the cavity of the blastocyst, which is known as the blastocoel, and 10–15 ES cells are released (Figure 14.7). After injection, the embryos are cultured for a few hours to allow them to re-expand slowly before being transferred to the uterus of a pseudo-pregnant foster mother. Pups should be born 17 days later.

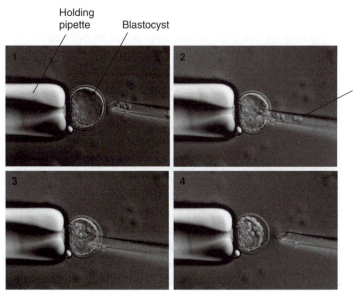

Figure 14.7   Injection of targeted ES cells into blastocysts. The blastocyst is held on the holding pipette by gentle suction (1). The injection needle containing ES cells is advanced into the blastocyst cavity (blastocoel) (2), where the ES cells are released (3) and the injection needle is removed (4).

14.3.2.5   Step 5. Identification of Chimeric Mice and Breeding to
*   Generate Homozygous Mutant (Knockout) Mice*

Approximately 1 week after mouse pups are born, their coat colour becomes apparent. At this stage, it is possible to identify agouti from non-agouti coat colour. It is therefore possible to identify chimeric mice by their coat colour if ES cells from the 129 mouse strain (agouti) have contributed to the development of a C57Bl/6 embryo (non-agouti). Embryos in which the ES cells had made no contribution would appear as wild-type C57Bl/6 (black), whereas those pups in which the 129 ES cells had made a contribution would contain a certain level of agouti coat colouring. Chimeric mice therefore contain some cells carrying the targeted mutation on one allele and other cells which are wild type.

To generate a gene knockout mouse, it is essential that some of the germ cells carry the targeted mutation. To test for germline transmission of the mutation, chimeric mice are bred to wild-type mice; should germline transmission occur, a proportion of the pups will be heterozygous for the targeted mutation. Heterozygous mice can then be bred to produce mice homozygous for the targeted mutation – gene knockout mice.

14.3.3   Summary of Advantages and Disadvantages of Generating Gene Knockout Mice

As will be evident from the section above, the generation of a knockout mouse by gene targeting in ES cells is not a process to be undertaken lightly. It is labour intensive, technically demanding and fraught with problems; however, with good experimental design and know-how, this technique will provide enormous amounts of invaluable information about the function of the target gene. Table 14.2 summarises the major advantages and disadvantages of this process.

14.4   CONDITIONAL GENE TARGETING

As has been described above, gene targeting has rapidly become a highly developed technology that facilitates the study of gene function. However, 15–20% of germline mutations result in embryonic lethality,[30] thereby precluding analysis of gene function in the adult tissues. Moreover, gene targeting disrupts the function of the gene within all tissues, creating a complex phenotype in which it may be difficult to distinguish direct function of the gene of interest in a particular tissue from any secondary effects. To resolve these critical problems, Gu *et al.* developed the conditional inactivation of gene expression using the Cre–loxP system.[31]

Table 14.2 Advantages and disadvantages of generating a knockout mouse by gene targeting in embryonic stem cells.

Advantages

The integration site and therefore the gene modification are highly specific

A variety of mutations can be achieved including null mutations (gene knockout), deletion/rearrangement of large regions of chromosomes, site-specific mutations, gene knock-in

Recessive alleles can be studied

Disadvantages

Microinjection requires specialist, expensive equipment and highly trained personnel

Process is very time consuming, taking 1.5–2 years to generate a targeting vector, target ES cells, identify homologous recombination events, microinject ES cells and test chimeric pups for germline transmission of mutation

Process is expensive as it is labour intensive, requires expensive equipment and the mouse husbandry costs will be high

Embryonic lethality – if the target gene is essential for development of the embryo, then it will not be possible to study the role of the gene in the adult mouse

Sometimes difficult to determine if the phenotype observed is primarily due to the deletion of the target gene or is a secondary consequence of the deletion of the target gene on a downstream pathway

5'-ATAACTTCGTATA**ATGTATGC**TATACGAAGTTAT-3'
3'-TATTGAAGCATAT**TACATACG**ATATGCTTCAATA-5'

Figure 14.8 Scheme of Cre binding to the loxP site. Cre specifically recognises and binds to inverted repeats surrounding an asymmetric 8 bp core region.

Cre (Cause recombination) recombinase is a 38 kDa enzyme isolated from the P1 bacteriophage. The loxP site is a 34 bp sequence comprising two 13 bp inverted repeats separated by an asymmetric 8 bp core which indicates the orientation of the loxP site (Figure 14.8). The Cre recombinase efficiently binds the two loxP sites and catalyses the recombination, resulting in the deletion, inversion or translocation of the sequences between the two recombination sites, according to the orientation of two loxP sites (Figure 14.9).

Cre is not the only recombinase that has been described for use in conditional gene targeting; Flp recombinase from *Saccharomyces cerevisiae* is also used.[32] Similarly to the Cre–loxP system, Flp recognises two FRT sites (34 bp DNA sequence) and catalyses the DNA recombination between the two recombination sites. The Flp recombinase is not as effective as Cre at 37 °C; its optimum function is at 25–30 °C.[33] The use of the Flp–FRT system is not as widespread as that of the Cre–loxP system, due to the limited availability of mouse lines expressing the

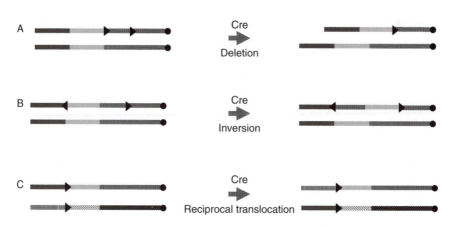

Figure 14.9   Cre-mediated recombination. The orientation of the loxP sites influences
the type of recombination that occurs. (A) Deletion of DNA occurs
between two loxP sites in the same orientation; following recombination
one loxP site remains in the genome. (B) Inversion of DNA occurs
between two loxP sites in the opposite direction. (C) Reciprocal translocation occurs between loxP sites placed in non-homologous chromosomes. The arrow heads indicate the orientation of the loxP sites.

Flp recombinase. This section will therefore focus on the description of
the Cre–loxP recombination system.

14.4.1   Generation of a Conditional Knockout Mouse Using the Cre–loxP System

The generation of conditional knockout mice is a multi-step process,
which involves mating two separate mouse lines generated by DNA
pronuclear injection and/or gene targeting by homologous recombination in ES cells.

1. Homologous recombination in embryonic stem (ES) cells is used to
   generate mutant mice (Flox mice) in which the essential exon(s) of
   the gene is flanked by two loxP sites. These sites do not interfere
   with the normal expression of the gene but constitute a binding
   domain for the Cre recombinase which will excise the DNA fragment between the loxP sites.
2. Generating inducible and/or tissue-specific Cre-expressing mouse
   lines by pronuclear injection or the knock-in method using
   embryonic stem cells.
3. Crossing the Flox mouse with the specific Cre-expressing mouse to
   achieve the deletion of the gene in a particular cell lineage or tissue
   type at a certain time.

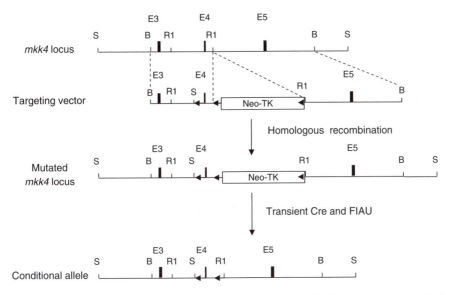

Figure 14.10 Conditional disruption of the *mkk4* gene. The figure shows the *mkk4* wild-type locus, the targeting vector containing the three loxP sites, represented by black triangles, the predicted structure of the mutated *mkk4* locus and the selected conditional allele obtained by Cre-mediated recombination. In the conditional *mkk4* allele, loxP sites are located either side of exon 4. Restriction enzyme sites are indicated (B, *Bam*HI; R, *Eco*RI; S, *Spe*I).

14.4.1.1 Step 1. Generation of Flox Mouse Models

The *tri-loxP* targeting vector has been successfully used to generate numerous Flox mouse lines. For example, to produce *mkk4* Flox mice (mkk4$^{\text{flox/flox}}$), a thymidine kinase neomycin resistance (Neo-TK) cassette containing two loxP sites was inserted behind exon 4, and a third loxP site was placed in front of exon 4 (Figure 14.10). Exon 4 of this gene contains the important kinase function domain; the removal of this exon causes a frame shift in the open reading frame, resulting in the loss of expression of MKK4. The targeting construct was electroporated into ES cells and then subjected to selection using the antibiotic G418. The resulting positive clones were transfected with a Cre-expression plasmid to remove the Neo-TK selection cassette; it is important to remove this cassette because it may interfere with the normal expression of the target gene. The ES cells were counter-selected with 1-(2-deoxy-2-fluoro-β-D-arabinofuranosyl)-5-iodouracil (FIAU), which results in the death of cells which retained the positive/negative selection cassette. The ES clones that retained two loxP sites flanking exon 4 were identified by Southern blotting and were subsequently used to generate the *mkk4* Flox mice.[34]

14.4.1.2   Step 2. Generation of the Cre-expressing Mouse Lines
Over the past few years, many tissue-specific Cre mouse lines have been developed which have greatly facilitated the study of gene function. Established Cre lines can be identified using a database (http://www.mshri.on.ca/nagy/Cre-pub.html). The core part of this approach is to choose a tissue-specific promoter which drives Cre expression specifically and efficiently. Three methods are widely used for generating Cre mouse models:

1. *Standard transgenesis by pronuclear injection.* The generation of a Cre-expressing mouse in which a transgene vector consisting of Cre driven by a tissue-specific promoter is inserted into the genome by standard transgenesis is the most popular method. It is a relatively rapid procedure with a straightforward concept and in many cases has generated very efficient Cre lines. However, this method involves random integration of the Cre-transgene into the genome, which sometimes results in unwanted Cre expression or low Cre expression levels which prohibits any further analysis of the function of the gene of interest.

2. *Bacterial artificial chromosome (BAC) transgenesis.* BAC clones contain large fragments of mouse genomic DNA, including most regulatory elements of the chosen promoter; hence introducing Cre into the BAC presents an advantage that the Cre expression pattern reflects truthful endogenous gene expression. The main drawbacks of this technique are that an additional loxP site exists in the BAC vector which needs to be removed prior to pronuclear injection and that other genes and their promoter regions may exist within the BAC vector and may therefore interfere with the desired Cre expression pattern.[35]

3. *Knock-in approach using homologous recombination in ES cells.* The knock-in strategy requires homologous recombination in ES cells to insert the Cre recombinase construct downstream of the endogenous targeted promoter. The advantage of this approach is that Cre expression is closely correlated to the expression profile of the chosen promoter; however, as has already been highlighted in this chapter, this is a long and expensive procedure.

Many genes have different functions, from developmental stages to adulthood. Using an early promoter (*i.e.* the promoter from a gene that is expressed very early in development) to drive Cre activity in a tissue-specific gene-knockout system sometimes results in embryonic lethality. To overcome this limitation, inducible gene targeting was first developed by Kühn *et al.*[36] This method can induce gene inactivation in mice at a given time point using modification of the Cre recombinase system. For example, the Cre–ER line is widely used to achieve such temporal control of genetic recombination; in this mouse line, Cre has been fused to a mutated form of the ligand binding domain of the estrogen receptor.[37] The mutant ER does not bind its natural ligand at physiological concentrations but binds to the synthetic ligand tamoxifen and several of its metabolites (*e.g.* 4-hydroxytamoxifen). Thus, Cre–ER activity can be induced following tamoxifen injection, thereby allowing the temporal (time-specific) deletion of the loxP flanked gene. At present, a number of tissue-specific inducible Cre lines have been reported, such as the αMHC–MerCreMer line for the study heart function[38] and the Nestin–CreERT2 line used in brain function studies.[39] The availability of these elegant and novel Cre lines has provided us with a unique opportunity to study gene function and regulation precisely.

14.4.1.3 Step 3. Breeding Strategies to Generate Temporal- and/or Tissue-specific Gene Knockout Mice

Two rounds of breeding are required to produce temporal- and/or tissue-specific gene-knockout mice (Figure 14.11). The first breeding step between the floxed mouse and the Cre-expressing mouse results in progeny which carry a single floxed allele and the Cre vector (Gene $A^{fl/+}$: Cre). The second breeding step crosses these mice with homozygous flox mice (Gene $A^{fl/fl}$); pups with a number of different genotypes will be born, including the desired conditional knockouts (Gene $A^{fl/fl}$:Cre).

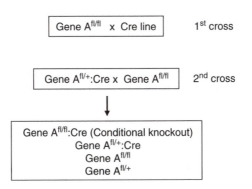

Figure 14.11 Breeding scheme to generate tissue/temporal-specific knockout mice. The floxed mouse is crossed with the Cre-expressing mouse, which after two breeding steps will result in a conditional gene knockout.

14.4.2   Chromosomal Engineering Using the Cre–loxP System

Chromosomal abnormalities cause many human genetic disorders and foetal loss, such as DiGeorge syndrome, which is due to the deletion of a large region within chromosomal 22 and is characterised by congenital heart defects, immunodeficiency and abnormal facial appearance. Due to the restriction on working with human material and the lack of experimental models, progress in dissecting the genetic basis underlying congenital diseases has been slow. In 1999, Lindsay *et al.* made a revolutionary step when they successfully created the DiGeorge mouse model bearing the 1.2 Mb chromosomal deletion using homologous recombination in ES cells and the Cre–loxP technique.[17] Due to the unique properties of the Cre–loxP system, the manipulation of chromosomal rearrangements (deletion, inversion or translocation) in order to model human diseases has become feasible and this will greatly empower scientists in studying human inherited diseases.

14.4.3   Summary of Advantages and Disadvantages of Conditional Gene Targeting

Conditional gene targeting is an exceptionally elegant extension of the classical gene targeting approach which permits the expression of a gene modification at a given time and/or in a particular cell type or tissue. The system can be exploited to introduce subtle mutations into the genome and to make large chromosomal modifications. However, its disadvantages must be fully considered before embarking on the study of gene function using this system. Table 14.3 summarises the main advantages and disadvantages of generating a conditional knockout mouse.

Table 14.3   Advantages and disadvantages of generating a conditional knockout mouse.

Advantages

Avoids embryonic lethality and complicated secondary effects which can be induced by conventional gene targeting

Gene function can be studied in a specific cell type or tissue at any given time point

Chromosomal abnormalities causing human diseases can be modelled in mice

Disadvantages

Two transgenic lines (Flox mice and Cre mice) are required to generate conditional knockout mouse models

Efficiency of gene deletion varies depending on the gene locus position and the Cre activity

Time consuming and expensive

14.5 PHENOTYPIC ANALYSIS OF GENETICALLY MODIFIED MICE

Genetically modified mice, generated by any of the techniques described within this chapter, are merely a sophisticated *tool* with which to work when determining the function of a gene. The mutant mouse must be fully analysed at the molecular, biochemical and *in vivo* levels to determine the function of that gene.

Initially all mice will be analysed to determine their *genotype*; for example, PCR or Southern blot analysis will be carried out to determine whether the mouse carries the desired gene mutation and whether it is heterozygous or homozygous for the mutation.

It is the effect that the *genotype* has on the *phenotype*, the characteristics and traits of the mouse, which is of key interest.

Having ascertained the genotype of the mouse, it is essential to determine that the introduced genetic modification has resulted in the desired altered gene regulation; for example, that disruption of the target gene or the introduction of a DNA construct has resulted in the ablation or overexpression of the gene product, respectively. RNA analysis by reverse transcriptase PCR (RT-PCR) or real-time PCR, Northern blotting or *in situ* hybridisation can be used to establish the expression pattern of the transgene or confirm disruption/deletion of the gene of interest. Modified levels/localisation of the gene product (protein) can then be identified by Western blot analysis and immunohistochemistry.

Having identified the direct effect that the gene modification has had on the gene of interest, it is important to investigate the downstream effects of the gene modification by determining expression levels of genes in the same pathways or known interaction partners, *etc*. DNA microarray technologies are extremely valuable in identifying novel genes whose expression may be influenced by modification of the gene of interest.[40]

The functional tests that must be carried out to determine the effect of the gene modification will depend on each particular gene; for example, it would be important to measure intracellular Ca^{2+} levels in a mouse model in which a calcium handling protein is disrupted or the activity of an enzyme should be tested where the gene encoding that enzyme has been modified.

To determine the role of the gene of interest in normal development, physiology and pathology, there are numerous assessments that can be carried out. The first thing to assess is whether the gene of interest is vital to embryonic development and therefore whether its modification leads to embryonic lethality. The simplest assessment is to analyse the genotyping results; for example, are homozygous gene knockout offspring born when heterozygous mice are crossed? If the gene modification does lead to a lethal phenotype, it will then be necessary to assess the point of lethality and ultimately how and why the modification leads to embryonic death.[41]

There is a vast array of tests that can be carried out to determine the function of the gene in normal physiology, ranging from very simple observations of the gross appearance of the mouse to extremely sophisticated screens. For example, a comprehensive screening protocol named SHIRPA has been developed to identify and characterise phenotypic changes associated with a gene mutation.[42] This protocol includes assessments of muscle and lower motor neurone function, spinocerebellar function, sensory function, neuropsychiatric function and autonomic function using a variety of functional, biochemical and histological techniques. The effect of the gene mutation on other organ systems including the heart can also be measured using specialised tests and equipment.[6,43] Simple breeding tests can be used to determine if the gene modification affects fertility, which can then be followed up, for example, with in depth analysis of sperm function.[44]

It is therefore the analysis of the functional, biochemical and molecular changes that occur due to a gene modification that will help us to determine the function of the genes in the genome.

14.6 ETHICAL AND ANIMAL WELFARE CONSIDERATIONS

The mouse has become an extremely popular model for the study of mammalian gene function for a number of reasons. Although it has long been known that the mouse has a similar physiological structure to humans, we now know that approximately 99% of mouse genes have human homologues.[45] Mice are mammals and subsequently undergo similar developmental processes to humans; they also have many physiological and behavioural similarities. In addition, mice develop some of the same diseases as humans, including cancer, cardiovascular disease, hypertension, obesity, diabetes, glaucoma, neurological defects, deafness, osteoporosis and asthma. For these reasons, mice are considered by many as an essential biomedical tool in our armoury to combat human disease through increased understanding of disease mechanisms and development of therapeutic agents.

It is not only the physiological and genetic aspects of the mouse that make it a suitable model for studying gene function; on a practical level, the mouse is very convenient to work with as it has a short gestation period (19–21 days), reaches breeding age within 6–8 weeks and colonies are relatively inexpensive to maintain.

It is clear, however, that the mouse is not without its disadvantages when using it as a model in which to study human gene function and the development of human disease. It is obvious that even though the two species share 99% gene homology, millions of years of evolution have resulted in major differences between the two species.

The mouse does not naturally develop all diseases exhibited by humans; for example, one of the most important diseases of the elderly population is Alzheimer's disease, which mice do not normally develop. However, genetic manipulation of the mouse genome through gene targeting and transgenesis has enabled many aspects of this disease to be modelled.[46] In other instances, reproducing mutations in the mouse genome which are known to cause disease in humans does not lead to mimicking of the human phenotype.[47,48] A recent development in extending the use of the mouse as a model for human disease is by 'humanising' the mouse; this is a process which includes replacing large segments of the mouse genome with the syntenic human sequence (syntenic regions are areas of chromosomes in which genes occur in the same order in different species), allowing both the gene itself and surrounding regulatory regions to be replaced.[49] This new adaptation of transgenic technology promises to enable more accurate models of human disease to be developed in the mouse. It is clear, therefore, that the mouse is not a perfect model in which to study mammalian gene function and for use as a model for human disease; however, the application of transgenic techniques and genetic mutation in the mouse genome has greatly informed our knowledge of gene function.

As research scientists using an animal system as a model, it is essential that we consider the welfare of our animal models and must closely weigh the benefit we gain from their use as part of our research against the cost to their welfare. This chapter has so far highlighted the benefits to research of transgenesis and of using the mouse as a model of human disease; however, the cost to animals used in this research may include suffering due to the phenotype caused by the genetic modification, embryonic or post-natal death as a result of the effect of the genetic modification or animal welfare may be comprised during the production of the GM mice, *e.g.* by superovulation, vasectomy, embryo collection and transfer.

There are many ways to ensure that a minimal number of animals are used to gain the maximum amount of information. One essential method is to ensure that animal experiments are carried out in combination with *in vitro*, cellular and bioinformatic techniques and are fully informed by the findings of these other research methods. The number of animals used can be reduced through good technical practice. It is also important that a research project is continually assessed in the light of new published data and that animal requirements are regularly reassessed to ensure minimal animal usage to maximise statistical analyses throughout the course of a project.

Transgenesis is a powerful technology which has been and will continue to be used successfully to increase our knowledge of gene function in health and disease. It is therefore incumbent on the researcher to ensure that exemplary practice regarding animal welfare is used to ensure that the cost:benefit ratio is correctly balanced.

14.7 CONCLUSIONS

The generation of mouse models using a variety of transgenesis techniques in order to study gene function has been immensely popular amongst researchers since the introduction of the techniques in the 1980s. This chapter gives an outline of several of the fundamental technologies involved, but there are many aspects of transgenesis which have not been addressed here, including the use of other model systems from plants, non-vertebrates, other vertebrates and even large mammals – all of which have an important place in research to determine gene function.

Over recent years, we have gathered immense amounts of information regarding mammalian gene function; however, we are still a long way from knowing the function of the majority of the genes in the genome.[50] The next major breakthrough in the use of transgenesis to understand mammalian gene function will be the development of large-scale projects which will systematically mutate every one of the 20 000–25 000 genes in the genome. Several such projects are now under way, including EUCOMM (European Conditional Mouse Mutagenesis Programme; www.eucomm.org), which is a European Union-funded consortium, KOMP (Knock-out Mouse Project; www.nih.gov/science/models/mouse/knockout/index.html), which is funded by the National Institutes of Health (NIH) in the USA, and NorCOMM (North American Conditional Mouse Mutagenesis Programme; www.norcomm.org), which is a Canadian project funded by Genome Canada, A number of other consortia are also preparing to join the international effort to knock out every gene in the genome.

14.8 ACKNOWLEDGEMENTS

We would like to thank Graham Morrissey and Lynnette Knowles for some of the photographs used in this chapter.

REFERENCES

1. J. D. McPherson, M. Marra, L. Hillier, R. H. Waterston, A. Chinwalla and J. Wallis, *et al., Nature*, 2001, **409**, 934.
2. International Human Genome Sequencing Consortium, *Nature*, 2004, **431**, 931.
3. J. W. Gordon, G. A. Scangos, D. J. Plotkin, J. A. Barbosa and F. H. Ruddle, *Proc. Natl. Acad. Sci. USA*, 1980, **77**, 7380.
4. R. L. Brinster, H. Y. Chen, M. Trumbauer, A. W. Senear, R. Warren and R. D. Palmiter, *Cell*, 1981, **27**, 223.
5. R. D. Palmiter, H. Y. Chen and R. L. Brinster, *Cell*, 1982, **29**, 701.
6. D. Oceandy, E. J. Cartwright, M. Emerson, S. Prehar, F. M. Baudoin, M. Zi, N. Alatwi, L. Venetucci, K. Schuh, J. C. Williams, A. L. Armesilla and L. Neyses, *Circulation*, 2007, **115**, 483.
7. K. Schuh, T. Quaschning, S. Knauer, K. Hu, S. Kocak, N. Roethlein and L. Neyses, *J. Biol. Chem.*, 2003, **278**, 41246.
8. J. F. Schiltz, A. Rustighi, M. A. Tessari, J. Liu, P. Braghetta, R. Sgarra, M. Stebel, G. M. Bressan, F. Altruda, V. Giancotti, K. Chada and G. Manfioletti, *Biochem. Biophys. Res. Commun.*, 2003, **309**, 718.
9. N. Iguchi, H. Tanaka, S. Yamada, H. Nishimura and Y. Nishimune, *Biol. Reprod.*, 2004, **70**, 1239.
10. A. A. Migchielsen, M. L. Breuer, M. S. Hershfield and D. Valerio, *Hum. Mol. Genet.*, 1996, **5**, 1523.
11. T. Choi, M. Huang, C. Gorman and R. Jaenisch, *Mol. Cell. Biol.*, 1991, **11**, 3070.
12. L. Kjer-Nielsen, K. Holmberg, J. D. Perera and J. McCluskey, *Transgenic Res.*, 1992, **1**, 182.
13. R. L. Brinster, H. Y. Chen, M. E. Trumbauer, M. K. Yagle and R. D. Palmiter, *Proc. Natl. Acad. Sci. USA*, 1985, **82**, 4438.
14. B. Hogan, R. Beddington, F. Costantini and E. Lacy, *Manipulating the Mouse Embryo: a Laboratory Manual*, Cold Spring Harbor Laboratory Press, Cold Spring Harbor, NY, 1994.
15. F. Costantini and E. Lacy, *Nature*, 1981, **294**, 92.
16. F. Buchholz, Y. Refaeli, A. Trumpp and J. M. Bishop, *EMBO Rep.*, 2000, **1**, 133.
17. E. A. Lindsay, A. Botta, V. Jurecic, S. Carattini-Rivera, Y. C. Cheah, H. M. Rosenblatt, A. Bradley and A. Baldini, *Nature*, 1999, **401**, 379.
18. J. A. Cearley and P. J. Detloff, *Transgenic Res.*, 2001, **10**, 479.
19. P. Dickinson, W. L. Kimber, F. M. Kilanowski, S. Webb, B. J. Stevenson, D. J. Porteous and J. R. Dorin, *Transgenic Res.*, 2000, **9**, 55.

20. Y. Geng, W. Whoriskey, M. Y. Park, R. T. Bronson, R. H. Medema, T. Li, R. A. Weinberg and P. Sicinski, *Cell*, 1999, **97**, 767.

21. N. Hosen, T. Shirakata, S. Nishida, M. Yanagihara, A. Tsuboi, M. Kawakami, Y. Oji, Y. Oka, M. Okabe, B. Tan, H. Sugiyama and I. L. Weissman, *Leukemia*, 2007, **21**, 1783.

22. M. Ikeya, M. Kawada, Y. Nakazawa, M. Sakuragi, N. Sasai, M. Ueno, H. Kiyonari, K. Nakao and Y. Sasai, *Int. J. Dev. Biol.*, 2005, **49**(7), 807.

23. F. L. Lin, K. Sperle and N. Sternberg, *Proc. Natl. Acad. Sci. USA*, 1985, **82**, 1391.

24. O. Smithies, R. G. Gregg, S. S. Boggs, M. A. Koralewski and R. S. Kucherlapati, *Nature*, 1985, **317**, 230.

25. E. Robertson, A. Bradley, M. Kuehn and M. Evans, *Nature*, 1986, **323**, 445.

26. R. L. Gardner and F. A. Brook, *Int. J. Dev. Biol.*, 1997, **41**, 235.

27. T. Doetschman, R. G. Gregg, N. Maeda, M. L. Hooper, D. W. Melton, S. Thompson S and O. Smithies, *Nature*, 1987, **330**, 576.

28. K. R. Thomas and M. R. Capecchi, *Cell*, 1987, **51**, 503.

29. H. te Riele, E. R. Maandag and A. Berns, *Proc. Natl. Acad. Sci. USA*, 1992, **89**, 5128.

30. B. P. Zambrowicz, A. Abuin, R. Ramirez-Solis, L. J. Richter, J. Piggott and H. BeltrandelRio, *et al., Proc. Natl. Acad. Sci. USA*, 2003, **100**, 14109.

31. H. Gu, J. D. Marth, P. C. Orban, H. Mossmann and K. Rajewsky, *Science*, 1994, **265**, 103.

32. S. M. Dymecki, *Proc. Natl. Acad. Sci. USA*, 1996, **93**, 6191.

33. F. Buchholz, L. Ringrose, P. O. Angrand, F. Rossi and A. F. Stewart, *Nucleic Acids Res.*, 1996, **24**, 4256.

34. X. Wang, B. Nadarajah, A. C. Robinson, B. W. McColl, J. W. Jin, F. Dajas-Bailador, R. P. Boot-Handford and C. Tournier, *Mol. Cell. Biol.*, 2007, **27**, 7935.

35. C. S. Branda and S. M. Dymecki, *Dev. Cell*, 2004, **6**, 7.

36. R. Kuhn, F. Schwenk, M. Aguet and K. Rajewsky, *Science*, 1995, **269**, 1427.

37. R. Feil, J. Brocard, B. Mascrez, M. LeMeur, D. Metzger and P. Chambon, *Proc. Natl. Acad. Sci. USA*, 1996, **93**, 10887.

38. D. S. Sohal, M. Nghiem, M. A. Crackower, S. A. Witt, T. R. Kimball, K. M. Tymitz, J. M. Penninger and J. D. Molkentin, *Circ. Res.*, 2001, **89**, 20.

39. I. Imayoshi, T. Ohtsuka, D. Metzger, P. Chambon and R. Kageyama, *Genesis*, 2006, **44**, 233.

40. J. D. Horton, N. A. Shah, J. A. Warrington, N. N. Anderson, S. W. Park, M. S. Brown and J. L. Goldstein, *Proc. Natl. Acad. Sci. USA*, 2003, **100**, 12027.

41. X. Wang, A. J. Merritt, J. Seyfried, C. Guo, E. S. Papadakis, K. G. Finegan, M. Kayahara, J. Dixon, R. P. Boot-Handford, E. J. Cartwright, U. Mayer and C. Tournier, *Mol. Cell. Biol.*, 2005, **25**, 336.

42. D. C. Rogers, E. M. Fisher, S. D. Brown, J. Peters, A. J. Hunter and J. E. Martin, *Mamm. Genome*, 1997, **8**, 711.

43. X. L. Tian, S. L. Yong, X. Wan, L. Wu, M. K. Chung, P. J. Tchou, D. S. Rosenbaum, D. R. Van Wagoner, G. E. Kirsch and Q. Wang, *Cardiovasc. Res.*, 2004, **61**, 256.

44. K. Schuh, E. J. Cartwright, E. Jankevics, K. Bundschu, J. Liebermann, J. C. Williams, A. L. Armesilla, M. Emerson, D. Oceandy, K. P. Knobeloch and L. Neyses, *J. Biol. Chem.*, 2004, **279**, 28220.

45. R. H. Waterston, K. Lindblad-Toh, E. Birney, J. Rogers, J. F. Abril and P. Agarwal, *et al.*, *Nature*, 2002, **420**, 520.

46. J. A. Richardson and D. K. Burns, *Ilar J.*, 2002, **43**, 89.

47. W.H. Colledge, B. S. Abella, K. M. Southern, R. Ratcliff, S. H. Cheng, L. J. MacVinish, J. R. Anderson, A. W. Cuthbert and M. J. Evans, *Nat. Genet.*, 1995, **10**, 445.

48. S. J. Engle, D. E. Womer, P. M. Davies, G. Boivin, A. Sahota, H. A. Simmonds, P. J. Stambrook and J. A. Tischfield, *Hum. Mol. Genet.*, 1996, **5**, 1607.

49. H. A. Wallace, F. Marques-Kranc, M. Richardson, F. Luna-Crespo, J. A. Sharpe, J. Hughes, W. G. Wood, D. R. Higgs and A. J. Smith, *Cell*, 2007, **128**, 197.

50. C. P. Austin, J. F. Battey, A. Bradley, M. Bucan, M. Capecchi and F. S. Collins, *et al.*, *Nat. Genet.*, 2004, **36**, 921.

CHAPTER 15

Protein Engineering

JOHN ADAIR[a] AND DUNCAN MCGREGOR[b]

[a] Ithaka Life Sciences Ltd, NETPark Incubator, Thomas Wright Way, Sedgefield, Co. Durham TS21 3FD, UK; [b] Cyclogenix Ltd, Crombie Lodge, Aberdeen Science and Technology Park, Balgownie Drive, Aberdeen AB22 8GU, UK

15.1 INTRODUCTION

Protein engineering is the process of constructing novel protein molecules, by design from first principles or by altering an existing structure. There are two main reasons for wishing to do this. First, there is the desire to understand for its own sake how proteins are assembled and what elements of the primary sequence contribute to folding, stability and function. These features can be probed by altering one or more specific amino acids in a directed manner within a protein and observing the outcome after production of the altered version. Often, related proteins with similar but not identical sequences exist in Nature that have slightly different properties and these differing sequences can be used as guides for the alterations.

A second reason for wishing to change a protein is that the protein may be suitable, in principle, for a particular technology purpose but the version found in Nature does not have the optimal properties required for the task. For example, an enzyme may be considered as part of an industrial process but a feature of the protein, such as the temperature stability or pH optimum for the catalytic activity or the need for a co-factor may not

Molecular Biology and Biotechnology, 5th Edition
Edited by John M Walker and Ralph Rapley
© Royal Society of Chemistry 2009
Published by the Royal Society of Chemistry, www.rsc.org

be compatible with the process. Amino acid changes can be made that can tailor the enzyme so that it functions better in the new environment. There are many other examples of how proteins can be altered to make them better suited to commercial and technological activities and some of these are noted in Section 15.3.

To engineer a protein implies an understanding of protein structural principles, knowledge of proteins as a material and an appreciation of the limits of the material so that rational design or alteration, of the properties can be achieved. In addition, the engineer should have to hand the tools to produce and analyse the desired protein. The tools and an increasing understanding of the underlying principles have been developing in a parallel but intertwined manner over the past two decades (see Refs 1–3 for reviews of early developments in the field). In this chapter we will provide a background to protein engineering and summarise some recent developments.

15.1.1 Protein Structures

Our current knowledge of how proteins look, how they behave and the principles that determine how a primary structure folds derives from two sources. First, studies on the structures of proteins using physical techniques, along with biochemical studies on the properties and physical interactions of the amino acids within proteins and with their environment, have shown that the amino acids of proteins adopt particular secondary structures such as the α-helix or β-strand and these secondary structures in turn are folded into tertiary structure motifs. Small proteins may comprise only one such tertiary structure motif; however, larger proteins are often comprised of a number of motifs, which may also be termed domains, that are themselves folded into a particular arrangement with specific interactions between the domains. In some cases separate proteins are organised into larger complexes by inter-domain interactions to form quaternary structures. Within the tertiary structures, particular amino acid groupings combine to provide the function of the protein. For example, certain amino acid side-chains located at different points in the primary amino acid sequence may be brought into close association by the tertiary structure to generate a catalytic function or a specific ligand-binding surface.

Second, and drawing on the first, theorising and calculation have provided insights into the ways proteins may assemble and function. For example, early observations on the likelihood of particular amino acids participating in different secondary structure motifs provided the initial impetus for protein structure prediction algorithms (*e.g.* Chou and

Fasman[4]). Many basic questions about the properties of proteins remain unresolved. For example the mechanics of protein folding remains a subject of debate.[5] This may not be unexpected, given that the number of sequences that have currently been obtained and structures that have been determined, are still a small fraction of the available repertoire occurring in Nature. Fortunately, Nature, via selection, repeats successful structural motifs and this redundancy allows an insight into the way in which different amino acid sequences can adopt similar structures. As further sequences become available and structures determined and as methods become more efficient at placing new sequences within existing sequence and structure families,[6] then the 'who, what and when' of protein structures may become more accessible. The 'why' and the 'how' still require a lot more effort. Protein engineering seeks to accelerate our understanding of the 'why' and the 'how'.

In addition, and adding a further layer of complexity, correct protein folding *in vivo* appears to be assisted by accessory protein complexes. The role of these folding chaperones and accessory proteins is still being elucidated,[7] but it appears that *in vivo* mechanisms have developed that allow incorrectly folded proteins to be rescued or slow folding domains to be assisted in their folding.

In the next section, the basic tools that are required for protein engineering are described, followed by a series of examples of successful protein engineering.

15.2 TOOLS OF THE TRADE

15.2.1 Sequence Identification

Sequence identification, by protein or gene sequencing, is now a relatively straightforward process. Large databases now exist containing many thousands of sequences, *e.g.* the Protein Data Bank.[8] In addition, genome-sequencing projects are providing new DNA sequences and open reading frames at an ever-increasing rate. The functions of many of these sequences are known, from biochemical or genetic data or by sequence homology to other known sequences. This gives the protein engineer an approach to rational design.

15.2.2 Structure Determination and Modelling

High-resolution structural information, determined by X-ray crystallography or nuclear magnetic resonance (NMR) techniques, is at the core of understanding of protein biochemistry. There are a large number of proteins for which high-resolution structural information is available, but this number remains well below the number of sequences identified by protein or gene sequencing. It has been suggested that a 'relatively' small number (around 1000–2000) of unique fold topologies may exist in nature.[6,9–11] Proteins with little primary sequence homology have been observed to adopt similar folds and can be grouped into protein superfamilies. These groupings may reflect ancient evolutionary relationships (but care needs to be taken in inferring functional activities to regions that may be conserved for purely physical reasons).

Structural genomics approaches and international cooperation to determine structures of representative proteins may allow better linkage of sequences to structures.[11,12] In the meantime, while this comprehensive structural database is being assembled, attempts are being made to predict protein structures by a variety of means.

Prediction of protein structures from primary sequence has a long history.[4] Significant progress has been made in predicting structures by modelling new sequences on to the known structures of homologous sequences or sub-sequences.[6,13] *Ab initio* calculation for structure prediction is still the subject of investigations.[14–16] A question that recurs is how close does a model approach reality and what confidence can be placed in the detail, before protein engineering experiments can begin. As will be seen below, some of the manipulation methods seek to circumvent this problem.

15.2.3 Sequence Modification

Modification of an existing protein by alteration of the gene sequence, from whole domains down to single amino acids, is now a routine process and this area is now the least challenging part of the engineering process. Procedures for preparing novel DNA sequences have been available since the early days of recombinant DNA technology. Large sequence segments were initially generated by synthesis of the desired double-stranded (ds) DNA sequence from short oligonucleotide segments that are annealed and ligated.[17,18] Currently, polymerase chain reaction (PCR)[19] amplification methods for gene synthesis are the most often used, to fill in and amplify partially overlapping oligonucleotides (see Figure 15.1, summarised in Ref. 10).

Individual nucleotide changes or alterations of short segments of sequence in a pre-existing sequence can be performed by oligonucleotide-directed site-specific mutagenesis or the use of the now ubiquitous PCR (see below).

GENE ASSEMBLY BY PCR

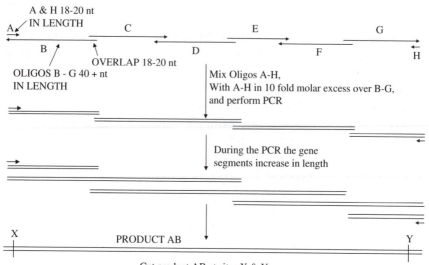

Cut product AB at sites X & Y,
Isolate sequence and ligate to vector

Figure 15.1 Gene assembly PCR. The required amino acid sequence is determined and a suitable nucleotide coding sequence is generated. The sequence may incorporate codon bias to assist with expression in a host cell and will usually also take advantage of the redundancy of the genetic code to include several useful restriction sites for later manipulation of the sequence. The sequence is then marked off into segments as shown (A–H), in which the sequences used alternate from the top (usually the sense strand) and bottom (antisense) strand. Oligonucleotide A has the sequence of the 5′ end of the top strand of the planned nucleotide sequence and will anneal to the 3′ region of oligonucleotide B. Oligonucleotide H has the sequence of the 5′ end of the bottom strand planned nucleotide sequence and will anneal to the 3′ region of oligonucleotide G. Oligonucleotides A and H are usually short oligonucleotides and will be present in the PCR in molar excess to drive the amplification of the full-length sequence once this has assembled in the early stages of the PCR. Oligonucleotides B–G cover the whole sequence to be assembled and can be of any length desired; the shorter the length the more that are required. With ever increasing fidelity of long oligonucleotide synthesis, lengths in excess of 100 nucleotides can be used. The 5′ region of oligonucleotide B can anneal to the 5′ region of oligonucleotide C and the 3′ region of oligonucleotide C can anneal to the 3′ region of oligonucleotide D, and so on, until the terminal oligonucleotide is reached. Oligonucleotides B–G are mixed with a molar excess of A and H and PCR is performed. In the early rounds of the reaction individual pairs of sequence anneal, *e.g.* C to D, and are extended. Similarly, A anneals to B and converts it to a dsDNA sequence. In the next and subsequent rounds, the CD product can now anneal to the AB product to give product covering the AD region. The CD product can also anneal to the EF product and so on. Similar annealing and extension events occur during subsequent rounds until a full-length sequence is assembled. The presence of excess A and H then ensures that this product is rapidly amplified. In the nucleotide sequence design, the terminal oligonucleotides (B and G) will encode restriction sites (X and) suitable for insertion into a cloning vector. Adapted from Ref. 20 with permission of the publisher.

These methods are restricted to gene-coded amino acids. Sequences may also be generated by chemical protein synthesis, which also allows the incorporation of non-coded amino acids into the protein sequence *in vitro*.[21,22]

15.2.3.1 *Defined Sequence Alterations*

In many situations, specific alteration of one or a small number amino acids is required. This can conveniently be done by site-specific mutagenesis procedures. There are two basic site-specific mutagenesis procedures. Both involve annealing of one or more oligonucleotides to a region of (at least temporarily) single-stranded DNA (ssDNA) followed by *in vitro* DNA polymerase-directed extension of the oligonucleotide(s).

Non-PCR Methods. In the older, non-PCR, procedures the DNA sequence to be modified is linked to a replication origin (*e.g.* in a plasmid, bacteriophage or 'phagemid'), allowing *in vivo* amplification of the modified and parental genotypes. A synthetic oligonucleotide harbouring the required mutation is annealed either to circular ssDNA template, for example the genome of an ssDNA phage such as M13 or ΦX174 or the ssDNA form of a phagemid (Figure 15.2a) or to a partially single-stranded dsDNA template (Figure 15.2b). Partial ssDNA templates can be generated by a variety of enzymic methods.

The annealed oligonucleotide acts as a primer for *in vitro* DNA synthesis using DNA polymerase, usually T4 or T7 DNA polymerase, and in the presence of DNA ligase, closed circular double-stranded DNA (dsDNA) molecules are generated (Figure 15.2c). The dsDNA is introduced into a suitable host cell where replication and hence segregation of parental and mutant daughter strand occur. The desired modification is then identified by one of a number of screening or selection procedures. The product Betaseron[23] (see Section 15.4.1.1) was generated in this manner.

PCR-based Methods. While the non-PCR methods are well known and have been optimised over a period of many years and commercial kits are available for the various procedures, PCR-based methods have now become more widespread.

In the PCR-based approaches, the desired modification is generated and amplified *in vitro* by annealing to denatured target DNA a synthetic oligonucleotide harbouring the required mutation along with an oligonucleotide that can act as a primer for replication of the complementary DNA strand (Figure 15.3).

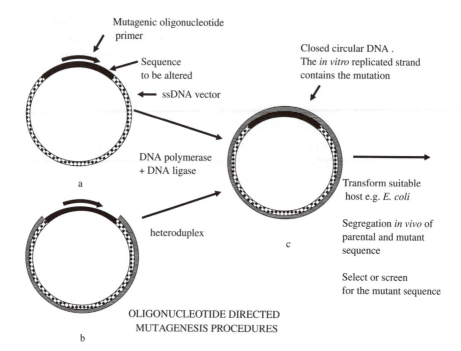

Mutagenic oligonucleotide primer

Sequence to be altered

ssDNA vector

Closed circular DNA .
The *in vitro* replicated strand
contains the mutation

DNA polymerase
+ DNA ligase

a

heteroduplex

c

Transform suitable
host e.g. *E. coli*

Segregation *in vivo* of
parental and mutant
sequence

Select or screen
for the mutant sequence

OLIGONUCLEOTIDE DIRECTED
MUTAGENESIS PROCEDURES

b

Figure 15.2 Strategy for non-PCR mutagenesis of ssDNA. (a) A single-stranded
DNA (ssDNA) template containing the sequence to be mutated is
obtained. This is usually achieved by cloning the required sequence into a
phage or phagemid vector and generating the ssDNA form. For a single
point mutation an oligonucleotide of between 15–20 nucleotides with the
proposed mutant sequence located centrally in the sequence is synthesised and is mixed in molar excess with, and annealed to, the ssDNA
template. For more complex mutations it is generally useful to have 15–
18 matched nucleotides at either side of the sequence mismatch. A DNA
polymerase is added along with DNA ligase and the remaining ssDNA
regions are converted to dsDNA and closed by the ligase (c). T7 DNA
polymerase is often used which has efficient 3′–5′ proof-reading activity
and good processivity but which lacks the detrimental 5′–3′ exonuclease
activity. (b) As an alternative to the procedure in (a), the vector containing the parental is rendered partially single stranded across the region
to be mutated. A very simple way to achieve this is by cloning the
required sequence into a phage or phagemid vector and generating the
ssDNA form. Then the dsDNA vector without the insert is annealed to
the ssDNA form generating a heteroduplex in which the cloned sequence
is exposed as ssDNA. The mutation is then generated using a synthetic
oligonucleotide, DNA polymerase and DNA ligase as in (a). (c) The
closed circular DNA that is generated has the mutation coded in the
in vitro generated strand and the parental sequence in the vector strand.
At the site of the mutation there is a DNA mismatch. The mutant and
parental sequences are separated by transformation into a suitable host
and allowing replication to occur. The mutant sequence can then be
identified by one of a number of screening or selection procedures
(reviewed in Ref. 20). In either procedure multiple mutations can be
incorporated within one oligonucleotide or by annealing several oligonucleotides to the ssDNA region. Adapted from Ref. 20 with permission
of the publisher.

There are a large variety of variants to the generalised method outlined in Figure 15.3, some of which involve including restriction sites into the mutagenic oligonucleotides. This can reduce the number of reaction involved. (For example, if in Figure 15.3 primers A and D have useful restriction sites in the sequences and in D the site is 3′ to the mutagenic sequence, the first reaction products can then be directly

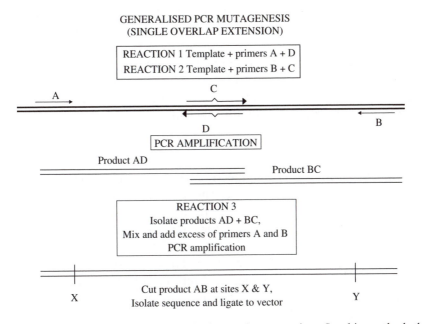

GENERALISED PCR MUTAGENESIS
(SINGLE OVERLAP EXTENSION)

REACTION 1 Template + primers A + D
REACTION 2 Template + primers B + C

C

A

D

B

PCR AMPLIFICATION

Product AD

Product BC

REACTION 3
Isolate products AD + BC,
Mix and add excess of primers A and B
PCR amplification

X

Cut product AB at sites X & Y,
Isolate sequence and ligate to vector

Y

Figure 15.3 PCR mutagenesis by single overlap extension. In this method the mutation to be introduced is coded into two oligonucleotides shown as C and D. These are designed to be complementary to each of the DNA strands at the site to be mutated and are capable of stably annealing. For a single point mutation the oligonucleotides are likely to be around 15–20 nucleotides. This usually allows the selection of an incubation temperature for the PCR that maintains the initial stability of the annealed oligonucleotide. Oligonucleotides A and B are often complementary to sequences within the vector and can be used for DNA sequence confirmation of the mutation when it is made. In a first round of parallel reactions, oligonucleotides A and D are mixed and annealed in a molar excess over the dsDNA template and similarly for B and C. PCR is performed and new dsDNA fragments AD and BC are produced that each have the mutation. After purification of the fragments an aliquot of each is mixed with molar excess of oligonucleotides A and B and PCR is gain performed. The result is the fragment AB. This fragment can then be cloned into a suitable vector using sites within the sequence or quite frequently, sites introduced to the sequence in the oligonucleotides A and B. This method has been termed single overlap extension (SOE) PCR.[137] Adapted from Ref. 20 with permission of the publisher.

cloned). The amplified DNA fragment is then ligated into a bacterial replication origin in a plasmid or 'phage and cloned by introduction into bacteria.

A variant of the procedure allows the linking together of separate sequences, which may code for domains from different proteins or allow the reorganisation of domains within a protein. In this procedure, outlined in Figure 15.4, the primers C and D are hybrids containing sequences that can anneal to both of the domains in question. The first reactions then generate dsDNA fragments that now overlap in sequence and the desired product can then be generated in a third reaction using the primers A and B.

In all of these cases, the design of the oligonucleotide sequences and the reaction conditions need to be considered carefully, along with the correct choice of thermostable DNA polymerase. However, these procedures are very rapid and highly efficient.

15.2.3.2   *Molecular Evolution*

The difficulty with protein engineering has been, and will continue to be, knowing what to alter. In many circumstances, directed modification of a sequence is not a suitable procedure for obtaining a desired outcome, because it is often not clear where the true target amino acid(s) reside(s) and to what to alter it/them. A number of strategies have been developed to produce and test members of large libraries or repertoires of variants of a particular sequence. Methods of this kind are rapidly becoming the workhorse for the protein engineer and are termed directed evolution.[24,25]

The approach relies on three main features: first, that a method exists to generate the large number of variants; second, that the nucleic acid that codes for the protein sequence of interest remains physically associated with the protein during the selection process; third, a screening or selection strategy is available that will enrich from among the collection of variants those novel protein sequences that have the phenotype of interest.

Generation of Variants. Molecular evolution methods require that a number of variants of a parental sequence be generated. The collection of variants is called a library. There are a variety of ways of generating this library of variants. For example, if selection among all possible amino acids at one location is required then the codon at that point can be replaced by the triplet NNN, where N means each of the four bases. This can be achieved either by direct replacement of a

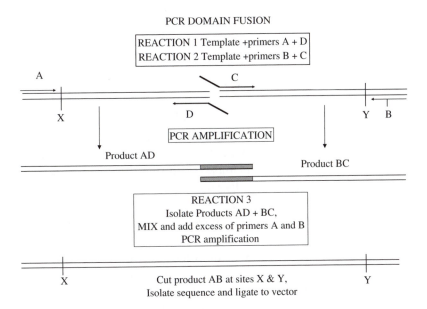

PCR DOMAIN FUSION

| REACTION 1 Template +primers A + D |
| REACTION 2 Template +primers B + C |

A   C

X   D   Y   B

PCR AMPLIFICATION

Product AD

Product BC

| REACTION 3 |
| Isolate Products AD + BC, |
| MIX and add excess of primers A and B |
| PCR amplification |

X   Y

Cut product AB at sites X & Y,
Isolate sequence and ligate to vector

Figure 15.4 PCR domain fusion, The domain fusion PCR is similar to the operation described in Figure 15.3. In this case the design of the central primers is more complex. Oligonucleotide A has the sequence of the sense strand at the N-terminal end of the first domain and so can anneal to the antisense strand of the coding sequence of the first domain. Oligonucleotide B has the sequence of antisense strand at the C terminal end of the second domain. Oligonucleotides A and B are usually 18–20 nucleotides in length and these lengths can be adjusted to ensure that the oligonucleotides remain annealed at the temperature of the PCR extension reaction. Oligonucleotide C is a hybrid of two sequences. At the 5′ end the first 20 (approximately) nucleotides are the same as the coding sequence at the C-terminus of the first domain. This part of the oligonucleotide sequence ends at the point where the fusion will occur. The second portion of the oligonucleotide sequence (also often around 20 nucleotides) has the sequence of the sense strand at the N-terminal end of the second domain. Therefore, oligonucleotide C can anneal to the antisense strand of the coding sequence of the second domain, leaving a non-annealed 'tail'. Oligonucleotide D is a similar hybrid sequence but can anneal to the sense strand of the first domain, also leaving a non-annealed 'tail'. Two PCR reactions are done. In the first, oligonucleotides A and D are in molar excess over the coding sequence for the first domain. In the second reaction, oligonucleotides B and C are in molar excess over the coding sequence for the second domain. At the end of the reactions, products AD and BC are formed. Product AD codes for the first domain and also codes for a number of amino acids from the N-terminal region of the second domain, with the point of fusion exactly as dictated by the sequence in oligonucleotide D. In product BC, the DNA sequence codes for the C-terminal amino acids of the first domain and the amino acids of the second domain. Importantly, the products AD and BC now have a significant nucleotide sequence homology. Products AD and BC are purified and an aliquot of each is mixed with molar excess of oligonucleotides A and B in a third PCR reaction. The resultant product AB now has the coding sequence of both domains in-frame and joined exactly at the desired point. This fragment can then be cloned into a suitable vector using sites near the ends of the sequence or, frequently, using sites introduced to the sequence in the oligonucleotides A and B. Adapted from Ref. 20 with permission of the publisher.

double strand section with NNN in both strands or by PCR using a priming oligonucleotide with NNN at the desired position. This strategy can be extended for a number of amino acid locations. More typically, rather than using NNN codons, libraries are encoded by NNK (K = G or T) to reduce the number of stop codons (TGA, TAA and TAG) appearing in libraries. A more recent approach, using trinucleotide codons in place of single base additions during oligonucleotide synthesis, results in an even distribution of all amino acid codons, while eliminating all stop codons.[26,27]

Larger scale variation can be achieved in certain circumstances by a technique known generically as DNA shuffling.[28,29] In the basic procedure, the nucleotide sequences of a number of closely related proteins are used. Breaks are introduced randomly in the sequences and the pooled DNA is then denatured and annealed. Gaps are filled by PCR and then the resultant new sequences are cloned. The procedure provides a means for crossover between multiple original sequences and generates new sequences in which the sequence differences are recombined.

Since the original procedure was described, a number of other procedures for generating the libraries have been published, with a variety of acronyms, each seeking to improve the procedure for particular circumstances: StEP,[30] ITCHY,[31] SCRATCHY,[32] RACHITT[33] and SHIPREC.[34]

Even with such procedures, only small segments of a protein can be altered at one time because of the number of possible combinations that are involved. Strategies the require the presentation and amplification in a cellular system are limited by the ability of the host to take up individual members and libraries of 10^9–10^{10} members are considered large numbers here, *in vitro* libraries can exceed these numbers so that libraries of 10^{12}–10^{14} can be formed. However, the possible combinations of all residues in even small protein domains far exceed this.

Linking Coding Sequence to Product. The physical linkage of populations of peptides or proteins to their encoding nucleic acids to create large, diverse display libraries has provided a rich source of ligands to a wide range of target molecules. Enrichment of ligands from these libraries is achieved by the co-selection of target-binding peptides along with their associated encoding nucleic acids, which allows the subsequent identification of the selected peptide sequences. This approach has been most widely exemplified using the phage display system. In this system, a range of options exists that provide for retaining a link between the coding sequence and the expressed product.[35,36]

In the original procedure, the coding sequence of interest is inserted into the gene sequence for one of the coat proteins of a bacteriophage ('phage), (usually the filamentous 'phages, as used for the site-directed mutagenesis methods noted above). When expressed during the course of infection the fusion is assembled into the virus coat. The hybrid 'phage can then be selected for by binding to specific ligand, for example an antibody. The hybrid protein remains linked to the coding sequence inside the 'phage. The 'phage genome can then be rescued and amplified and further rounds of binding can be performed. This procedure provides the basis for a means of selection among variants.

In the original peptide system, libraries of peptide were screened for binding to antibody to identify binding epitopes. However, the same principle applies to protein: protein interactions.

As an alternative to a 'phage, a 'phagemid' can be used. Here, the 'phage coat protein that is used to generate the presenting fusion has been inserted into a plasmid. The host bacterium, containing the phagemid vector, must then be co-infected with autonomously replicating bacteriophage, termed helper 'phage, to provide the full complement of proteins necessary to produce mature phage particles. Today this technique provides the basis for most library screening methods.

A variant of this procedure involves generating a fusion of the protein or peptide of interest with a DNA binding protein and arranging for the recognition motif to be included in the 'phage genome along with the coding sequence for the fusion. When the fusion is expressed it binds to its cognate DNA sequence on the 'phage genome and becomes packaged. The protein of interest protrudes through the 'phage coat and is available fro ligand binding and selection. The subsequent steps of selection and amplification are the same as for phage display.[37]

In addition to presentation on the surface of a 'phage, proteins and peptides can also be presented on the surface of bacteria, baculovirus or eukaryotic cells.[38–43]

Another *in vivo* approach is exemplified by the 'peptides on plasmids' approach, where peptides are fused to the *lacI* DNA-binding protein and expressed in the cytoplasm of bacteria, resulting in a protein–DNA complex library that can be selected against targets after bacterial lysis to release the complexes.[44]

Alternatively, *in vitro* procedures have been devised. The challenge here is to be able to maintain the specific coding sequence–product link and to separate out the individual products. The *in vivo* procedures described above require the insertion of the library DNA into bacterial cells and the efficiency of bacterial transformation restricts library sizes to the 10^9–10^{10} range. Similarly, the display of peptides on bacterial or yeast surfaces or the display of peptides directly on their encoding plasmids all require a transformation step which imposes a limit on the library sizes that can be constructed.

It is generally accepted that there is a correlation between library size and the affinity of ligands that can be isolated from them.[45] This has prompted the development of display technologies in which the size-limiting transformation step is unnecessary, allowing ever-larger display libraries to be constructed. These new technologies enable higher affinity ligands to be obtained through the sampling of an increased structural repertoire, which is made possible through the generation of libraries which are up to four orders of magnitude larger than those that can be constructed for phage display. These can be separated into RNA- and DNA-based procedures.

RNA-based procedures were developed first. The best known of these are known as 'ribosome display' and 'mRNA display'.[45-51] In both of these systems, protein-coding mRNA is transcribed and translated *in vitro*. In ribosome display, the translated protein folds while emerging from the ribosome and remains linked to the mRNA during selection for function, after which the coding mRNA sequence is reverse transcribed to DNA, amplified and transcribed to make more mRNA for further rounds of selection. This is achieved through linking the library protein to a spacer protein (a number of different spacer proteins can be used) to allow space for the library protein to fold correctly unhindered by proximity to the ribosome and by deleting the stop codon from the coding sequence. This these latter change causes ribosome stalling and hence linkage between the protein and its mRNA molecule. Both prokaryotic and eukaryotic cell-free translation systems can be used, although the prokaryotic methods are more prevalent. The key is to ensure that the folded protein does not disengage from the ribosome while the selection for function is occurring. This requires that selections are performed at low temperature and in the presence of magnesium ions.

In mRNA display, mRNA molecules encoding protein or peptide libraries are tagged at the 3'-end with the antibiotic puromycin.[50,52] During translation, when the ribosome reaches the puromycin molecule, the antibiotic is covalently transferred to the elongating peptide chain (while still attached to the mRNA), thereby linking the mRNA to the encoded protein. Although this is more robust than ribosome display, there is no method for ensuring attachment between the correct elongating peptide -chain and its encoding mRNA molecule. As both methods are mRNA based, there are stability issues that must be considered when using these procedures.

DNA-based methods can be divided into non-covalent (CIS display[53]), covalent[54,55] and emulsion-based systems.[56]

CIS display exploits the high-fidelity *cis*-activity that is exhibited by a group of bacterial plasmid DNA-replication initiation proteins typified by RepA of the R1 plasmid. In this context, *cis*-activity refers to the property of the RepA family of proteins to bind exclusively to the template DNA from which they have been expressed. By genetically fusing peptide libraries to the N-terminus of the RepA protein, Odegrip *et al.*[53] demonstrated that it is possible to achieve a direct linkage of peptides to the DNA molecules that encode them *in vitro*. The rationale of the system is that *in vivo*, R1 plasmid replication is initiated through the binding of RepA to the plasmid origin of replication (*ori*). *Ori* is separated from the RepA-coding sequence by a DNA element termed *CIS*. The consensus model for *cis*-activity is that the *CIS* element, which contains a rho-dependent transcriptional terminator, causes the host RNA polymerase to stall. This delay allows a nascent RepA polypeptide emerging from a translating ribosome to bind transiently to *CIS*, which in turn directs the protein to bind to the adjacent *ori* site, thus linking genotype to phenotype.

DNA-based covalent display approaches are essentially similar to *CIS* display, but require a covalent interaction between protein and DNA to retain the linkage of genotype to phenotype, through the *cis* action of the cross-linking protein. Two requirements are needed for successful use of this technique. First, proteins are required which interact *in vitro* with the DNA sequence which encodes them (cis action), and second, the said proteins must establish a covalent linkage to their own DNA template. This method suffers from the fact that the DNA is chemically modified, which can prevent the recovery and identification of the binding peptide of interest. Despite this, Reiersen *et al.*[55] demonstrated the isolation of scFv antibody fragments from an immunised library using a covalent display approach based on the *cis*-acting P2 A DNA binding protein.

Tawfik and Griffiths[56] demonstrated that directed evolution procedures could be undertaken in cell-free emulsions in which transcription and translation occur in water microdroplets in an oil emulsion. This allows the physical co-localisation of coding sequence and product. In the original procedure, optimisation of an enzyme was performed and selection could be detected by the presence of product. More recently, the procedure has been generalised to allow for selection among ligand binding variants[57] [see Rothe *et al.*[58] for a recent review of these *in vitro* compartmentalisation (IVC) methods].

A further *in vitro* procedure combines emulsion compartmentalisation with a *cis*-display method.[59] Here, in an elegant experiment, a biotinylated DNA fragment library composed of random sequence fused to DNA encoding streptavidin is compartmentalised in emulsions (one DNA molecule per compartment), then transcribed and translated. The resulting fusion protein can then bind to the biotinylated DNA to maintain the link between coding sequence and expressed product.

Identification of the Required Phenotype. Third, these methods require a screening or selection strategy that will enrich from among the library those novel protein sequences that have the phenotype of interest. It has been suggested that the First Law of Directed Evolution is 'you get what you screen (or select) for'.[60,61] Screens which involve binding to a ligand can easily be devised.[62] Procedures for screening or selection for enzyme function or protease resistance have been summarised.[63,64] Selection systems often involve complementation of an essential function within the host organism,[65] but more novel *in vivo* selection systems have also been proposed[66,67] where selection takes place within a whole organism.

15.2.4 Production

Once a novel sequence has been identified and designed, it must be expressed to determine that the function is as required. A range of expression systems are available, from microbial systems (*Escherichia coli* being the most obvious),[68] yeasts (*e.g. Saccharomyces cerevisiae* and *Pichia pastoris*), insect cells, filamentous fungi and mammalian cell cultures through to transgenic animals and plants. Cell-free systems for the expression of relatively small quantities of product for initial analysis are also being developed. These can now produce sufficient quantities of product for early stage analysis.[69]

Chemical *in vitro* synthesis of polypeptides and small proteins was noted earlier as a means of introducing non-natural amino acids into the sequence, but might also be considered as an alternative expression option, assuming that it can be done cost-effectively.[70,71]

The scale of production also varies. In the early stages of development of a novel protein, for example where a protein has been identified among a number from a display library, only small quantities (micrograms) of the protein may be needed to confirm a biological property. In the second stage, larger quantities (tens to hundreds of milligrams) of purer material are required, typically to obtain structural information and to perform more demanding *in vitro* and *in vivo* bioassays. In some situations, larger quantities (grams to the multi-kilogram or tonne scale) of material purified by rigorous (and usually costly) procedures may be required if the protein is required for commercial purposes (for example, for industrial enzymes or for therapeutic purposes).

Often a number of approaches may need to be tried to find a suitable expression host, as it cannot safely be assumed that an altered protein will express in the same manner as the parental sequence and in the case of novel sequences no precedent exists. These factors are becoming of increasing importance as more strategies rely on library display systems, where members of a library may be lost or under-represented because of poor expression or folding properties.

15.2.5 Analysis

Methods must exist for the analysis of the properties of the modified proteins. Where a function is being altered (introduced, modified or removed), a biological assay may already be available to measure the function and activity of the parental protein or can usually be devised to measure the ability of the newly produced protein (this assay may well be the same as the screening assay). The issue is often of ensuring that the assay can function with very small amounts of material, such as that obtained from library amplification procedures. Methods that include physical adsorption or binding may be better suited to this than those that rely on detection of a catalytic product. In the latter case, diffusible product must remain close to the point of production and so water–oil emulsion compartmentalisation or single well techniques are better suited here.

In many cases, acquiring and demonstrating the new function may be all that is required. However, in addition to functional assays, interpretation of the results may require a structural understanding of the alteration, particularly to gain further insights into protein folding. Assuming that a reasonable quantity of the novel protein can be prepared, gross properties can readily be detected by spectroscopic or other large-scale measurements (*e.g.* circular dichroism, turbidity, enthalpic, sedimentation or chromatographic properties). However, detailed structural information about the end product is still the rate-limiting step in rational protein engineering and design. Casimiro *et al.*[72] showed that PCR-based gene synthesis, expression in *E. coli* of milligram quantities of isotopically labelled protein and NMR spectroscopy can be achieved in a reasonably short time, with first NMR spectra available for a novel sequence as little as 2 months after the initiation of the gene synthesis steps. More recently, there has been interest in developing methods for rapid determination, on a small scale, of the structural information about coded sequences identified in genomic projects, which can be applied to engineered proteins.[73]

15.3 APPLICATIONS

Applications of protein engineering are everywhere in biology, as a means of producing novel molecules and as a means of understanding the basic properties of proteins. Therefore, to attempt a comprehensive survey is fruitless. However, it is useful to note some specific examples of the different types of molecules that can be achieved, from the simplest point mutations and domain rearrangements to more demanding multi-site alterations and *de novo* designs. The examples given in the following sections reflect our personal bias towards biopharmaceuticals, but hopefully will prove of general interest. The examples indicate that protein engineering procedures have been used to develop potentially beneficial medicines for over 20 years now. Walsh[74] has recently tabulated approved recombinant protein medicines, several of which are the result of protein engineering experiments.

15.3.1 Point Mutations

Individual point mutations in proteins can be readily achieved once the gene sequence is available, using the techniques noted earlier. Numerous examples of amino acid sequence variants are known. Many of the early proteins considered and approved as pharmaceuticals have now been modified in various ways and modified forms are now available as drug products. Some key examples are given below.

15.3.1.1 Betaseron/Betaferon (r-Interferon β-1b)
One of the earliest examples of pharmaceutical protein engineering was the production of interferon β-1b. This novel protein was generated by the substitution of a cysteine (Cys) for serine (Ser) at residue 17 of the 154 amino acid interferon β.[23] This substitution reduces the possibility of incorrect disulfide bridge formation during synthesis in *E. coli* and also removes a possible site for post-translational oxidation. The resultant protein is expressed in *E. coli* at a specific activity approaching that of native fibroblast-derived interferon β. The molecule has been licensed since 1993 for use for the reduction of frequency and degree of severity of relapses in ambulatory patients with relapsing remitting multiple sclerosis.

15.3.1.2 Novel Insulins
Insulin is a pancreatic hormone that acts to regulate blood glucose levels. Insulin is given to used in the treatment of diabetes to help

regulate blood glucose levels. Recombinant insulin (humulin) was the first biotechnology therapeutic product in 1982. Since the early approval of recombinant insulin, a number of altered forms have been developed that seek to improve the effectiveness of the delivered insulin. The first of these, Humalog (insulin lispro), was approved in 1996. Humalog is an engineered, fast-acting analogue of insulin. In Humalog, the two residues at the C-terminus of the B chain, Pro28 and Lys29, have been reversed in their order. The C-terminal alterations were designed based on structural and sequence homology to insulin-like growth factor 1 (IGF-1) and the sequence reversal reduces the dimerisation of the B subunit. This reduced self-association means that the monomer is more readily available to function after food uptake and therefore can be administered very shortly before meals.

Since 1996, a number of recombinant forms have been licensed.[74]

15.3.1.3 *Aranesp (Darbopoietin α) Modified Erythropoietin*
Erythropoietin (EPO) is a hormone that stimulates the formation of red blood cells.[75] It is approved for treatment of anaemia. Since its approval in 1989, EPO has become the largest selling biotechnology drug with multi-billion dollar sales. In 2001, a modified version, Aranesp, was approved. Aranesp is a hyper-glycosylated form of EPO in which the amino acid sequence has been altered at two locations so as to introduce two new N-linked carbohydrate attachment motifs. Aranesp therefore bears five N-linked carbohydrate structures compared with the three in the native form. This results in a longer *in vivo* half-life for the product and allows for less frequent dosing of the patient [76] Aranesp is fast on its way to being a blockbuster drug.

15.3.2 Domain Shuffling (Linking, Swapping and Deleting)

Most large proteins are composed of smaller, independently folding domains, generally corresponding to the protein folds mentioned earlier. These domains are often linked together by short peptide sequences and in many, but not all, cases the domains are identifiable as separate exons in the gene sequence. Using standard molecular biology techniques, it is possible to add, remove or swap domains from one protein to another to rebuild proteins.

15.3.2.1 *Linking Domains*
Domain Fusions for Cell Targeting. Among the very earliest examples of protein engineering, the binding site region of an antibody was

genetically linked to an enzyme. The IgG antibody is a Y- or T-shaped structure comprising two identical 50 kDa 'heavy' chains and two identical 25 kDa 'light' chains. The heavy chain comprises an N-terminal variable domain (VH) and three constant domains (CH1, CH2, CH3) with a short hinge region between CH1 and CH2. The light chain comprises an N-terminal variable domain, VL and a constant domain CL. The antigen-binding region, known as the Fab region, comprises the VH and CH1 domain of one heavy chain and one complete light chain. The C-terminal CH2 and CH3 domains of the two heavy chains associate together to form the Fc region. The Fc region is involved with interactions with various cells of the immune system and with complement and is also important in determining the serum half-life of antibodies, at least of the IgG type.[77]

The antibody–enzyme gene fusions were constructed by taking the DNA sequence that codes for the heavy chain component of Fab of the antibody and linking this either to staphylococcal nuclease[78] or to *E.coli* DNA polymerase coding sequences.[79] Introduction of these fusion genes in a cell that produces the antibody light chain and expression of the fusion gene reconstituted both antigen binding and enzymic activity.

These early examples provided the foundation for many other examples where a binding function [antibody, cytokine, growth factor or extracellular ligand binding domain (ECD) of a receptor] is linked to an effector function (toxin, enzyme, cytokine). Also, a non-antibody binding domain can be attached to the Fc region of an antibody to take advantage of the long serum half-life of antibodies to improve the pharmacokinetic properties of a designed molecule.

Some of these domain fusions are now licensed for human use (Table 15.1). For example, Enbrel (etanercept) comprises the extracellular domains of the receptor for the cytokine tumour necrosis factor α (TNFα) genetically fused to the Fc regions of IgG. Enbrel is used to block the activity of the TNFα.[80] Enbrel is currently licensed for the reduction in signs and symptoms of moderately to severely active rheumatoid arthritis in patients who have an inadequate response to one or more disease-modifying anti-rheumatic drugs. Other approved fusions with similar architecture include Amevive and Orencia (Table 15.1).

Ontak is a different form of genetic fusion in which the cytokine interleukin-2 (IL-2) is used to replace the receptor-binding domain of diphtheria toxin.[81] Ontak is used to target the diphtheria A and B domain to cells expressing the receptor for IL-2, which is overexpressed in leukaemia cells. Ontak was approved in 1999 to treat certain patients with advanced or recurrent cutaneous T-cell lymphoma (CTCL) when other treatments have not worked.

Table 15.1 Selected engineered protein products.

Product (international non-proprietary name)	Format	First approval date	Target	Application
Approved fusion proteins				
Enbrel (etanercept)	Tumour necrosis factor receptor (TNFR) p75–Ig Fc fusion (dimeric)	1998	TNFα	Rheumatoid arthritis
Ontak (denileukin diftitox)	Recombinant interleukin (rIL)-2–diphtheria toxin fusion protein	1999	IL-2 receptor	Cutaneous T-cell lymphoma
Amevive (alefacept)	CD2-binding region of LFA-3 fused to IgG1 Fc	2003	CD2	Treatment of adult patients with moderate to severe chronic plaque psoriasis who are candidates for systemic therapy or phototherapy
Orencia (abatacept)	CTLA4–Ig Fc fusion (dimeric)	2005	CD80, CD86	Rheumatoid arthritis

Fusions to Stabilise Dimeric Proteins. In other situations, it is useful to fuse together genetically domains which are normally associated by non-covalent interactions or disulfide bridges. This allows more convenient production and ensures that the domains remain in proximity even at the low concentrations that would be observed during *in vivo* dosing. Examples include joining the VH and VL domains of antibodies together with a linker peptide to form molecules called single chain Fvs (scFvs) and linking the domains of heterodimeric cytokines (*e.g.* IL-12 to produce Flex-12.[82] The scFv molecule is another veteran of protein engineering, having first been disclosed during the 1980s.[83,84]

15.3.2.2 Swapping Protein Domains
Another simple procedure used in protein engineering is whole domain swapping, in which analogous domains from different sources are swapped into a multi-domain protein to provide a novel functionality.

Chimeric Mouse–Human Antibodies. Here the antigen binding domains from a mouse monoclonal antibody are linked to the constant regions (which provide immune effector functions and dictate biological half-life) from a human antibody.[85] This switching can markedly reduce unwanted immunogenicity compared with the original mouse antibody.[86] Several chimeric antibody products have been licensed as pharmaceuticals (Table 15.2).

Polyketide Synthases (PKSs). Polyketides are a class of chemical compounds which include many pharmaceutical compounds, including antibiotics, antifungals and immunosuppressants, and which together account for billions of pounds of sales per annum. These chemicals are produced in various soil microorganisms and fungi. Their synthesis involves the regulated action of a sequence of enzymes, the polyketide synthases (PKSs). The PKSs occur either as small proteins that have a small number of distinct non-repeating catalytic domains (iterative PKSs) or as large multi-domain polypeptides. In the latter, while each domain has a specific enzyme function, the same function may be present a number of times (modular PKSs). In both cases, the product of one catalytic domain forms the substrate for the reaction of a neighbouring domain, the modular polyketides producing the more complex molecules. By selectively adding, deleting and rearranging the order of the catalytic domains to generate novel PKSs, a novel order of catalytic reactions is established and hence new products can be generated.[87–91]

Table 15.2 Approved recombinant antibodies

Product (international non-proprietary name)	Format	First approval date	Target	Application
Approved recombinant mouse–human chimeric antibodies				
ReoPro (abciximab)	Fab fragment	1994	gpIIb/IIIa	Adjunctive therapy of percutaneous coronary interventions (PCI) Adjunctive therapy in refractory unstable angina when PCI is planned within 24 h
Rituxan (also marketed as MabThera (rituximab)	IgG	1997	CD20 antigen found on the surface	Non-Hodgkin's lymphoma Rheumatoid arthritis
Remicade (infliximab)	IgG	1998	Tumour necrosis factor α (TNFα)	Crohn's disease Rheumatoid arthritis Ankylosing spondylitis Psoriatic arthritis Ulcerative colitis Plaque Psoriasis
Simulect (basiliximab)	IgG	1998	CD25	Prophylaxis of acute organ rejection in de novo allogeneic renal transplantation
Erbitux (cetuximab)	IgG	2003	EGFR	Colorectal cancer Squamous cell carcinoma of the head and neck
Approved recombinant humanised antibodies				
Zenapax (daclizumab)	IgG	1997	CD25	Prophylaxis of acute organ rejection in patients receiving renal transplants
Synagis (palivizumab)	IgG	1998	Respiratory syncytial virus (RSV) F protein	Prevention of serious lower respiratory tract disease caused by RSV in paediatric patients at high risk of RSV disease

Table 15.2 (*Continued*).

Product (international non-proprietary name)	Format	First approval date	Target	Application
Herceptin (trastuzumab)	IgG	1998	HER2 (human epi-dermal growth factor receptor 2) protein	Breast cancer
Mylotarg (gemtuzumab, ozogamicin)	IgG–calicheamicin drug conjugate	2000	CD33	Acute myeloid leukaemia (AML)
CAMPATH, MAB-CAMPATH (alemtuzumab)	IgG	2001	CD52	Chronic lymphocytic leukaemia
Xolair (olizumab)	IgG	2003	IgE	Asthma
RAPTIVA (efalizumab)	IgG	2003	CD11a	Plaque psoriasis
Avastin (bevacizumab)	IgG	2004	VEGF	Colorectal carcinoma Non-squamous, non-small-cell lung cancer
TYSABRI (natalizumab)	IgG	2004	α4 subunit of α4β1 (VLA-4) and α4β7 integrins	Multiple sclerosis
Actemra (tocilizumab)	IgG	2005	IL-6R	Castleman's disease
TheraCIM (nimotuzumab)	IgG	2004	EGFR	Head and neck cancer
LUCENTIS (ranibizumab)	Fab fragment	2006	VEGF	Neovascular (wet) age-related macular degeneration
Soliris (eculizumab)	IgG	2007	C5a	Paroxysmal nocturnal haemaglobinuria
CIMZIA (certolizumab, pegol)	Fab fragment–poly (ethylene glycol) (PEG) conjugate	2008	TNFα	Crohn's disease

15.3.2.3 Deleting Domains

In some situations, only a subset of the domains in a naturally occurring protein are required for the development of a functional therapeutic. Tissue plasminogen activator (tPA) is a serine protease secreted by endothelial cells. Following binding to fibrin, tPA activates plasminogen to plasmin which then initiates local thrombolysis. Reteplase is a variant form in which three of the five domains of tPA have been deleted. One of the domains that confers fibrin selectivity and the catalytic domain are retained.[92] Reteplase is licensed as Retavase for the treatment of acute myocardial infarction to improve blood flow in the heart.

15.3.3 Whole Protein Shuffling

Many proteins exist as multi-member families in which the homologues display slightly different biological activities determined by the sequence variation. In the early days of protein engineering, hybrid genes were generated by swapping sequence stretches using convenient common restriction sites[93] or by *in vivo* recombination between homologous genes in a more random manner.[94] These methods produced small numbers of novel genes that could be examined individually. More recently, protein homologues have been used to generate very large libraries of novel variants by sequence shuffling. New variants, such as thermostable enzyme mutants, can then be identified by selection or screening procedures, as noted earlier.[95]

15.3.4 Protein–Ligand Interactions

15.3.4.1 Enzyme Modifications

Many examples of protein engineering involve changes to enzymes to examine and modify enzyme–substrate interactions. These experiments can be traced back to the earliest attempts at protein engineering, using tyrosyl-tRNA synthetase.[96–98]

The types of changes include the following: enhancing catalytic activity; modification of substrate specificity, including the *de novo* generation of novel catalytic functions; alterations to pH profiles, so that an enzyme can function in non-physiological conditions; improving oxidation resistance by replacing oxidation sensitive amino acids such as cysteine (Cys), tryptophan (Trp) or methionine (Met) by sterically similar non-oxidizable amino acids such as serine (Ser), phenylalanine (Phe) or glutamate (Glu), respectively; improving stability to heavy metals by replacing Cys and Met residues and surface carboxyl groups; removing protease cleavage motifs; and removing sites at which catalytic product might otherwise bind to induce allosteric feedback inhibition. Leisola and Turunen[99] have provided a recent brief summary with pertinent references; see also Ref. 100.

15.3.4.2 Substitution of Binding Specificities

A number of examples are available where the specificity for a ligand has been transferred from one protein background to another. Some of the examples require small changes, some require large-scale transfer of many residues. For example, the specificity of the hormone prolactin was modified by substitution of eight amino acids at the receptor binding surface so that the modified hormone now binds to the receptor for growth hormone.[101]

Protein domains often have loop regions that span between other secondary structure motifs (α-helix, β-strand). In some cases, the functionality of the protein resides in these loops and they can be the target for relatively simple substitution experiments. For example, the specificity of basic fibroblast growth factor was altered to that of acidic fibroblast growth factor by the substitution of one particular loop region.[102]

On a more adventurous scale, antibody humanisation involves transfer of up to six loops that form the antigen binding surface, from the β-sheet frameworks of the antigen binding domains of a non-human antibody to that of a human antibody. This feat has taken some years to become routine, since the first description in 1986.[18] However, it has now progressed to the point where a significant number of humanised antibodies are marketed and profitable drugs (Table 15.2).

15.3.5 Towards *De Novo* Design

15.3.5.1 Development of Novel Binding Molecules

An intermediate step between experiments of the kind described above and 'true' *de novo* design is to take a particular protein scaffold and modify it so as to introduce new functions or retain function on a radically altered scaffold. This is often done by taking a small, compact domain of known structure and using phage display procedures to modify part of the surface of the protein and to screen or select for the introduction of a new function. Early examples included examination of the structural plasticity of a protein fold by redesigning the N- and C-termini within the sequence of the IL-4 cytokine, circularly permuting the sequence without affecting the overall fold.[103] In another early example, Martin *et al.*[104] described how a 'minibody' scaffold derived from an immunoglobulin binding domain and retaining only two loops and six β-strands was used to obtain a novel antagonist for IL-6.

Over the last several years, there has been a growing use of naturally occurring, small, stably folding domains as the basis for developing novel proteins with specific functions, usually as binding domains.[105–111] The desired specificity and affinity features can be introduced using the molecular evolution procedures described earlier, and significant progress has been made in preparing novel molecules. These newer molecules aim to provide high-affinity binding and to avoid some of the practical issues surrounding the antibody molecules (*e.g.* engineering and production and not least the crowded intellectual property arena). A non-exhaustive list of examples being commercialised is given in Table 15.3. A long-term aim of examining such molecules is to bring protein or peptide biopharmaceutical closer to the 'normal' pharmaceutical paradigm of an orally available, synthetic, low molecular weight compound. Beyond this there is the possibility of true *de novo* or *ab initio* design.

15.3.5.2 De Novo *Design*
Discussion of and attempts to design protein domains from first principles have been under way for some time.[112–114] *Ab initio* design brings different challenges. In principle, for any protein of *n* residues there are 2×10^n different possible sequences. As noted earlier, however, known structure space is populated by a relatively small number of stable fold types. The structures and sequences databases demonstrate that a range of sequences can be accommodated in similar protein folds and a range of sequences can fulfil similar functions. One approach to designing a functional protein is to decide what fold may be appropriate and then identify what sequence(s) would be needed to generate the desired fold and function.[115–118] The next decade should bring significant developments in this area.

15.4 CONCLUSIONS AND FUTURE DIRECTIONS

Protein engineering, as a methodology, is now a mature technology. Through the pragmatic approach allowed by molecular evolution methods, novel proteins are being produced for research and commercial applications at an ever-increasing rate. In the near future, there will be an increasing effort to exploit the technology in areas that have been less tractable to date: in understanding the structure and function of membrane proteins that are the focus for many pharmaceutical strategies,[119,120] and in the developing field of nanotechnology, including biosensors, bioelectronics and biocomputing.

Table 15.3 Selected antibody alternatives and non-antibody protein scaffolds under investigation as potential therapeutic entities.

Scaffold	Organisation	Application indication	Latest clinical stage	Reference
Nanobody: single-domain antibodies from the camelid family	Ablynx (Ghent, Belgium); http://www.ablynx.com	Therapy: Autoimmune diseases Thrombosis (arterial stenosis) Cancer Alzheimer's disease	ALX-0081 Phase I Target: Von Willebrand's factor	Hulstein et al., 2005[121]
Affibody: three-helix bundle from Z-domain of Protein A from *S. aureus*	Affibody (Bromma, Sweden); http://www.affibody.com	Therapy: Cancer Apheresis: Alzheimer's disease Chromatography: industrial-scale separation	Preclinical	Nord et al., 1997[122] Nilsson and Tolmachev, 2007[123] Tolmachev et al., 2007[124]
Maxibody: LDL receptor monomers	Avidia (Mountain View, CA, USA); http://www.avidia.com (now Amgen)	Therapy: Autoimmunity Inflammation Cancer	C326 Phase I Target: IL-6	Silverman et al., 2005[125]
Trans-body: human transferrin	BioRexis (King of Prussia, PA, USA); http://www.biorexis.com (now Pfizer)	Therapy: Cancer	Preclinical	–
Tetranectin: monomeric or trimeric human C-type lectin domain	Borean Pharma (Aarhus, Denmark); http://www.boreanpharma.com	Therapy: Autoimmune diseases HIV	Preclinical	Gunnarsson et al., 2006[126]

Peptides: EPO agonist	Affymax (Palo Alto, CA, USA); http://www.affymax.com	Therapy: Anaemia	Hematide Phase III Target: EPO receptor	Woodburn et al., 2007[127]
AdNectin: human tenth fibronectin type III domain	Adnexus (Waltham, MA, USA; http://www.adnexustx.com/ (now BristolMyersSquibb	Therapy: Cancer Autoimmune disease	Angiocept Phase I Target: VEGFR-2	Lipovsek et al., 2007[128]
Bispecific scFv BiTes	Micromet (Munich, Germany) http://www.micromet.de	Therapy: Cancer	MT-103 (MEDI-538) Phase II Target: CD19	Mølhøj et al., 2007[129]
Domain antibody: variable domain of human light or heavy chain	Domantis (Cambridge, UK); http://www.domantis.com (now GlaxoSmithKline)	Therapy: Autoimmune diseases Asthma	Preclinical	Holt et al., 2003[130]
Kunitz-type domain of human and bovine trypsin inhibitor	Dyax (Cambridge, MA, USA); http://www.dyax.com	Therapy: Hereditary angio-oedema Open heart surgery	DX-88 Target: kallikrein Phase III Target: kallikrein Phase II	Williams and Baird, 2003[131] Devy et al., 2007[132]
Evibody: human cyto-toxic-associated anti-gen (CTLA-4)	Evogenix (Sydney, Australia); http://www.evogenix.com	Therapy: Cancer	Preclinical	Nuttall et al., 1999[133]
Ankyrin repeat protein	Molecular Partners (Zurich, Switzerland); http://www.molecularpartners.com	Therapy: Diagnostic	Preclinical	Binz et al., 2004[134]

Table 15.3 (Continued).

Scaffold	Organisation	Application indication	Latest clinical stage	Reference
Anticalin: human lipo-calins/lipocalins from *P. brassicae* butterfly	Pieris Proteolab (Freising, Germany): http://www.pieris.org	Therapy: Cancer Cardiovascular diseases	Preclinical	Schlehuber and Skerra, 2002[135]
Affilin molecule: human g-crystallin/human ubiquitin	Scil Proteins (Halle, Germany); http://www.scilproteins.com	Therapy: Cancer Ophthalmology Inflammation.	Preclinical	Ebersbach et al., 2007[136]
SMIPs	Trubion Inc. (Seattle, WA, USA) http://www.trubion.com	Therapy: Rheumatoid arthritis	TRU-015 Phase IIb Target: CD20	See Trubion website information: http://www.trubion.com/pdf/Trubion_TRU-015_Fact_Sheet.pdf [Accessed 22 October 2007]

Rational, *ab initio*, protein design is still at an early stage and remains the domain of experts. Much will need to be done to achieve the goal of rational design and it might be argued that evolutionary methods, which provide means for finding the desired needle from among a stack of needles, are the way forward. It is likely each route will find a niche. There seems little doubt that protein engineering will provide the basis for enormous opportunities in healthcare and the materials sciences.

REFERENCES

1. R. Wetzel, *Protein Eng.*, 1986, **1**, 3.
2. W. W. Shaw, *Biochem. J.*, 1987, **246**, 1.
3. J. A. Brannigan and A. J. Wilkinson, *Nat. Rev. Mol. Cell. Biol.*, 2002, **3**, 964.
4. P. Y. Chou and G. D. Fasman, *Biochemistry*, 1974, **13**, 222.
5. A. N. Naganathan, U. Doshi, A. Fung, M. Sadqi and V. Muñoz, *Biochemistry*, 2006, **45**, 8466.
6. L. H. Greene, T. E. Lewis, S. Addou, A. Cuff, T. Dallman, M. Dibley, O. Redfern, F. Pearl, R. Nambudiry, A. Reid, I. Sillitoe, C. Yeats, J. M. Thornton and C. A. Orengo, *Nucleic Acids Res.*, 2007, **35**, D291.
7. S. Lee and F. T. Tsai, *J. Biochem. Mol. Biol.*, 2005, **38**, 259.
8. H. Berman, K. Henrick, H. Nakamura and J. L. Markely, *Nucleic Acids Res.*, 2007, **35**, D301.
9. C. Chothia, *Nature*, 1992, **357**, 543.
10. R. I. Sadreyev and N. V. Grishin, *BMC Struct. Biol.*, 2006, **6**, 6.
11. M. Levitt, *Proc. Natl. Acad. Sci. USA*, 2007, **104**, 3183.
12. R. L. Marsden, J. A. G. Ranea, A. Sillero, O. Redfern, C. Yeats, M. Maibaum, D. Lee, S. Addou, G. A. Reeves, T. J. Dallman and C. A. Orengo, *Philos. Trans. R. Soc. London, Ser. B*, 2006, **361**, 425.
13. K. Ginalski, N. V. Grishin, A. Godzik and L. Rychlewski, *Nucleic Acids Res.*, 2005, **33**, 1874.
14. S. Wu, J. Skolnick and Y. Zhang, *BMC Biol.*, 2007, **5**, 17.
15. J. Moult, *Philos. Trans. R. Soc. London, Ser. B*, 2006, **361**, 453.
16. D. Baker, *Philos. Trans. R. Soc. London, Ser. B*, 2006, **361**, 459.
17. M. D. Edge, A. R. Green, G. R. Heathcliffe, P. A. Meacock, W. Schuch, D. B. Scanlon, T. C. Atkinson, C. R. Newton and A. F. Markham, *Nature*, 1981, **292**, 756.
18. P. T. Jones, P. H. Dear, J. Foote, M. S. Neuberger and G. Winter, *Nature*, 1986, **321**, 522.

19. R. K. Saiki, S. Scharf, F. Faloona, K. B. Mullis, G. T. Horn, H. A. Erlich and N. Arnheim, *Science*, 1985, **230**, 1350.
20. J. R. Adair and T. P. Wallace, in *Molecular Biomethods Handbook*, ed. R Rapley and J. Walker, Humana Press, Totowa, NJ, 1998 pp. 347–360.
21. F. Albericio, *Curr. Opin. Chem. Biol.*, 2004, **8**, 211.
22. T. Durek, V. Y. Torbeev and S. B. Kent, *Proc. Natl. Acad. Sci. USA*, 2007, **104**, 4846.
23. D. F. Mark, S. D. Lu, A. A. Creasey, R. Yamamoto and L. S. Lin, *Proc. Natl. Acad. Sci. USA*, 1984, **81**, 5662.
24. C. Neylon, *Nucleic Acids Res.*, 2004, **32**, 1448.
25. T. Matsuura and T. Yomo, *J. BioSci. Bioeng.*, 2006, **101**, 449.
26. J. Sondek and D. Shortle, *Proc. Natl. Acad. Sci. USA*, 1992, **89**, 3581.
27. J. Yáñez, M. Argüello, J. Osuna, X. Soberón and P. Gaytán, *Nucleic Acids Res.*, 2004, **32**, e158.
28. W. P. Stemmer, *Nature*, 1994, **370**, 389.
29. W. P. Stemmer, *Proc. Natl. Acad. Sci. USA*, 1994, **91**, 10747.
30. H. Zhao, L. Giver, Z. Shao, J. A. Affholter and E. H. Arnold, *Nat. Biotechnol.*, 1998, **16**, 258.
31. M. Ostermeier, J. H. Shim and S. J. Benkovic, *Nat. Biotechnol.*, 1999, **17**, 1205.
32. S. Lutz, M. Ostermeier, G. L. Moore, C. D. Maranas and S. J. Benkovic, *Proc. Natl. Acad. Sci. USA*, 2001, **98**, 11248.
33. W. M. Coco, W. E. Levinson, M. J. Crist, H. J. Hektor, A. Darzins, P. T. Pienkos, C. H. Squires and D. J. Monticello, *Nat. Biotechnol.*, 2001, **19**, 354.
34. V. Sieber, C. A. Martinez and F. H. Arnold, *Nat. Biotechnol.*, 2001, **19**, 456.
35. G. P. Smith, *Science*, 1985, **228**, 1315.
36. J. K. Scott and G. P. Smith, *Science*, 1990, **249**, 386.
37. D. McGregor and S. Robins, *Anal. Biochem.*, 2001, **294**, 108.
38. Y. Boublik, P. DiBonito and I. M. Jones, *Bio/technology*, 1995, **13**, 1079.
39. G. Georgiou, C. Stathopoulos, P. S. Daugherty, A. R. Nayak, B. L. Iverson and R. Curtiss III, *Nat. Biotechnol.*, 1997, **15**, 29.
40. S. Ståhl and M. Uhlén, *Trends Biotechnol.*, 1997, **15**, 185.
41. M. J. Feldhaus, R. W. Siegel, L. K. Opresko, J. R. Coleman, J. M. Feldhaus, Y. A. Yeung, J. R. Cochran, P. Heinzelman, D. Colby, J. Swers, C. Graff, H. S. Wiley and K. D. Wittrup, *Nat. Biotechnol.*, 2003, **21**, 163.

42. S. A. Gai and K. D. Wittrup, *Curr. Opin. Struct. Biol.*, 2007, **17**, 467.
43. P. S. Daugherty, *Curr. Opin. Struct. Biol.*, 2007, **17**, 474.
44. M. G. Cull, J. F. Miller and P. J. Schatz, *Proc. Natl. Acad. Sci. USA*, 1992, **89**, 1865.
45. A. S. Perelson and G. F. Oster, *J. Theor. Biol.*, 1979, **81**, 645.
46. L. C. Mattheakis, R. R. Bhatt and W. J. Dower, *Proc. Natl. Acad. Sci. USA*, 1994, **91**, 9022.
47. J. Hanes and A. Pluckthun, *Proc. Natl. Acad. Sci. USA*, 1997, **94**, 4937.
48. M. He and M. J. Taussig, *Nat. Methods*, 2007, **4**, 281.
49. G. Coia, L. Pontes-Braz, S. D. Nuttall, P. J. Hudson and R. A. Irving, *J. Immunol. Methods*, 2001, **254**, 191.
50. R. W. Roberts and J. W. Szostak, *Proc. Natl. Acad. Sci. USA*, 1997, **94**, 12297.
51. C. Zahnd, P. Amstutz and A. Plückthun, *Nat. Methods*, 2007, **4**, 269.
52. N. Nemoto, E. Miyamoto-Sato, Y. Husimi and H. Yanagawa, *FEBS Lett.*, 1997, **414**, 405.
53. R. Odegrip, D. Coomber, B. Eldridge, R. Hederer, P. A. Kuhlman, C. Ullman, K. FitzGerald and D. McGregor, *Proc. Natl. Acad. Sci. USA*, 2004, **101**, 2806.
54. J. Bertschinger and D. Neri, *Protein Eng. Des. Sel.*, 2004, **17**, 699.
55. H. Reiersen, I. Lobersli, G. A. Løset, E. Hvattum, B. Simonsen, J. E. Stacy, D. McGregor, K. Fitzgerald, M. Welschof, O. H. Brekke and O. J Marvik, *Nucleic Acids Res.*, 2005, **33**, e10.
56. D. S. Tawfik and A. D. Griffiths, *Nat. Biotechnol.*, 1998, **16**, 652.
57. A. Sepp, D. S. Tawfik and A. D. Griffiths, *FEBS Lett.*, 2002, **532**, 455.
58. A. Rothe, R. N. Surjadi and B. E. Power, *Trends Biotechnol.*, 2006, **24**, 587.
59. M. Yonezawa, N. Doi, Y. Kawahashi, T. Higashinakagawa and H. Yanagawa, *Nucleic Acids Res.*, 2003, **31**, e118.
60. C. Schmidt-Dannert and F. H. Arnold, *Trends Biotechnol.*, 1999, **17**, 135.
61. F. H. Arnold, http://eands.caltech.edu/articles/arnold/arnold.html.
62. H. Zhao and F. H. Arnold, *Curr. Opin. Struct. Biol.*, 1997, **7**, 480.
63. E. G. Hibbert and P. A. Dalby, *Microb. Cell Fact.*, 2005, **4**, 29.
64. C. Chiarabelli, J. W. Vrijbloed, R. M. Thomas and P. L. Luisi, *Chem. Biodivers.*, 2006, **3**, 827.
65. P. Kast and D. Hilvert, *Curr. Opin. Struct. Biol.*, 1997, **7**, 470.

66. R. Pasqualini and E. Ruoslahti, *Nature*, 1996, **380**, 364.
67. D. R. Christianson, M. G. Ozawa, R. Pasqualini and W. Arap, *Methods Mol. Biol.*, 2007, **357**, 385.
68. K. Graumann and A. Premstaller, *Biotechnol. J.*, 2006, **1**, 164.
69. F. Katzen, G. Chang and W. Kudlicki, *Trends Biotechnol.*, 2005, **23**, 150.
70. B. L. Bray, *Nat. Rev. Drug Discov.*, 2003, **2**, 587.
71. T. Bruckdorfer, O. Marder and F. Albericio, *Curr. Pharm. Biotechnol.*, 2004, **5**, 29.
72. D. R. Casimiro, P. E. Wright and H. J. Dyson, *Structure*, 1997, **5**, 1407.
73. W. Peti, R. Page, K. Moy, M. O'Neil-Johnson, I. A. Wilson, R. C. Stevens and K. Wüthrich, *J. Struct. Funct. Genomics*, 2005, **6**, 259.
74. G. Walsh, *Nat. Biotechnol.*, 2006, **24**, 769.
75. M. Joyeux-Faure, *J. Pharmacol. Exp. Ther.*, 2007, **323**, 759.
76. A. M. Sinclair and S. Elliott, *J. Pharm. Sci.*, 2005, **94**, 1626.
77. P. J. Carter, *Nat. Rev. Immunol.*, 2006, **6**, 343.
78. M. S. Neuberger, G. T. Williams, E. B. Mitchell, S. S. Jouhal, J. G. Flanagan and T. H. Rabbitts, *Nature*, 1985, **314**, 268.
79. G. T. Williams and M. S. Neuberger, *Gene*, 1986, **43**, 319.
80. K. M. Mohler, D. S. Torrance, C. A. Smith, R. G. Goodwin, K. E. Stremler, V. P. Fung, H. Madani and M. B. Widmer, *J. Immunol.*, 1993, **151**, 1548.
81. D. P. Williams, K. Parker, P. Bacha, W. Bishai, M. Borowski, F. Genbauffe, T. B. Strom and J. R. Murphy, *Protein Eng.*, 1987, **1**, 493.
82. G. J. Lieschke, P. K. Rao, M. K. Gately and R. C. Mulligan, *Nat. Biotechnol.*, 1997, **15**, 35.
83. R. E. Bird, K. D. Hardman, J. W. Jacobson, S. Johnson, B. M. Kaufman, S. M. Lee, T. Lee, S. H. Pope, G. S. Riordan and M. Whitlow, *Science*, 1988, **242**, 423.
84. J. S. Huston, D. Levinson, M. Mudgett-Hunter, M. S. Tai, J. Novotný, M. N. Margolies, R. J. Ridge, R. E. Bruccoleri, E. Haber, R. Crea and H. Oppermann, *Proc. Natl. Acad. Sci. USA*, 1988, **85**, 5879.
85. S. L. Morrison, M. J. Johnson, L. A. Herzenberg and V. T. Oi, *Proc. Natl. Acad. Sci. USA*, 1984, **81**, 6851.
86. A. F. LoBuglio, R. H. Wheeler, J. Trang, A. Haynes, K. Rogers, E. B. Harvey, L. Sun, J. Ghrayeb and M. B. Khazaeli, *Proc. Natl. Acad. Sci. USA*, 1989, **86**, 4220.
87. J. Staunton and K. J. Weissman, *Nat. Prod. Rep.*, 2001, **18**, 380.

88. H. G. Menzella, R. Reid, J. R. Carney, S. S. Chandran, S. J. Reisinger, K. G. Patel, D. A Hopwood and D. V. Santi, *Nat. Biotechnol.*, 2005, **23**, 1171.

89. H. G. Menzella, J. R. Carney and D. V. Santi, *Chem. Biol.*, 2007, **14**, 143.

90. D. H. Sherman, *Nat. Biotechnol.*, 2005, **23**, 1083.

91. H. Jenke-Kodama, T. Börner and E. Dittmann, *PLoS Comput. Biol.*, 2006, **2**, e132.

92. T. K. Nordt and C. Bode, *Heart*, 2003, **89**, 1358.

93. A. G. Porter, L. D. Bell, J. Adair, G. H. Catlin, J. Clarke, J. A. Davies, K. Dawson, R. Derbyshire, S. M. Doel, L. Dunthorne, M. Finlay, J. Hall, M. Houghton, C. Hynes, I. Lindley, M. Nugent, G. J. O'Neil, J. C. Smith, A. Stewart, W. Tacon, J. Viney, N. Warburton, P. G. Boseley and K. G. McCullagh, *DNA*, 1986, **5**, 137.

94. H. Weber and C. Weissmann, *Nucleic Acids Res.*, 1983, **11**, 5661.

95. J. Hecky and K. M. Muller, *Biochemistry*, 2005, **44**, 12640.

96. G. Winter, A. R. Fersht, A. J. Wilkinson, M. Zoller and M. Smith, *Nature*, 1982, **299**, 756.

97. A. J. Wilkinson, A. R. Fersht, D. M. Blow, P. Carter and G. Winter, *Nature*, 1984, **307**, 187.

98. A. Fersht and G. Winter, *Trends Biochem. Sci.*, 1992, **17**, 292.

99. M. Leisola and O. Turunen, *Appl. Microbiol. Biotechnol.*, 2007, **75**, 1225.

100. F. H. Arnold, www.che.caltech.edu/groups/fha/directed_evolution, 2007. Accessed 9 November 2007.

101. B. C. Cunningham, D. J. Henner and J. A. Wells, *Science*, 1990, **247**, 1461.

102. A. P. Seddon, D. Aviezer, L.-Y. Li, P. Bohlen and A. Yayon, *Biochemistry*, 1995, **34**, 731.

103. R. J. Kreitman, R. K. Puri and I. Pastan, *Proc. Natl. Acad. Sci. USA*, 1994, **91**, 6889.

104. F. Martin, C. Toniatti, A. L. Salvati, G. Ciliberto, R. Cortese and M. Sollazzo, *J. Mol. Biol.*, 1996, **255**, 86.

105. I. Tomlinson, *Nat. Biotechnol.*, 2004, **22**, 521.

106. H. K. Binz, P. Amstutz and A. Pluckthun, *Nat. Biotechnol.*, 2005, **23**, 1257.

107. K. J. Jeong, R. Mabry and G. Georgiou, *Nat. Biotechnol.*, 2005, **23**, 1493.

108. D. S. Gill and N. K. Damle, *Curr. Opin. Biotechnol.*, 2006, **17**, 653.

109. R. J. Hosse, A. Rothe and B. E. Power, *Protein Sci.*, 2006, **15**, 14.

110. A. Rothe, R. J. Hosse and B. E. Power, *FASEB J.*, 2006, **20**, 1599.
111. P. M. Watt, *Nat. Biotechnol.*, 2006, **24**, 177.
112. K. E. Drexler, *Proc. Natl. Acad. Sci. USA*, 1981, **78**, 5275.
113. C. Pabo, *Nature*, 1983, **301**, 200.
114. L. Regan and W. F. deGrado, *Science*, 1988, **241**, 976.
115. B. I. Dahiyat and S. L. Mayo, *Science*, 1997, **278**, 82.
116. P. B. Harbury, J. J. Plecs, B. Tidor, T. Alber and P. S. Kim, *Science*, 1998, **282**, 1462.
117. G. Dantas, B. Kuhlman, D. Callender, M. Wong and D. Baker, *J. Mol. Biol.*, 2003, **332**, 449.
118. B. Kuhlman, G. Dantas, G. C. Ireton, G. Varani, B. L. Stoddard and D. Baker, *Science*, 2003, **302**, 1364.
119. N. Hurwitz, M. Pellegrini-Calace and D. T. Jones, *Philos. Trans. R. Soc. London, Ser. B*, 2006, **361**, 465.
120. P. Barth, J. Schonbrun and D. Baker, *D. Proc. Natl. Acad. Sci. USA*, 2007, **104**, 15682.
121. J. J. Hulstein, P. G. de Groot, K. Silence, A. Veyradier, R. Fijnheer and P. J. Lenting, *Blood*, 2005, **106**, 3035.
122. K. Nord, E. Gunneriusson, J. Ringdahl, S. Ståhl, M. Uhlén and P. A. Nygren, *Nat. Biotechnol.*, 1997, **15**, 772.
123. F. Nilsson and V. Tolmachev, *Curr. Opin. Drug Discov. Dev.*, 2007, **10**, 167.
124. V. Tolmachev, A. Orlova, F. Y. Nilsson, J. Feldwisch, A. Wennborg and L. Abrahmsén, *Expert Opin. Biol.Ther.*, 2007, **7**, 555.
125. J. Silverman, Q. Liu, A. Bakker, W. To, A. Duguay, B. M. Alba, R. Smith A. Rivas, P. Li, H. Le, E. Whitehorn, K. W. Moore, C. Swimmer, V. Perlroth, M. Vogt, J. Kolkman and W. P. Stemmer, *Nat. Biotechnol.*, 2005, **23**, 1556.
126. L. C. Gunnarsson, L. Dexlin, E. N. Karlsson, O. Holst and M. Ohlin, *Biomol Eng.*, 2006, **23**, 111.
127. K. W. Woodburn, Q. Fan, S. Winslow, M. J. Chen, R. B. Mortensen, N. Casadevall, R. B. Stead and P. J. Schatz, *Exp Hematol.*, 2007, **35**, 1201.
128. D. Lipovsek, S. M. Lippow, B. J. Hackel, M. W. Gregson, P. Cheng, A. Kapila and K. D. Wittrup, *J. Mol. Biol.*, 2007, **368**, 1024.
129. M. Mølhøj, S. Crommer, K. Brischwein, D. Rau, M. Sriskandarajah, P. Hoffmann, P. Kufer, R. Hofmeister and P. A. Baeuerle, *Mol Immunol.*, 2007, **44**, 1935.
130. L. J. Holt, C. Herring, L. S. Jespers, B. P. Woolven and I. M. Tomlinson, *Trends Biotechnol.*, 2003, **21**, 484.
131. A. Williams and L. G. Baird, *Transfus. Apher. Sci.*, 2003, **29**, 255.

132. L. Devy, S. A. Rabbani, M. Stochl, M. Ruskowski, I. Mackie, L. Naa, M. Toews, R. van Gool, J. Chen, A. Ley, R. C. Ladner, D. T. Dransfield and P. Henderikx, *Neoplasia*, 2007, **9**, 927.

133. S. D. Nuttall, M. J. Rousch, R. A. Irving, S. E. Hufton, H. R. Hoogenboom and P. J. Hudson, *Proteins*, 1999, **36**, 217.

134. H. K. Binz, P. Amstütz, A. Kohl, M. T. Stumpp, C. Briand, P. Forrer, M. G. Grütter and A. Plückthun, *Nat. Biotechnol.*, 2004, **22**, 575.

135. S. Schlehuber and A. Skerra, *Biophys. Chem.*, 2002, **96**, 213.

136. H. Ebersbach, E. Fiedler, T. Scheuermann, M. Fiedler, M. T. Stubbs, C. Reimann, G. Proetzel, R. Rudolph and U. Fiedler, *J. Mol. Biol.*, 2007, **372**, 172.

137. S. N. Ho, H. D. Hunt, R. M. Horton, J. K. Pullen and L. R. Pease, *Gene*, 1989, **77**, 51.

CHAPTER 16

Immobilisation of Enzymes and Cells

GORDON F. BICKERSTAFF

School of Science, University of the West of Scotland, Paisley Campus, Paisley PA1 2BE, UK

16.1 INTRODUCTION

The number of new biocatalysts required to support the rapid growth of biotechnology applications is increasing substantially. Although many biocatalysts can be used as free enzymes or as whole cells, immobilisation adds features that can significantly improve the commercial viability and stability. Immobilisation methodology has expanded greatly in the past 40 years in a wide range of analytical, biotransformation and medical applications.[1] A consequence of the explosion of this technology is that there is now a bewildering array of permutations for immobilisation of biological material.[2] Biological catalysts have a high degree of individual variability and, although many immobilisation techniques have wide applicability, it is impossible for one or even a few methods to cater for the great diversity of requirements inherent in biological material. This is especially so when the aim is to produce an optimum system in which an immobilised biocatalyst can function at high levels of efficiency, stability and economy.[2]

Molecular Biology and Biotechnology, 5th Edition
Edited by John M Walker and Ralph Rapley
© Royal Society of Chemistry 2009
Published by the Royal Society of Chemistry, www.rsc.org

16.2 BIOCATALYSTS

The range of biocatalysts available for industrial and medical applications increases as advances in biological sciences reveal new understanding of the components of living organisms.[3] Enzymes/cells/organelles that were previously overlooked because of high cost/poor stability have benefited from advances in downstream processing and genetic engineering.[4] New catalyst types such as abzymes, ribozymes and multicatalytic complexes have added to the repertoire of biocatalysts in the biotechnologist's toolkit and stimulated new ideas on the future development and application of biocatalysts.[4]

16.2.1 Enzymes

Enzymes are a group of proteins that are synthesised by living cells to function as catalysts for the many thousands of biochemical reactions that constitute the metabolism of a cell.[3] More than 2500 different enzymes are known and there are still many organisms that have not been screened for novel enzymes or metabolic pathways. It is likely that there are many enzymes awaiting discovery. A new addition to the designated list of enzymes in 2007 was the enzyme chlorophyllase (EC 3.1.1.14), which is part of the chlorophyll degradation pathway that is responsible for de-greening processes that occur during fruit ripening and flowering. The enzyme hydrolyses chlorophyll to phytol + chlorophyllide. Living cells require enzymes to drive metabolic pathways because at physiological temperature and pH, uncatalysed reactions would proceed at too slow a rate for the vital processes necessary to sustain life.[3] A list of the reaction types catalysed by enzymes is provided in Table 16.1. In common with all catalysts, enzymes are subject to the normal laws concerning catalysis of reactions:

- A catalyst cannot speed up a reaction that would not occur in its absence, because it is not thermodynamically possible.
- A catalyst is not consumed during the reaction and so relatively few catalyst molecules are capable of catalysing a reaction many times.
- A catalyst cannot alter the equilibrium position of a given reaction.

Most reactions proceed eventually to a state of equilibrium, in which the rate of the forward reaction is equal to the rate of the reverse reaction. At equilibrium the substrate and product have specific equilibrium concentrations that are a special characteristic of the reaction. For example, isomerisation (see Table 16.1) of glucose (substrate)

Table 16.1 Reactions types catalysed by enzymes.[a]

EC No.	Reaction type	General example of the reaction type	
EC 1	Oxidoreductase reactions	$X–H + Y \rightarrow X + Y–H$	(reduction)
		$X + 1/2O_2 \rightarrow XO$	(oxidation)
EC 2	Transferase reactions	$X–CH_3 + Y \rightarrow X + Y–CH_3$	(transfer)
EC 3	Hydrolase reactions	$X–Y + H_2O \rightarrow X–OH + Y–H$	(splitting)
EC 4	Lyase reactions	$X–CO_2–Y \rightarrow X–Y + CO_2$	(removal/addition)
EC 5	Isomerase reactions	$X–Y–X \rightarrow Y–X–X$	(rearrangement)
EC 6	Ligase reactions	$X + Y + ATP \rightarrow$ $X–Y + ADP + Pi$	(joining)

[a]Enzyme reaction types were grouped into six main families by an Enzyme Commission (EC) and underneath each main family there are sub-classes and sub-sub-classes that further describe details of the reaction catalysed. Each enzyme receives an EC number that provides details of the reaction catalysed. For example, the EC number for the enzyme lactate dehydrogenase is EC 1.1.1.27:
The first number(1) indicates the enzyme is an oxidoreductase.
The second number(1) indicates a sub-class that specifies CH–OH as a group involved in the reaction.
The third number(1) indicates a sub-section that specifies NAD as the coenzyme used in the reaction.
The fourth number(27) indicates the 27th enzyme in this sub-section that specifies L-lactate as the substrate for the reaction (note there are over 290 enzymes in this sub-section).

to produce fructose (product) is catalysed by the enzyme glucose isomerase:

$$\text{glucose} \underset{\text{substrate}}{\overset{\text{glucose isomerase}}{\rightleftharpoons}} \text{fructose}_{\text{product}}$$

From a starting solution of 100% glucose, the reaction proceeds to equilibrium and for this reaction the relative proportions at equilibrium are 45% fructose and 55% glucose.[3] The enzyme cannot change the equilibrium position of the reaction, but it can substantially reduce the time that the reaction normally takes to reach equilibrium. In these respects, enzymes are no different from other catalysts. However, enzymes do possess two special attributes that are not found to any great extent in other non-biological catalysts: specificity and high catalytic power.

16.2.1.1 Specificity
Specificity is a distinctive feature of enzymes. Chemical catalysts have limited specificity, whereas enzymes have strong specificity for reactants

(substrates), and for the susceptible bond involved in a reaction. The degree of specificity can vary from absolute to fairly broad. For example, the enzyme urease is absolutely specific for its substrate urea (NH_2CONH_2) and an almost identical structural analogue such as thiourea (NH_2SONH_2) is not hydrolysed (see Table 16.1). In contrast, the enzyme hexokinase is less specific and has group specificity for a small set of related sugar molecules:

$$\text{glucose} + \text{ATP} + Mg^{2+} \overset{\text{hexokinase}}{\rightleftharpoons} \text{glucose-6-phosphate} + \text{ADP} + Mg^{2+}$$

Glucose is the principal substrate. Hexokinase will also catalyse phosphorylation of several other sugars such as mannose and fructose, but *not* galactose, xylose, maltose or sucrose. In addition to substrate specificity, enzymes display remarkable product specificity, which ensures that the final product is not contaminated with by-products. Thus, in the above phosphorylation of glucose, the product is exclusively glucose-6-phosphate and no other phosphoglucose (*e.g.* glucose-1-phosphate or glucose-2-phosphate) is produced during the reaction. The formation of by-products by side reactions is a significant problem associated with most of the less specific catalysts. In many enzymes, specificity also extends to selective discrimination between stereoisomers of a substrate molecule. This stereospecificity is shown by the enzyme D-amino acid oxidase, which is specific for D-amino acids only and will not catalyse the oxidation of the L-amino acid stereoisomers.

Specificity is an inherent feature of enzymes because catalysis takes place in a particular region of the enzyme, which is designed to accommodate the substrates involved in a reaction (Figure 16.1). This

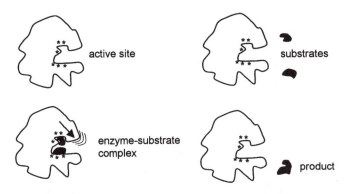

Figure 16.1 Schematic representation of an enzyme active site with amino acid residues (*) located at the active site, binding of substrates and change in enzyme shape after substrates are bound.

region or active site is normally a small pocket, cleft or crevice on the surface of the enzyme. It is designed to bring a few of the enzyme amino acid residues into contact with the substrate molecule. The active site has strong affinity for the substrate because site amino acid residues are primed for interaction with groups or atoms on the substrate molecule. Consequently, a substrate molecule must have the correct shape with groups/atoms in the correct position to fit into the active site and participate in the interactions (Figure 16.1). Enzymes with absolute specificity have very precise shape/interaction requirements, which are only found in one particular substrate molecule. Enzymes with broad specificity have active site requirements that are more flexible and therefore accept a wider range of substrate molecules. The active site amino acid residues participate directly in a catalytic reaction and are largely responsible for the high catalytic power associated with enzyme reactions.

16.2.1.2 Catalytic Power

During any reaction, the reactants briefly enter a state in which susceptible substrate bonds are not completely broken and new bonds in the product are not completely formed. This temporary condition is called a transition state and is energy dependent because energy is needed to make and break chemical bonds (*e.g.* $350\,kJ\,mol^{-1}$ for each C–C covalent bond) in substrate and product molecules. Formation of a transition state represents an energy barrier to successful reaction and is the reason why most reactions proceed extremely slowly in the absence of external help.

Substrates can be helped towards a transition state by addition of heat energy, high pressure or extreme pH to weaken bonds or by addition of catalysts. Enzymes are very effective at reducing the energy barrier to allow the formation of a transition state and increased the rate of a reaction. The efficiency of enzyme catalysis varies, but most enzymes can enhance the rate of an uncatalysed reaction by a factor in the range 10^5–10^{14}. One of the most efficient enzymes is carbonic anhydrase, which catalyses the hydration of up to 600 000 molecules of CO_2 per second under optimal conditions:

$$CO_2 + H_2O \quad \overset{\text{carbonic anhydrase}}{\rightleftharpoons} \quad HCO_3^- + H^+$$

Carbonic anhydrase has a vital role in maintaining blood pH at 7.4 and in removal of CO_2 from blood by gas exchange in the lungs. In medical conditions/emergencies when breathing has stopped or is

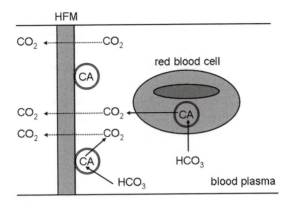

Figure 16.2   Schematic of a CO_2 removal assist device based on a hollow-fibre membrane (HFM) containing bound carbonic anhydrase (CA) for conversion of bicarbonate (HCO_3) to CO_2 for removal through the membrane.

difficult and intervention to remove CO_2 is required, it is normally achieved with assistance of mechanical ventilators. When lung damage/dysfunction is apparent, then ventilator assistance can harm weakened lungs and in such circumstances there is a need for a breathing assistance device that is independent of the lungs.[5] Core to the development of such a device is immobilisation of carbonic anhydrase and recent studies have demonstrated the use of hollow-fibre membranes (HFMs) with immobilised carbonic anhydrase to enhance CO_2 exchange in a respiratory assist device as depicted in Figure 16.2.

16.2.2   Ribozymes, Deoxyribozymes and Ribosomes

Ribozymes and deoxyribozymes are RNA and DNA molecules with endonuclease catalytic activity and were probably evolutionary forerunners of enzymes. A variety of ribozymes have been discovered, such as the hammerhead ribozyme, which is one of the smallest ribozymes that can catalyse the site-specific hydrolysis of a phosphodiester bond in RNA.[6] Although described as hydrolysis (see Table 16.1), the autocatalytic reaction does not involve water; rather, the cleavage reaction is an isomerisation rearrangement of phosphodiester bonds in the RNA.

Production of these new biological catalysts, which can be tailored to suit a specific biotransformation, is a major development that has the potential to increase the usefulness of biological catalysts. The biocatalysts currently available for development are those existing in Nature and applications evolve around the existing set of biocatalysts. New applications will be possible if bespoke biocatalysts can be designed to suit an application. Ribozymes have attracted great attention because comparative studies of the properties of enzymes and ribozymes can provide new fundamental information on the chemical/physical principles of biological catalysis.[6] Ribozymes also recognise their target substrate (RNA) in a highly sequence-specific manner and this degree of specificity should permit the development of important therapeutic applications for both inherited and acquired diseases.[6] For example, ribozymes may be used to counteract harmful gene expression by attacking gene products (mRNA) produced by faulty, damaged or uncontrolled genes.[6] Such targets might be the mRNA produced by oncogenes (cancer) and viral genomes from human immunodeficiency virus type 1 (HIV-1).

In another area of development, a 49 nucleotide long ribozyme has been immobilised on an agarose support material and used to catalyse formation of C–C bonds in a Diels–Alder reaction, which is one of the most important C–C bond formation reactions used by organic chemists.[7] An important point with this development is that a number of limitations imposed by protein-based enzymes that catalyse these reactions can be overcome using a ribozyme.[7]

Ribosomes have been known for a long time as the structures in cells that are responsible for protein synthesis. Recent studies on bacterial ribosomes have revealed that the active site of a ribosome where the peptide bond is formed between two amino acids is composed of RNA. Thus although a ribosome is a complex multi-component ribonucleotide and protein structure; it has a ribozyme at its core active site.[8]

16.2.3 Splicesomes

In the late 1970s, it was found that eukaryotic genes are transcribed into mRNA sequences that contain conserved coding sequences (exons) and non-conserved non-coding sequences (introns). Given the notion of a chromosome as a single long length of DNA then a process of alternative splicing allows for coding sequences to be copied out from various places in a chromosome and spliced together to form different coding sequences of mRNA that generate proteins with specific characteristics.[9] This mechanism has provided evolution with a capacity for genetic diversity, to produce families of enzymes by varying the splicing process, and a capacity to produce many more copies of functional protein than would be possible if one gene coding sequence was totally responsible for only one protein.

The splicesome is a multi-component ribonucleotide protein structure that catalyses the excision of intron non-coding sequences from transcribed immature lengths of mRNA.[9] Further understanding of the role and properties of the splicesome will have a great impact on genetic engineering and construction of new and novel enzymes. The enzymes in the biotechnologist's current toolkit are those that evolution has shaped and refined to meet its needs over countless millennia. It may be that there are earlier constructs that were tried and discarded or new ones that have not yet been tested and that these can be produced using mechanisms employed by a splicesome. It is also apparent from studies on the effects of mutations that errors in the splicing process can result in disease. A mutation responsible for mental retardation associated with lactic acidosis (Leigh's syndrome) was found to cause binding of a human splicing regulatory factor that then caused retention rather than excision of intron sequences in the coding sequence for a subunit of the enzyme pyruvate dehydrogenase.[9] A fuller understanding of splicesome mechanisms will help to guide therapeutic solutions for diseases caused by malfunction of splicesome activity.

16.2.4 Abzymes

Antibodies are proteins and, like enzyme and receptor proteins, they have binding domains (regions) where they bind other molecules. Antibodies bind molecules called antigens and can bind small molecules (haptens) or small parts (determinants) of high molecular weight antigens, which may be attached to cells. Catalytic antibodies or abzymes are an additional class of biocatalyst because they have distinct catalytic properties.[10] They catalyse several reactions, including hydrolysis, cyclisation reactions, elimination reactions and synthesis.

Comparative studies between enzymes and abzymes indicate that enzymes have varying levels of specificity built into their structure, whereas antibodies have much greater recognition qualities. It seems likely that enzymes and antibodies are 'molecular cousins' that evolved from a common ancestor with catalytic and recognition properties. Enzymes probably evolved to concentrate on catalysis and their recognition property adapted to provide specificity. Antibodies evolved the recognition property to a very high degree and retained useful elements of catalysis. Comparison of the properties of enzymes and abzymes will provide new understanding of the selectivity process inherent in enzyme specificity,[10]

Abzymes could have exciting applications in many areas of biotechnology, medicine, analytical biochemistry and molecular biology.[11] In particular, enzyme specificity is limited to that already provided by Nature and often such limitation restricts the range of possible catalysis. With protein engineering, it will be possible to alter and improve the specificity of key enzymes and improve the catalytic power of useful abzymes. Antibodies have a vast repertoire of highly specific binding functions, which can be used to increase further the portfolio of bio-catalysts for the biotechnologist's toolkit.

Biosensors are an obvious example where abzymes could provide new improvement in recognition of analyte (molecule to be measured) and thereby increase the range of analytes that can be detected and measured. Enzyme-based biosensors are limited in scope and tend to discriminate between low molecular weight molecules. Abzyme-based biosensors, on the other hand, have considerable scope and will be able to discriminate between toxins, viruses, microbial cells, native and foreign proteins, nucleic acids, human cells, cancer cells, *etc.* Evolution has shaped anti-bodies for a particular role in Nature so it is not surprising that abzymes have less catalytic power than enzymes and have a tendency not to release product quickly after catalysis.[11] However, unlike enzymes, antibodies have common gross protein structural features so a single immobilisation method may be more readily applied to different abzymes.

16.2.5   Multienzyme Complexes

Application of extracellular enzyme-based catalysis has mostly been con-cerned with single enzymes performing single-step reactions. Although there is still much work to be done with single enzyme biotransformations, other biotechnologists have focused attention on the next phase of extracellular biocatalysis, *i.e.* sequential multiple-step biotransformations as currently performed in biochemical pathways by cells and organelles. Multienzyme complexes are ordered assemblies of functionally and structurally different enzymes and proteins, which catalyse successive steps in a biochemical reaction pathway. Greater understanding of such com-plexes has provided scope to take enzyme-based catalysis beyond single-step reaction processes to more sophisticated reaction schemes. A number of multienzyme structures have been studied and examples below reveal some of the opportunities and challenges that can be obtained from research on these complexes.

16.2.5.1   Apoptosomes

Individual cells have a capacity to self-terminate in a controlled manner, if required for the benefit of the organism as a whole, in a process called apoptosis.[12] This process has an important function in processes that involve cell turnover such as development of an embryo, cell recycling, cell-based diseases and in removal or recycle of surplus, damaged or

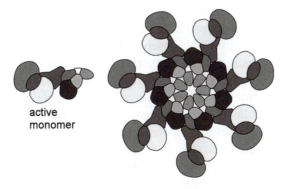

active
monomer

Figure 16.3 Apoptosome schematic showing the complex organisation of seven active monomers to form a wheel-like structure. Adapted from Ref. 13.

faulty cells. The process involves a suite of protease enzymes called *caspases* (*c*ysteine, *asp*artic acid specific prote*ases*).[13] Normally these enzymes are in a dormant precursor state, but when activated they produce a cascade of proteolytic activity that destroys intracellular protein structures, enzymes and other vital proteins in an organised shutdown mechanism that results in cell death and subsequent removal of the dead cell by phagocytes.

Work on apoptosome structure is progressing and recent work reveals a model in which various protease and non-protease components are assembled in a wheel-like oligomeric structure (Figure 16.3).[13] This model is based on an assembly of recombinant proteins because extraction of a functional apoptosome is still a major challenge for current isolation/extraction technology.[13] The role of apoptosis in disease is also under investigation and it is clear that disturbance of the natural role of apoptosis can influence disease and dysfunction. In some diseases such as cancer, apoptosis would be antagonistic to cell proliferation and so it may be that oncogenes produce inhibitors that block normal apoptosis, whereas in degenerative diseases it may be a feature of this type of disease that premature activation of apoptosis is stimulated.[12] There is clearly a great wealth of information waiting to be discovered both at the local level on how protease enzymes are orchestrated and at the higher cell and tissue level on how tissues are sculptured to meet the needs of the organism.

16.2.5.2 Proteosomes

These are large (typically 2000 kDa) multienzyme complexes consisting of more than 25 different proteins, which range in size from 22 to 110 kDa. They are found in both cytoplasm and nucleus of eukaryotic cells and structurally appear (side-on-view) as a cylindrical stack of four rings with six or seven proteins in each ring.[14] The function of the proteosome is rapid hydrolysis (see Table 16.1) of intracellular protein to enable fast turnover of protein. Proteins destined for hydrolysis by proteosomes are tagged by the regulatory protein ubiquitin and working together proteosomes & ubiquitin have a central role in the function of many critical cellular processes. Fast recycle of protein is important in many cellular processes such as embryogenesis, cell growth, cell differentiation, cell-cycle control, response to environmental stress and removal of damaged/faulty protein. It is also vital for continuous supply of amino acids for new protein synthesis and for immune responses, some key functions of the nervous system and acquisition of memory.[14]

The complex contains both endopeptidase (cut the inside of protein to produce large fragments) and exopeptidase (cut at the ends of polypeptide fragments to produce peptides and amino acids) and is therefore capable of complete hydrolysis of protein to amino acids. Proteosome also have a role in the presentation of antigens as part of the immune defence system and proteosome inhibitors have been investigated as possible tools to combat viral infections.[15]

16.2.5.3   Cellulosomes

Cellulose is the most abundant source of carbon and chemical energy on Earth and around two-thirds of the billions of kilograms of industrial and domestic waste generated by individual countries around the world is potentially biodegradable because of the cellulose content.[16] However, it is a very stable polymer of glucose, which is usually inter-mixed with other polysaccharides such as xylan and lignin to produce a complex structure that is resistant to standard mechanisms of chemical or biological degradation. Cellulolytic microorganisms have evolved a multifunctional multienzyme complex called a cellulosome, which provides a systematic hydrolysis of cellulose to release glucose sugars for growth and energy supply.[16] The complexity of cellulose structure that plants have evolved to try to resist tissue loss often requires concerted cooperation between various cellulolytic wood-degrading microorganisms.

Typically, a cellulosome is composed of a central protein skeleton called scaffoldin on to which are attached link proteins called dockerins as depicted in Figure 16.4. Attached to dockerin proteins are various enzymes associated with cellulose hydrolysis and also attached are binding proteins that attach to the cellulose substrate and physically locate the complex on to cellulose. A raft of enzymes have been identified including endoglucanases, exoglucanases, β-glucosidase, xylanase,

Figure 16.4 Cellulosome schematic showing various enzymes attached to cohesin proteins, which bind to dockerins attached to a central backbone called scaffoldin. Also attached to scaffoldin is a cellulose binding module (cbm). Adapted from Ref. 16.

ligninase, pectinase and the amount and pattern of enzymes can be varied by a given microorganism to suit the different structured architectural forms of cellulose produced by a given plant.[17] Flexibility and adaptability are key features that are highly desirable in biocatalysis and it is likely that there are some very clever principles that can be derived from detailed studies on the cellulosome. Given an abundance of cellulose on Earth, it is not surprising that several thousand cellulolytic strains have been found among microorganisms. Recent work has evaluated the potential of engineering cellulosomes for waste management by selecting particular enzyme components from different cellulolytic organisms for combination in a designer cellulosome for specified applications. Microorganisms have evolved cellulosomes to degrade cellulose for their own particular energy and growth requirements. It should be possible to extract the best set of catalytic features from the spectrum of different organisms and design cellulosomes that will reproduce the processes currently operating in landfill sites around the world by suites of cellulolytic organisms, but in a faster, more controlled process, that will also release the energy trapped in cellulose.[17]

16.2.5.4 Multienzyme Complexes and Future Immobilisation Technology

An exciting prospect for biocatalysis is that studies on multienzyme complexes such as apoptosomes, proteosomes and cellulosomes will undoubtedly shape future methods of biocatalyst immobilisation technology. It may be possible to make immobilisation a more active participant in biotransformation reactions by introducing specificity to support materials via immobilised specific binding domains that improve accurate binding of particular substrate materials so increasing substrate specificity and catalytic power. It will be possible to mimic the cellulosome and immobilise an array of specific enzymes and regulatory proteins together for sophisticated multi-step biotransformations (Figure 16.4).

The future will provide one modular universal support material on to which can be attached any number of binding, regulatory and catalytic protein modules that are required for a given biotransformation. In the future, successful development of these systems will allow the production of cell-free complex processes and will allow more sophisticated opportunities for enzyme-based biocatalysis such as versatile fuel cells, recycling of complex waste materials, multianalyte analytical biosensors and cell-free biosynthesis of biopolymers such as proteins and carbohydrates. A development that will support this approach to biocatalysis is nanobiotechnology[18] and the availability of highly defined nanoparticles (see Section 16.3.1.1).

16.2.6   Cells

Cells are a different class of biocatalyst in that they contain considerable arrays of molecular biocatalysts (such as enzymes), which they harness in sequential step reactions to catalyse extensive biotransformations (*e.g.* glucose $\rightarrow CO_2 + H_2O$). Therefore, the range and scale of biocatalysis within a particular cell is vastly greater than that of particular extracellular enzymes, ribozymes, *etc.* Cell-based biocatalysis can be achieved with animal, plant and microbial sources. Cell-based systems are particularly valuable for multiple-step reactions and reactions that involve complex energy transduction such as ATP hydrolysis.[19] At first sight it may seem that the need for extracellular enzymes is questionable. However, cells are living organisms and their priority is life support, rather than completion of a biotransformation required by a biotechnologist. Consequently, biochemical resources and energy are used by cells for growth and so cells are not efficient biocatalysts for many simple biotransformations.[20] Sometimes cells need to be 'tricked' into switching on a metabolic pathway, which is dormant because the cells have no need for it, when the biotechnologist has a need for the pathway to complete a biotransformation. Cells are important biocatalysts for complex and multistep reactions and the merits of various cells as biocatalysts are discussed briefly below.

16.2.6.1 Animal Cells

Commercial applications for animal cells are increasing and they are now routinely used for expression of proteins from recombinant DNA. They are also used as hosts for attenuated strains of important viruses in the production of vaccines for foot-and-mouth disease, polio, rabies, measles and rubella, *etc.*[21]

A particularly important catalytic biotransformation of animal cells is post-translational modification of commercial proteins. Many key proteins receive biochemical modification after biosynthesis (translation of mRNA to protein) and such modifications are essential for proper functioning of a mature protein.[22] Bacterial cells can be used to express commercial human proteins, but bacteria cannot complete the post-translational modifications needed to make a protein fully functional. Post-translational modifications include glycosylation (adding sugar residues), formation of disulfide bridges, amidation, carboxylation or phosphorylation of amino acid residues and highly specific protease-based cutting of a protein chain to produce a particular protein shape.[23] However, animal cells are fragile, have special growth requirements, have low product yields and are susceptible to infection by bacteria and viruses. Animal cell culture is therefore comparatively expensive and normally reserved for high-value medical products.

16.2.6.2 Plant Cells

In addition to crops, many plants produce compounds that have commercial value. Over three-quarters of the 32 000+ known natural products are derived from plants.[24] These include medicines and drugs such as atropine, morphine and digoxin, essential oils and fragrances such as menthol, strawberry, vanilla and camphor, pigments such as anthocyanin, beta-cyanin and saffron, and speciality products such as enzymes, fungicides, pesticides, peptides, vitamins and pigments.[25] Perhaps the most notorious plant products are narcotics such as opium, cocaine and morphine.

Greater development of plant biocatalysis has been slow due to the disadvantages associated with whole plant cultivation, such as weather requirements, low product yield, geographical complications, pesticide/herbicide requirements and expensive extraction processes. A particular problem relates to plant vacuoles that accumulate waste products. During cell disruption to release plant protein products, waste materials are also released that contaminate the product and often cause disruption/inactivation of the final protein product.[26]

Advances in plant genetic engineering will enable a whole range of products, which were previously difficult to obtain from plants, to be produced in quantity from domesticated crop plants. Transgenic plants are beginning to compete with microbial cell systems for bulk production of biomolecules. Other improvements in plant cell tissue culture have permitted processes that provide a suitable alternative to whole plant cultivation for speciality low-volume products.[27]

16.2.6.3 Microorganisms (Bacteria, Yeast and Filamentous Fungi)

Microorganisms are without question the most versatile and adaptable forms of life and this ability has enabled them to survive on this planet for over 3 billion years. A key feature in this remarkable success is an immense capacity that microorganisms have for biocatalysis. We have made extensive use of the biocatalytic properties of microorganisms in production of beverages and foodstuffs, and this is a common factor shared by all communities in the world.

Advances in microbiological sciences over the past 70 years have revealed an enormous biocatalytic potential inherent in microorganisms and this has stimulated new contributions to medicine, agriculture, waste (water and hazardous) management and animal feed. It is clear that microorganisms have considerably more potential for biotransformation of a wide range of organic and biochemicals and current opinion suggests that less than 3% of the total range of microorganisms on Earth have been thoroughly characterised in terms of biocatalytic potential. As biocatalysts, some microorganisms are producers of organic material (autotrophs), whereas some are consumers of organic material (heterotrophs), and although bacteria are prokaryotic (*i.e.* they lack nuclear membrane, mitochondria, endoplasmic reticulum, Golgi apparatus and lysosomes) they possess sophisticated and flexible catalytic systems for both biosynthesis and biodegradation. Although easier to cultivate than animal or plant cells, they are not without some drawbacks as biocatalysts. A critical concern is safety and assurance that products are free from residual bacterial/fungal toxins/cells.

16.2.7 Biocatalyst Selection

The choice of catalyst for a particular biotransformation will be influenced by a number of factors, but three are fundamental.[2] The first is the availability of biocatalysts, and this may range from pure enzyme, partially purified enzyme, crude enzyme preparation, cell/organelle extract, dead whole cells, to living whole cells, which may be growing or resting cultures. The next factor is a suitable bioreactor together with optimum configuration and operating conditions for efficient biocatalysis. Many configurations are available and these considerations are influenced by biocatalyst characteristics, substrate properties and requirements of the finished product. The third key factor is the need or not for the additional features provided by immobilisation, and this will depend on choices made for the first two factors.[2]

16.3 IMMOBILISATION

In a solution, biocatalysts behave as any other solute in that they are readily dispersed in solution or solvent and have complete freedom of movement in solution (Figure 16.5). Immobilisation may be viewed as a procedure specifically designed to limit freedom of movement of a biocatalyst. Immobilisation normally involves attachment of biocatalyst to or location within, an insoluble support material.[3] In effect, a biocatalyst is separated from the bulk of the solution to create a heterogeneous two-phase system (Figure 16.5). Immobilisation provides three basic advantages over soluble-based systems:

- repeated use of the biocatalyst in batch reactions;
- simple separation of biocatalyst from product after reaction;
- continuous use of a biocatalyst in a suitable reactor system.

An immobilised biocatalyst is easily recovered after a batch reaction and this facility allows repeated use of the biocatalyst in a fresh batch.

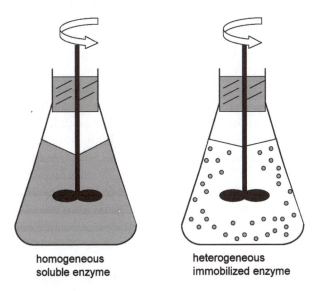

homogeneous
soluble enzyme

heterogeneous
immobilized enzyme

Figure 16.5 Diagram showing how immobilisation separates a biocatalyst from the bulk solution phase to produce a heterogeneous two-phase mixture.

This advantage can improve the commercial viability of a reaction process that was previously too expensive in terms of biocatalyst costs. Removal of biocatalyst from a product solution after reaction is complete means that the final product is not contaminated with biocatalyst and fewer expensive extraction procedures are required in finishing steps. Immobilisation of a biocatalyst will increase the options for bioreactor selection. An immobilised biocatalyst can be incorporated into cylindrical flow-through columns or other devices to produce continuous bioreactors in which a continuous flow of substrate enters at one end of the bioreactor and a continuous flow of product emerges from the other end.[28] Such systems can operate continuously for weeks or months.

Recent expansion of biotechnology and expected developments that will accrue from advances in genetic technology have stimulated enthusiasm for immobilisation of biocatalysts.[29] Research and development work has provided an extensive array of support materials and methods for biocatalyst immobilisation and much of the new immobilisation may be attributed to specific improvements for given applications rather than a concerted attempt to build a set of industrial standards for immobilisation methodology. Consequently, there have been few detailed and comprehensive comparative studies on immobilisation methods and supports. Therefore, no ideal support material or method of immobilisation has emerged to provide a standard for a given type of immobilisation. Some work using functionalised mesoporous silica with open pores has extended the understanding of enzyme behaviour in an entrapment model and has contributed to the design of confining matrices to optimise the catalytic process by developing a model that accounts for the influence of a matrix environment on concentration effects, catalytic activity, turnover, structural integrity, enzyme denaturation/renaturation cycles and solvent effects.

Selection of support material and method of immobilisation is made by weighing the various characteristics and required features of a biocatalyst application against the properties/limitations/characteristics of a combined immobilisation/support (Figure 16.6). A number of practical aspects should be considered before embarking on experimental work to ensure that the final immobilised biocatalyst is fit for a planned application and will operate with optimum effectiveness.[30]

16.3.1 Choice of Support Material

A decision on type of support material for immobilisation will require careful evaluation of the proposed use of the biocatalyst and characteristics of the intended application.[31] Virtually any inorganic, organic

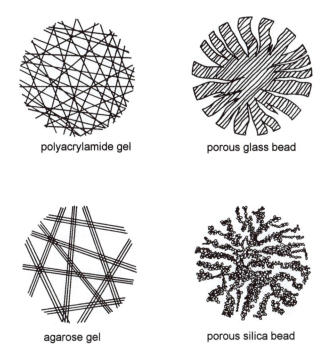

polyacrylamide gel   porous glass bead

agarose gel   porous silica bead

Figure 16.6 Microstructure of some popular support materials.

or biological material can be used or modified for use as a support material for immobilisation of biocatalysts (see Table 16.2). The choice of support is influenced by some of the support properties listed below and an evaluation process should help to inform which of the properties are important for the immobilised biocatalyst under consideration:

- Various physical properties of a support may be important, such as strength, non-compression of particles, available surface area, shape/form (beads/sheets/fibres, *etc.*), non-porous, porous (degree of porosity), pore volume, permeability, density, space for increased biomass, flow rate and pressure drop (Figure 16.6).
- Chemical features such as hydrophilicity (water binding), inertness towards a biocatalyst and available functional groups for modification, regeneration and re-use of the support may be important.
- Stability characteristics of a support material may influence storage, residual enzyme activity, cell productivity, regeneration of biocatalyst activity, maintenance of cell viability and mechanical integrity of the support.
- A resistance to bacterial/fungal attack, disruption by chemical/ pH/temperature/organic solvent, cellular defence mechanisms

Table 16.2 Some support materials for biocatalyst immobilisation.

Inorganic	Organic	Biological	Nanoparticles
Aluminium oxide	Polyethylene	Cellulose	Gold nanoparticles
Nickel oxide	Polystyrene	Dextran	Quantum dots
Stainless steel	Polyacrylate	Agarose	Nanobodies
Porous glass	Nylon	Starch	Liponanoparticles
Porous silica	Polyacrylamide	Alginate	Nanofibres
Fused silica	Polymethacrylate	Carrageenan	Nanoporous silica
Diatomaceous earth	Polypyrrole	Chitin	Nanoprisms
Iron oxide	Polyaniline	Bone	Nanotubes
Titanium oxide	Polyphenol	Chitosan	Nanoparticles
Pumice stone	Polyester	Collagen, gelatin	
Zirconium oxide	Poly(vinyl alcohol)	Liposome	
Silicon dioxide	PTFE	Cells (yeast)	
Activated carbon			

(antibodies/killer cells) may be required if a biocatalyst is to operate in a harsh or challenging environment.

- Operational safety considerations such as biocompatibility (immune response), toxicity of component or residual chemical reagents may be important if a support will come into direct or indirect contact with food, biological fluids, *etc.*

- There may be health and safety concerns with a support itself or support modification procedures and these may be important for process workers and end product users, and an immobilised biocatalyst may be subject to GRAS (generally recognised as safe) requirements for FDA approval (for food/pharmaceutical/medical applications).

- Key economic concerns may be raised concerning the availability and cost of the support material and/or chemicals, special equipment/reagents, technical skill required, environmental impact, industrial-scale chemical preparation, feasibility for scale-up, continuous processing, effective working life, reusable support and contamination (enzyme/cell-free product).

- Overall reaction characteristics influenced by a support will include flow rate, enzyme/cell loading, catalytic productivity, reaction kinetics, side reactions, multiple enzyme and or cell systems, batch or continuous operation system, reactor type and diffusion limitations on mass transfer of cofactors/substrates/products.

16.3.1.1 Nanotechnology and Nanoparticle Support Materials

The microelectronics industry and the biosensor industry have both sought greater levels of miniaturisation in their respective technology development and both have now reached a scale where it is possible to work together to produce new products.[31,32] Engineers have shrunk the dimensions of microfabricated structures to provide faster high-density electronic chips for the electronics industry. Biotechnologists have reduced the volumes required in analytical devices from millilitre to microlitre and nanolitre levels. A convergence can now take place that will allow each sector to take advantage of the unique benefits and technology that each side has developed and create new innovation in the area of nanobiotechnology.[33] Biological materials offer impressive molecular recognition and self-assembly capacity and can be immobilised on non-biological electro/mechanical components to produce new hybrid systems. New research is seeking to use the natural self-assembly characteristics of some biological molecules to assist microfabrication of nanowires for production of two- and three-dimensional nanocircuits for electronics development. A notable benefit of biological intervention in microfabrication is that biological materials allow a greater range of assembly options that can enhance traditional photolithography methods for production of electronic circuits.[33]

Innovation in molecular detection has a focus on moving towards highly multiplexed molecular recognition systems using preconfigured arrays and departing from arrays with fixed recognition sites printed on solid surfaces. This requires new methods of registering and quantifying a specific binding event through electrochemical and/or electronic events and nanotechnology will be instrumental in delivering this objective.[33] The incorporation of biological components into microelectromechanical systems (MEMS) and implantable devices is achievable, but must be built with biocompatible materials or coated with biocompatible materials.[34]

Immobilisation methods and materials have moved into a new phase to benefit from the opportunities provided by nanotechnology. New support materials such as polymeric microspheres (nanoparticles) have been produced with size ranges from 50 to 10 000 nm and used to covalently bind enzymes, antibodies and proteins.[35] Various nanosized objects have been produced, including nanotubes, nanochannels, nanoparticles, nanopores, nanocapacitors, quantum dots, nanorods and nanoprisms.[35] These have increased the range of potential support materials and support configurations that can be used for immobilisation. Further indications are that nanoparticles may be engineered to provide cell-specific delivery of therapeutic agents such as enzymes, proteins and drugs after injection into the bloodstream.[36]

Until relatively recently, most enzymes and cells were immobilised on/ in relatively large structures (micrometre to millimetre scale) to ease the collection, recovery and handling of immobilised enzyme preparations. New procedures have immobilised biocatalysts to nanoparticles with diameters on the nanometre scale. Reduction in size of support material will generally reduce some of the negative support material effects that limit the effectiveness of a catalytic reaction, particularly in relation to diffusion of substrates and products to and from an enzyme.[37] In addition, nanoparticles composed of or covered with biocompatible materials have potential for *in vivo* delivery of therapeutic systems. For example, nanoparticles of gold have very low toxicity.

Nanotechnology can provide exciting opportunities for a wide range of significant developments in biotechnology. However, there is one obstacle that requires a solution if the opportunities are to be fully realised and that is energy. Biological systems require and can transduce very specific chemical energy to drive biological systems. Microelectronic systems require and can transduce electrical energy to drive microelectromechanical devices. The two are not easily compatible. An efficient, workable solution that provides a common energy currency to both systems will provide a foundation for new generations of bio-microelectromechanical devices. A solution will facilitate production of a wider range of efficient functional hybrid devices composed of active biological and active non-biological electronic materials each drawing energy from a single energy source.

16.3.2   Choice of Immobilisation Procedures

There are five principal methods for immobilisation of biocatalysts: adsorption, covalent binding, entrapment, encapsulation and cross-linking (Figure 16.7), and the relative merits of each are discussed below.

16.3.2.1   Adsorption

Immobilisation by adsorption is the simplest method and involves reversible non-covalent interactions between biocatalyst and support material (Figure 16.7).[38] The forces involved are mostly weak electrostatic such as van der Waals forces and ionic and hydrogen bonding interactions, although hydrophobic bonding can be significant. These forces are individually very weak, but sufficiently large in number to provide good binding. For example, it is known that yeast cells have a surface chemistry that is substantially negatively charged so that use of a positively charged support material will allow immobilisation. Existing surface chemistry on a biocatalyst and on a support is utilised for adsorption, so no chemical activation or modification of support is

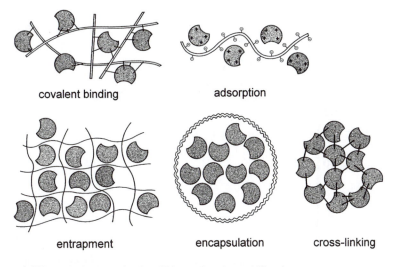

Figure 16.7 Principal methods of biocatalyst immobilisation.

required and little damage is inflicted on biocatalysts in this method of immobilisation.[39]

The procedure consists of mixing together biocatalyst and support under suitable conditions of pH, ionic strength, *etc.*, for a period of incubation, followed by collection of immobilised biocatalyst and extensive washing to remove non-bound biocatalyst.

Advantages:

- little or no damage to a biocatalyst;
- simple, cheap and quick to obtain immobilisation;
- no chemical changes required for support or biocatalyst;
- easily reversed to allow regeneration with fresh biocatalyst.

Disadvantages:

- leakage of biocatalyst from support/contamination of product;
- non-specific binding of other material to the support;
- overloading of a support with biocatalyst;
- steric hindrance of a biocatalyst by a support.

The most significant disadvantage is leakage or desorption of a biocatalyst from support material. This can occur under circumstances such as environmental changes in pH, temperature and ionic strength. Sometimes a firmly adsorbed enzyme is readily desorbed during reaction

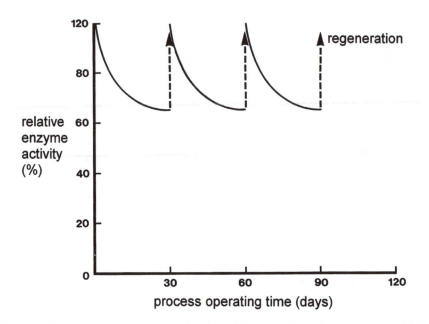

Figure 16.8   Graph showing the steady loss of immobilised enzyme with time and regeneration of activity at monthly intervals by addition of fresh enzyme.

as a result of substrate binding, binding of contaminants present in the substrate, product production or other condition leading to change in conformation (molecular shape) of the adsorbed enzyme.

Physical characteristics of bioreactors such as flow rate, bubble agitation, particle–particle abrasion and scouring effect of particulate materials on reactor vessel walls can lead to desorption.[40] Desorption can be turned to advantage if regeneration of support is built into the operational regime to allow rapid expulsion of 'exhausted' biocatalyst and replacement with fresh biocatalyst. In some commercial biotransformations, the economics of a process require the use of crude substrates containing contaminants that 'poison' the biocatalyst and so regular regeneration of biocatalyst is an essential part of the operation (Figure 16.8).

Non-specific binding can become a problem if substrate, product and or residual contaminants possess charges that allow interaction with a support material. This can lead to diffusion limitations and reaction kinetics problems, with consequent alteration in enzyme kinetic parameters (see Section 16.4). Further, binding of hydrogen ions to a support material can result in an altered pH microenvironment around a support with a consequent shift in pH optimum (1–2 pH units) of a biocatalyst, and this may be detrimental to enzymes with precise pH requirements. Unless carefully controlled, overloading a support with excess biocatalyst can lead to low catalytic activity and in some circumstances a spacer molecule situated between an enzyme molecule and a support may be required to overcome problems of steric hindrance (see Section 16.4).

16.3.2.2 Covalent Binding

This method of immobilisation involves the formation of a covalent bond between a biocatalyst and support material (Figure 16.7). The bond is normally formed between functional groups present on the surface of a support material and functional groups on amino acid residues on the surface of an enzyme. A number of amino acid functional groups are suitable for participation in covalent bond formation. Those most often involved are the amino group (NH_2) of lysine or arginine, the carboxyl group (CO_2H) of aspartic acid or glutamic acid, the hydroxyl group (OH) of serine or threonine and threshold group (SH) of cysteine.[3]

Many varied support materials are available for covalent binding and the extensive range of supports available reflects the fact that there is no ideal support material (see Table 16.1). Therefore, the advantages and disadvantages of a given support are taken into account when considering possible procedures for a given biocatalyst immobilisation. Many factors may influence the selection of a particular support and research has shown that hydrophilicity (water attraction) is the most important factor for maintaining enzyme activity in a support material environment. Consequently, polysaccharide polymers, which are very hydrophilic, are popular support materials for enzyme immobilisation.[41] For example, cellulose, dextran, starch and agarose are often used for enzyme immobilisation. The sugar residues in these polymers contain hydroxyl (OH) groups that are suitable for chemical activation to provide covalent bond formation (Figure 16.9). The OH groups also form

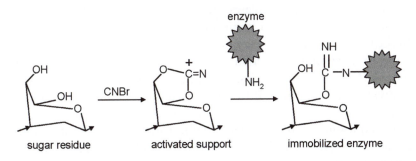

Figure 16.9 Covalent immobilisation by activation of a carbohydrate support material with CNBr and coupling of an enzyme (via free amino group) to the activated support to form an isourea bond.

hydrogen bonds with water molecules and thereby create a hydrophilic environment within a support material.

Polysaccharide supports do have a weakness in that they are susceptible to microbial/fungal disintegration and organic solvents used in some reaction processes can cause shrinkage of polysaccharide gels. Support materials are usually used in bead form (Figure 16.6) and there are many procedures for coupling an enzyme and a support in a covalent bond. However, most reactions fall into the following main categories:

- formation of an isourea linkage;
- formation of a diazo linkage;
- formation of a peptide bond;
- an alkylation reaction.

It is important to choose a reaction method that will not inactivate an enzyme by reacting with amino acid residues at the enzyme active site. Hence if an enzyme employs a carboxyl group at its active site (see Figure 16.1) for participation in catalysis, then it is wise to choose a reaction method that involves amino groups for covalent bond formation with a support.[42] Basically two steps are involved in covalent binding of enzymes to support materials.

First, functional groups on a support material are activated by a specific chemical reagent such as cyanogen bromide (CNBr), and second, an enzyme is added in a coupling reaction to form a covalent bond with the activated support material. Normally an activation reaction is designed to make functional groups on a support strongly electrophilic (electron deficient). In the coupling reaction, these groups will react with strong nucleophiles (electron donating), such as the amino (NH_2) functional groups of certain amino acids on the surface of an enzyme, to form a covalent bond. It is possible to vary the coupling conditions to allow single-point or multiple-point attachment (see Figure 16.11) and influence the stability of an immobilised enzyme (see Section 16.4).

In the CNBr method, an enzyme and support are joined via an isourea linkage. In carbodiimide activation, a support material should have a carboxyl (CO_2H) functional group and an enzyme and support are joined via a peptide bond. If the support material contains an aromatic amino functional group, it can be diazotised using nitrous acid. Subsequent addition of enzyme leads to the formation of a diazo linkage between a reactive diazo group on a support and the ring structure of an aromatic amino acid such as tyrosine.

It is important to recognise that no method of immobilisation is restricted to a particular type of support material and that an extremely large number of permutations are possible between methods of immobilisation and support materials. This is possible by chemical modification of normal functional groups on a support material to produce a range of derivatives containing different functional groups. For example, the normal functional group in cellulose is the OH of glucose and chemical modifications of these groups have produced a range of cellulose derivatives such as AE-cellulose (aminoethyl), CM-cellulose (carboxymethyl) and DEAE-cellulose (diethylaminoethyl). Thus chemical modification increases the range of immobilisation methods that can be used for a given support material.[43]

16.3.2.3 Entrapment

Immobilisation by entrapment differs from adsorption and covalent binding in that enzyme molecules are free in solution, but restricted in movement by the lattice structure of a gel (see Figure 16.6, agarose). The porosity of a gel lattice is controlled to ensure that the structure is tight enough to prevent leakage of enzyme or cells and at the same time allow free movement of substrate and product. Inevitably the support will act as a barrier to mass transfer and, although this can have serious implications for reaction kinetics, it can have useful advantages as harmful cells, proteins and enzymes are prevented from interaction with the immobilised biocatalyst. There are several methods of entrapment:

- ionotropic gelation with multivalent cations (*e.g.* alginate);
- temperature-induced gelation (*e.g.* agarose, gelatin);
- organic polymerisation by chemical/photochemical reaction (*e.g.* polyacrylamide);
- precipitation from an immiscible solvent (*e.g.* polystyrene).

Entrapment can be achieved by mixing an enzyme with a polyionic polymer material and then cross-linking the polymer with multivalent cations in an ion-exchange reaction to form a lattice structure that traps the biocatalyst (ionotropic gelation).[44] Temperature change is a simple method of gelation by phase transition using 1–4% solutions of agarose or gelatin. However, the gels formed are soft and unstable. A significant development in this area has been an introduction of κ-carrageenan polymers, which can form gels by ionotropic gelation and by temperature-induced phase transition, and this has introduced a greater degree of flexibility in gelation systems for immobilisation.

Alternatively, it is possible to mix an enzyme with chemical monomers that are then polymerised to form a cross-linked polymeric network, trapping the enzyme in the interstitial spaces of a lattice (see Figure 16.6, polyacrylamide). The latter method is more widely used and a number of acrylic monomers are available for the formation of hydrophilic copolymers. For example, acrylamide monomer is polymerised to form polyacrylamide and methyl acrylate is polymerised to form polymethacrylate. In addition to a monomer, a cross-linking agent is added during polymerisation to form cross-linkages between polymer chains and help to create a three-dimensional network lattice. The pore size of a gel and its mechanical properties are determined by the relative amounts of monomer and cross-linking agent. It is possible to vary these concentrations to influence the porosity of a lattice structure. The polymer formed may be broken up into particles of a desired size or polymerisation can be arranged to form beads of defined size to suit particular bioreactor requirements. Precipitation occurs by phase separation rather than by chemical reaction, but does bring a biocatalyst into contact with a water-miscible organic solvent and most biocatalysts are not tolerant of such solvents. Hence this method is limited to highly stable/previously stabilised enzymes or non-living cells.

16.3.2.4 Encapsulation

Encapsulation of enzymes and or cells can be achieved by enveloping biological components within various forms of semi-permeable membrane (see Figure 16.7). It is similar to entrapment in that enzymes/cells are free in solution, but restricted in space. Large proteins or enzymes cannot pass out of or into the capsule, but small substrates and products can pass freely across the semipermeable membrane. Many materials have been used to construct microcapsules varying from 10 to 100 µm in diameter and nylon membranes and cellulose nitrate membranes have proved popular.[45]

Problems associated with diffusion of substrates in/products out are more acute and can result in rupture of a membrane if products from a reaction accumulate rapidly (Figure 16.10). A further problem may arise if an immobilised biocatalyst particle has a density fairly similar to that of the bulk solution and therefore 'floats' in the solution. This can cause process problems and may require reassessment of reactor configuration, flow dynamics, *etc*.

It is also possible to use biological cells as capsules and a notable example of this is the use of erythrocytes (red blood cells). The membrane of an erythrocyte is normally only permeable to small molecules. However, when erythrocytes are placed in a hypotonic solution they swell up and the cell membrane stretches and increases membrane

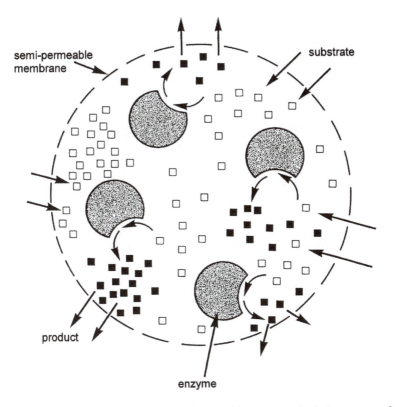

semi-permeable membrane

substrate

product

enzyme

Figure 16.10   Immobilisation in a semipermeable enzyme depicting some of the problems associated with diffusion of substrate into and product out of the structure.

permeability. In this condition, erythrocyte proteins diffuse out of the cell and enzymes can diffuse into the cell. Returning swollen erythrocytes to an isotonic solution allows the cell membrane to return to normal and trapped enzymes inside the cell do not leak out. Recent work has used a filamentous fungus to entrap yeast cells in an arrangement that represents a forced symbiosis with modification of growth media to supply a particular carbon source that can only be used by the fungus.[46]

16.3.2.5   Cross-linking
This type of immobilisation is support free and involves joining cells (or enzymes) to each other to form a large three-dimensional complex structure and it can be achieved by chemical or by physical methods (see Figure 16.7). Chemical methods of cross-linking normally involve

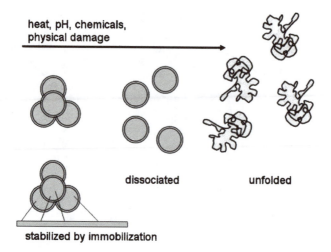

heat, pH, chemicals,
physical damage

dissociated   unfolded

stabilized by immobilization

Figure 16.11 Schematic depicting the stages in denaturation of an enzyme and stabilisation of an enzyme against protein denaturation by multi-point attachment to a support material.

covalent bond formation between proteins or cells by means of a bi- or multifunctional reagent such as glutaraldehyde and toluene diisocyanate.[47] However, the toxicity of such reagents is a limiting factor in applying this method to living cells and many enzymes. Both albumin and gelatin have been used to provide additional protein molecules as spacers to minimise the close proximity or crowding problems that can be caused by cross-linking an enzyme.

Physical cross-linking of cells by flocculation is well known in the biotechnology industry and does lead to high cell densities. Flocculating agents such as polyamines, polyethylenimine, polystyrene sulfonates and various phosphates have been used extensively and are well characterised. Cross-linking is rarely used as the only means of immobilisation because absence of mechanical properties and poor stability are severe limitations. Cross-linking is most often used to enhance other methods of immobilisation by assisting the reduction of cell or enzyme leakage in other systems.

A potentially useful development is the production of cross-linked enzyme crystals (CLECs).[48] Production of enzyme crystals has long been seen as a final step in enzyme purification procedures as it signifies high purity. The principal use of enzyme crystals *per se* has been in X-ray crystallography to study enzyme structures and recent recognition that some enzyme crystals were catalytically active has led to the development of CLECs.

Indications are that CLECs have superior stability characteristics and may enable novel enzyme-catalysed biotransformations to occur in difficult environments such as in organic solvents, in a gas phase or in supercritical fluids. Enzyme crystals might be more easily incorporated into microelectronic devices such as a CHEMFET (chemically sensitive field effect transistor) and ISFET (ion-selective field effect transistor) to produce novel 'bio-chips' for the next generation of biosensors.

16.4 PROPERTIES OF IMMOBILISED BIOCATALYSTS

When a biocatalyst is immobilised, its fundamental characteristics are usually changed in one way or another and the change may be a drawback or an improvement. The nature of the alteration depends on the inherent properties of the biocatalyst and additional characteristics imposed by the support material on the biocatalyst, substrate and product.[49] It is very difficult to quantify these properties and characteristics given the diversity of biocatalysts, support materials and methods of immobilisation. Consequently, it has proved impossible to predict completely what effect a particular immobilisation will have on an enzyme or cell or on the reaction that it catalyses, and the only recourse is to evaluate a number of methods to discover and develop a system that provides the greatest positive improvement for the application under consideration.[50]

16.4.1 Stability

The two most important properties that may be changed by immobilisation are stability and catalytic activity. Stability is defined as an ability to resist alteration and in the context of biocatalyst stability it is important to distinguish several different types of stability:

- inactivation by heat;
- disruption by chemicals;
- digestion by proteases or cells;
- inactivation by change in pH;
- loss of catalytic activity during storage;
- loss of catalytic activity due to process operations.

The various types are not necessarily interdependent and an observed increase in heat stability does not indicate that there will be a corresponding increase in storage stability or operational stability. Although immobilisation does not guarantee an improvement in stability, it is generally recognised that it does represent a strategy that can be used as a means of developing more stable enzyme preparations.[51]

Generally, it is found that covalent immobilisation is more effective than the other methods at improving enzyme resistance to heat, chemical disruption and pH changes. Disruptants normally induce loss of catalytic activity by causing a considerable alteration in the protein structure of an enzyme (Figure 16.11). In particular, disruptants disperse the many non-covalent bonds responsible for holding the enzyme polypeptide chain in its highly specific shape or conformation, thus causing the polypeptide chain to unfold with a consequent loss of active site structure and catalytic activity. Given that unfolding is associated with loss of activity, it is probable that multi-point attachment of an enzyme polypeptide chain to a support material provides extra rigidity to the folded protein chain and therefore greater resistance to protein unfolding (Figure 16.11).

16.4.2 Catalytic Activity

Immobilisation almost invariably changes the catalytic activity of an enzyme, and this is clearly reflected in alterations in the characteristic kinetic constants of the enzyme-catalysed reaction. In particular, the maximum reaction velocity (V_{max}) obtained with an immobilised enzyme is usually lower than that obtained with the corresponding soluble enzyme under the same reaction conditions. The Michaelis constant (K_m), which reflects the affinity that the enzyme has for its substrate, is usually changed upon immobilisation, indicating that binding of substrate to the active site has been altered. Four principal factors influence the catalytic activity of immobilised enzymes: (a) conformation, (b) steric, (c) micro-environment and (d) diffusion.

The conformation of an enzyme refers to the particular shape adopted by the polypeptide chain, which is essential for maintaining the active site structure (see Figure 16.1). Immobilisation procedures that involve modification or interaction with amino acid residues on the polypeptide chain can sometimes disturb protein structure and thereby affect the enzyme activity. Covalent immobilisation is most likely to cause an alteration in the protein conformation of an enzyme. A steric problem arises if the enzyme is immobilised in a position that causes the active site to be less accessible to the substrate molecules. In solution, a free enzyme molecule is surrounded by a homogeneous micro-environment in which the enzyme is fully integrated with all components of the solution.

Immobilisation creates a heterogeneous micro-environment consisting of two phases, *i.e.* the immobilised enzyme and the bulk of the solution from which the immobilised enzyme is separated (see Figure 16.5). Therefore, all components of the reaction, substrate, products, activators, ions, *etc.*, are partitioned between the immobilised enzyme phase and the bulk solution phase. This feature can significantly alter the characteristics of an enzyme reaction even if the enzyme molecule itself is not changed by immobilisation. The support material may influence the partitioning effect. If the support material attracts substrate then this can improve catalytic activity. Reaction rate is also reduced by diffusion restriction. As the substrate is consumed, more substrate must diffuse into the enzyme from the bulk solution and product must diffuse away from the active site. This is normally a problem for all forms of immobilised enzymes, but particularly so for encapsulated enzymes (see Figure 16.10).

Diffusional limitations may be divided into two types, external diffusion restriction and internal diffusion restriction. The external type refers to a zone or barrier that surrounds the support material, called the Nernst layer. Substrate molecules can diffuse into this layer by normal convection and by a passive molecular diffusion. If substrate molecules pass through this layer slowly, then this may limit the rate of enzyme reaction. External diffusion restriction can be improved by speeding up the flow of solvent over and through the immobilised enzyme by increasing the stirring rate.

Internal diffusion restrictions are due to a diffusion limitation inside the immobilised enzyme preparation (see Figure 16.10). In this case, diffusion of substrate molecules occurs by a passive molecular mechanism only, which may be more difficult to overcome if it is a seriously limiting factor. The overall rate of diffusion is markedly influenced by the method of immobilisation. Covalent and adsorption procedures cause less diffusion limitation than do entrapment and encapsulation procedures.

16.4.3 Coenzyme Regeneration

Around 25% of enzymes are oxidoreductases (see Table 16.1) and they catalyse many reactions that are of great interest to biotechnologists. However, most of these enzymes require participation of a coenzyme such as NAD, NADH, NADP, NADPH or ATP.[51] Coenzymes are chemically changed during reaction and so in effect become an expensive consumable. Hence efficient regeneration of coenzyme is essential

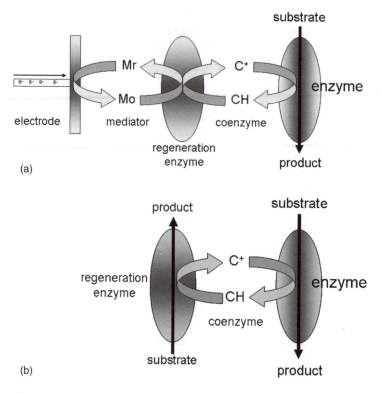

(a)

(b)

Figure 16.12   Diagram depicting coenzyme regeneration using (a) an electrode, regeneration enzyme and chemical mediator (Mr, reduced; Mo, oxidised) to recycle the coenzyme and (b) a regeneration enzyme in a reaction that utilises a secondary substrate.

for applications that utilise these enzymes. There are basically two mechanisms to regenerate coenzyme, as depicted in Figure 16.12. In the first method (Figure 16.12a), the coenzyme is regenerated electrochemically using electrons supplied from an electrode.[51] To improve efficiency, a regeneration enzyme and chemical mediators are added to more effectively interact with the electrode. The second method (Figure 16.12b) does not use an electrode and instead uses a second enzyme that catalyses a reverse reaction to regenerate coenzyme. Use of a second enzyme reaction requires close coupling of the reaction kinetics of the secondary enzyme to match that of the primary enzyme reaction. For example, if the rate of catalysis of the primary reaction is $500\,\mu\mathrm{mol\,min^{-1}}$ and the secondary enzyme catalyses at a rate of $200\,\mu\mathrm{mol\,min^{-1}}$, then the slower reaction will drag the primary reaction and lower the efficiency. On the other hand, if the secondary reaction is too fast (*e.g.* 2000 µmol min^{-1}), then there would be inefficient and uneconomical waste of material. There are enzymes available that have been adapted for coenzyme regeneration to improve the efficiency of the process.[52]

Many attempts have been made to immobilise coenzymes and enzymes together to try to produce a universal efficient biocatalytic system with continuous coenzyme regeneration and, although some systems work well, further innovation is still required to produce a fully efficient and effective system that will realise the full potential of all the valuable enzymes that have coenzyme requirements.

16.5 APPLICATIONS

Immobilisation technology is now firmly established[53] among the tools used by biotechnologists to improve the effectiveness of biocatalysts for biotransformations in industry, analysis and medicine. Consequently, there are many examples of applications that use immobilised biocatalysts and further information can been obtained from reviews and books. A few examples are given in Table 16.3 to indicate the wide range. The future for immobilised biocatalysts is promising and exciting developments in this field will support the expected expansion and progress of biotechnology in various industries and medicine in the 21st century. Advances in nanotechnology have provided engineering tools to facilitate the development of enzyme-immobilised microchannel reactors that utilise microlitre and nanolitre fluidic systems for enzyme reactions.[53]

Material engineers have produced highly defined nanoparticles that enable immobilised enzymes to be situated in defined fluidic microchannels for microscale reaction processes. Advances in the understanding and application of microfluidics will increase the scope for microscale reactions that have many advantages including substantial reduction of mass transfer effects and high enzyme loading.[37,54] However, the debate surrounding the impact of bio-nanotechnology on society has started and, in the light of public understanding and reaction to genetically modified (GM) food, it is likely that biotechnologists will have an important role to play in ensuring that applications and impacts associated with environmental and health issues have a stronger profile in general discussions of the impact of bio-nanotechnology on society.[55]

Table 16.3   Some applications of immobilised biocatalysts.

Application	Biocatalyst	Immobilised biocatalyst
Medicine		
Metabolic deficiencies	Fibroblast cells	Cell/alginate
Diabetes	Pancreatic islet cells	Cell/agarose
Bone regeneration	Stem cells	Cell/alginate
Liver transplant	Hepatocyte cells	Cell/alginate
Parkinson's disease	Adrenal chromaffin cells	Cell/alginate
Cancer therapy	Neuraminidase	Enzyme/gelatin membrane
Cystic fibrosis lung decongestion	DNAase	Enzyme/dextran
Burn and ulcer treatment	Collagenase	Enzyme/cotton fibres
Drug detoxification (overdose)	Cytochromes P450	Enzymes/hollow fibres
Biosensors		
Serum cholesterol level	Cholesterol esterase	Enzyme/nylon membrane
Serum glucose level	Glucose dehydrogenase	Enzyme/collagen membrane
Serum alcohol level	Alcohol oxidase	Enzyme/cross-linked polymer
Diamine level in foodstuff	Diamine oxidase	Enzyme/cross-linked with glutaraldehyde
NADH level	NADH oxidase	Enzyme/Immobilon AV membrane
Glucuronide-drug levels	Animal cell microsomes	Microsomes/organic membrane
Food industry		
Glucose → fructose	Bacterial cells	*Arthrobacter* cells/polyelectrolyte complex
Glucose → fructose	Glucose isomerase	Enzyme/polystyrene beads
Lactose → glucose + fructose	Lactase enzyme	Enzyme/alginate beads
Aspartame biosynthesis	Thermolysin	Enzyme/hydrophobic beads
Starch → maltose	α-Amylase	Enzyme/hollow fibres
Pharmaceutical industry		
6-APA → ampicillin	Bacterial cells	*Bacillus magaterium*/DEAE-cellulose
Glucose → penicillin G	Bacterial cells	*Penicillium chrysogenum*/poly-acrylamide gel
Digitoxin → digoxin	Plant cells	*Digitalis lanata*/alginate beads
Human gamma interferon	Animal cells	Chinese hamster ovary/alginate
Biotechnology		
Urocanic acid biosynthesis	Bacterial cells	*Achromobacter liquidum*/poly-acrylamide gel
Coenzyme A biosynthesis	Bacterial cells	*Brevibacterium ammoniagenes*/Cellophane
Hydrolysis of palm oil	Lipase	Enzyme/zeolite type Y
Nitrate removal (groundwater)	Bacterial cells	*Scenedesmus obliquus*/polyvinyl foam
Cellulose hydrolysis	Cellulosome complex	Multienzyme complex/calcium phosphate gel

REFERENCES

1. G. F. Bickerstaff, *Genetic Eng. Biotechnol.*, 1995, **15**, 13–30.
2. G. F. Bickerstaff, *Immobilisation of Enzymes and Cells*, Humana Press, Totowa, NJ, 1997.
3. G. F. Bickerstaff, *Enzymes in Industry and Medicine*, Edward Arnold, London, 1987.
4. D. J. Pollard and J. M. Woodley, *Trends Biotechnol.*, 2006, **25**, 66–73.
5. J. L. Kaar, H.-I. Oh, A. J. Russell and W. J. Federspiel, *Biomaterials*, 2007, **28**, 3131–3139.
6. S. Schubert and J. Kurreck, *Curr. Drug Targets*, 2004, **5**, 667–681.
7. J. C. Schlatterer, F. Stuhlmann and A. Jäschke, *ChemBioChem*, 2003, **4**, 1089–1092.
8. M. V. Rodnina, M. Beringer and W. Wintermeyer, *Trends Biochem. Sci.*, 2006, **32**, 20–26.
9. J. Tazi, S. Durand and P. Jeanteur, *Trends Biochem. Sci.*, 2005, **30**, 469–478.
10. A. G. Gabibov, *Immunol. Lett.*, 2006, **103**, 1–2.
11. A. G. Gabibov, N. A. Ponomarenko, E. B. Tretyak, M. A. Paltsev and S. V. Suchkov, *Autoimmun. Rev.*, 2006, **5**, 324–330.
12. C. Adrain, G. Brumatti and S. J. Martin, *Trends Mol. Med.*, 2006, **31**, 565–570.
13. X. Yu, D. Acehan, J.-F. Ménétret, C. R. Booth, S. J. Ludtke, S. J. Riedl, Y. Shi, X. Wang and C. W. Akey, *Structure*, 2005, **13**, 1725–1735.
14. A. C. Ciechanover and M. G. Masucci, *The Ubiquitin–Proteasome Proteolytic System*, World Scientific, Singapore, 2002.
15. A. J. Rivett and A. R. Hearn, *Curr. Protein Pept. Sci.*, 2004, **5**, 153–161.
16. E. A. Bayer, R. Lamed and M. E. Himmel, *Curr. Opin. Biotechnol.*, 2007, **18**, 237–245.
17. E. A. Bayer, J.-P. Belaich, Y. Shoham and R. Lamed, *Annu. Rev. Microbiol.* 2004, **58**, 521–554.
18. G. B. Sukhorukov and H. Möhwald, *Trends Biotechnol.*, 2006, **25**, 93–98.
19. G. Orive, R. M. Hernández, A. R. Gascón and J. L. Pedraz, in *Immobilisation of Enzymes and Cells*, 2nd edn, ed. J. M. Guisan, , Humana Press, Totowa, NJ, 2006, pp. 427–437.
20. E. E. Kheadr, J. C. Vuillemard and S. A. El-Deeb, *Food Res. Int.*, 2003, **36**, 241–252.

21. V. Nedovic and R. Willaert, *Applications of Cell Immobilisation Biotechnology*, Springer, Berlin, 2005.
22. V. Nedovic and R. Willaert, *Fundamentals of Cell Immobilisation Biotechnology*, Springer, Berlin, 2004.
23. H. Lodish, A. Berk, P. Matsudaira, C. A. Kaiser, M. Krieger, M. P. Scott, S. L. Zipursky and J. Darnell, *Molecular Cell Biology*, 5th edn, Freeman, New York, 2004.
24. V. Gomod, P. Chamberlain, R. Jefferis and L. Faye, *Trends Biotechnol.*, 2005, **23**, 559–565.
25. F. Bourgaud, A. Gravot, S. Milesi and E. Gontier, *Plant Sci.*, 2001, **161**, 839–851.
26. R. Y. K. Yang, O. Bayraktar and H. T. Pu, *J. Biotechnol.*, 2003, **100**, 13–22.
27. O. S. M. Goddijin and J. Pen, *Trends Biotechnol.*, 1995, **13**, 379–387.
28. L. Cao, Carrier-bound Immobilized Enzymes, Wiley InterScience, New York, 2006.
29. P. Gadler and K. Faber, *Trends Biotechnol.*, 2006, **25**, 83–88.
30. G. F. Bickerstaff, in *Immobilisation of Enzymes and Cells*, ed. G.F. Bickerstaff, Humana Press, Totowa, NJ, 1997, pp. 1–11.
31. A. Curtis and C. Wilkinson, *Trends Biotechnol.*, 2001, **19**, 98–101.
32. B. M. Cullum and T. Vo-Dinh, *Trends Biotechnol.*, 2000, **18**, 388–393.
33. L.-Q. Wu and G. F. Payne, *Trends Biotechnol.*, 2004, **22**, 593–599.
34. P. Fortina, L. J. Kricka, S. Surrey and P. Grodzinski, *Trends Biotechnol.*, 2005, **23**, 168–173.
35. K. K. Jain, *Trends Biotechnol.*, 2006, **24**, 143–145.
36. S. S. Davis, *Trends Biotechnol.*, 1997, **15**, 217–223.
37. J. Kim, J. W. Grate and P. Wang, *Chem. Eng. Sci.*, 2006, **61**, 1017–1026.
38. J. M. Guisán, in *Immobilisation of Enzymes and Cells*,2nd edn, ed. J. M. Guisan, , Humana Press, Totowa, NJ, 2006, pp. 1–13.
39. N. Öztürk, A. Akgöl, M. Arisoy and A. Denizli, *Sep. Purif. Technol.*, 2007, **58**, 83–90.
40. R. Reshmi, G. Sanjay and S. Sugunan, *Catal. Commun.*, 2006, **7**, 460–465.
41. M. Paterson and J. F. Kennedy, in *Immobilisation of Enzymes and Cells*, ed. G.F. Bickerstaff, Humana Press, Totowa, NJ, 1997, pp. 153–165.
42. M. F. Cardosi, in *Immobilisation of Enzymes and Cells*, ed. G.F. Bickerstaff, Humana Press, Totowa, NJ, 1997, pp. 217–227.

43. C. Mateo, J. M. Palomo, G. Fernández-Lorente, J. M. Guisán and R. Fernández-Lafuente, *Enzyme Microb. Technol.*, 2007, **40**, 1451–1463.
44. J. E. Fraser and G. F. Bickerstaff, in *Immobilisation of Enzymes and Cells*, ed. G.F. Bickerstaff, Humana Press, Totowa, NJ, 1997, pp. 61–66.
45. G. Orive, M. R. Hernández, A. R. Gascón, R. Calafiore, T. M. S. Chang, P. deVos, G. Hortelano, D. Hunkeler, I. Lacík and J. L. Pedraz, *Trends Biotechnol.*, 2004, **22**, 87–92.
46. R. A. Peinado, J. J. Moreno, J. M. Villalba, J. A. González-Reyes, J. M. Ortega and J. C. Mauricio, *Enzyme Microb. Technol.*, 2006, **40**, 79–84.
47. A. L. Margolin, *Trends Biotechnol.*, 1996, **14**, 223–230.
48. H. Noritomi, A. Sasanuma, A. Kato and K. Nagahama, *Biochem. Eng. J.*, 2007, **33**, 228–231.
49. A. Pandey, C. Webb, C. R. Soccol and C. Larrochem (eds.), *Enzyme Technology*, Springer, Berlin, 2006.
50. J. Rudge and G. F. Bickerstaff, *Biochem. Soc. Trans.*, 1984, **12**, 311–312.
51. J. Toher, A. M. Kelly and G. F. Bickerstaff, *Biochem. Soc. Trans.*, 1990, **18**, 313–314.
52. W. Liu and P. Wang, *Biotech. Adv.*, 2007, **25**, 369–384.
53. S. Venkatasubbarao, *Trends Biotechnol.*, 2004, **22**, 630–637.
54. M. Miyazaki and H. Maeda, *Trends Biotechnol.*, 2006, **24**, 463–470.
55. D. Parr, *Trends Biotechnol.*, 2005, **23**, 395–398.

CHAPTER 17

Downstream Processing

DANIEL G. BRACEWELL,[a] MOHAMMAD ALI S. MUMTAZ[a] AND
C. MARK SMALES[b]

[a] Department of Biochemical Engineering, University College London,
Torrington Place, London WC1E 7JE, UK; [b] Protein Science Group,
Department of Biosciences, University of Kent, Canterbury, Kent CT2 7NJ,
UK

17.1 INTRODUCTION

The production of high-value biological products has intensified many
times over in the last few decades. Since the original sourcing of natural
products from the likes of plants, animals and microbes, the advent of
recombinant DNA technology has allowed researchers and industrialists
alike the ability to produce almost any biological product that one might
desire to a relatively large scale and with consistent reproducibility.
Whether using 'natural' sources of biological products (*e.g.* blood,
plants, microorganisms) or recombinant material expressed intracellularly or secreted into culture medium and feedstocks, in almost all cases
it is necessary to purify (at least partially) the product of interest from
other biological contaminants. This process is termed downstream
processing and refers to the recovery of target biological products such
as proteins, peptides, DNA and virus particles from other contaminants.

With the growing demand for biological products and increased protein yields from fermentation processes, particularly from mammalian
expression systems, downstream bioprocessing has become both a major

Molecular Biology and Biotechnology, 5th Edition
Edited by John M Walker and Ralph Rapley
© Royal Society of Chemistry 2009
Published by the Royal Society of Chemistry, www.rsc.org

expense and a bottleneck in the production of biological products at large scale. Indeed, since the 1980s, developments in the use of mammalian cell culture for the expression of high-value therapeutic proteins have seen productivities and yields increase by over 100-fold,[1] and this has resulted in pressure on the ability to deliver efficient and cost-efficient downstream processing of these molecules. Although improvements in product yields per unit volume are economically beneficial, unfortunately this does not directly correlate with downstream processing. In most cases, where chromatography is central to downstream processing, it is the total mass of the product that determines the amount of chromatography resin required and therefore cost.[2] Currently it is estimated that industrial downstream bioprocessing of biopharmaceuticals constitutes over 40% of the manufacturing cost,[3] and this is sure to rise as yields are further increased. The design of downstream processing systems on an industrial scale must therefore be carefully undertaken and optimised, although this is not necessarily as crucial for academic or small-scale industrial laboratory-based processes.

Traditionally, downstream processing has been, and continues to be, heavily reliant upon adsorptive chromatographic procedures, although high-performance liquid chromatography (HPLC) and size-exclusion chromatography (SEC) also play key roles. The actual approach and series of steps undertaken during the downstream processing for any given biomolecule are determined by a number of factors, including the nature of the molecule, the source and how the material is presented for processing. Further, on an industrial scale, the number of steps involved, recoveries, requirements and cost of the process as a whole must be carefully considered before designing the approach to be taken. Hence the selection of all unit operations to be utilised throughout a processing workflow must be assessed before designing and undertaking any system. The variables involved in such systems and the interactions between these have recently been investigated using two case studies.[4] We note that throughout the work process it is necessary to monitor continually the integrity and authenticity of the molecule of interest using an array of analytical technologies, and this must be integrated into the workflow. Here we restrict ourselves to the purification of proteins from *in vitro* cultured expression systems (*e.g. E. coli*, mammalian cells) to illustrate the typical workflow and processes involved in downstream bioprocessing.

17.2   INITIAL CONSIDERATIONS AND PRIMARY RECOVERY

With respect to the use of recombinant DNA technology and cell expression systems, a significant difference between the *E. coli* and mammalian cell expression systems used both academically and industrially is that whereas *E. coli* proteins are expressed intracellularly, industrially relevant protein targets expressed in mammalian cells are typically processed through the secretory pathway and secreted into the surrounding culture media. Therefore, the recombinant material from such *in vitro* mammalian expression systems may be recovered directly from the culture medium whereas in the case of expression in *E.coli* it is necessary to break open the cells (lysis) in order to release the intracellular recombinant protein.

17.2.1 Centrifugation and Filtration

Centrifugation and filtration are often the first step employed in any downstream bioprocessing workflow. Centrifugation is used either to remove or to collect cells and debris (collect cells where intracellular material needs to be released; remove cell debris where the biomolecule of interest is in the culture medium) before further processing is undertaken. Due to the limitations of size, batch centrifugation on a large scale is not feasible and in this case continuous-flow centrifugation can be used. Both centrifugation and filtration can be used to clarify cell culture supernatants and cell extracts; however, on a large scale traditional filtering is not possible due to the membrane size that would be required and fouling of the membrane. Cross-flow filtration has been employed to reduce both fouling of the membrane and improve the flow rate, and this approach has been successfully used to remove soluble host proteins during the recovery of inclusion bodies in *E. coli* cell lysates.[5]

17.2.2 Cell Lysis

In the case of biologics that are expressed intracellularly (*e.g.* in bacterial and plant cell expression systems), it is necessary to release the material so that it can be recovered and purified. The three main approaches for achieving this are enzymatic, chemical and physical lysis. Both chemical and enzymatic approaches can be used to help selectively release target products, but this requires the optimisation of each procedure for individual target molecules.[6] Chemical approaches generally rely on the use of detergents or high pH/alkaline solutions to disrupt and lyse the cell, both of which can adversely affect the molecule of interest if these approaches are not optimised for individual targets. Enzymatic lysis can also be used to give specificity to both the process of lysis and release of the target product and therefore offers a more controlled method of cell lysis.[7] The recent advances in the development of enzyme systems for the rupturing of bacterial and yeast cells have been well described by Salazar and Asenjo.[8] Physical approaches to cell lysis at the laboratory scale include grinding (*e.g.* in a mortar), high pressure (*e.g.* in a French press) and osmotic shock. In the case of heat-stable products, heat lysis methods may also be applicable, a readily scaleable approach. Large-scale physical disruption is, however, most commonly performed by either high-speed agitator bead mills or high-pressure industrial homogenisers. The equipment used derives from the paint and dairy industries, respectively.[9]

17.2.3 Recovery of Material from Inclusion Bodies

The high-level expression of recombinant protein using *E. coli* expression systems often results in the formation of inclusion bodies, insoluble aggregates of material that form *in vivo*.[10] Inclusion bodies usually result due to the inability of the *E. coli* host to fold and process correctly the polypeptide chain(s) of the recombinant protein and/or via intermolecular interactions between partially folded polypeptide chains. Although the protein material found within inclusion bodies has no biological activity and is therefore usually of little value as is, inclusion bodies can be easily isolated from *E. coli* cells at high purities, thereby acting as a useful purification means.[10] Further, it is often possible to recover the biological activity of the protein of interest once the inclusion bodies have been purified, making this a potential means of expressing and purifying target recombinant proteins. However, recovery of protein from inclusion bodies can be awkward, recoveries can be low and the process can be expensive. Further, inclusion bodies are not usually absolutely pure and there may be significant loss of protein during the recovery of material from inclusion bodies. Hence careful attention must be paid to the design of the strategies employed for the recovery of material from inclusion bodies to ensure good recovery and purification of the protein of interest.

There are four steps typically involved in the purification and biological recovery of proteins within inclusion bodies. These are the isolation of the inclusion bodies from the host *E. coli* cells, resolubilisation of the protein material within the inclusion bodies, refolding of the resolubilised material into a bioactive form and further purification or polishing of the protein of interest.[10] The most important steps are the solubilisation and refolding. Inclusion bodies and therefore the proteins contained within, are usually resolubilised using high concentrations (6–8 M) of chaotropic agents such as urea. The resolubilised proteins are then refolded by slowly removing the resolubilising agent while transferring the protein to an appropriate buffer system and concentration in order to reach and retain the correct biological three-dimensional structure and activity. Unfortunately, for those proteins with disulfide bonds, recovery of correctly folded and bioactive protein can often be extremely low.

17.3   PROTEIN PRECIPITATION

Precipitation of protein from culture medium or following cell lysis is commonly used in laboratory separations as a first step in the purification of a protein, and a quick survey of where this has been used and reported in the literature suggests that the yield of recombinant material returned is often around 80% with a purification factor in the region of 4-fold on average, although this can vary widely for specific proteins. The most commonly used precipitating agent is ammonium sulfate, although other agents such as acetone and polyethylene glycol are also used (Figure 17.1). Ammonium sulfate is preferred due to its relatively low cost, high solubility and low toxicity.

Recent studies have investigated the use of affinity precipitation to allow the selective precipitation of a specific target of interest. Affinity precipitation works by linking an affinity ligand (*e.g.* nickel) with a stimuli responsive material which is precipitated by altering the environmental conditions.[11] This method has so far not been widely used or developed for large-scale use; however, as it links the simplicity of precipitation approaches with the biospecific capturing ability of affinity-based

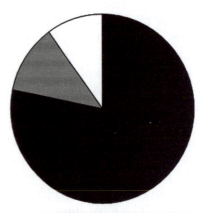

Figure 17.1   Example of the relative levels of use for different precipitation agents from a survey of the literature. Ammonium sulfate (black) is the most common precipitating agent utilised, although organic solvents such as acetone (grey) and polymers (*e.g.* polyethylene glycol) (white) are also utilised.

separations, it shows some promise. A further emerging technology which is yet to be fully explored or adopted is centrifugal precipitation, which was described in detail by Ito.[12]

17.4 CHROMATOGRAPHY

The major step(s) in most downstream processes is purification of the target molecule by chromatography. There are four main chromatographic techniques used for the purification of proteins: ion-exchange, affinity, gel filtration and hydrophobic interaction. The principle of each, examples and the usual stage at which they are used during downstream processing are detailed in Table 17.1. Of the four, ion-exchange, affinity and hydrophobic interaction chromatography are adsorptive processes whereby binding and elution are undertaken. The use of adsorptive chromatographic processes for downstream processing, the constraints upon this type of chromatography and future outlook have been reviewed by Lydiatt.[13]

The selection of which type of chromatography/approach to use, and when, is much more complicated than perhaps might be expected from Table 17.1. The scalability of the system, binding capacity for adsorptive approaches, flow rates, resolution, pH, ionic strengths, costs and

Table 17.1 Details of the four major chromatographic methods used for downstream processing of biologics.

Chromatographic method	Basis of separation	Example ligands/ systems	Principle	Stage of purification procedure applied
Ion exchange	Charge	Cation exchange, *e.g.* carboxymethyl (CM) and sulfopropyl (SP)	Positively charged molecules bind	Usually early in purification process, giving high recovery and purification factor
		Anion exchange, *e.g.* diethylaminoethyl (DEAE) and quaternary amine (Q)	Negatively charged molecules bind	
Affinity	Biospecific affinity	Protein A (for IgGs)	Specific interaction between protein and ligand	Usually at early stage, high resolution and purification factor
Hydrophobic interaction	Polarity	Phenyl-Sepharose	Polarity, interaction increases with ionic strength	Often after salt precipitation
Gel filtration	Size	Sephadex	Large molecules elute first	Polishing step late in purification

monitoring of the process are among a few of the variables that must be considered before design of the system and ultimate scale-up operation. Another important consideration is the chromatography matrix itself, and ideally this should be:

- reusable;
- chemically stable (not leach during processing);
- physically stable (*i.e.* withstand the required flow conditions);
- have appropriate internal structure/porosity to give suitable mass transfer and capacity;
- be commercially available from a reliable source to meet regulatory requirements;
- be cost effective.

Typical matrix types often encountered include agarose and cross-linked agaroses, cross-linked dextran, composites of dextran and poly-acrylamide or dextran and agarose, cellulose and organic polymer-based beads.

A survey of approximately 100 papers in the primary literature where the four chromatographic approaches outlined above have been utilised confirms that affinity chromatography achieves the highest purification factors of all these approaches (Table 17.2). However, ion-exchange chromatography is the most commonly utilised chromatographic approach, comprising approximately 41% of all purification steps used, and gel filtration chromatography (17%) is the second most widely used approach. Precipitation is used almost exclusively as a first step (Figure 17.2), whereas ion exchange is used as a bulk purification step and as a polishing step late in the workflow (Figure 17.2).

Table 17.2 The average and maximum purification factors of the four major chromatographic methods utilised during downstream processing as determined from a survey of the primary literature.

Chromatography method	*Purification factor*	
	Average	*Maximum*
Ion exchange	7	143
Affinity	57	499
Hydrophobic interaction	10	79
Gel filtration	5	26
Precipitation[a]	4	33

[a]Precipitation is not a chromatographic method but is included for comparison.

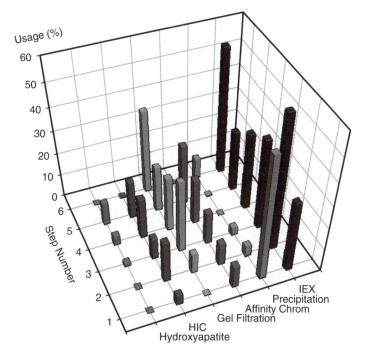

Figure 17.2 The use of chromatography and precipitation methods as a percentage of the total number of unit operations at a given step throughout downstream processing as determined from a survey of the primary literature. IEX, ion exchange; HIC, hydrophobic interaction chromatography.

17.4.1 Ion-exchange Chromatography (IEX)

Ion-exchange chromatography is the most commonly used chromatographic step in downstream bioprocessing and is an adsorptive binding and elute method. Ion exchange can be subdivided into cation exchange, whereby positively charge molecules bind, and anion exchange, whereby negatively charged molecules bind the resin. In both cases, the bound material is usually eluted by increasing the salt concentration. The target protein of interest must therefore be in a buffer of the appropriate pH and salt concentration before being applied to the column. As outlined in Table 17.1, commonly used ligands include carboxymethyl and diethylaminoethyl (DEAE) for cation and anion exchange, respectively. Further, as can be seen in Figure 17.2, ion exchange dominates as the method of choice in most steps due to the fact it is used as a bulk purification method and a polishing step and the fact that a weak binding high capacity ion-exchange step is followed by a lower capacity strong affinity binding step.

17.4.2 Affinity Chromatography

Applied correctly and with the appropriate ligand, affinity chromatography is the most powerful chromatographic purification method
available to yield high purification factors in a single step. Due to the
high biospecificity of this approach, it is used as a fundamental capture
step in most processes where an appropriate ligand is available and this
is one of the limitations in using the approach. The approach is thus
based upon the affinity of the target molecule for a ligand immobilised
on an appropriate resin. Protein A capture of monoclonal antibodies
expressed from *in vitro* cultured mammalian cells is one of the most
widely used affinity chromatography steps. Indeed, industrially the gold
standard purification of monoclonal antibodies is to use protein A
chromatography as the first step in the purification process after primary
recovery steps.[14] For example, in 2002 the demand for protein A resins
exceeded 10 000 L and an already $50 million a year market was growing
at a rate of 50% per year.[15] However, increased yields of recombinantly
produced monoclonal antibodies coupled with the high cost of protein A
resins mean that investigators are currently focused upon improving the
capacity and throughput of this approach in order to continue making
this approach economical. A further area of concern with the use of
protein A is the stability of these resins upon cleaning, and therefore
there is much interest in finding alternatives to protein A chromatography for antibody purification.[14]

A number of other affinity ligands are also used at the industrial and
laboratory scale. Dye-based ligands are perhaps the largest industrially
utilised approach, largely for the purification of industrially produced
enzymes. Immobilised metal affinity chromatography is also widely used
at the laboratory scale, particularly the use of poly His-tags, whereby the
target protein of interest is engineered to contain a His-tag usually of
six consecutive His residues at either the C- or N-terminal end of the
protein. This approach is popular due to the high binding capacity,
stability, robustness and recovery offered by poly His-tags.[16] However, a
drawback of such an approach is that it is often necessary to remove the
His-tag during bioprocessing to maintain the authenticity of the biomolecule, adding further steps and costs to the process. It is therefore most widely applied to high-throughput research approaches
where there is a need to purify a large number of different proteins for
further study.

Although the development of affinity separations is one of enormous
potential, the growth of regular packed bed affinity chromatography
seems to have plateaued over the past few years. The limitations do not

lie in affinity–based separations but in packed bed chromatography columns and the lack of cheap ligands available for use at an industrial scale. The recent development of the Capture Select technology by BAC and GE Healthcare is a new approach for designing/developing novel ligands. This technology uses a unique antibody found in Camelidae, specifically a small 12 kDa antigen–binding fragment, to develop customised ligands which can be used for purifying recombinant proteins, scavenging impurities and removing viral load from complex media using an animal–free system. This approach offers optimal binding affinity coupled with a high specificity over a broad range which allows the step to be fine tuned for a given process by incorporating specific elution conditions and stability.[17] The technology has not yet become widely adopted at the industrial scale but offers an alternative method to affinity chromatography development. Other alternatives, such as affinity precipitation, expanded bed adsorption and affinity membrane chromatography, are therefore currently being investigated as possible replacements.

17.4.3 Hydrophobic Interaction Chromatography (HIC)

Hydrophobic interaction chromatography is a useful tool for the purification of proteins, providing a powerful complementary separation step to ion-exchange and gel filtration chromatography. It has also shown much promise as a simultaneous protein purification and refolding step.[18] As hydrophobic interactions are strongest at high ionic strengths, this approach is often used fairly early in downstream bioprocessing, directly after precipitation steps[19] when salt levels are high, or in combination with gel filtration[20] or ion-exchange chromatography.[21] Although hydrophobic interaction chromatography is an adsorptive process, the interactions are generally weaker than in affinity and ion-exchange chromatography, thus resulting in minimal structural damage and loss of activity to the target protein.[22] Hydrophobic interaction chromatography is also very useful in the removal of DNA impurities and for purifying DNA-based products.[23]

17.4.4 Gel Filtration Chromatography

Gel filtration chromatography does not involve bind–elute chromatography but rather separates proteins on the basis of size through porous beads. Large molecules are not able to enter the pores and elute first whereas small molecules elute later. In order for gel filtration chromatography to separate entirely on the basis of size alone, there must be no interaction between the matrix and the molecules in the mobile phase. Gel filtration chromatography has been used for over 50 years and is primarily used as a polishing step late in downstream processing where the feedstream is relatively pure compared with earlier purification steps. Gel filtration is mainly used at the industrial scale as a final chromatography step to remove protein aggregates and small impurities and has a low loading capacity and low throughput. A vast array of different columns and matrices are now commercially available with pore sizes as small as $3\,\mu m$.[24] The majority of the developments in the use of gel filtration in recent years have been in its utilisation as an analytical tool and for the separation of biopolymers[25] rather than proteins, especially at an industrial scale.

17.5 ALTERNATIVES TO PACKED BED CHROMATOGRAPHY

Packed bed chromatography has been, and continues to be, the most widely used mode of chromatography used on both a large and small scale, industrially and academically. Despite this approach being on the whole of high cost and low throughput, it has predominated, largely due to the very high resolution that can be achieved and a lack of alternatives that can deliver similar resolution. However, the high cost of this approach coupled with increased product yields/concentrations in feedstocks and the practicalities and limitations that these bring mean that there is now a real need to investigate and develop alternatives to this traditional approach. Such alternatives include, but are not limited to, membrane chromatography, expanded bed adsorption, aqueous two-phase extraction and protein crystallisation. These potential approaches and the likelihood of their success have been discussed in detail by Przybycien *et al.*[26]

17.5.1 Expanded Bed Adsorption

Expanded bed adsorption combines the essential product capture step with the separation of solids and liquids in one step, thus reducing the number of steps in a process, and has been well described and reviewed by Anspach *et al.*[27] This approach can result in increased yield and reduced overall process time and cost.[28] The first stabilised expanded bed adsorption column was developed in the early 1990s,[13,29] and this approach has now been used in the processing of whole cell mammalian cell culture broth and has found successful application in affinity chromatography. This approach has also been used for the purification of monoclonal antibodies, providing higher recovery at lower cost and shorter processing time compared with packed bed chromatography.[30] Expanded bed adsorption has also been used as an alternative process for the purification of an insulin precursor, MI3.[31] Despite the apparent advantages, expanded bed adsorption currently remains an alternative technique and has not yet proven to be a threat to the dominance of packed bed chromatography in the biopharmaceutical industry.

17.5.2 Aqueous Two-phase Extraction

The process of aqueous two-phase extraction has recently attracted the attention of investigators wishing to develop alternatives to traditional packed bed chromatographic procedures. This has been particularly evident in the purification of plant-manufactured biopharmaceuticals.[32] Aqueous two-phase systems integrate the steps of cell disruption with product recovery using either polymer-based systems, hydrophobic and affinity precipitation or the use of surfactants.[33] The theory is that the protein of interest and cellular debris and other material will exhibit differential solubility between two aqueous phases. The two phases are usually created by mixing solutions of polyethylene glycol with either dextran or salts to form two immiscible layers or phases. Despite the fact that aqueous two-phase extraction can be applied to very crude mixtures while providing high capacity and easy scale-up, its use to date has been limited to selected industrial applications. This is largely due to the high cost of the polymers used and a lack of fully developed design approaches. Novel polymers are currently being developed that may overcome such limitations, especially the development of low-cost starch- and cellulose-based polymers for some of the more attractive uses of aqueous two-phase systems such as affinity precipitation.

17.5.3 Membrane Chromatography and Filtration

A number of recent reviews on the use of membrane chromatography have suggested that theoretically this approach could have a number of advantages over traditional packed bed chromatography.[34] Membrane chromatography can exhibit dynamic binding capacities of up to 10 times higher and flow rates of up to a 100 times faster than packed bed chromatography, especially for selected applications such as the purification of large particles.[35] Membrane chromatography is also significantly cheaper per cycle if hidden costs, including validation, cleaning and the manpower involved in packed bed chromatography, are taken into account. Despite these perceived advantages, membrane chromatography has not been widely adopted by the bioprocessing industry, mainly because it is limited by the throughput as a result of the available surface area per unit compared with packed bed chromatography.

Membrane filtration is also used as an alternative method of clarifying cell culture media and cell extracts and techniques such as ultrafiltration and high-performance tangential flow filtration (HPTFF) have been investigated as integral steps in downstream processing schemes. Indeed, there have been reports where HPTFF has been successfully used as a polishing step.[36] Perhaps one of the more significant advances in this area has been the use of pH and charged membranes to manipulate the product. These approaches have been shown to be useful for the removal of viruses from mammalian cell systems and endotoxins from bacterial systems. Despite such advances, these approaches have not been fully adopted and membrane fouling remains a problem during such processing, and these approaches are usually used prior to chromatographic steps.

17.5.4 Crystallisation

Crystallisation as a purification technique has long been used by investigators and industrialists in the small-molecule field, but has largely been ignored until recently in the biologics field. With increasing product yields and concentrations, this approach is now attracting much interest as an alternative to some chromatographic steps, thereby reducing the number of these required. Crystallisation of proteins has long been used in structural analysis studies where screens have been developed to find conditions that allow the successful crystallisation of a target protein. Now the potential of this approach in downstream processing is being seriously considered and has been used during the purification of glucose isomerase and insulin. A number of investigators have now reviewed the principles of protein crystallisation and the technical aspects of process development for its large-scale use.[37] Despite the promise of this approach, there are still scale-up issues involved in further developing this technique and a lack of process development in the area.[38] The development of novel microanalysis methods to determine large-scale conditions that will drive crystallisation of the target protein of interest at high yield is currently under way and may help overcome such limitations.[39] Hence there is the potential to develop low-cost methods of purification that utilise crystallisation for many biologics, although currently this potential is largely unrealised.

17.5.5 Monolith Columns

Monoliths are continuous stationary phases available in a variety of shapes and sizes and have shown potential to be developed with chromatographic ligands.[40] Monoliths are polymerised directly into a single unit although, for biotechnology purposes, many regular monolith tubes are prepared and interlaced into a single column with a limited pore size of around 5 µm.[41] Monoliths show great potential for providing high mass transfer through convective flow, thus enabling both capacity and resolution to be independent of flow rate. A further perceived advantage over conventional chromatographic beds is that they do not require packing and therefore reduce validation and labour costs. However, despite the fact that monoliths are being developed for affinity and ion-exchange chromatography, they have as yet not been able to make significant inroads into industrial-scale protein purification processing, largely due to technical problems involved in developing large-scale monolith columns. One use of monoliths that has been commercially validated is for the purification of pharmaceutical-grade plasmid DNA.[42] Success has been more readily achieved in this case as there are clear capacity advantages over conventional chromatographic resins which have porous internal structures designed for smaller protein-sized molecules. This capacity advantage may also translate to purifications of other very large molecular structures such as virus-like particles and viruses.

17.6 DESIGN OF BIOMOLECULES FOR DOWNSTREAM PROCESSING

There is much interest in how high-value biological products, and particularly proteins, can be designed for downstream bioprocessing. Although it has long been recognised that the properties of a target molecule influence its behaviour and the approaches taken during downstream processing, until relatively recently there has not necessarily been a concerted and integrated approach to link the two together for any particular molecule. This requires designing the molecule for the downstream processing envisaged and integrates the upstream and downstream investigations such that the two processes are seamless.

To date, the design of molecules for downstream processing has largely been limited to the use of affinity tags and fusion proteins for purification. These have been particularly adopted at the small laboratory scale but are often not practical, too costly or potentially have legislative issues associated with them, which have prevented such approaches becoming more widely adopted at large scale. However, the design of molecules to aid in more efficient traditional chromatographic approaches, such that fewer steps and greater yields are achieved, and/or for aiding alternative approaches such as protein crystallisation and aqueous phase partitioning is an attractive proposition. Further, careful design of biomolecules may also be used in the future to prevent or limit aggregation and harmful post-translation modifications occurring (*e.g.* deamidation) during the expression of the protein of interest in fermenters and while undergoing downstream processing.[43] This area of designing molecules for manufacturability, from expression to formulation, is likely to be one in which there will be much interest and investigation in the coming years and may ultimately dictate the direction that industrial-scale downstream processing follows.

17.7 SCALEDOWN METHODS

Scaledown methods are critical to assist decision making when the intention is ultimately to operate a purification process at large scale. This is because when increasing the scale of a purification process from what is typically a laboratory-scale sequence of operations, many changes occur. Some might impact on the cells, such as the change from a benchtop centrifuge to a continuous centrifuge for cell harvest, others may impact on the final product such as the pumping and vialing operation in a large-scale process, all of which can have unpredictable effects that can seriously delay the development of large-scale operations. Scaledown methods seek to mimic these critical changes to measure the effect (if any) of these changes but to allow this to be achieved with a minimal amount of material, far less than would be required in the pilot plant environment. This provides scope to assess more processing options and the manufacturability of different molecules and host cell lines (as described in the previous section) at realistic scale and hence cost. Methods have been established for operations such as filtration[44,45] and centrifugation[46,47] making bench-scale studies which predict full-scale operation feasible. Such methodology has also been developed for chromatography.[48]

17.8 VALIDATION AND ROBUSTNESS

A traditional mantra used in the sector is 'the process is the product'. This statement goes a long way to explaining how complex biologics can be manufactured in a safe manner. A strict definition of the process is made which has validated limits within which the process must run.

By operating in this fashion, one seeks to produce a consistent product which has proven efficacy from clinical trials. The US Food and Drug Administration (FDA) defines this act of process validation as 'establishing documented evidence which provides a high degree of assurance that a specific process will consistently produce a product meeting its predetermined specifications and quality attributes'.[49] This reliance on defining the process to such a high level arises from the difficulty in defining by analytical techniques biological products, particularly biopharmaceuticals, to the necessary degree of accuracy.

The key end-point of process development is demonstration to the regulatory authorities that the final large-scale process can indeed be operated consistently (typically three repeats are required). To ensure this and that subsequent manufacture is reliable, it makes logical sense that the process is designed such that the validated limits of operation are in robust regions, that is, the variability natural to such processes will be tolerated by the process and it will remain within its validated limits. To assist in developing such robust processes, many companies now embrace the principles of 'quality by design'. Key to this strategy is the judicious use of monitoring and control strategies within the process.[50] Traditionally, this involves approaches such as absorbance measurements, gel electrophoresis and HPLC analysis; however, with the advent of more state-of-the-art technologies such as mass spectrometry, a greater degree of monitoring and hence confidence in the process is now envisaged. These more advanced approaches are being actively encouraged by the FDA via their Process Analytical Technology (PAT) initiative.

17.9 FORMULATION AND ANTIVIRAL TREATMENTS

17.9.1 Formulation

In order that biomaterials, and particularly biopharmaceutical protein-based therapeutic drugs, can be maintained and delivered in a biologically active state, it is often necessary to formulate purified materials in excipients and additives that protect the product from deleterious reactions and degradations that compromise the authenticity of such products. This can be achieved in either the solid (lyophilised) or solution state and both are routinely used for the preparation of such products. Current methods for determining the best formulation(s) for preservation, stability and delivery rely heavily upon trial-and-error approaches. Preservation is then usually investigated using higher temperature and varying pH studies in order to 'force' stability issues (often described as accelerated stability studies) and stability is then investigated routinely using isoelectric focusing, SDS-PAGE and fluorescence methodology. Particularly with regard to antibody-based therapies where relatively large amounts of material are required per dose,[51] there is interest in the development of high-concentration liquid formulations ($> 100 \, mg \, mL^{-1}$) destined for clinical use. In particular, there is much academic and industrial interest in developing our ability to predict appropriate formulation strategies and compositions for particular proteins that would negate the need for time-consuming trial-and-error studies.[52]

17.9.2 Antiviral Treatments

Ever since the transmission of HIV infection to haemophiliac patients through contaminated blood-based products in the early 1980s, there have been stricter controls and scrutiny of the methods of viral inactivation utilised during bioprocessing. This is of particular concern when using mammalian cells to produce biotherapeutic products, as there is clear potential to host harmful viruses. Current good practice therefore dictates that products destined for use as therapeutics are submitted to antiviral treatments with two different viral-inactivation steps each of which gives a four to five log reduction in virus load. To achieve this level of reduction, several steps with viral removal functionality are required, although some of these steps may have dual functions. The methods employed to achieve this are varied and the regulatory bodies do not require specific viral inactivation treatments but rather judge the validity of the approach used for each product on its own merits.[53]

The most commonly used methods to reduce viral load are the heating of protein solutions and lyophilisates, nanofiltration, low-pH hold steps, UV irradiation, solvent/detergent treatments and those process removal steps that are part of the purification methodology or the production of the protein of interest. Of these treatments, heat treatment is the current 'gold standard' and many therapeutic protein preparations are treated in this way (*e.g.* albumin, Factor VIII, Factor IX, thrombin, fibrinogen, IgG). Further, heat treatment has the advantage of being one of the few methods that can be applied to a given product in the vial without the need for disruption of the product. Typical heat treatment in the liquid state consists of 60 °C for 10 h, whereas in the lyophilised state samples are typically subjected to temperatures of 90 °C for 10 h or 80 °C for 72 h. The disadvantage of heat treatment is that as proteins denature and undergo deleterious reactions upon heat stress, it is necessary to formulate the protein of interest in stabilising excipients such as sucrose to prevent such reactions. It has been shown that the excipients themselves can break down and react with the protein of interest under certain heat treatment conditions, resulting in disadvantageous side reactions that compromise the authenticity of the protein product.[53]

17.10 CURRENT DEVELOPMENTS AND FUTURE DIRECTIONS

Downstream processing of high-value biological products has long been dominated by adsorptive packed bed chromatographic approaches, despite a number of limitations to this technique, including the high cost, low throughput and complexity of scale-up. However, the ability of currently available expression technologies to produce ever higher yields and concentrations of high-value biological products means that the downstream bioprocessing community has begun to consider alternatives to this traditional approach that may at least reduce the extent of its use. Many consider that the traditional approach of small incremental improvements to operations already working at the limit is proving ineffective at resolving downstream processing bottlenecks. Consequently, there is a drive to investigate unit operations that might replace some of the current dependence on packed bed chromatography. Recent reviews suggest that the most notable trends in this area are to integrate steps (thereby reducing the number), the development of platform technologies (so as to move towards a more systemic approach), further process optimisation and the increased use of disposable equipment.

Process integration is a logical approach towards reducing costs as a reduced number of steps will allow greater throughput and potentially yield. This approach has already delivered a number of new technologies, including affinity precipitation and expanded bed adsorption. Unsurprisingly, the focus tends towards eliminating intermediate purification steps and moving directly from product capture to product polishing. The aim is to develop an integrated process which uses the minimum number of steps in a logical way, with each step focusing on the removal of a certain impurity. This has resulted in some polishing steps being employed before bulk separation processes, for instance, the removal of specific proteases early on in the process to preserve product stability. However, to implement such a concept, a systematic approach has to be taken towards process development. Increased regulatory issues, and the fact that integrating new developments into the process can often yield haphazard results, have necessitated that a more organised approach be taken. This must be undertaken to ensure that individual steps during downstream processing all work together and are managed in such a way that optimum processing is achieved at each step. A current lack of systematic approaches highlights the need to design platform technologies for specific types or classes of biological products so as to make process optimisation a simpler task. This will be aided by developments in microscale bioprocess optimisation and process simulation software, which have allowed the formulation of a process very early in the development of a product.[54] Finally, the trend towards disposable technology is likely to continue in the foreseeable future and any novel developments in downstream processing must be able to incorporate such a move. For example, recent studies have shown that smaller disposable membranes use up to 95% less buffer than conventional systems and can reduce monoclonal antibody polishing costs by 70%. Disposables are also encroaching upon upstream bioprocessing with disposable bioreactors as large as 1000 L being developed. All of these areas and approaches, particularly systematic approaches, are likely to play key roles in future developments for optimising downstream processing of high-value biological products.

REFERENCES

1. F. M. Wurm, *Nat. Biotechnol.*, 2004, **22**, 1393–98.
2. D. Lowa, R. O'Leary and N. S. Pujar, *J. Chromatogr. B*, 2007, **848**, 48–63.
3. Frost and Sullivan Research Analyst Pharmaceuticals and Biotechnology Research and Consulting Practice (2004) Report on The strategic analysis of downstream processing in biopharmaceuticals production.
4. J. M. King, N. J. Titchener-Hooker and Y. Zhou, *Bioprocess Biosyst. Eng.*, 2007, **30**, 123–34.
5. A. Venkiteshwaran, P. Heider, S. Matosevic, A. Bogsnes, A. Staby, S. Sharfstein and G. Belfort, *Biotechnol. Prog.*, 2007, **23**, 667–72.
6. H. Schutte and M. R. Kula, *Biotechnol. Appl. Biochem.*, 1990, **12**, 599–620.
7. J. A. Asenjo and B. A. Andrews, *Bioprocess Technol.*, 1990, **9**, 143–75.
8. O. Salazar and J. A. Asenjo, *Biotechnol. Lett.*, 2007, **29**, 985–94.
9. M. R. Kula and H. Schutte, *Biotechnol. Prog.*, 1987, **3**, 31–42.
10. S. M. Singh and A. K. Panda, *J. Biosci. Bioeng.*, 2005, **99**, 303–10.
11. R. Freitag, I. Schumacher and F. Hilbrig, *Biotechnol. J.* 2007, **2**, 685–90.
12. Y. Ito, *Anal. Biochem.*, 2000, **277**, 143–53.

13. A. Lyddiatt, *Curr. Opin. Biotechnol.*, 2002, **13**, 95–103.
14. D. Low, R. O'Leary and N. S. Pujar, *J. Chromatogr. B*, 2007, **848**, 48–63.
15. J. Curling, *Biopharm. Int.* 2004, August 1.
16. P.-O. W. C. K. I. Y. G. B. M. Ashok Kumar, *Biotechnol. Bioeng.*, 2003, **84**, 494–503.
17. F. Detmers, P. Hermans and M. ten Haaft, *Genet. Eng. News*, 2007, **27**, 8.
18. X. D. Geng and Q. Bai, *Sci. China, Ser. B*, 2002, **45**, 655–69.
19. M. L. Stracke, H. C. Krutzsch, E. J. Unsworth, A. Arestad, V. Cioce, E. Schiffmann and L. A. Liotta, *J. Biol. Chem.*, 1992, **267**, 2524–9.
20. C. T. Tomaz and J. A. Queiroz, *J. Chromatogr. A*, 1999, **865**, 123–8.
21. T. Unge, H. Ahola, R. Bhikhabhai, K. Backbro, S. Lovgren, E. M. Fenyo, A. Honigman, A. Panet, J. S. Gronowitz and B. Strandberg, *AIDS Res. Hum. Retroviruses*, 1990, **6**, 1297–303.
22. J. A. Queiroz, C. T. Tomaz and J. M. Cabral, *J. Biotechnol.*, 2001, **87**, 143–59.
23. M. M. Diogo, J. A. Queiroz, G. A. Monteiro, S. A. Martins, G. N. Ferreira and D. M. Prazeres, *Biotechnol. Bioeng.*, 2000, **68**, 576–83.
24. H. G. Barth, B. E. Boyes and C. Jackson, *Anal. Chem.*, 1998, **70**, 251R–78R.
25. H. G. Barth and G. D. Saunders, *LC GC N. Am.*, 2006, **24**, 38.
26. T. M. Przybycien, N. S. Pujar and L. M. Steele, *Curr. Opin. Biotechnol.*, 2004, **15**, 469–78.
27. F. B. Anspach, D. Curbelo, R. Hartmann, G. Garke and W.-D. Deckwer, *J. Chromatogr. A*, 1999, **865**, 129–44.
28. A. Suding and M. Tomusiak, *Abstr. Pap. Am. Chem. Soc.*, 1993, **205**, 61–BTEC.
29. N. M. Draeger and H. A. Chase, *Bioseparation*, 1991, **2**, 67–80.
30. Y. Gonzalez, N. Ibarra, H. Gomez, M. Gonzalez, L. Dorta, S. Padilla and R. Valdes, *J. Chromatogr. B*, 2003, **784**, 183–7.
31. P. Brixius, I. Mollerup, O. E. Jensen, M. Halfar, J. Thommes and M. R. Kula, *Biotechnol. Bioeng.*, 2006, **93**, 14–20.
32. D. Balasubramaniam, C. Wilkinson, K. Van Cott and C. Zhang, *J. Chromatogr. A*, 2003, **989**, 119–29.
33. R. M. Banik, A. Santhiagu, B. Kanari, C. Sabarinath and S. N. Upadhyay, *World J. Microbiol. Biotechnol.*, 2003, **19**, 337–48.
34. R. Ghosh, *J. Chromatogr. A*, 2002, **952**, 13–27.
35. H. N. Endres, J. A. C. Johnson, C. A. Ross, J. K. Welp and M. R. Etzel, *Biotechnol. Appl. Biochem.*, 2003, **37**, 259–66.

36. R. van Reis and A. Zydney, *Curr. Opin. Biotechnol.*, 2001, **12**, 208–11.
37. E. K. Lee and W. S. Kim, *Biotechnol. Bioprocess.*, 2003, **27**, 277–320.
38. S. Schmidt, D. Havekost, K. Kaiser, J. Kauling and H. K. Henzler, *Eng. Life Sci.*, 2005, **5**, 273–76.
39. V. Klyushnichenko, *Curr. Opin. Drug Discov. Dev.*, 2003, **6**, 848–54.
40. M. Vodopivec, A. Podgornik, M. Berovic and A. Strancar, *J. Chromatogr. B*, 2003, **795**, 105–13.
41. A. Jungbauer and R. Hahn, *J. Sep. Sci.*, 2004, **27**, 767–78.
42. J. Urthalera, R. Schlegl, A. Podgornik, A. Strancar, A. Jungbauer and R. Necina, *J. Chromatogr. A*, 2005, **1065**, 93–106.
43. M. E. Cromwell, E. Hilario and F. Jacobson, *Aaps J.*, 2006, **8**, E572–9.
44. N. B. Jackson, J. M. Liddell and G. J. Lye, *J. Membr. Sci.*, 2006, **276**, 31–41.
45. T. Reynolds, M. Boychyn, T. Sanderson, M. Bulmer, J. More and M. Hoare, *Biotechnol. Bioeng.*, 2003, **83**, 454–64.
46. M. Boychyn, S. S. Yim, M. Bulmer, J. More, D. G. Bracewell and M. Hoare, *Bioprocess Biosyst. Eng.*, 2004, **26**, 385–91.
47. N. Hutchinson, N. Bingham, N. Murrell, S. Farid and M. Hoare, *Biotechnol. Bioeng.*, 2006, **95**, 483–91.
48. M. D. Wenger, P. DePhillips, C. E. Price and D. G. Bracewell, *Biotechnol. Appl. Biochem.*, 2007, **47**, 131–9.
49. G. Sofer and L. Hagel, *Handbook of Process Chromatography: a Guide to Optimisation, Scale-up and Validation*, Academic Press, San Diego, 1997.
50. G. K. Sofer and D. W. Zabriskie, *Biopharmaceutical Process Validation*, Marcel Dekker, New York, 2000.
51. S. Kozlowski and P. Swann, *Adv. Drug Deliv. Rev.*, 2006, **58**, 707–22.
52. A. L. Daugherty and R. J. Mrsny, *Adv. Drug Deliv. Rev.*, 2006, **58**, 686–706.
53. C. M. Smales, D. S. Pepper and D. C. James, *Biotechnol. Bioeng.*, 2000, **67**, 177–88.
54. M. Micheletti and G. J. Lye, *Curr. Opin. Biotechnol.*, 2006, **17**, 611–8.

CHAPTER 18

Biosensors

MARTIN F. CHAPLIN

London South Bank University, Borough Road, London SE1 0AA, UK

18.1 INTRODUCTION

Biosensors are analytical devices that convert biological actions into electrical signals in order to quantify them.[1–3] In this chapter, only biosensors that make use of the specificity of biological processes are described, that is, the recognition of enzymes for their substrates or other ligands, antibodies for their antigens, lectins for carbohydrates and nucleic acids or peptide nucleic acids for their complementary sequences. Biosensor science is interdisciplinary, bringing together chemistry, physics, biology, electronics and engineering to solve real-world analytical problems. The parts of a typical biosensor are shown in Figure 18.1.[4]

The primary advantage of using biologically active molecules as part of a biosensor is due to their high specificity and, hence, high discriminatory power. Thus, they are generally able to detect particular molecular species from within complex mixtures of other materials with similar structure that may be present at comparable or substantially higher concentrations. Often, samples can be analysed with little or no prior clean-up. In this respect they show distinct advantages over most 'traditional' analytical methods; for example, colorimetric assays such as the Lowry assay for proteins.

Molecular Biology and Biotechnology, 5th Edition
Edited by John M Walker and Ralph Rapley
© Royal Society of Chemistry 2009
Published by the Royal Society of Chemistry, www.rsc.org

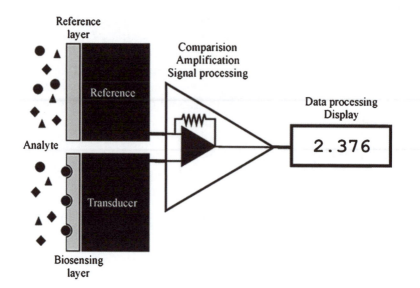

Figure 18.1   The functional units of a biosensor. The biological reaction usually takes place in close contact with the electrical or optical transducer, here shown as a 'black-box'. This intimate arrangement ensures that most of the biological reaction is detected. The resultant electrical or optical signal is compared with a reference signal that is usually produced by a similar system without the biologically active material. The difference between these two signals, which optimally is proportional to the material being analysed (*i.e.* the analyte), is amplified, processed and displayed or recorded. The reference electrode stabilises the output, so reducing signal drift and improving reproducibility.

   Biosensors may serve a number of analytical purposes. In some applications, for example in clinical diagnosis, it is sometimes important only to determine whether the analyte is above or below some predetermined threshold, whereas in process control there often needs to be a continual accurate but imprecise feedback of the level of analyte present. In the former case, the biosensor must be designed to give the minimum number of false positives and, more importantly, false negatives. In the latter case, it is generally the rapid response to changes in the analyte that is more important than its absolute precision. Minimising false negatives is often more important in clinical analyses than minimising false positives, as in the former case a diagnosis may be missed and therapy delayed whereas in the latter case further tests may show the error. Other biosensor analytical applications may require accuracy and precision over a wide analytical range.

   Biosensors must show advantages over the use of the free active biomolecules, which possess at least equal specificity and discriminatory power, if they are to be acceptable. Their main advantages usually involve rapid response, the reduced need for sample pretreatment, their ease of use and often their reusability and portability. Repetitive reuse of the same biologically active sensing material generally ensures that similar samples give similar responses as any need for accurate aliquoting of such biological materials to contain precisely similar activity is avoided. This circumvents the possibility of introducing errors by inaccurate pipetting or dilution and is a necessary prerequisite for automated and on-line monitoring. Repetitive and reagentless methodology offers considerable savings in terms of reagent costs, so reducing the cost per assay. In addition, the increased operator time per assay and the associated higher skill required for 'traditional' assay methods also involves a cost penalty which is often greater than that due to the reagents. These advantages must be sufficient to encourage the high investment necessary for the development of a biosensor and the purchase price to the end-user. Table 18.1 lists a number of important attributes that a successful biosensor may be expected to possess. In any particular case only some of these may be achievable.

Table 18.1 The properties required of a successful biosensor.

Required property	Achievable with ease?
Specificity	Yes
Discrimination	Yes, except for very low concentrations in complex biological fluids
Repeatability	Yes
Precision	Yes
Safe	Yes
Accuracy	Yes, as easily calibrated
Appropriate sensitivity	Yes, except in trace analysis
Fast response	Yes, usually
Miniaturisable	Yes
Small sample volumes	Yes
Temperature independence	Yes, may be electronically compensated
Low production costs	Yes, but only if mass-produced
Reliability	Yes, if to be useful
Marketable	Difficult, if there is pre-existing competing methodology
Drift free	Difficult but possible; may be compensated
Continuous use	Yes, for short periods (days)
Robust	Getting there, but generally needs careful handling
Stability	No, except on storage (months) or in the short term (weeks)
Sterilisable	No, except on initial storage
Autoclavable	Not currently achievable

Table 18.2 The properties of some important biosensor configurations. Constant development reduces the more negative features when large-scale production is planned.

Biosensor	Amperometric	Potentiometric	Piezoelectric	Optical
Cost	Low	Low	Medium	High
Reliability	High	Medium	Varies	Medium
Complexity	Medium	Low	Low	High, becoming less
Selectivity	High	High	Often low	High
Sensitivity	High	Medium	Often low	Medium
Speed of response	Medium	Slow	Fast	Sometimes fast
Applicability	High	Medium	Narrow	Medium
Present usage	Highest	Medium	Small	High and growing
Future prospects	High	Medium	Medium	High

The different types of biosensors have their own advantages and disadvantages that are summarised in Table 18.2. Apart from the important related area of colorimetric test strips, the most important commercial biosensors are electrochemical, but the principal area of new development is optical. There are amperometric biosensors available for glucose, lactate, glycerol, ethanol, lactose, L-amino acids, cholesterol, fish freshness and microorganisms. Potentiometric devices have been marketed for glucose, other low molecular weight carbohydrates and alcohols. Receptor-based and immunosensors have been marketed utilising surface plasmon resonance devices and piezocrystals.

Most of the market share for biosensor applications is for glucose analysis for use in the monitoring and control of diabetes.[5] Only a few years ago, market research projections for the growth of the biosensor industry predicted expansion at an impressively fast pace of almost 50% per year. These were over-optimistic, with current growth being about 10%. However, industry sales as a whole were expected to be worth about $5000 million in 2007. Although currently over 1000 scientific papers on biosensors are published each year, biosensor science has not progressed as significantly as other areas of modern biotechnology.[6] A vast number of biosensor systems have been published, some being very imaginative, but with just a few reaching the marketplace and even these often utilise well-established ideas. It is now accepted that there are substantial and investment-intensive difficulties involved in producing such robust and reliable commercial analytical devices that are able to operate under authentic real-life conditions, even where a novel and highly promising prototype device has been produced. The cost of the biosensor is often of importance and considerable effort has been expended in the production of disposable devices using cheap integrated chip technology; disposable technology offering opportunities for continuing and increased revenue. Such silicon-based methodology[7] has opened up the possibility of having a number of different sensors on one device, allowing multi-parameter assays. A realistic target for the future is a density of a million sensors per square centimetre. However, there must be a demand for very large numbers of such devices in order for their production to be economically viable. Relatively unsophisticated processes, involving screen printing or ink-jet technologies, are currently preferred, due to their lower start-up costs, whereas photolithography can be used for the production of sub-micron structures with well-defined geometry.

By far the largest biosensor application area is in clinical diagnostics. This includes monitoring of critical metabolites during surgery. The major target markets are concerned with use for self-testing at home, within the physician's office for screening and within casualty and other hospital departments for point-of-care diagnosis. These application areas are potentially very wide. The use of rapid biosensor techniques in doctors' surgeries and in hospitals avoids the need for expensive and, most importantly, time-consuming testing at central clinical laboratories. Thus diagnosis and treatment may start during the first visit of a patient. This removes the need to wait for a return visit after the clinical tests have been completed elsewhere and allowing time, perhaps, for the patient's condition to deteriorate somewhat. Also, there is less likelihood of the sample being mishandled or contaminated. As an example, testing for diabetes using glucose biosensors is now routine even if a patient has an apparently unrelated complaint. Centralised clinical analytical facilities remain a necessity due to the need, in many cases, for multiple different analyses on the same sample and difficulties such as regulatory compliance. Legislation and the increasing possibility of medical malpractice claims often impose stringent quality assurance and control standards on clinical analyses. This makes it much more expensive to bring novel clinical biosensors to the marketplace today than in the past, but generally allows those biosensors that are already established, giving them a distinct competitive edge.

Home diagnosis and testing is an area that is being opened up by, for example, pregnancy and ovulation test kits, blood cholesterol testing and diabetes control.[8] Clearly, there are risks and problems involved with their more widespread use, but many people prefer to use them as indications for whether a trip to the doctor's surgery is really necessary or not. As counselling may be necessary with some potential home diagnosis applications (for example, cancer and AIDS), controversy exists over their development.

One of the major potential uses for biosensors is for *in vivo* applications such as for the close control of diabetes. The purpose here is to monitor continuously the levels of metabolites such that corrective action can be employed immediately when necessary. Clearly, such biosensors must be biocompatible and miniaturised so that they are implantable. In addition, they should be reagentless, the reaction being controlled only by the presence of the metabolite and the stabilised bioreagent. The signal generated must be drift free over the period of interest. At the present time, such biosensors have a relatively short lifespan of a few days at most, due mainly to problems which arise from the body's response.

Industrial analyses involve food, cosmetics and fermentation process control and quality control and monitoring. The defence industry is interested in detectors for explosives, nerve gases and microbial spores and toxins. Environmental uses of biosensors are mainly in areas of water quality and pollution control. A typical application might be to detect parts per million of particular molecular species such as an industrial toxin within the highly complex mixtures produced as process effluent.

18.2 THE BIOLOGICAL REACTION

An important factor in most biosensor configurations is the sensing surface. This normally consists of a thin layer of biologically active material in intimate contact with the electronic or optical transducer. In some cases the biological material may be covalently or non-covalently attached to the surface, but often in electrochemical biosensors it forms part of a thin membrane covering the sensing surface. Generally, the conversion of the biological process into an electronic or optical signal is most efficient where there is minimal distance between where the biological reaction or binding takes place and where the electronic transduction takes place. In addition, it is important for the retention of biological activity that the biological material is not lost into analyte solutions. The immobilisation technology for holding the biocatalyst in place is extensive, with examples of immobilisation methods summarised in Table 18.3. Much current research is directed at stabilising enzymes for use in biosensors.[9] At a suitable pH, polyelectrolytes such as diethylaminoethyldextran wrap around enzymes, restricting their movement and reducing their tendency to denature. Polyelectrolytes in combination with polyalcohols, such as sorbitol and lactitol, have been shown to stabilise enzymes considerably against thermal inactivation for use in biosensors. More recently, enzymes have been specifically engineered for their use in biosensors.

Table 18.3 Examples of biosensor immobilisation methods.

Physical entrapment
Biologically active material held next to the sensing surface by a semipermeable membrane that prevents it from escaping to the bulk phase but allows passage of the analyte. Often this membrane is synthesised *in situ* and sometimes it is made such that it increases the specificity of the sensing or reduces unwanted side-reactions. This is usually a simple and inexpensive method

Non-covalent binding
Adsorption of proteins on a porous carbon or silicon electrode. This may suffer from gradual leaching of the enzyme to the bulk phase

Covalent binding
Treatment of the biosensor surface with 3-aminopropyltriethoxysilane followed by coupling of biologically active material to the reactive amino groups remaining on the cross-linked siloxane surface. Proteins may be attached by use of carbodiimides, which form amide links between amines and carboxylic acids. Such methods permanently attach the biological material but are difficult to reproduce exactly and sometimes cause a large reduction in activity

Membrane entrapment
Cross-linking of proteins with glutaraldehyde within a cellulose or nylon supporting net. Due to limiting diffusion, this method can only be used for low molecular weight analytes

18.3 THEORY

In the absence of diffusion effects (see later), most biological reactions and binding processes can be described in terms of saturation kinetics:

$$\text{biological material} + \text{analyte} \rightleftharpoons \text{bound analyte} \tag{1}$$

The bound analyte is then detected optically or by weight or gives rise to a biological action, so generating the electronic response. This electronic or optical response varies with the extent of binding or biological response, which, in turn, varies with the concentration of the bound analyte. Apart from the logarithmic relationship in potentiometric biosensors, the optical, biological and electronic responses are often proportional:

bound analyte → biological or optical response → electronic response

$$\text{electronic response} = \frac{(\text{maximum electronic response possible}) \times (\text{analyte concentration})}{(\text{half- saturation constant}) + (\text{analyte concentration})} \tag{2}$$

where the half-saturation constant is equal to the analyte concentration which gives rise to half the maximum electronic response possible (Figure 18.2). The response is linear, to within 95%, at analyte concentrations up to about one-twentieth of the half-saturation constant.

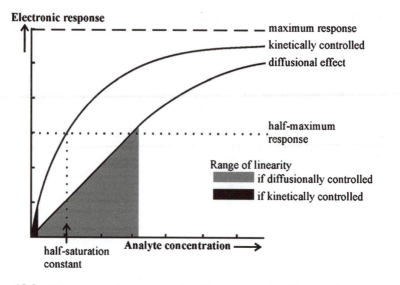

Figure 18.2 The range of response of a biosensor under biocatalytic kinetic and diffusional control.

A biosensor may be used over a wider, non-linear range if it has compensatory electronics.

This relationship holds for both reacting (e.g. enzymes) and non-reacting (e.g. immunosorption) processes. Where the analyte reacts as part of the biological response [e.g. during a biocatalytic reaction utilising enzyme(s) and/or microbial cells], an additional factor is the diffusion of the analyte from the bulk of the solution to the reactive surface. If this rate of diffusion is less than the rate at which the analyte would otherwise react, there follows a reduction in the local concentration of analyte undergoing reaction. The rate of diffusion increases as the concentration gradient increases:

$$\text{rate of diffusion} = (\text{diffusivity constant}) \\ \times (\text{analyte concentration gradient}) \qquad (3)$$

where the analyte concentration gradient is given by the difference between the bulk analyte concentration and the local (microenvironmental) analyte concentration on the sensing surface of the biosensor divided by the distance through which the analyte must diffuse.

As most biocatalytic biosensor configurations utilise a membrane-entrapped biocatalyst, this concentration gradient depends not only on the analyte concentration in the bulk and within the membrane but also on the membrane's thickness. The thicker the membrane, the greater is the diffusive distance from the bulk of the solution to the distal sensing surface of the biosensor and the greater is the amount of biocatalyst encountered. Both effects increase the likelihood that the overall reaction will be controlled by diffusion. Hence such biosensors can be designed to be under diffusional or kinetic control by varying the membrane thickness. When the rate of analyte diffusion is slower than the rate at which the biocatalyst can react, the electronic response decreases due to the lower level of analyte available for reaction. A steady state is rapidly established when the rate of arrival of the analyte equals its rate of reaction. This steady-state condition may be determined wholly by the rate of diffusion (diffusional control) or wholly by the rate of reaction (kinetic control) or by an intermediate dependence. Where the reaction rate depends solely on the rate of diffusive flux of the analyte, this determines the electronic response:

$$\text{electronic response} \propto \text{rate of diffusive flux} \qquad (4)$$

As the rate of diffusion depends on the bulk concentration of the analyte, this electronic response is linearly related to the bulk analyte concentration and, most importantly and intriguingly, is independent of the properties of the enzyme. Thus, the biosensor is linear over a much wider range of substrate concentrations (see Figure 18.2) and relatively independent of changes in the pH and temperature of the biocatalytic membrane, provided that the system remains diffusion controlled. It should be noted, however, that under these conditions the response is reduced relative to a system containing the same amount of biocatalyst but not diffusionally limited. Maximum sensitivity to analyte concentration would be accomplished by the utilisation of thin membranes containing a high biocatalyst activity and a well-stirred analyte solution. The overall kinetics of most biosensor configurations are difficult to predict. They depend on the diffusivities in the bulk phase and within the biocatalytic volume, the nature, porosity and physical properties of any membrane, the intrinsic biocatalytic kinetics, the electronic transduction process and kinetics, the way in which the analyte is presented and on other non-specific factors. Generally, such overall kinetics are determined experimentally using the complete biosensor and, hence, it is very important that the biosensor configuration is reproducible.

In biosensors utilising binding only, such as immunosensors, the major problem encountered is non-specific absorption that blocks the binding sites. There is need to minimise this and maximise the specific binding. As binding is an equilibrium process, high sensitivity necessitates a very high affinity between the analyte and the sensor surface.

18.4 ELECTROCHEMICAL METHODS

Electrochemical biosensors are generally fairly simple devices. There are three types utilising electrical current, potential or resistive changes:

1. amperometric biosensors, which determine the electric current associated with the electrons involved in redox processes;
2. potentiometric biosensors, which use ion-selective electrodes to determine changes in the concentration of chosen ions (*e.g.* hydrogen ions); and
3. conductimetric biosensors, which determine conductance changes associated with changes in the ionic environment.

There has been much progress in miniaturising these devices using microfabrication technologies developed by the electronics industry. These include the use of screen-printing and the deposition of nano-litre volumes of enzymes using advanced ink-jet printing and conducting inks.

18.4.1 Amperometric Biosensors

Enzyme-catalysed redox reactions can form the basis of a major class of biosensors if the flux of redox electrons can be determined. Normally a constant potential is applied between two electrodes and the current, due to the electrode reaction, is determined. The potential is held relative to that of a reference electrode and is chosen such that small variations do not affect the rate of the electrode reaction. Direct electron transfer between the electrode and the redox site associated with the enzyme is kinetically hindered due to the distance between them or the presenta-tion of an unfavourable pathway for the electrons. This can usually be addressed by the use of mediators and/or special electrode materials.

The first and simplest biosensor was based on this principle. It was for the determination of glucose and made use of the Clark oxygen elec-trode.[10] Figure 18.3 shows a section through such a simple ampero-metric biosensor. A potential of $-0.6\,V$ is applied between the central platinum cathode and the surrounding silver/silver chloride reference electrode (the anode). Dissolved molecular oxygen at the platinum cathode is reduced and the circuit is completed by means of the satu-rated KCl solution. Only oxygen can be reduced at the cathode due to its covering by a thin Teflon or polypropylene membrane through which the oxygen can diffuse but which acts as a barrier to other electroactive species.

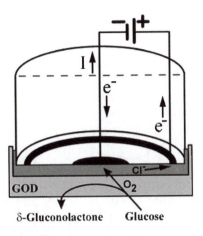

Figure 18.3 Amperometric glucose biosensor based on the oxygen electrode utilising glucose oxidase (GOD).

$$\text{Pt cathode reaction } (-0.6\,\text{V}): \quad O_2 + 4H^+ + 4e^- \rightarrow 2H_2O \qquad (5)$$

$$\text{Anode reaction}: \quad 4Ag^0 + 4Cl^- \rightarrow 4AgCl + 4e^- \qquad (6)$$

The biocatalyst is retained next to the electrode by means of a membrane, which is permeable only to low molecular weight molecules including the reactants and products.

Glucose may be determined by the reduction in the dissolved oxygen concentration when the redox reaction, catalysed by glucose oxidase, occurs (Figure 18.4a):

$$\text{glucose} + O_2 \xrightarrow{\text{glucose oxidase}} \delta\text{-gluconolactone} + H_2O_2 \qquad (7)$$

It is fortunate that this useful enzyme is also one of the most stable oxidoreductases found.

Conditions can be chosen such that the rate at which oxygen is lost from the biocatalyst-containing compartment is proportional to the bulk glucose concentration. Other oxidases can be used in this biosensor configuration and may be immobilised as part of a membrane by treatment of the dissolved enzyme(s), together with a diluent protein, with glutaraldehyde on a cellulose or nylon support. An alternative method of determining the rate of reaction is to detect the hydrogen peroxide produced directly by reversing the polarity of the electrodes (Figure 18.4b). This is the principle used in YSI (Yellow Springs)

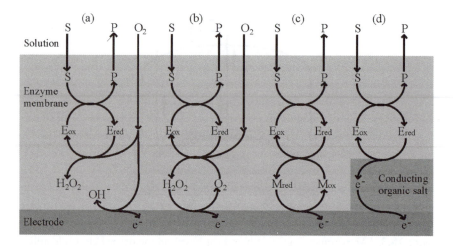

Figure 18.4   The redox mechanisms for various amperometric biosensor configurations.

analysers. The use of a covering, such as the highly anionic Nafion membrane, prevents electroactive anions such as ascorbate reaching the electrode without restricting glucose. Internally, a cellulose acetate membrane replaces the Teflon membrane to allow passage of the hydrogen peroxide. This arrangement has a higher sensitivity than that utilising an oxygen electrode but, in the absence of highly selective membranes, is more prone to interference at the electrode surface.

$$\text{Cathode reaction:} \quad 2AgCl + 2e^- \rightarrow 2Ag^0 + 2Cl^- \tag{8}$$

$$\text{Pt anode reaction}(+0.6\,\text{V}): \quad H_2O_2 \rightarrow O_2 + 2H^+ + 2e^- \tag{9}$$

These electrodes can be developed further for the determination of substrates for which no direct oxidase enzyme exists. Thus sucrose can be determined by placing an invertase layer over the top of the glucose oxidase membrane in order to produce glucose (from the sucrose), which can then be determined. Interference from glucose in the sample can be minimised by including a thin anti-interference layer of glucose oxidase and peroxidase over the top of both layers, which removes the glucose without significantly reducing the oxygen diffusion to the electrode. A clever alternative approach to assay sucrose and glucose together[11] makes use of the lag period in the response due to the necessary inversion of the sucrose that delays its response relative to the glucose (Figure 18.5). This arrangement shows the typical craft required for using existing biosensors to analyse alternative analytes.

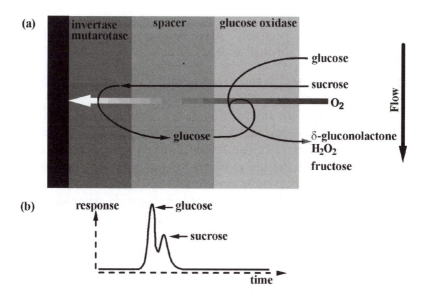

Figure 18.5 A kinetically controlled anti-interference membrane for sucrose. The sample is presented as a rapid pulse of material in the flowing stream. Use of phosphate buffer, which catalyses mutarotation, removes the need for the mutarotase enzyme.

Many biochemicals may be analysed using similar biosensors. One interesting example is the biosensor for determining the artificial sweetener aspartame in soft drinks, where three enzymes are necessary in the sensing membrane in order to produce the H_2O_2 for the electrode reaction:

$$\text{aspartame} + H_2O \xrightarrow{\text{peptidase}} \text{L-aspartate} + \text{L-phenylalanine methyl ester} \quad (10)$$

$$\text{L-aspartate} + \alpha\text{-ketoglutarate} \xrightarrow{\text{aspartate aminotransferase}} \text{L-glutamate} + \text{oxaloacetate} \quad (11)$$

$$\text{L-glutamate} + O_2 + H_2O \xrightarrow{\text{glutamate oxidase}} \alpha\text{-ketoglutarate} + NH_4^+ + H_2O_2 \quad (12)$$

Fish freshness can be determined using a similar concept; the nucleotides in fish changing due to a series of reactions after death. Fish freshness can be quantified in terms of its K value, where

$$K = \frac{(\text{HxR} + \text{Hx}) \times 100}{\text{ATP} + \text{ADP} + \text{AMP} + \text{IMP} + \text{HxR} + \text{Hx}} \quad (13)$$

where HxR, IMP and Hx represent inosine, inosine-5′-monophosphate and hypoxanthine, respectively. After fish die, their ATP undergoes catabolic degradation through a series of reactions:

$$ATP \rightarrow ADP \rightarrow AMP \rightarrow IMP \rightarrow HxR \rightarrow Hx \rightarrow xanthine \rightarrow uric\ acid \quad (14)$$

The accumulation of the intermediates inosine and hypoxanthine relative to the nucleotides is an indicator of how long the fish has been dead and its handling and storage conditions and hence its freshness. A commercialised fish freshness biosensor has been devised which utilises a triacylcellulose membrane containing immobilised nucleoside phosphorylase and xanthine oxidase over an oxygen electrode:

$$inosine + phosphate \xrightarrow{\text{nucleoside phosphorylase}} hypoxanthine \\ + ribose\ phosphate \quad (15)$$

$$hypoxanthine + O_2 \xrightarrow{\text{xanthine oxidase}} xanthine + H_2O_2 \quad (16)$$

$$xanthine + O_2 \xrightarrow{\text{xanthine oxidase}} uric\ acid + H_2O_2 \quad (17)$$

The electrode may be used to determine the reduction in oxygen[12] or the increase in hydrogen peroxide. The inosine content may be determined after the hypoxanthine content by the addition of the necessary phosphate. The nucleotides can be determined using the same electrode and sample, subsequent to the addition of nucleotidase and adenosine deaminase. Typically, K values below 20 show the fish is very fresh and may be eaten raw. Fish with a K value between 20 and 40 must be cooked but those with a K value above 40 are not fit for human consumption. Critical K values vary amongst species, but can be a reliable indicator applicable to frozen, smoked or fish stored under modified atmospheres. Clearly, a relatively simple probe that accurately and reproducibly determines fish freshness has significant economic importance to the fish industry. In its absence, freshness is determined completely subjectively by inspection.

Although such biosensors are easy to produce, they do suffer from some significant drawbacks. The reaction is dependent on the concentration of molecular oxygen, which precludes its use in oxygen-deprived environments such as *in vivo*. Also, the potential used is sufficient to cause other redox processes to occur, such as ascorbate oxidation/reduction, which may interfere with the analyses. Much research has been undertaken on the development of substances that can replace oxygen in these reactions.[13] Generally, oxidases are far more specific for the oxidisable reactant than they are for molecular oxygen itself as the oxidant. Many other materials can act as the oxidant. The optimal properties of such materials include fast electron transfer rates, the ability to be easily regenerated by an electrode reaction and retention within the biocatalytic membrane. In addition, they should not react with other molecules, including molecular oxygen that may be present. Many such oxidants, now called mediators, have been developed. Their redox reactions, which together transfer the electrons from the substrate to the electrode, so producing the electrical signal, are summarised in Figure 18.6. As they allow reduced working potentials, interference from natural redox materials is reduced.

The mediated biosensor reaction consists of three redox processes:

$$\text{substrate}_{\text{(reduced)}} + \text{enzyme}_{\text{(oxidised)}} \xrightarrow{\text{enzyme reaction}} \text{product}_{\text{(oxidised)}} \\ + \text{enzyme}_{\text{(reduced)}} \quad (18)$$

$$\text{enzyme}_{\text{(reduced)}} + \text{mediator}_{\text{(oxidised)}} \xrightarrow{\text{enzyme reaction}} \text{enzyme}_{\text{(oxidised)}} \\ + \text{mediator}_{\text{(reduced)}} \quad (19)$$

$$\text{mediator}_{\text{(reduced)}} \xrightarrow{\text{electrode reaction}} \text{mediator}_{\text{(oxidised)}} + e^- \quad (20)$$

When a steady-state response has been obtained, the rates of all of these processes and the rate of the diffusive flux in must be equal (see Figure 18.4c). Any of these, or a combination, may be the controlling factor. Although the level of response is reproducible, it is therefore difficult to predict its magnitude.

Several blood glucose biosensors, for the control of diabetes, have been built and marketed based on this mediated system (Figure 18.7). The sensing area is a single-use disposable electrode, typically produced by screen-printing on to a plastic strip and consisting of an Ag/AgCl reference electrode and a carbon working electrode containing glucose oxidase and a derivatised ferrocene mediator. Both electrodes are covered with wide-mesh hydrophilic gauze to permit even spreading of a blood drop sample and to prevent localised cooling effects due to uneven evaporation. The electrodes are kept dry until use and have a shelf-life of 6 months when sealed in aluminium foil. They can detect glucose concentrations of 2–25 mM in single tiny drops of blood (1–4 µl) and accurately display the result within 5–40 s. The pricing is such that the biosensor is sold cheaply with the major profit arising from the sale of the necessary disposable electrodes. Such biosensors are currently used

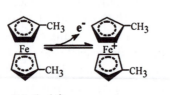

Methyl ferrocene

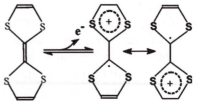

Tetrathiafulvalene

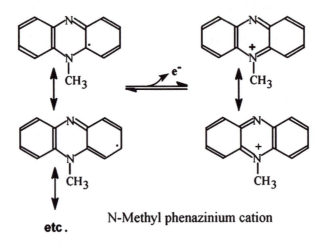

etc.   N-Methyl phenazinium cation

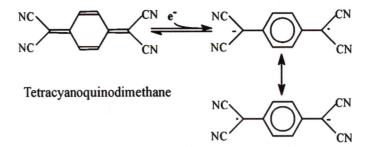

Tetracyanoquinodimethane

Figure 18.6   The redox reactions of amperometric biosensor mediators. Tetracyanoquinodimethane acts as a partial electron acceptor whereas ferrocene,
tetrathiafulvalene and *N*-methylphenazinium can all act as partial electron donors.

by millions of people with diabetes in more than 50 countries worldwide. The profits from this area, although by far the greatest in the
biosensor industry, are levelling off as, while the world diabetic population stands at over 200 million and is steadily expanding, diabetic
patients use their biosensors less often when their diabetes appears under

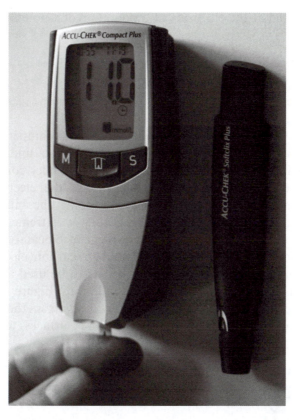

Figure 18.7   A biosensor for glucose used in diabetes monitoring. A drop of blood is
applied to the disposable electrode at the bottom with its glucose concentration being produced on the read-out a few seconds later. On the
right is a system for the controlled pricking of a finger.

control. Although the subject of much research, the use of continuous
subcutaneous *in vivo* glucose sensing, using electrode-based biosensors,
has not overcome the difficulties caused by the body's response and poor
patient acceptability. New non-biosensor technologies are being developed,[14] involving near-infrared light which can penetrate tissue and
detect glucose in the blood, albeit at low sensitivity and in the face of
almost overwhelming interference.

   When an oxidase is unable to react rapidly enough with available
mediators, horseradish peroxidase, which rapidly reacts with ferrocene
mediators, can be included with the enzyme. This catalyses the reduction
of the hydrogen peroxide produced by the oxidase and consequent
oxidation of the mediator. In this case, the mediator is acting as an
electron donor rather than acceptor. The oxidised mediator then can be

rapidly reduced at the electrode at moderate redox potential:

$$\text{mediator}_{\text{(reduced)}} + \tfrac{1}{2}\text{H}_2\text{O}_2 + \text{H}^+ \xrightarrow{\text{peroxidase}} \text{mediator}_{\text{(oxidised)}} + \text{H}_2\text{O} \qquad (21)$$

$$\text{mediator}_{\text{(oxidised)}} + \text{e}^- \xrightarrow{\text{electrode reaction}} \text{mediator}_{\text{(reduced)}} \qquad (22)$$

A major advance in the development of microamperometric biosensors came with the discovery that pyrrole can undergo electrochemical oxidative polymerisation (Figure 18.8) under conditions mild enough to entrap enzymes and mediators at the electrode surface without denaturation. A membrane, entrapping the biocatalyst and mediator, can be formed at the surface of even extremely small electrodes by polymerising pyrrole in the present of biocatalyst. The polypyrrole adheres tightly to platinum, gold or carbon electrodes. This allows silicon chip microfabrication methods to be used and for many different sensors to be laid down on the same chip (Figure 18.9). Similar membranes may be prepared less wastefully *in situ* from the oxidation of pyrrole by chloroauric acid to give sensing surfaces with excellent

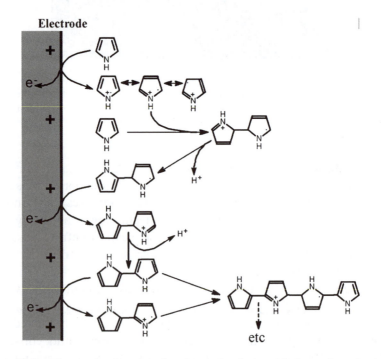

Figure 18.8 The mechanism for the electrochemical oxidative polymerisation of pyrrole.

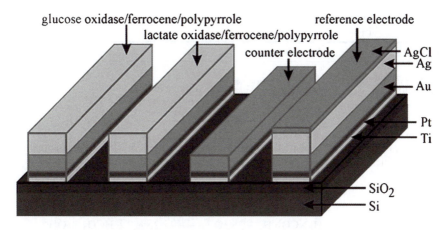

Figure 18.9 A combined microelectrode for glucose and lactate.

electron transfer properties formed from polypyrrole-immobilised enzymes and gold nanoparticles.[15]

Another advance has been the use of conducting organic salts on the electrode. These allow the direct transfer of electrons from the reduced enzyme to the electrode without the use of any (other) mediator (Figure 18.4d). Conducting organic salts consist of a mixture of two types of planar aromatic molecules, electron donors and electron acceptors (see Figure 18.6), which partially exchange their electrons. These molecules form segregated stacks, containing either the donor or acceptor molecules, with some of the electrons from the donors being transferred to the acceptors. These electrons, which have been partially transferred, are mobile up and down the stacks, giving the organic crystals a high conductivity. There must not be a total electron transfer between the donor and acceptor molecules or the crystal becomes an insulator through lack of electron mobility. These electrodes give the somewhat misleading appearance of direct electron transfer to the electrode. As both the components of the organic salts, in the appropriate redox state, are able to mediate the reaction, it is highly probable that these electrodes are behaving as a highly insoluble mediator prevented from large-scale leakage by electrostatic effects.

18.4.2 Potentiometric Biosensors

Changes in ionic concentrations are easily determined by use of ion-selective electrodes. This forms the basis for potentiometric biosensors.[16] Many biocatalysed reactions involve charged species, each of which will

Table 18.4 Biocatalytic reactions that can be used with ion-selective electrode biosensors.

Electrode	Reactions
Hydrogen ion	
Penicillin	penicillin $\xrightarrow[\text{lipase}]{\text{penicillinase}}$ penicilloic acid $+ H^+$
Lipid	triacylglycerol $\xrightarrow{}$ glycerol $+$ fatty acids $+ 3H^+$
Urea	$H_2NCONH_2 + H_2O + 2H^+ \xrightarrow{\text{urease(pH 6)}} 2NH_4^+ + CO_2$
Ammonia	
L-Phenylalanine	L-phenylalanine $\xrightarrow{\text{phenylalanine ammonia-lyase}} NH_4^+ + $ *trans*-cinnamate
L-Asparagine	L-asparagine $+ H_2O \xrightarrow{\text{asparaginase}} NH_4^+ + $ L-aspartate
Adenosine	adenosine $+ H_2O + H^+ \xrightarrow{\text{adenosine deaminase}} NH_4^+ + $ inosine
Creatinine	creatinine $+ H_2O \xrightarrow{\text{creatininase}}$ creatine $\rightarrow$
Creatine	creatine $+ H_2O \xrightarrow{\text{creatinase}} H_2NCONH_2 + $ sarcosine $\rightarrow$
Urea	$H_2NCONH_2 + 3H_2O \xrightarrow{\text{urease (pH 7)}} 2NH_4^+ + HCO_3^- + OH^-$
Iodide	
Peroxide	$H_2O_2 + 2I^- + 2H^+ \xrightarrow{\text{peroxidase}} 2H_2O + I_2$

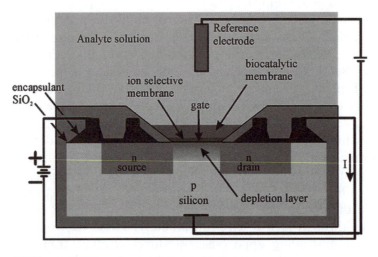

Figure 18.10   A FET-based potentiometric biosensor.

absorb or release hydrogen ions according to their pK_a and the pH of the environment. This allows a relatively simple electronic transduction using the commonest ion selective electrode, the pH electrode. Table 18.4 shows some biocatalytic reactions that can be utilised in potentiometric biosensors. Potentiometric biosensors can be miniaturised by the use of field effect transistors (FETs).

Ion-selective field effect transistors (ISFETs) are low-cost devices that are in mass production. Figure 18.10 shows a diagrammatic cross-section through an npn hydrogen ion-responsive ISFET with an approximately 0.025 mm^2 biocatalytic membrane covering the ion-selective membrane. The build-up of positive charge on this surface (the gate) repels the positive holes in the p-type silicon, causing a depletion layer and allowing the current to flow. The reference electrode is usually an identical ISFET without any biocatalytic membrane. A major practical problem with the manufacture of such enzyme-linked FETs (ENFETs) is protection of the silicon from contamination by the solution, hence the covering of waterproof encapsulant. Because of their small size, they only require minute amounts of biological material and can be produced in a form whereby they can determine several analytes simultaneously. A further advantage is that they have a more rapid response rate compared with the larger, sluggish ion-selective electrode devices. The enzyme may be immobilised to the silicon nitride gate using polyvinylbutyral deposited by solvent evaporation and cross-linked with glutaraldehyde. Such devices still present fabrication problems such as reproducibility, drift, sensitivity to light and the need for on-chip temperature compensation.

Use of membranes selective for ions other than hydrogen ions, such as ammonium, allows many related biosensors to be constructed. Potentiometric biosensors for DNA have been developed which use anti-DNA monoclonal antibodies conjugated with urease. DNA is bound to a membrane, placed on the electrode and quantified by the change in pH on addition of urea (see Table 18.4).

18.4.3 Conductimetric Biosensors

Many biological processes involve changes in the concentrations of ionic species. Such changes can be utilised by biosensors that detect changes in electrical conductivity. A typical example of such a biosensor is the urea sensor, utilising immobilised urease,[17] and used as a monitor during renal surgery and dialysis (Figure 18.11). The reaction gives rise to a large change in ionic concentration at pH 7.0, making this type of biosensor particularly attractive for monitoring urea concentrations.

$$NH_2CONH_2 + 3H_2O \xrightarrow{\text{electrode reaction}} 2NH_4^+ + HCO_3^- + OH^- \qquad (23)$$

An alternating field between the two electrodes allows the conductivity changes to be determined while minimising undesirable electrochemical processes. The electrodes are interdigitated to give a relatively long track length (~ 1 cm) within a small sensing area (0.2 mm^2). A steady-state response can be achieved in a few seconds, allowing urea to be determined within the range 0.1–10 mM. The output

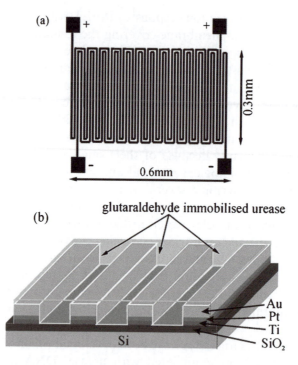

(a)

0.3mm

0.6mm

(b)

glutaraldehyde immobilised urease

Au
Pt
Ti
SiO₂

Si

Figure 18.11 Parts of a conductimetric biosensor electrode arrangement. (a) Top view; (b) cross-sectional view. The tracks are about 5 μm wide and the thicknesses of the various layers are approximately SiO₂ 0.55 μm, Ti 0.1 μm, Pt 0.1 μm, Au 2 μm.

is corrected for non-specific changes in pH and other factors by comparison with the output of a non-enzymic reference electrode pair on the same chip. The method can easily be extended to use other enzymes and enzyme combinations that produce ionic species, for example amidases, decarboxylases, esterases, phosphatases and nucleases. As molecular iodine molecules change the conductivity of iodine-sensitive phthalocyanine films, they may be used in peroxidase-linked immunoassays where the peroxidase converts iodide ions to iodine. Conductimetric biosensors have also be devised to determine DNA hybridisation.[18]

18.5 PIEZOELECTRIC BIOSENSORS

The piezoelectric effect is due to some crystals containing positive and negative charges that separate when the crystal is subjected to a stress, causing the establishment of an electric field. As a consequence, if this crystal is subjected to an electric field it will deform. An oscillating electric field of a resonant frequency will cause the crystal to vibrate with a characteristic frequency dependent on its composition and thickness and also the way in which it has been cut. As this resonant frequency varies when molecules adsorb on the crystal surface, a piezoelectric crystal may form the basis of a biosensor. Even small changes in resonant frequencies are easy to determine with precision and accuracy using straightforward electronics. Differences in mass, even as small as 1 ng cm^{-2}, can be measured when adsorbed on the sensing surface. Changes in frequency are generally determined relative to a similarly treated reference crystal but without the active biological material. As an example, a biosensor for cocaine in the gas phase may be made by attaching cocaine antibodies to the surface of a piezoelectric crystal. This biosensor changes frequency by about 50 Hz for a 1 ppb atmospheric cocaine sample and can be reused after flushing for a few seconds with clean air. The relative humidity of the air is important, because if it is too low the response is less sensitive and if it is too high the piezoelectric effect may disappear altogether. Cocaine in solution can be determined after drying such biosensors.[19]

Enzymes with gaseous substrates or inhibitors can also be attached to such crystals, as has been proved by the production of biosensors for formaldehyde incorporating formaldehyde dehydrogenase and for organophosphorus insecticides incorporating acetylcholinesterase.

One of the drawbacks preventing the more widespread use of piezoelectric biosensors is the difficulty in using them to determine analytes in solution. The frequency of a piezoelectric crystal depends on the liquid's viscosity, density and specific conductivity. Under unfavourable conditions, the crystal may cease to oscillate completely. There is also a marked effect of temperature due to its effect on viscosity. The binding of material to the crystal surface may be masked by other intermolecular effects at the surface and bulk viscosity changes consequent upon even small concentration differences. There is also the strong possibility of interference due to non-specific binding.

Antibody–antigen binding can be determined by measuring the frequency changes in air after drying the crystal. Such procedures, although sensitive, are difficult to reproduce repetitively, as the antibody layer may be partially lost when the antigen is removed. However, one-shot biosensors have been developed, using this principle, for the detection of several food contaminants such as enterobacteria.[20]

Piezoelectricity is also utilised in surface acoustic wave (SAW) devices where a set of interdigitated electrodes is microfabricated at each end of a rectangular quartz plate (Figure 18.12). Binding of molecules to the surface affects the propagating wave, generated at one end, such that its

AC signal send   receiver

coated sensor

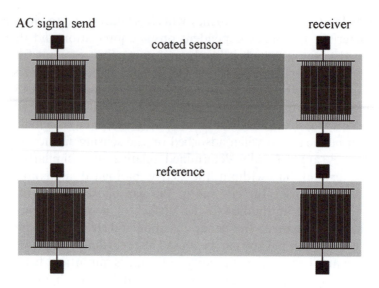

reference

Figure 18.12   A surface acoustic wave (SAW) biosensor.

frequency is reduced before reception at the other. The sensitivity of these devices is proportional to their frequency and dependent on minimising the interference from non-specific binding.

18.6   OPTICAL BIOSENSORS

Optical biosensors (also called optodes) have generated considerable interest, particularly with respect to the use of fibre optics and optoelectronic transducers.[21] These allow the safe non-electrical remote sensing of materials in hazardous or sensitive (*i.e. in vivo*) environments. An advantage of optical biosensors is that no reference sensor is needed; a comparative signal is generally easily generated by splitting the light source used by the sampling sensor. A simple example of an optical biosensor is the fibre optic lactate sensor (Figure 18.13) which senses changes in molecular oxygen concentrations by determining its quenching of a fluorescent dye:

$$O_2 + \text{lactate} \xrightarrow{\text{lactate monooxygenase}} CO_2 + \text{acetate} + H_2O \qquad (24)$$

The presence of oxygen quenches (reduces) the amount of fluorescence generated by the dyed film. An increase in lactate concentration reduces the oxygen concentration reaching the dyed film so alleviating the

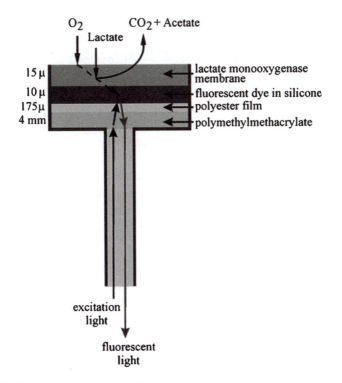

Figure 18.13 A fibre optic lactate biosensor.

quenching and consequentially causing an increase in the fluorescence output.

Simple colorimetric changes can be monitored in some biosensor configurations. A lecithin biosensor has been developed containing phospholipase D, choline oxidase and bromothymol blue. The change in pH, due to the formation of the acid betaine from the released choline, causes a change in the bromothymol blue absorbance at 622 nm.[22] Gasphase reactions can also be monitored.[23] For example, alcohol vapour can be detected by the colour change of a dry dispersion of alcohol oxidase and peroxidase plus the redox dye 2,6-dichloroindophenol.

One of the most widely established biosensor technologies is the low-technology single-use colorimetric assay based on a paper pad impregnated with reagents. This industry revolves mainly round blood and urine analysis with test strips costing only a few cents. A particularly important use for these colorimetric test strips is the detection of glucose. In this case, the strips contain glucose oxidase and horseradish peroxidase together with a chromogen (e.g. *o*-toluidine) which changes colour when oxidised by the peroxidase-catalysed reaction with the hydrogen peroxide produced by the aerobic oxidation of glucose:

$$\text{(reduced)chromogen(2H)} + H_2O_2 \xrightarrow{\text{peroxidase}} \text{(oxidised)dye} + 2H_2O \qquad (25)$$

The colour produced can be evaluated by visual comparison with a test chart or by the use of a portable reflectance meter. Many test strips incorporate anti-interference layers to produce more reproducible and accurate results.

It is possible to link up luminescent reactions to biosensors, as light output is a relatively easy phenomenon to transduce to an electronic output. As an example, the reaction involving immobilised (or free) luciferase can be used to detect the ATP released by the lysis of micro-organisms:[24]

$$\text{luciferin} + \text{ATP} + O_2 \xrightarrow{\text{luciferase}} \text{oxyluciferin} + CO_2 + \text{AMP}$$
$$+ \text{pyrophosphate} + \text{light} \qquad (26)$$

This allows the rapid detection of urinary infections by detecting the microbial content of urine samples.

18.6.1 Evanescent Wave Biosensors

A light beam will be totally reflected when it strikes an interface between two transparent media, from the side with the higher refractive index, at angles of incidence (θ) greater than the critical angle (Figure 18.14a). This is the principle that allows transparent fibres to be used as optical waveguides. At the point of reflection an electromagnetic field is induced which penetrates into the medium with the lower refractive index, usually air or water. This field is called the evanescent wave and it rapidly decays exponentially with the penetration distance and generally has effectively disappeared within a few hundred nanometres. The exact depth of penetration depends on the refractive indices and the wavelength of the light and can be controlled by the angle of incidence. The evanescent wave may interact with the medium and the resultant electromagnetic field may be coupled back into the higher refractive index medium (usually glass) by essentially the reverse process. This gives rise to changes in the light emitted down the waveguide. Thus, it can be used to detect changes occurring in the liquid medium. The necessary surface interactions impose a limitation on the sensitivity of such devices at about $10\,\text{pg}\,\text{mm}^{-2}$ and the requirement to limit non-specific absorption.

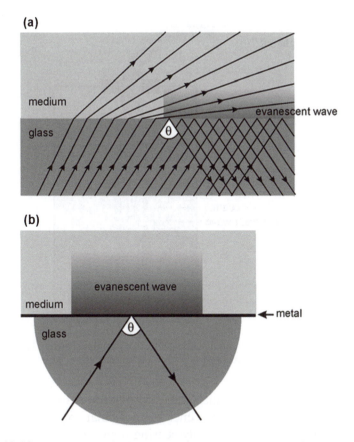

Figure 18.14 Production of (a) an evanescent wave and (b) surface plasmon resonance. At acute enough angles of incidence the light is totally internally
reflected at the glass surface. In (a) an evanescent wave extends from
this surface into the air or water medium. This process is amplified in
(b) by the presence of the thin metal film.

Various effects, dependent on the biological sensing processes, can be
determined, including changes in absorption, optical activity, fluorescence and luminescence. Because of the small degree of penetration, this
system is particularly sensitive to biological processes in the immediate
vicinity of the surface and independent of any bulk processes or changes.
Due to the small pathlength through the solution, it can even be used for
the continuous monitoring of apparently opaque solutions.

This biosensor configuration is particularly suitable for immunoassays
as there is no need to separate bulk components since the wave only
penetrates as far as the antibody–antigen complex. Surface-bound
fluorophores may be excited by the evanescent wave and the excited light
output detected after it is coupled back into the fibre (Figure 18.15).

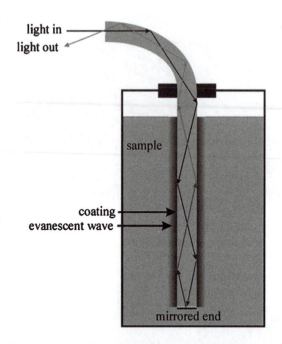

Figure 18.15 The principle behind evanescent wave immunosensor. The light output
is reduced by absorption within the evanescent wave.

Sensors can be fabricated which measure oxidase substrates using the
principle of quenching of fluorescence by molecular oxygen as described
earlier. Another advantage of only sensing a surface reaction less than
1 μm thick is that the volume of analyte needed may be very small indeed,
that is, less than 1 nl using suitable fluid transfer microfluidics.

Protein A, an important immunoglobulin-binding protein from
Staphylococcus aureus, has been determined by this method using a
plastic optical fibre coated with its antibody. Detection was by the
fluorescence of a fluorescein-bound anti-protein A immunoglobulin
which was subsequently bound, sandwiching the protein A.[25]

18.6.2 Surface Plasmon Resonance

The evanescent field generated by the total internal reflection of
monochromatic plane-polarised light within a fibre optic or prism may
be utilised in a different type of optical biosensor by means of the
phenomenon of surface plasmon resonance (SPR).[26] If the surface of
the glass is covered with a very thin layer of metal (usually pure gold,
silver or palladium, just a nanometre or so thick) then the electrons
at the surface may be caused to oscillate in resonance with the photons

(as surface plasmon polaritons). This generates a surface plasmon wave and amplifies the evanescent field on the far side of the metal (Figure 18.14b). If the metal layer is thin enough to allow penetration of the evanescent field to the opposite surface, the effect is critically dependent on the 100 nm or so of medium that is adjacent to the metal. This effect occurs only when the light is at a specific angle of incidence dependent on its frequency, the thickness of the metal layer and the refractive index of the medium immediately adjacent the metal surface within the evanescent field.[27] The generation of this surface plasmon resonance adsorbs some of the energy of the light so reducing the intensity of the internally reflected light (Figure 18.16). Changes occurring in the medium caused by biological interactions may be followed using the consequential changes in the intensity of the reflected light or the resonance angle. Figure 18.16 shows the change in the resonance angle of a human chorionic gonadotrophin (hCG) biosensor on binding hCG to surface-bound hCG antibody.[28] The sensitivity in such devices is limited by the degree of uniformity of the surface and the bound layer and the more-sensitive devices minimise light scattering. Under optimal conditions just 20 or 30 protein molecules bound to each square micrometre of surface may be detected. As with other immunosensors, the main problem occurring in such devices is non-specific absorption.

The biological sensing can be achieved by attaching the bioactive molecule to the medium side of the metal film. Physical adsorption may be used but, because this may lead to undesired denaturation and weak

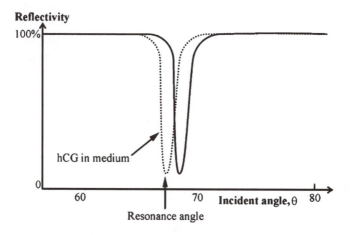

Figure 18.16 The change in absorption due to surface plasmon resonance.

binding, covalent binding is often preferred. Gold films can be coated with a monolayer of long-chain 1,ω-hydroxyalkylthiols, which are copolymerised to a flexible un-cross-linked carboxymethylated dextran gel, allowing the subsequent binding of bioactive molecules. This flat plate system, marketed as Biacore,[29] allows the detection of parts per million of protein antigen where the appropriate antibody is bound to the gel. Similarly, biosensors for DNA detection can be constructed by attaching a DNA or RNA probe to the metal surface when as little as a few femtograms of complementary DNA or RNA can be detected[30] and, as a bonus, the rate of hybridisation may be determined. Such biosensors retain the advantages of the use of evanescent fields as described earlier. They can also be used to investigate the kinetics of the binding and dissociation processes.[31] In spite of the relatively low costs possible for producing biosensing surfaces, the present high cost of the instrumentation is restricting developments in this area. However, developments in the use of low-cost light-emitting diodes and lens arrays together with the use of porous surfaces to increase surface area, microfluidics and integrated devices with in-built waveguided inter-ferometry of reference beams are allowing the field to progress rapidly.

18.7 WHOLE CELL BIOSENSORS

As biocatalysts, whole microbial cells can offer some advantages over pure enzymes when used in biosensors.[32] Generally, microbial cells are cheaper, have longer active lifetimes and are less sensitive to inhibition, pH and temperature variations than the isolated enzymes. Against these advantages, such devices usually offer longer response and recovery times, a lower selectivity and they are prone to biocatalytic leakage. They are particularly useful where multistep or coenzyme-requiring reactions are necessary. The microbial cells may be viable or dead. The advantage of using viable cells is that the sensor may possess a self-repair capability, but this must be balanced against the milder conditions necessary for use and problems that might occur due to membrane permeability and cellular outgrowth. Different types of whole cell biosensors are shown in Table 18.5.

Biochemical oxygen demand (BOD) biosensors most often use a single selected microbial species. Although rapid, linear and reproducible, they give a different result to conventional testing, which involve incubation over 5 days, and reflect the varying metabolism of a mixed microbial population. Thermophilic organisms have been used in such a biosensor for use in hot wastewater.

Table 18.5 Whole cell biosensors.

Analyte	Organism	Biosensor
Ammonia	*Nitrosomonas sp.*	Amperometric (O_2)
Biological oxygen demand (BOD)	Many	Amperometric (O_2/mediated), polarographic (O_2) or potentiometric (FET/H_2)
Cysteine	*Proteus morganii*	Potentiometric (H_2S)
Glutamate	*Escherichia coli*	Potentiometric (CO_2)
Glutamine	*Sarcina flava*	Potentiometric (NH_3)
Herbicides	Cyanobacteria	Amperometric (mediated)
Nicotinic acid	*Lactobacillus arabinosus*	Potentiometric (H^+)
Sulfate	*Desulfovibrio desulfuricans*	Potentiometric (SO_3^-)
Thiamine	*Lactobacillus fermenti*	Amperometric (mediated)

Table 18.6 A selection of receptor-based sensors.

Analyte	Sensing method	Biosensor
Anthrax spores (warfare)	Antibody	Surface acoustic wave
Cholera toxin (warfare)	Antibody	Potentiometric
Verotoxin-producing *E. coli*	Peptide nucleic acid	Surface plasmon resonance
Cocaine	Antibody/protein A	Piezoelectric
Human chorionic gonadotrophin	Antibody/catalase	Amperometric (O_2)
	Antibody	Surface plasmon resonance
Hepatitis B surface antigen	Antibody/peroxidase	Potentiometric (I^-)
Insulin	Antibody/catalase	Amperometric (O_2)
Ricin (warfare)	Antibody	Evanescent wave
Trinitrotoluene (TNT)	Antibody/labelled antigen	Fluorescence

18.8 RECEPTOR-BASED SENSORS

Receptor-based sensors include immunosensors[33] (Table 18.6) and some of these have been mentioned earlier. Most biosensor configurations may be used, with Figure 18.17 showing some of those possible. Direct binding of the antigen to immobilised antibody (Figure 18.17a) or antigen–antibody sandwich (Figure 18.17b) may be detected using piezoelectric or SPR devices,[34] as can antibody release due to free antigen (Figure 18.17c). Binding of enzyme-linked antigen (Figure 18.17d) or antibody can form the basis of all types of immunosensors but has proved particularly useful in amperometric devices. The amount of enzyme activity bound in these receptor-based sensors is dependent on the relative concentrations of the competing labelled and unlabelled

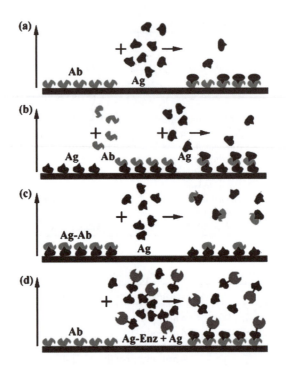

Figure 18.17   Different configurations for biosensor immunoassays: (a) antigen binding to immobilised antibody; (b) immobilised antigen binding antibody
which binds free second antigen; (c) antibody bound to immobilised
antigen partially released by competing free antigen; (d) immobilised
antibody biding free antigen and enzyme-labelled antigen in competition.

ligands and so it can be used to determine the concentration of unknown
antigen concentrations.

The main problems involved in developing receptor-based sensors
centre on non-specific binding and incomplete reversibility of the
binding process, both of which reduce the active area and hence sensitivity, on repetitive assay. Single-use biosensing membranes are a way
round this but they require strict quality control during production.

DNA chips have been produced that contain tens of thousands of
different but known single-stranded polynucleotides immobilised on a
silicon chip. Thus, fluorescently labelled single-stranded DNA can be
probed for structure due to the specificity of DNA–DNA binding.[35]
Introduction of redox materials that can intercalate between double-
stranded DNA, but not bind to single-stranded DNA allows the
development of electrochemical versions of these chips without the need
for fluorescent probes or fluorescence readers (Figure 18.18).[36] A simple
example involves ferrocene-linked naphthalene diimide.[37]

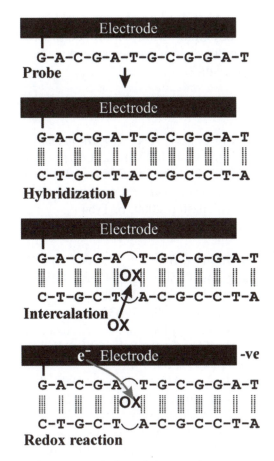

Figure 18.18 DNA biosensor. Binding of the complementary strand allows certain redox molecules to intercalate into the DNA fragment to allow detection via amperometry.

Proteins, such as antibodies, may also be attached to DNA chips by means of covalently attached complementary probes, so allowing thousands of differently binding proteins to be specifically attached to single chips. The high negative charge on the complementary linkers almost totally removes the chance of non-specific binding. This area is likely to expand considerably over the next few years.

18.9 CONCLUSION

Biosensors form an interesting and varied part of biotechnology. They have been applied to solve a number of analytical problems and some have achieved notable commercial success. They have been slow to

evolve from research prototype to the marketplace and have not yet
reached their full potential. Many more commercial products are
expected over the next few years, particularly in medical diagnostics and
the war against terrorism.[38]

REFERENCES

1. J. D. Newman and S. J. Setford, *Mol. Biotechnol.*, 2006, **32**, 249.
2. M. N. Velasco-Garcia and T. Mottram, *Biosyst. Eng.*, 2003, **84**, 1.
3. H. Nakamura and I. Karube, *Anal. Bioanal. Chem.*, 2003, **377**, 446.
4. G. A. Urban, in *Encyclopedia of Analytical Chemistry*, ed. R. A.
   Meyers, Wiley, Chichester, 2000, p. 1164.
5. J. D. Newman and A. P. F. Turner, *Biosens. Bioelectron.*, 2005,
   20, 2435.
6. P. T. Kissinger, *Biosens. Bioelectron.*, 2005, **20**, 2512.
7. Z. Zhu, J. Zhang and J. Zhu, *Sens. Lett.*, 2005, **3**, 71.
8. J. Wang, *Electroanalysis*, 2001, **13**, 983.
9. T. D. Gibson, *Analusis*, 1999, **27**, 632.
10. L. C. Clark and C. H. Lyons, *Ann. N. Y. Acad. Sci.*, 1962, **102**, 29.
11. E. Watanabe, M. Takagi, S. Takei, M. Hoshi and C. Shu-gui,
    Biotechnol. Bioeng., 1991, **38**, 99.
12. S. S. Hu and C. C. Liu, *Electroanalysis*, 1997, **9**, 1174.
13. F. W. Scheller, F. Schubert, B. Neumann, D. Pfeiffer, R. Hintsche,
    I. Dransfeld, U. Wollenberger, R. Renneberg, A. Warsinke,
    G. Johansson, M. Skoog, X. Yang, V. Bogdanovskaya, A. Bückmann
    and S. Y. Zaitsev, *Biosens. Bioelectron.*, 1991, **6**, 245.
14. H. E. Koschwanez and W. M. Reichert, *Biomaterials*, 2007,
    28, 3687.
15. J. Njagi and S. Andreescu, *Biosens. Bioelectron.*, 2007, **23**, 168.
16. R. Koncki, *Anal. Chim. Acta*, 2007, **599**, 7.
17. N. F. Sheppard, D. J. Mears and A. Guiseppielie, *Biosens. Bio-
    electron.*, 1996, **11**, 967.
18. S.-J. Park, A. Taton and C. A. Mirkin, *Science*, 2002, **295**, 1503.
19. B. S. Attili and A. A. Suleiman, *Microchem. J.*, 1996, **54**, 174.
20. M. Plomer, G. G. Guilbault and B. Hock, *Enzyme Microb. Tech-
    nol.*, 1992, **14**, 230.
21. A. Leung, P. M. Shankar and R. Mutharasan, *Sens. Actuators B*,
    2007, **125**, 688.
22. V. P. Kotsira and Y. D. Clonis, *J. Agric. Food Chem*, 1998, **46**, 3389.
23. S. Lamare and M. D. Legoy, *Trends Biotechnol.*, 1993, **11**, 413.
24. M. F. Chaplin, in *Physical Methods for Microorganisms Detection*,
    ed. W. M. Nelson, CRC Press, Boca Raton, FL,1991, p. 81.

25. Y. H. Chang, T. C. Chang, E. F. Kao and C. Chou, *Biosci. Biotechnol. Biochem.*, 1996, **60**, 1571.
26. X. D. Hoa, A. G. Kirk and M. Tabrizian, *Biosens. Bioelectron.*, 2007, **23**, 151.
27. J. Homola, *Anal. Bioanal. Chem.*, 2003, **377**, 528.
28. J. W. Attridge, P. B. Daniels, J. K. Deacon, G. A. Robinson and G. P. Davidson, *Biosens. Bioelectron.*, 1991, **6**, 201.
29. R. L. Rich and D. G. Myszka, *Drug Discov. Today: Technol.*, 2004, **1**, 301.
30. T. Schwarz, D. Yeung, E. Hawkins, P. Heaney and A. McDougall, *Trends Biotechnol.*, 1991, **9**, 339.
31. M. Fivash, E. M. Towler and R. J. Fisher, *Curr. Opin. Biotechnol.*, 1998, **9**, 370.
32. Y. I. Korpan and A. V. Elskaya, *Biochemistry (Moscow)*, 1995, **60**, 1517.
33. O. Lazcka, F. J. Del Campo and F. X. Muñoz, *Biosens. Bioelectron.*, 2007, **22**, 1205.
34. C. L. Morgan, D. J. Newman and C. P. Price, *Clin. Chem.*, 1996, **42**, 193.
35. R. C. Anderson, G. McGall and R. J. Lipshutz, *Top. Curr. Chem.*, 1997, **194**, 117.
36. A. L. Ghindilis, M. W. Smith, K. R. Schwarzkopf, K. M. Roth, K. Peyvan, S. B. Munro, M. J. Lodes, A. G. Stöver, K. Bernards, K. Dill and A. McShea, *Biosens. Bioelectron.*, 2007, **22**, 1853.
37. J. Wang, *Anal. Chim. Acta*, 2002, **469**, 63.
38. J. J. Gooding, *Anal. Chim. Acta*, 2006, **559**, 137.

CHAPTER 19

Biofuels and Biotechnology

JONATHAN R. MIELENZ

Oak Ridge National Laboratory, Oak Ridge TN 37831, USA

19.1 INTRODUCTION

The world obtains 86% of its energy from fossil fuels, 40% from petroleum, a majority of which goes to the transportation sector (www.IEA.gov). Well-recognized alternatives are fuels derived from renewable sources known as biofuels. There are a number of biofuels useful for transportation fuels, which include ethanol, biobutanol, mixed alcohols, biodiesel and hydrogen. These biofuels are produced from biologically derived feedstock, almost exclusively being plant materials, either food or feed sources or inedible plant material called lignocellulose or biomass. This chapter discusses technologies for production of liquid transportation biofuels from renewable feedstocks that can use existing infrastructure, so hydrogen will not be included. In addition, a specific emphasis will be placed upon the research opportunities and potential for application of system biology tools to dissect and understand the biological processes central to production of these biofuels from biomass and biological materials.

There are a number of technologies for production of each of these biofuels that range from fully mature processes such as grain-derived ethanol, the emerging technologies for cellulose-derived ethanol, biobutanol and thermochemical conversion technologies and immature processes such as production of hydrogen from renewable biological

Molecular Biology and Biotechnology, 5th Edition
Edited by John M Walker and Ralph Rapley
© Royal Society of Chemistry 2009
Published by the Royal Society of Chemistry, www.rsc.org

feedstocks. Conversion of biomass by various thermochemical and combustion technologies to produce thermochemical biodiesel or steam and electricity provide growing sources of bioenergy. However, like hydrogen, these technologies are outside of the scope of this chapter, as is the use of biological processing for upgrading and conversion of fossil fuels. Therefore, this chapter will focus on the current status of production of biofuels produced from biological-derived feedstocks using biological processes.

Regardless of the status of development of the biological process for production of the biofuels, each process can benefit from research and resulting development activities using the latest biological research tools and techniques. Among the most recently evolving research tools is what is collectively known as 'omics' techniques, such as genomics, transcriptomics, proteomics, metabolomics and fluxomics, plus an ever growing omics word generation.[1] These and other similar methodologies are central to understanding the interactive functioning of gene expression, resulting protein/enzyme production, which impacts the cellular metabolism and carbon and metabolite flow. These emerging system biology 'omics' tools are just beginning to be applied to understand and improve the biological processes involved in conversion of renewable plant and animal material to biofuels, and further opportunities will be discussed in this chapter.

19.2 PRODUCTION OF THE MAJOR BIOFUELS

Ethanol is currently (March 2008) being produced from over 140 commercial biorefineries throughout the USA and numerous processing plants in Brazil, yielding over 8.0 billion gallons in the USA (www.e-thanolrfa.org) and nearly 4.5 billion gallons in Brazil (2007) (www.iea.org). In the USA, most of the feedstock for production of ethanol currently is corn, whereas Brazil exclusively uses sucrose from sugar cane.[2,3] This latter process is the simplest as sucrose is readily metabolized by yeast to ethanol in a highly efficient mature process that provides minimal areas for improvement of the fermentation process and will not be discussed further. However, the conversion processes for corn starch-based ethanol is considerably more complex, whether it uses the simpler dry mill process or the complex corn wet milling, and both biological conversion technologies for production of ethanol from corn offer opportunities for process improvement. Due to their simplicity, dry mills comprise a majority of the corn ethanol plants.

Biodiesel use is growing rapidly in the USA using primarily soybean oil as the feedstock with additional sources coming from waste fats and oils from food processing and restaurants. Total production in the USA is only 250 million US gallons (2008), although the production plant capacity is 2.24 billion US gallons, so the production capacity greatly exceeds sales (www.biodiesel.org). In Europe, an estimated 2.3 billion US gallons of biodiesel were produced in 2007, with 84% of the biodiesel coming from canola/rapeseed sources (www.eubia.org). Over half of the production comes from Germany, with the nearest producer country, France, producing only about 30% as much biodiesel. Alteration of the crop source offers good opportunity for expansion in the USA to replace food sources such as soybean oil.

19.2.1 Corn Processing and Ethanol

The corn dry milling process is a simpler process than the wet mill process and has fewer unit operations. The corn is initially milled with a hammer mill or grinder to crush the corn kernels, followed by addition of water and heating of the corn slurry to liquefy the corn starch. During the heating process a thermostable α-amylase[4] is added to initiate the starch hydrolysis at high temperature (90 °C) and decrease microbial contaminants. This hydrolysis process generates partially hydrolyzed starch maltodextrins and little if any free glucose. Upon cooling to 50–60 °C and adjustment of the pH to about 5, glucoamylase is added to initiate the cleavage of single glucose units off of the starch oligomers produced by the α-amylase action. A critical process parameter is maximizing glucose generation while minimizing reversion sugars formed by reverse enzyme action as conversion approaches over 95% glucose. The corn slurry–enzyme mixture is cooled and, when the temperature reaches below 40 °C, fermentation yeast, *Saccharomyces cerevisiae* is added. This yeast is either obtained by on-site production or purchased from an inoculum producer. The fermentation continues over a few days until a majority of the starch-derived glucose is converted to ethanol, typically yielding a beer with about 9% (w/w) ethanol.[5,6]

By contrast, the corn wet mill process is a true refinery process initiated by steeping the corn with water and sulfur dioxide to swell the corn and make it easily disrupted by mechanical processing. This steeping process lasts at 40–60 h and supports active lactic acid fermentation by resident *Lactobacillus*. Subsequently, the milling process permits the separation of the corn pericarp, corn starch, germ and soluble portion of the corn as a result of the steeping process. This steep slurry is refined by a complex series of separation processes yielding highly purified corn starch and fractions that yield corn oil after further purification, pericarp and germ residues that provide high-protein feed products. The corn steep liquor containing nearly all amino acids, minerals, other nutritional components along with typically about 50% lactic acid after extensive evaporation of the water. The highly purified corn starch is the largest product by weight and is used to produce a number of starch products such as cooking starch and derivatives destined for a variety of uses including adhesives, gums and coatings. Additionally, the starch can be further processed by a series of hydrolysis processes that include acid hydrolysis to produce a variety of specialty syrups for the baking, candy and food processing industries. Additional products employ enzymes to take advantage of their high degree of specificity that yield highly selective materials. The starch conversion process usually is initiated at high temperature (approximately 90–95 °C) in a jet cooker dosed with a thermostable α-amylase,[4] which rapidly cleaves the starch slurry to high molecular weight oligomers. Like the dry mill process, the maltodextrin syrup is then converted to over 95% glucose by the action of glucoamylase, but with much higher purity compared with the dry mill process. The glucose is purified for the incomplete hydrolysis products and is destined for various uses, including intravenous glucose, fermentation grade glucose used to produce light beer and sold to the food industry. One of the largest markets is the production of high-fructose corn syrup (HFCS), which is produced by conversion of glucose to fructose by the action of xylose/glucose isomerase, typically sold as 42% fructose, or via enrichment processes, 55% fructose HFCS. A chromatographic enrichment process is needed to attain 55% fructose due the natural enzymatic equilibrium between glucose and fructose at about 42% fructose. HFCS is sold largely as a sugar substitute, especially to the soft drinks market.

Purified glucose is highly versatile and can be used as a carbon source for a number of fermentative conversion processes. Over the last two decades, large volumes of ethanol have been produced by yeast fermentation in high purity and yield using wet mill-produced glucose. This highlights the benefit of the capital intensive wet mill process. For example, wet mill operations can produce ethanol from the glucose or divert any specified amount of the glucose stream to produce alternative higher value products. Examples of existing higher value products include antibiotics, amino acids and other food and feed additives such as vitamins. Glucose is also used to produce industrial products such as enzymes, organic acids such as citric acid and chemical intermediates such as 2,3-propanediol. This chemical intermediate is being produced by the DuPont/Tate and Lyle joint venture, located in Loudon, TN, using glucose obtained from the adjacent Tate and Lyle corn wet mill.[7]

19.2.2 Biomass Conversion for Ethanol

Multifaceted bioconversion technologies are emerging for the production of ethanol from lignocellulosic biomass. A high degree of complexity is required because, unlike corn kernels, which are over 70% starch on a dry basis, biomass contains three primary structural components, 33–50% cellulose, 17–35% hemicellulose and 12–24% lignin, with the remainder being minerals, protein and other minor materials.[8] In addition, since in Nature plants must withstand normal degradative processes as part of their survival, plant cell walls and the exterior of plant are highly resistant to enzymatic and microbial attack. As a result, over the last 30 + years a number of conversion processes have evolved that vary in complexity. The initial disruptive process needed to initiate the structural breakdown of biomass is called pretreatment and at least seven processes have been developed, ranging from highly alkaline to acidic environments and organic solvent based.[9–11] They all accomplish the partial disruption of the plant structure to allow the internal components, in particular the carbohydrates, to be exposed to enzymatic and microbial attack under controlled environments of a fermentation process. These processes vary from multi-enzyme and multi-fermentation approaches called separate hydrolysis and fermentation (SHF)[12] to simpler simultaneous cellulose hydrolysis or saccharification and fermentation (SSF).[13–15] In the SHF process, the biomass cellulose is hydrolyzed with cellulases to liberate fermentable glucose followed by a separate step for fermentation to ethanol. The SSF process combines the enzymatic hydrolysis with the fermentation, simultaneously reducing the process complexity. A natural extension is simultaneous saccharification and cofermentation (SSCF) with microorganisms that are able to covert both hexose and pentose sugars to ethanol. This process simplification culminates with the development of fermentation microorganism produces its own enzymes for cellulose hydrolysis, called consolidated bioprocessing (CBP). A fermentation technology that is the most versatile is the combination of gasification of biomass and other organic materials to a gaseous form followed by fermentation of the resulting gases to ethanol. All these conversion options have significant potential to benefit from use of advanced molecular and biotechnology analysis tools offered by systems biology tools.

19.2.2.1 Enzymatic Hydrolysis and Conversion

The enzymes that are typically employed in the hydrolysis of the biomass carbohydrates are cellulases including endoglucanases or β-1, 4-endoglucan hydrolase (EC 3.2.1.4) and exoglucanases which are glucan 1,4-β-glucosidases (EC 3.2.1.74). An additional required enzymatic activity is 1,4-β-cellobiosidases (EC 3.2.1.91) and β-glucosidases (EC 3.2.1.21), that collectively hydrolyze cellobiose disaccharide and other oligosaccharides to glucose. These combined enzyme mixtures are fully capable of hydrolyzing both crystalline and amorphous cellulose to glucose, making it available for fermentation to ethanol.[16,17] Depending on the pretreatment process used, additional enzymes may be required, such as hemicellulase enzymes such as *endo*-1,4-β-D-xylanase (EC 3.2.1.8) and β-xylosidase (EC 3.2.1.37), to hydrolyze hemicellulose (xylose) polymers that are poorly hydrolyzed under alkaline conditions. Biomass sources also contain limited amounts of starch and pectin and so addition of á-amylase, glucoamylase and pectinase enzymes may be beneficial for selected substrates. Finally, biomass contains between 12 and 24 wt% lignin, a complex polyphenolic phenylpropanoid polymer built up from *p*-coumaryl, coniferyl and sinapyl alcohols, resulting in complex arrangement of *p*-hydroxyphenyl (H), guaiacyl (G) and syringyl (S) units.[18] There are a number of enzymes that are known to partially degrade this complex material, including lignin peroxidase (EC 1.11.1.7), manganese peroxidase (EC 1.11.1.13) and laccase (EC 1.10.3.2), but they have not been involved in ethanol production from biomass due to cost and conversion considerations.

Enzymatic hydrolysis processes ideally produce a mixed sugar stream with predominantly glucose and xylose sugars along with a number of minor sugars such as arabinose, mannose and galactose, which collectively must be converted to ethanol by fermentation to maximize ethanol yield and specific productivity. Table 19.1 shows examples of specific analyses for three selected plants which shows the considerable variation

Table 19.1 Selected examples of detailed composition of selected biomass.[19]

Component	Composition (mass %)		
	Corn stover	Switchgrass	Hybrid poplar (debarked)
Glucan (glucose sugar)	37.12	33.59	44.65
Xylan (xylose sugar)	20.31	23.08	18.55
Araban (arabinose sugar)	2.46	2.91	0.59
Mannan (mannose sugar)	0.48	0.37	2.63
Galactan (galactose sugar)	0.92	1.15	0.82
Uronic acids (sugar acids)	1.72	1.12	0.85
Lignin	18.15	19.45	23.91
Extractives	3.89	10.82	3.22
Ash	12.54	5.84	0.85
Mass closure	97.59	98.33	95.22

in cellulose (glucan), xylan, lignin and ash content.[19] However, the minor sugars plus sugar acids together comprise 5–6% of the mass of biomass and should not be ignored as they contribute to ethanol production. The production of microorganisms that are able to ferment these various sugars has been the subject of active research over the last three decades and has included yeast (*Saccharomyces, Pichia*) and bacteria such as *E coli, Zymomonas* and *Klebsiella*.[20–23] Particularly challenging has been the goal of adding the capability to ethanol-producing (ethanologen) microorganisms the ability to ferment the five carbon sugars along with glucose, since many microorganisms lack this dual fermentation capability or ferment xylose with significant diauxic lag extending the fermentation time excessively.[24–27]

Significant progress has been made with a number of these microorganisms and a limited number have been chosen for commercialization, although details are often scarce. In addition, through the use of genetic engineering, research efforts are under way to combine the need for hydrolytic enzymes to be synthesized by a versatile multi-sugar-fermenting microorganism. Ingram's group, using both *Escherichia coli* and *Klebsiella*, is actively pursuing the addition of cellulases to versatile ethanologens.[21,28,29] Initial results have demonstrated the beneficial contribution of genetically adding a cellulolytic enzyme gene and capability to the genetic make-up of an ethanologen. Similar efforts are under way with yeast[30] and the development of an improved SSF ethanologen microorganism that produces one or more required complex carbohydrase enzymes appears likely.

19.2.2.2 Consolidated Bioprocessing

The possibility of the simultaneous production of hydrolytic enzymes that facilitate the degradation of complex carbohydrates to simple fermentable sugars along with fermentation to ethanol and other chemical byproducts is not new. Nature has provided anaerobic microorganisms that conduct this challenging task as a fundamental part of our ecosystem's carbon cycle. Both mesophilic and thermophilic microorganisms are known to degrade and ferment biomass carbohydrates producing ethanol, other alcohols, organic acids and hydrogen as common byproducts of their anaerobic fermentation.[31,32]

Intense research efforts have been made in a number of laboratories to harness selected individual candidate microorganisms that rapidly hydrolyze cellulose and/or hemicellulose and produce ethanol as part of their byproduct profile, through the aforementioned process of CBP.[33–35] Of particular interest have been the cellulase enzymes that facilitate the biomass cellulose breakdown. Through research over two decades, the structure of a highly evolved complex multi-unit multi-enzyme complex, called the cellulosome, has been elucidated and it is known to be directly responsible for biomass carbohydrate hydrolysis.[36-39] This enzyme complex, which can contain as many as 30 proteins with various functions, have been the subject of detailed structural biology research involving a number of clostridial species, including *Clostridium thermocellum*, *C. cellulovorans* and *C. cellulolyticum*, and the basic structure of he cellulosome is similar.[40] These anaerobic cellulose-utilizing microorganisms are able to produce simultaneously this highly complex cellulosome, which requires a significant level of metabolic energy, while actively fermenting the resulting cellulose and hemicellulose degradation products to ethanol and other byproducts such as organic acids. The present author remained skeptical regarding the excessive energy requirement needed for enzyme production and sugar transport and fermentation in an energy-deficient anaerobic environment. However, this changed when it was discovered that these microorganisms, such as *C. thermocellum*, preferentially produce and transport cellulose degradation units containing an average of four glucose units, thus conserving precious high-energy transport by obtaining three glucose molecules 'free' relative to phosphate-driven sugar transport.[41]

The discovery of an explanation for the rapid anaerobic growth and fermentation capabilities of *C. thermocellum* elevated this and similar microorganisms to possible candidates for a viable industrial process for the production of ethanol or other selected fermentation products, including hydrogen. However, thus far candidate CBP microorganisms all lack the preferred characteristics for the ideal ethanologen, as shown in Table 19.2. Research efforts are under way in a number of laboratories to address many of these limitations to take advantage of the inherent biomass carbohydrate degradative capacity of these anaerobes.[34-35,42]

Table 19.2 Preferred characteristics of an ideal CBP microorganism

Produces target fermentation product as sole product
Hydrolyzes cellulose to fermentable oligomers
Hydrolyzes hemicellulose to fermentable oligomers
Ferments cellulose oligomers
Ferments xylose or xylose oligomers
Produces target fermentation product in high titer (product resistance needed?)
Resistant to up to 1% acetic acid derived from hemicellulose
Grows at thermophilic temperatures at 55–80 °C
Moderately resistant to common pretreatment inhibitors (furans, lignin aromatics)
Produces a multi-carbohydrase portfolio on the cellulosome

19.2.2.3   Gasification for Ethanol

Thermochemical processing of biomass, as mentioned in the Introduction, is not the subject of this chapter, but it has the real advantage in that any carbon-containing material can be converted rapidly with no pretreatment. A process called gasification is a thermochemical conversion process conducted in the absence of oxygen. With this process, all carbon-based materials yield common molecular building blocks of CO_2, CO, H_2O and H_2. Gasification with oxygen produces synthesis gas with typically about 40% CO, 40% H_2, 3% CH_4 and 17% CO_2 on a volume basis.[43] Typical uses include generation of heat and power or as feedstock for chemicals.[44] However, the diversity of Nature has produced anaerobic microorganisms that are able to utilize the aforementioned gasified products to grow and produce byproducts of fermentation. One microorganism, *C. ljungdahlii*, is of particular interest because it produces primarily ethanol under certain conditions along with smaller amounts of acetic acid and hydrogen. Engineering research has been under way for over 15 years to develop a commercial process with this unique fermentation process[45–48] and commercialization is apparently moving forward (http://alicoinc.com/news_releases/2007/03-01-07A.html). However, little genetic improvement has occurred with these types of microorganism, providing a potentially fertile ground for biotechnological improvement.

19.2.3   Biodiesel

Biodiesel is either a derivative of triglycerides resulting in a fatty acid methyl ester (FAME) or the result of thermochemical conversion of biomass. The feedstock for FAME biodiesel is typically soybean oil in the USA or rapeseed/canola in Europe or Canada, although waste grease and fat from food operations can serve as feedstocks. However, the conversion technology to produce the FAME is the very mature, low-cost chemical transesterification process, which permits little opportunity for alternative processes such as enzymatic esterification.[49–51] The process entails blending a triglyceride with an alcohol, usually methanol, in molar excess with an alkaline catalyst such as methoxide that caused the methanol to replace glycerol ester to the fatty acid yielding free glycerol and a FAME. The most significant opportunity for process and product improvement lies in the selection of alternative crop species beyond soybeans, and canola fills this niche to some extent. However, the prime driver for alternative crop species is both high yield of oil per acre and an optimal degree of unsaturation. Soybeans typically yield about 376 lb per acre of oil whereas canola produces about twice that amount per acre.[52]

The degree of unsaturation is important in order to provide a degree of cold tolerance when blended with petroleum diesel at the 20% level, but oxidative instability is caused by the excessive double bond content. Canola is nearly 92% unsaturated but 63% of the fatty acids have only one double bond compared with 84% and 25% for soybean oil, respectively. Ideal plant species, such as an improved canola that has higher productivity and is able to be grown at in warmer mid-latitudes, should be the subject of plant breeding and bioengineering. Recent production expansion aims at producing 1.8 billion US gallons, but only 0.33 billion gallons are in production at present. Unfortunately, the current cost of soybean oil has risen from $0.31 to $0.55 per pound, causing plants to reconsider operation.[53] Germany, the largest producer of biodiesel, is facing market problems for very different reasons as of January 2008 due to large Federal tax increases, and this threatens the otherwise huge production from canola/rapeseed.

19.3 APPLICATION OF BIOTECHNOLOGY TOOLS TO BIOFUELS PROCESSES

Technologies for biofuel production vary in the degree of process maturity from highly refined corn ethanol and FAME biodiesel production to emerging technologies for production of ethanol and thermochemical biodiesel, not to mention nascent renewable gasoline production schemes. Therefore, these technologies provide varying degrees of potential improvement by advanced biotechnology tools collectively called system biology tools. A system biology approach essentially examines a biological process holistically at the system or total biological process level, including the system's genetic content (genome), gene expression, enzyme and protein production, flow of metabolic intermediates and resulting interaction with the genome and enzymatic processes. An example of application of system biology tools is the examination of differential expression of the genome by transcriptomic analysis using whole genome microarray methods where theoretically the degree of differential expression of every gene can be examined. Such an approach provides potential insights into the interaction of multiple gene systems and pathways. A companion biological tool is examination of the proteins present within a biological system, called a proteome, using advanced protein processing and analysis tools. Combining the transcriptome and the proteome data and analysis of results can provides unparalleled information needed to begin to understand the functioning of a biological system.

Critical to most of these system biology tools is access to the genomic sequence data as they form the foundation of gene expression and proteome analysis. Fortunately, genome sequence technology is advancing at an immense pace where bacterial genome sequencing is measured in hours and genome maps can be constructed and genes identified in weeks. This genomic information is used to construct microarrays using selected oligonucleotides for coding regions of the genome. The oligonucleotides are attached to a solid support and used for hybridization studies with labeled cDNA produced from RNA from selected biological samples. Due in part to the vagaries of nucleic hybridization, differential hybridization is often used with RNA from two samples from different biological conditions labeled with different molecular tags. The microarray is able to detect difference in expression of specific genes by the ratio of hybridization by competing labeled RNAs produced from two different experimental conditions. Use of this technology is growing rapidly, spawning numerous companies and supportive products with improved data sets.[1]

Analysis of proteins expressed by a biological system, called proteomics, can provide valuable information regarding the biological process functioning in a specific environment especially regarding the presence of specific enzymes for known biological processes. Like transcriptomics, proteomic analysis of protein expression requires knowledge of the system's genome sequence, whether it is a complex biological system with multiple (micro)organisms or a biological process functioning with a single (micro)organism. Early proteomic analysis examined proteins using two-dimensional gel analyses, which is a particularly challenging technique to obtain acceptable separation of the thousands of proteins present in microbial, let alone eukaryotic, cells. Gel-based proteomic analysis has largely been replaced with mass spectrometric analysis aided by various initial sample separation technologies such as liquid chromatography. With advanced separation and detection instrumentation, detection of nearly all the soluble proteins present in a microorganism, called the 'whole proteome', can be accomplished.[54] However, the proteome analysis can miss poorly expressed proteins and hard to capture hydrophobic and membrane proteins.

Of particular interest are the alterations in the enzymes and other proteins present in a biological system as conditions change and the biological system responds to a changing environment. However, both proteomics and transcriptomics are limited to the degree of accuracy of gene calling where the gene function has been assigned to open reading frames in the genome. Additionally, the function of numerous theoretical genes or open reading frames remains unknown even within the genome of *E. coli*, the most studied organism, limiting full understanding of a biological system. An additional challenge is that the reason for the change in thousands of genes seen by differential transcriptomic or proteomic analysis is often unknown due the lack of holistic knowledge of the gene expression cascades at the organism level.

During biofuel production, the cell's metabolic content, called metabolome, changes in response to the fermentation conditions. Through knowledge of the internal and external biochemicals present, which are the result of ongoing metabolic activity, much can be learned about the overall operation and functioning of the biological system under investigation. With temporal evaluation of metabolism, particularly after a perturbation or substrate addition, the flow of biological intermediates and precursor–product relationships can be determined. This analysis of the flow of metabolic intermediates, or fluxomics, is valuable both to understand the functions of a biological system and to determine metabolic responses to environmental or process changes or stimuli. A true systems biology analysis of a biological process will involve all the aforementioned 'omics' analysis tools to understand an ongoing biological process. A prime example of systems biology tools application is the process of bacterial sporulation, where known environmental stimuli (such a nitrogen deficiency) lead to a cascade of gene and protein expression culminating in the construction of biological survival system, the bacterial endospore.[55] All these 'omics' tools have the potential also to delineate the biological processes under way during biofuels production, thus potentially providing knowledge whose use may provide keys to making significant improvements in the microorganism and engineered processes, yielding more economical biofuels production.

19.3.1 Improved Production of Corn Ethanol

The production of ethanol from corn or grain starch is mature technology that was the target of early genetic engineering enzyme production in the early 1980s with the cloning and expression of the key enzymes required for glucose production. For example, α-amylase, glucoamylase and other accessory enzymes such as pullulanase and isomaltase needed for cleavage of α-1,4- and α-1,6-glyceride linkages in starch and elimination of reversion sugars[56–60] were genetically cloned and made commercially available. Production of many of these enzymes proceeded through classical protein engineering improvement strategies to improve thermostability, pH optimum/stability and specific activity in conjunction with genetic and process improvements to improve the volumetric productivity of the enzyme fermentation.[61,62] Improvements in characteristics and production levels of these enzymes have expanded applications, but the largest markets for enzymes, such as sweetener and detergent enzymes, has become a commodity business. This has made further research for incremental improvements difficult to support.[63]

The fermentation microorganism *Saccharomyces cerevisiae* is a highly productive and efficient producer of ethanol from glucose within both the wet and dry mill processes. However, opportunities remain for process improvement using new approaches to analyze the metabolic limitations that may arise. Fundamental analysis of the *S. cerevisiae* genome has been investigated for laboratory fermentation.[64,65] In addition, this technology has been applied in the beer and wine industry with the temporal analysis of the fermentation of simple sugars by *S. cerevisiae*.[66] Interesting differences have been detected among wine yeast strains that ferment primarily glucose and fructose from wine grapes *Vitis vinifera* L. For example. microarray analysis of natural vineyard populations of *S. cerevisiae* found significant differences in amino acid expressions, especially methionine, which showed that the natural population can vary significantly in gene expression due to segregation patterns of relatively few genes.[67] These differences presumably may impact flavor components in the fermentation broth, in this case wine. In addition, transcriptomic analysis has been used to determine the impact of high ethanol levels routinely reached during fermentation, especially with wine and champagne yeast where ethanol tolerance is required to meet product specifications.[68] A more holistic approach has been taken for beer production where the yeast must ferment hydrolyzed starch from added grains along with supplemented sugar. Distinct differences detected in the proteome of ale versus lager beer yeast have allowed researchers to conclude that the yeast involved has undergone significant interspecies hybridization and evolution as a 'genetic tree' can be derived with selected yeast.[49,69] To some extent it is encouraging that very traditional, long-standing industries such as wine and beer manufacturing have already applied the tools of systems biology to their craft, yielding potentially beneficial insights into product improvement.

Among all yeast fermentations, glycerol has long been accepted as an expected co-product of ethanol production, which must be removed during processing, especially for the fuel alcohol industry. Significant progress has been made in understanding why glycerol is produced, providing an avenue to strain improvement. It has been determined that glycerol is produced or induced as an osmotic stress response.[70] Indeed, mutant analysis has made progress in identifying specific mutant changes that impact glycerol production, permitting strategies for reduction of glycerol production including strategies to limit substrate, modify acid/base additions or other known engineering conditions present in commercial fermenters.[71,72] Further research is needed as additional *S. cerevisiae* glycerol mutants are produced. Through rapid sampling and processing, the yeast RNA can be acquired and processed for gene expression analysis, for example, providing information on the gene expression under way even from industrial-sized fermenters. This information, coupled with closely monitored fermentation broth ethanol, glycerol and other byproducts (external metabolome), will provide information on the fermentation temporal changes permitting engineering and possible genetic engineering solutions as needed. However, such investigations must be limited to on-going inefficiencies or difficulties that impose a sufficient economic burden to warrant research time and costs.

19.3.2 Ethanol Production from Biomass

19.3.2.1 SHF and SSF Processing

Enzyme manufacture for SHF and SSF/SSCF applications have grown in anticipation of biomass ethanol biorefineries that will require cellulases and other biomass enzymes. An important producer of cellulases for biomass hydrolysis is *Trichoderma reesei* and Genencor/Danisco has used microarray technology to evaluate and presumably improve the enzyme production.[73] Similar approaches are undoubtedly underway at Novozymes and other enzyme manufacturers, but much will remain unavailable until patent application publication. Target enzymes include 1,4-β-cellobiosidases and *endo*-1,4-β-D-xylanases. Indeed, Wilson has used microarray and proteomic analysis to detect induction of a xylanase with cellobiose by a biomass degrading bacterium *T. fusca*.[74] Finally, in the Fall of 2007, Genencor launched a new enzyme formulation intended for SSF conversion, called Accellerase, which is a blend of cellulases and cellobiosidases (genencor.com).

The process for production of ethanol from biomass described earlier contains two biological unit operations: enzymatic hydrolysis of the complex biomass carbohydrates cellulose and hemicellulose that remain after pretreatment by exogenously added enzymes, and fermentation of the resulting simple carbohydrates by an ethanologenic microorganism to ethanol. The latter fermentation operation can comprise two fermentations: conversion of cellulose-derived glucose to ethanol and fermentation of hemicellulose-derived pentose/xylose sugars to ethanol. Depending on the fermentation capabilities of the ethanologenic microorganism chosen, either one or two fermentation steps are required,

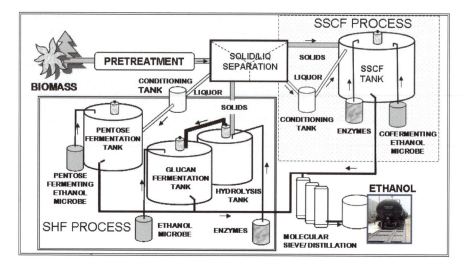

Figure 19.1   Diagram of biomass fermentation by either the separate hydrolysis and
fermentation process (SHF) (left) or simultaneous saccharification and
cofermentation (SSCF) process (right). SHF also shows conversion of
glucan solids and xylose hydrolyzates separately with different ethanol-
producing microorganisms, whereas SSCF requires an organism able to
ferment both carbohydrate sources. Use of a co-fermenting microorganism in the SHF process will reduce the tank number by one.
Conditioning is a process to remove fermentation inhibitors from the
pretreatment liquor. Waste streams not included.

since many natural ethanol-producing microorganisms such as *S. cerevisiae* do not ferment xylose/pentose sugars. As described earlier, separate hydrolysis and fermentation operations (SHF) permit the rapid
enzymatic hydrolysis of cellulose at elevated temperatures ($\sim 60\,^\circ$C),
followed by free sugar fermentation to ethanol at a lower temperature.
As shown in Figure 19.1, this process has been largely replaced in typical
plant designs by the simpler SSF process, in which cellulose hydrolysis
and ethanol production are conducted in one vessel. The ability to use
only one vessel requires the use of an ethanologic microorganism that
can ferment hexose and pentose sugars to avoid the double vessel configuration shown in the SHF schematic in Figure 19.1. Note the use of a
conditioning vessel that is needed to remove fermentation inhibitors
contributed by the pretreatment process, which will be discussed further
below. This technology, in various configurations, is the state-of-art
heading to commercialization.[75,76] Indeed, a pilot plant has been in
operation at the 660 000 gallons per year scale at Iogen Corporation
in Ottawa, Canada, using SSF technology.[77,78] Still no full-scale plant is
in operation at present but a number are planned, including some of the

six cellulosic ethanol plants the DOE selected for up to $385 million in US federal funding (http://www.energy.gov/news/4827.htm), which include SSF-like ethanol production (details are proprietary).

With the technologies for SHF and SSF approaching commercialization, there is still a dearth of detailed biological systems understanding regarding the optimal metabolic flow needed for most efficient production of ethanol from authentic biomass not laboratory sugars. This provides an excellent opportunity for the developing biological tools to elucidate the ethanologen strains' response to biomass carbohydrate hydrolysis and fermentation process. Indeed, the calculations as to the advantages of SHF, which separates hydrolysis and fermentation in two tanks, versus the combined SSF process have been limited to benefits of higher hydrolysis rates at higher temperatures (SHF) versus lower capital costs for SSF. There has been no investigation of the impact of these very different process designs on the fundamental biological or biochemical conversion processes except differences in volumetric ethanol productivity,[6,79] thus providing opportunities to apply advanced biology tools to near commercial biomass ethanol technology.

Separate hydrolysis and fermentation, SHF (Figure 19.1), process design was a logical first process design as the mixing actively fermenting microorganisms with exogenously added enzymes was not common. SHF continues to be evaluated as it has the advantage of operation with the hydrolysis at elevated temperatures, such as 50–60 °C, to enhance enzyme activity without concerns about killing common mesophilic ethanologens such as *S. cerevisiae* and *E coli*. Analysis of the best enzyme mixture to use has been the subject of detailed analysis,[16,17] as has the impact of ethanol production.[80,81] However, the impact of high concentrations of glucose, xylose and arabinose sugars derived from the biomass hydrolysis step on specific ethanologens is just beginning to be analyzed[82] and is ideal territory for further analysis with system biology tools. A transcriptomic analysis along with metabolomic analysis of the chemical intermediates in the fermentation could shed light on carbon flux limitations impacting the rate of ethanol production in authentic biomass fermentations. Depending on the specific process needs, additional detailed analysis of the fermentation portion of the SHF process may provide sufficient benefits to permit its commercialization instead of the SSF process under certain circumstances. One example is with biomass substrates such as eucalyptus and softwoods that produce excessive fermentation inhibitors that could be removed after hydrolysis, prior to ethanol production.

The SSF process is more complex than the SHF process due to the interaction of hydrolysis rates that provide the sugars for fermentation and the resulting response by the ethanologen microorganism. As the action of the cellulases (endogluconase, exogluconases) proceeds, significant levels of the cellobiose are produced which are known to competitively inhibit the cellulase action. As a result of this, commercial cellulase preparations, such as Accellerase, used in the SSF process include extra β-glucosidase.[16,17] Improper enzyme ratios, differential inactivation or overall insufficient enzyme addition could result in diminished sugar availability during the SSF process, producing a carbon-limited fermentation. Like the SHF process, both transcriptomic and metabolomic analysis should be able to detect depleted carbon flow by initial starvation response accompanied by decreased sugar intermediates in the metabolome. The gene response during carbon source depletion will likely be accompanied by increases in expression of signal transduction and sugar transport genes, for example, as genetic indications of hydrolytic imbalance, as has been seen with pure sugar fermentations.[83] If the ethanologen is capable of fermenting xylose and arabinose in SSCF mode, an additional level of complexity exists that can readily be evaluated by transcriptomic analysis of the pentose phosphate pathway during the SSF conversion. This examination may be particularly valuable if the xylose utilization capability is the result of addition of genetic modification,[84] including possible addition of exogenous foreign genes involved in pentose sugar conversion. In this regard, detection of the level of a exogenous enzyme, such as the anaerobic fungal xylose isomerase,[83,85–87] could be readily detected by both transcriptomic and proteomic analysis permitting estimation of whether the fermentation conditions supports sufficient levels of this or other critical enzymes throughout the SSF process.

Complete utilization of the pentose sugars (primarily xylose and arabinose) has been a vexing problem not easily accomplished with both yeast and some bacterial that negatively impacts SSF fermentations.[22,23] The pentose phosphate pathway (PPP) is very complex and this common but unexplainable incomplete fermentation may be the result of an imbalance of carbon flow, redox balance or net energy balance. Indeed, fermentation of high levels of glucose and xylose (50:50) by *S. cerevisiae* preferentially converted to xylose to xylitol and not ethanol after producing ethanol from glucose.[83] Determination of the pentose phosphate pathway and key glycolytic intermediates showed that increased xylose conversion was the result of higher levels of PPP intermediates, but whether this was due to flux rates or downstream pathway limitations could not be determined. A similar approach is needed for the more complex SSF fermentation during initial rapid pentose sugar fermentation compared with later during the fermentation where the pentose conversion wanes and often is not complete. Such analysis is likely to detect PPP gene expression changes, depletion of PPP enzymes in the proteome and particularly changes in the contents of the metabolome or fluxome as the pentose sugar fermentation process changes during the SSF process. Detection of the cause of the metabolic imbalance preventing full conversion of the pentose sugars would permit a directed genetic and/or fermentation engineering solution to be selected and improvements made in *S. cerevisiae*, *E. coli* and *Z. mobilis* ethanologens developed over the last decade.[88,89]

19.3.2.2 Impact of Enzyme and Fermentation Inhibitors

Fermentation for beer, wine and biomass ethanol does not benefit from a relatively pure source of simple carbohydrates that Brazilian sugar cane production enjoys.[2] These feedstocks contain a variety of soluble plant (flavor) components created during the processing. Wine fermentation contains limited fermentation inhibitors especially with white wine fermentation and additional inhibitors are common in beer production due to the cooking steps and the addition of hops. Neither of these processes approach the level of fermentation inhibitors present in the very complex milieu of partially hydrolyzed plant material after biomass has been pretreated, neutralized and conditioned prior to either SHF or more commonly SSF conversion. Among the most common inhibitory compounds for biomass conversion is free acetic acid liberated from acetylated hemicellulose during pretreatment. This acetate can reach typically ~ 2–$4\,\mathrm{g\,L^{-1}}$ levels depending upon the plant substrate involved, providing a significant source of fermentation inhibition.[90]

An additional source of inhibitors is the degradative products from lignin, which is a complex polyphenolic structural compound found in all plant matter.[18] Acidic processing of biomass can yield aromatic acids, aldehydes, ketones and other derivatives, all of which are highly inhibitory to metabolic activity.[91] Indeed, these compounds impact both cell growth and final ethanol concentration. While phenolic compounds from lignin breakdown provide significant inhibition to ethanologen's metabolic activity, lignin itself, in various forms after pretreatment, has been implicated in reducing the enzyme activities needed for either successful SHF or SSF production of ethanol. It has been hypothesized that lignin is partially solubilized by both alkaline and acidic pretreatment at elevated temperatures and then it is redeposited on the biomass upon cooling. This has been confirmed with the development of a protein/enzyme protection process using low-cost protein that minimizes lignin inactivation or binding to cellulase enzymes needed for cellulose hydrolysis.[92] Therefore, lignin poses a significant problem to biomass ethanol production, whether degraded or intact.

Dilute acid pretreatment and steam or hot water pretreatment result in an acidic environment during the initial biomass processing steps, whether from added acids or self-generated acetic acid from hemicellulose deacetylation. It is known that acid attack on sugars leads to the production of degradation products called furans, with xylose yielding furfural and glucose yielding hydroxymethylfurfural. This chemical reaction is accelerated at the high temperatures needed for effective pretreatment. These furan derivatives not only emanate from monomeric sugars, but also kinetic analysis has determined that intact cellulose and cellulose oligomers are subject to acid attack generating furan inhibitors.[10,93] Regardless of the source, the pretreatment process must be operated carefully to minimize the excessive acidic hydrolysis of both the hemicellulose and cellulose since these furans are highly toxic to all ethanologenic microorganisms tested.[94] Fortunately, there are a number of processes, collectively called conditioning (Figure 19.1), that can be used to minimize or eliminate the toxic byproducts of pretreatment of biomass, which include the over-liming process,[95–98] that permit fermentation processes to proceed with a variety of ethanologen microorganisms.

Addition of fermentation inhibitors during ethanol production is needed to elucidate the metabolic impact of the myriad of chemical inhibitors, each of which may have a different impact on the fermentation process. Transcriptomic analysis before and immediately after charging fermentation with the selected inhibitor would be a powerful approach to detecting directly the impact of that inhibitor on the biological processes that occur during the fermentation. Work is under way at the USDA laboratory in Peoria, IL, to investigate the impact of furfural and hydroxyfurfurals on *S. cerevisiae* fermentations and has resulting in the detection of specific genes associated with the pentose phosphate pathway for xylose utilization that mitigates the inhibitory response.[99,100] Their results using microarray analysis indicate that there is a nontransitory multigene response to furfural addition that eventually leads to adaptation to this potent metabolic inhibitor and provide ample opportunities for further evaluation of a very complex genetic response.

19.2.2.3 Consolidated Bioprocessing

A process for production of biomass ethanol without addition of exogenous cellulase enzymes has been under development for years using mesophilic and thermophilic anaerobes capable of growing and fermenting cellulose to ethanol and other fermentative byproducts. This process is called consolidated bioprocessing (CBP) due to the elimination of the requirement for exogenously added enzymes as needed by the

SHS or SSF processes.[35] CBP process relies on the production of carbohydrase enzymes by numerous anaerobic bacteria that can grow on complex carbohydrates such as cellulose and hemicellulose. In particular, certain clostridia produce a complex carbohydrase enzyme complex, the aforementioned cellulosomes that are expressed outside the cell but remain attached. These enzyme complexes attach to cellulose using a cellulose-binding module, which facilitates bringing the cell in close proximity to the substrate and permitting hydrolysis and consumption of the cellulose and other complex carbohydrates.[101] These cellulosomes are extremely complex and have been the subject of decades of research, especially with *C. thermocellum*. The research has centered on structural biology studies for many of these years and recent genome sequencing has augmented research considerably with the discovery of over 70 genes potentially associated with the cellulosome.[40] However, research has been slowed by the lack of widely available genetic transfer systems. Still, systems biology tools such as proteomic analysis can be critical to evaluating changes in cellulosome structure and composition both with and without genetic modification. For example, research has evaluated the expression of cellulosomal protein during growth on cellobiose, which does not require cellulosomal activity.[102] Stevenson and Weimer identified 17 genes differentially expressed during growth on cellobiose or cellulose in continuous culture, six involved in carbohydrate metabolism, and observed the majority had higher differential expression, but no direct correlation was possible based upon growth rate or fermentation products.[102] Doi's group has investigated the response of selected cellulosomal enzymes for *Clostridium cellulovorans* to growth on a variety of substrates such as hemicellulose, cellulose and pectin carbohydrates and analyzed the enzyme composition of the cellulosome by RNA Northern blots and PAGE and discovered changes in the enzyme profile of known carbohydrase enzymes.[103–106] Recently, Gold and Martin used metabolic labeling with *C. thermocellum* to identify changes in cellulosome protein subunit during growth on either cellobiose or cellulose using proteomic analysis with mass spectrometry.[107] They identified 16 new catalytic units but more than 23 putative cellulosomal proteins remained undetectable when grown on cellulose or cellobiose alone.

Recent work in our laboratory has investigated the linkage of complex carbohydrate growth substrate and expression of cellulosomal catalytic subunits using mass spectrometry-based proteomic analysis (to be published). *C. thermocellum* ATCC 27405 was grown in small fermenters with a variety of carbon sources including crystalline and amorphous cellulose, cellobiose and with various fixed mixtures of these materials with hemicellulose and/or pectin. These experiments were conducted with ^{15}N metabolic labeling and multi-dimensional LC–MS/MS proteomic analysis. Interesting patterns of expression of endogluconases were observed with fermentations of crystalline and amorphous cellulose and cellobiose at the individual enzyme family level. Xylanase enzyme expression was poorly responsive to cellulose and increased in expression in the presence of cellobiose. Not surprisingly, since *C. thermocellum* does not metabolize xylose and other pentose sugars, xylanase enzymes showed no specific pattern of induction in the absence of cellulose. However, more than 21 additional cellulosomal proteins were detected and specific responses to alternative non-cellulose substrates were observed. This study forms the foundation for the analysis of the cellulosomal composition present during the fermentation of natural biomass substrates such as switchgrass and poplar that are the subject of on-going research.

Genetic transformation systems have been developed for specific microorganisms capable of CBP, which opens heretofore closed doors for strain development aimed at the cellulosome and other metabolic processes. For example, genetic systems have been developed for a limited number of anaerobic cellulose-hydrolyzing microorganisms including *Clostridium cellulolyticum*,[108] *C. thermocellum*[109] and *Thermoanaerobacterium saccharolyticum*.[110] These genetic tool developments position the researchers to investigate the structure–function relationship of cellulosome gene components and the intricacies of the cellulose fermentation. An example of the power of a genetic system is seen with the recent genetic engineering of *T. saccharolyticum* into a homoethanologenic bacterium.[111] This thermophilic anaerobe ferments preferentially hemicellulose through the action of its portfolio of hemicellulases and produce by fermentation ethanol, lactic acid, acetic acid and hydrogen. Genetic engineering strategies were used specifically to block key genes in the acetate and lactate pathways, thus terminating the production of these common fermentative organic acids. Through successive genetic blocking strategies, lactic acid production was eliminate first,[112] followed by disruption of the acetate pathway, with an industrially important compensating increasing in ethanol production.[111] With the elimination of organic acid production, this genetically developed strain of *T. saccharolyticum* has the potential to be developed into a useful microorganism for conversion of the hemicellulose portion of biomass that is particularly present in the alkaline pretreatment methods due to the lack of acid-catalyzed hemicellulose depolymerization.[113]

Development of a holistic picture of a biological system requires that gene and protein expression be examined and analyzed. This can be further connected to any resulting metabolic products, either extra-cellular as is common for fermentation processes or for intracellular metabolites, both typically detected by advance mass spectrometric techniques. This combined approach has recently been applied to the complex process of fermentation of cellulose or cellobiose to metabolic products (ethanol, acetate and lactate) in our laboratory. The experimental approach was the fermentation of either cellulose or cellobiose by *C. thermocellum* using two identical fermenters for dual biological replicates. Fermentations were sampled throughout exponential and stationary phase to the point of full carbon source depletion.

The transcriptomic analysis was conducted using the complete genome sequence that was completed by the US DOE Joint Genome Institute (JGI) (http://genome.jgi-psf.org/finished_microbes/cloth/cloth.home.html) using microarrays prepared at ORNL with 70-mer oligonucleotides for predicted ORFs and employing differential dye tagging as standard protocols.[114] Typical fermentations yielded six sample points, so the differential microarray analysis was conducted in a variety of ways both within a single type of fermentation (cellobiose or cellulose) during exponential versus stationary growth or between fermentations (cellobiose versus cellulose). Analysis of the differential expression between cellobiose and cellulose fermentations yielded numerous differences, which included a strong induction of cellulosomal proteins as the culture reached stationary phase regardless of carbon source (to be published). Both cultures supported higher levels of energy and biosynthetic pathways as expected for a rapidly growing culture. Analysis between different carbon source cultures at either late exponential or stationary phase showed strong induction of cellulosomal proteins during active growth that tapered upon reaching the stationary phase. Although these are global analyses based upon functional groups, more detailed comparisons are under way with this total transcriptomic analysis. One approach that assists in this further analysis is cluster analysis for gene expression during six time points within fermentation. Cluster analysis for the cellulose fermentation permitted both clustering based on similar temporal comparisons and multiple categories of gene patters were observed with about 378 genes showing near level expression, while 192 genes showed significant decreasing expression throughout the fermentation. By comparison, 424 genes showed increasing genes expression throughout the fermentation process and 145 genes showed dramatic increased expression only upon reaching late stationary. Interestingly, only 11 of the expressed genes within this latter set were from the multi-gene sporulation pathway which contains 51 identified genes involved in sporulation. This is apparently because the minimal medium used for our research does not support sporulation of *C. thermocellum* (to be published).

Global or even functional group analysis provides information on general trends but analysis of specific gene patterns found within the huge database that microarrays provide can investigate specific questions. Among the logical targets for *C. thermocellum* are genes that impact ethanol production. *C. thermocellum* produces ethanol and organic acids acetate and lactate during fermentation, but these latter metabolites detract from the ethanol yield. Elimination of these organic acids by classical genetic engineering would be the preferred approach. Examination of the genome of *C. thermocellum* detected two structural genes for lactate dehydrogenase (LDH) in the pathway to lactic acid production. One of these genes has been cloned from *C. thermocellum* ATCC 27405.[115] It is not known if they were functional duplications or spurious crossovers, but knowledge of the expression of both genes would likely simplify the genetic engineering approach. Examination of the genetic expression of LDH by microarray analysis showed that indeed there was a significant difference in expression of two genes. One LDH gene (Genome ID Cthe0345) showed continuous temporal expression in cells grown on cellulose, whereas the other (Genome ID Cthe1053) did not demonstrate any significant expression under the conditions used. Apparently only one gene is strongly expressed and should be the initial target for genetic modification or knockout. Interestingly, the gene that was cloned by Özkan *et al.*[115] was Cthe1053 and it demonstrated LDH activity when put under expression of the *lac* promoter. Either the gene regulatory signals for the Cthe1053 were faulty or the regulatory signals prevented expression of this otherwise functional Cthe1053 LDH gene under the conditions used at a high enough level to be detected by the microarray.

In additional to gene expression patterns determined with microarrays, examination of the proteins present in cells for different conditions or different strains can shed additional light on the cell's metabolism. An important characteristic needed for *C. thermocellum* is tolerance to ethanol. The wild-type *C. thermocellum* is tolerant to less than 1% ethanol, but early researchers found that the microorganism can adapt to increasing ethanol added to the selection culture and have demonstrated growth in 2% ethanol.[116] Strobel and co-workers have developed a strain of *C. thermocellum* that can grow in as high at 7% ethanol and used proteomic analysis to detect protein differences between the wild-type and the stable adapted strains.[117] The study detected numerous differentially expressed membrane proteins that were down-regulated in the ethanol-adapted strain, providing potential targets for genetic improvement by protein overexpression to introduce tolerance in potential production strains. Due to the high homology of Nature, it highly likely that genes that cause ethanol tolerance will function in related microorganisms for an analogous homologous gene, for example in *T. saccharolyticum*. Also, selection of production strains resistant to inhibitors such as acetate, furfural and lignin monomers remains an important potential avenue for strain improvement.

19.2.2.4 Gasification Ethanol Production

One of the most versatile processes for the utilization of carbonaceous material is gasification, where the carbonaceous feedstock is converted to CO, CO_2, H_2 and H_2O. A gasifier can convert plant matter (biomass), plastics and rubber to these compounds due to the high temperature controlled anaerobic process. Interestingly, there are a group of anaerobic bacteria, some thermophiles, which are able to convert these gaseous compounds to cellular material and ethanol as a byproduct of fermentation.[48] The biochemistry uses the reductive acetyl-CoA pathway to permit conversion of the synthesis gas to ethanol, acetate and other chemical byproducts. The process uses high-density *Clostridium ljungdahlii* as the microbial catalyst and, depending on the fermentation conditions, either acetic acid or ethanol has been produced.[118] Indeed, production of acetic acid as a fermentation product has limited the yield of ethanol under certain fermentation conditions. The process requires gas to liquid to microorganism transfer, so improving mass transfer has been the primary target of process improvement over the past 20 years.[42–45] However, claims of up to 80% consumption of gases in a specially designed fermenter have been made by BRI Engineering, allowing potential commercialization of this process.

The potential for process improvement using genetic engineering of the key enzymes acetyl-CoA synthase and CO dehydrogenase could yield significant process improvements. Additionally, transcriptomic and proteomic analysis of wild-type strains during syngas fermentation would likely detect the level of expression of enzymes involved in ethanol synthesis and acetate production, shedding light on the importance of blocking acetate production. These include alcohol dehydrogenase for ethanol synthesis and acetate kinase and phosphotransacetylase for acetate production. Additionally, quantitative proteomics could detect potential pathway limitations, providing guidance for genetic engineering over expression of both predictable and unexpected proteins needed for syngas fermentation. To accomplish this, the genome sequence of *C. ljungdahlii* is required. This microorganism is in the pipeline for genome sequencing by the DOE JGI and has been under way in Gottschalk's laboratory funded by Celanese Chemicals, so it remains to be seen if this genome sequence will be available soon.

19.3.3 Biobutanol

Butanol produced from renewable materials is called biobutanol and is the result of fermentation of glucose from corn processing using *Clostridium acetobutylicum* and other anaerobic bacteria that produce butanol as a fermentative byproduct. Butanol is a potentially interesting biofuel because it is readily blended with gasoline, is poorly miscible with water, so it can be shipped via current pipeline infrastructure, and can be used in all current automobiles. The fermentation proceeds along the same pathway as for ethanol until pyruvate and acetyl-CoA. In butanol-producing bacteria acetyl-CoA is converted to acetoacetyl-CoA by the action of acetyl-CoA-acetyl transferase. Acetoacetyl-CoA is either metabolized to acetone and CO_2 or converted butyryl-CoA, which is the branch point to either butyrate or butanol. Butyrate is further metabolized to acetone and butanol enzymatically yielding multiple byproducts. Many bacteria also produce ethanol, hence yielding an acetone–butanol–ethanol (ABE) process.[119] Subsequently it was discovered that key genes for ABE production are located on large plasmids in *C. acetylbutylicum*.[120] Research has been aimed at manipulation of the pathway towards butanol at the expense of butyrate, but it is a difficult metabolic challenge.[121] However, one of the largest hurdles is the toxicity of butanol to the cellular membrane and metabolic processes. Resistant mutants have been created as a partial solution,[122] as have genetic engineering approaches.[123,124] Approaches for improvement include removal of the butanol solvent[125,126] and recent investigations of alternative substrates besides corn starch have been undertaken.[127]

The genomic sequence of *C. acetobutylicum* has been determined, allowing further sophisticated investigations to proceed.[128] Work in a number of laboratories has been examining gene expression during butanol fermentation to address the double problems of solvent toxicity and the delicate balance of butanol and butyric acid accumulation.[129] The genome data have been used to undertake transcriptomic analysis by the groups of both Blaschek[129] and Papoutsakis,[130,131] so significant progress might be made. Indeed, DuPont and British Petroleum (BP) announced in June 2006 that they are developing technology for biobutanol production,[132] and certainly they have the capability to apply all the new tools for systems biology to solve the low yield issue with this otherwise interesting biofuel. However, in-depth use of the new biological tool kit ('the omics') to this complex metabolic challenge will likely assist further in solving this difficult process problem.

19.3.4 Biodiesel

As stated previously, transesterification (TE) is a mature chemical process lending itself to minimal process improvement. As a chemical conversion process, it has the inherent benefit of functioning at high temperatures and pressures, providing very fast catalysis rates. Since the reactants are simple (alcohol and triglycerides), process separations after TE conversion are also relatively easy. This situation lends itself to minimal if any competitive advantage of an enzyme catalyst such as lipases,[49] unless specific products such as surfactant esters are desired. However, progress must be made to develop an alternative use for glycerol, a significant byproduct of biodiesel production, if the biodiesel business is to continue to expand.

19.3.5 New Concepts

At least two new companies have been formed recently that are aiming at the production of petroleum-like products that can serve as renewable gasoline or diesel. LS9 Inc. (www.LS9.com) in San Carlos, CA, was founded by Professors George Church and Chris Sommerville of Harvard and Stanford University, respectively. LS9 intends to develop and commercialize technology to convert renewable materials such as biomass into highly reduced petroleum equivalents using a biotechnology-based bioconversion process. Their website describes conversion of renewable sugars and fatty acids, obtained from plant material or natural triglycerides in plant or animal fats and oils. If a fatty acid feedstock is used the process apparently reduces the fatty acid C-terminus to a reduced form such as a methyl group. While the process is essentially the reverse of technology for production of dicarboxylic acids developed a decade ago,[133] this new process will likely face the same hurdles of high-priced feedstock as witnessed with the biodiesel industry currently and the cost of the chemical energy to drive the reductive processes. The company has attracted significant venture capital investment for the next few years.

The second company, Amyris Biotechnologies (www.amyrisbiotech.com), in Emeryville, CA, has similar business plan to produce renewable gasoline and petroleum fuels from biomass feedstocks through the application of synthetic biology. Their website describes synthetic biology as 'a new scientific discipline that involves the design and construction of new biological parts or systems, as well as the re-design of existing biological systems, for specific applications'. They are developing a fermentation process for production of both a renewable gasoline and a biodiesel and their website emphasizes technology demonstrated for production of terpenoids by fermentation.[134]

These and similar companies face significant challenges, essentially producing a highly reduced fuel from either approximately half-reduced carbon ($C_6H_{12}O_6$, hexose sugars) or expensive triglycerides and fatty acids. The use of sugars from plants as feedstock will face the same cost and conversion issues regarding efficient and effective hydrolysis of carbohydrate polymers while avoiding inhibitory byproducts discussed earlier. In addition, development of a microbial catalyst that provides its own cellulases, *etc.*, will be equally beneficial since the CBP process concept is not limited to ethanol and alternative fermentation products can be produced driven only by the microorganism's genetic and metabolic capabilities. Therefore, application of all the relevant systems biology tools will likely be very beneficial to identify, for example, biological processes that are inefficient regarding the use of valuable reducing equivalents in their fermentation processes as well as the understanding the impact of expected inhibitors generated during the biomass fermentation regimens.

19.4 FUTURE PERSPECTIVES

Long term, regardless of the conversion technology, the supply of renewable feedstock must fit the process for which it is destined. While thermochemical processes, of course, have much less restrictions regarding the biomass sources supplied, the biological conversion systems described here are dramatically impacted by the feedstock characteristics and quality. Corn-based ethanol processes require a high starch content and the seed grain suppliers have improved the corn yields with high starch content kernels. However, the nascent biomass energy industry is being built upon available residues from the food and feed industry and, for maximum cost effectiveness, this must change.

An important additional aspect to consider is that there are numerous types of plant biomass feedstocks available for biofuels production,[135] ranging from herbaceous corn stover and switchgrass to woody sources such as hardwoods and pine forest residues. Their composition varies considerably with differences in the levels of cellulose, hemicellulose and lignin, as shown in Table 19.3. Among the critical differences from a process perspective are the levels of acetic acid and lignin, as these are important contributors to fermentation inhibition.[94,96–98] Process

Table 19.3 Composition of selected biomass sources.[19]

Plant source	Composition (%)				
	Glucan	Xylan	Araban	Lignin	Ash
Switchgrass	33–34	21–22	2.6–2.8	17–21	5.2–6.2
Corn stover	37–38	21–22	2.4–2.5	18–19	**10–13**
Bagasse (sugarcane residues)	**39–43**	20–23	1.5–1.7	**23–24**	4
Wheat straw	33	19	2.4	17	**10**
Hybrid poplar	**40–44**	17–18	<1	**23–24**	1

modifications must be considered when feedstocks containing considerable acetylated hemicellulose, such as pine, or high levels of lignin, such as hybrid poplar or sugar cane bagasse, are the available biomass source because of their expected higher levels of inhibitor caused by the feedstock chemical composition. The bold entries in the table are particularly important characteristic for bioconversion, including high glucan as a benefit and high lignin or mineral content as detriments.

Current biomass ethanol processes are using either wheat straw, for example at the Iogen pilot plant,[77,78] or waste sugar cane residues, bagasse, corn plant residue or corn stover. While these materials have a beneficial composition (Table 19.3), wheat straw resources are limited,[135] sugar cane planting range is limited due to its current climatic range for growth and the US National Corn Growers Association estimates corn ethanol production will peak at approximately 15 billion gallons (www.ncga.com). Additionally, the environmental impacts of corn production and energy balance of corn ethanol have become important considerations before this upper limit is reached.[136,137] To change this situation, dedicated bioenergy crops must be selected, tested for bioconversion capabilities, domesticated and grown on a large scale.

In the mid-1990s, the US DOE Oak Ridge National Laboratory (ORNL) selected hybrid poplar and switchgrass as short rotation woody and herbaceous dedicated bioenergy crops needing development. Both are native North American species that have a wide grow range. In addition, hybrid poplar is a hardwood that grows very rapidly and, in spite of a relatively high lignin content, is readily conversed to ethanol with proper pretreatment and SSF processes.[138,139] For trees, storage is not an issue since they are a year-round source waiting to be harvested to support continuous biorefinery operation. Switchgrass is a high-yielding prairie grass that requires minimal fertilizer inputs, has deep root systems providing carbon sequestration benefits, grows in regions with low rainfall and can be planted and harvested reasonably well with existing planting equipment.[140] These two plant species are not the only potential dedicated crop selections as other native grasses, willow species and tropical energy cane have specific benefits for certain agronomic regions.[141-143]

To maximize productivity of biomass to ethanol and other chemicals selected dedicated energy crop species must be improved by plant genetic engineering and breeding with the aim of building in plant characteristics that are superior to the best available cultivar regarding ease of complete bioconversion to product. Such research has begun with a forage species, alfalfa, aimed at the reduction of lignin content. The Samuel Roberts Noble Foundation has shown genetic alteration of six different enzymatic steps involved in lignin production with reduced lignin content and proportionately improved enzymatic hydrolysis of the plant residues.[144] Similar breeding and genetic research on potential crops is under way aimed at improving not only the yield of the selected crop, but also modification of the plant characteristics by addition of hydrolytic enzymes involved in bioprocessing.[145,146] Improvement of hybrid poplar took a significant leap forward with the completion of its genome sequence.[147] Work has been under way at ORNL to utilize this genomic sequence for selected genetic modification and gene knockouts to improve this fast-growing tree for bioconversion. Indeed, the simultaneous improvements of switchgrass and hybrid poplar are central themes of the ORNL-centered multi-partner US DOE Bioenergy Centers awarded in the Fall of 2007 (www.bioenergycenter.org). The ideal eventual process will be comprised of dedicated, high-yielding biomass feedstocks specially bred for minimal impact on the environment while having composition and structural characteristics that facilitate rapid carbohydrate depolymerization and conversion to ethanol or other chemicals. This will be accomplished by concerted and linked research on both improvement of the plant species and optimization of the conversion process for this improved feedstock, permitting an economically sustainable biofuels industry to grow and displace significant levels of petroleum.

REFERENCES

1. W. D. Deckwer, D. Jahn, D. Hempel and A. P. Zeng, *Eng. Life Sci.*, 2006, **6**(5), 455–469.
2. G. M. Zanin, C. C. Santana, E. P. S. Bon, R. C. L. Giordano, F. F. de Moraes, S. R. Andrietta, C. C. De Carvalho Neto, I. C. Macedo, D. L. Fo, L. P. Ramos and J. D. Fontana, *Appl. Biochem. Biotechnol.*, 2000, **84–86**(1–9), 1147–1161.

3. J. Goldemberg, *Science*, 2007, **317**(5843), 1325.
4. G.E. Inglett (ed.), *Corn: Culture, Processing and Products*, AVI Publishing, Westport, CT, 1970, pp. 241–261.
5. S. Rajagopalan, E. Ponnampalam, D. McCalla and M. Stowers, *Appl. Biochem. Biotechnol.*, 2005, **120**(1), 37–50.
6. A. McAloon, F.Taylor, W. Yee, K. Ibsen and R. Wooley, Determinig the Cost of Producing Ethanol from Corn Starch and Lignocellulosic Feedstocks, National Renewable Energy Laboratory, Golden, Colorado, Report TP-580-28893. 2000.
7. C. E. Nakamura and G. M. Whited, *Curr. Opin. Biotechnol.*, 2003, **14**(5), 454–459.
8. C. E. Wyman, *Annu. Rev. Energy Environ.*, 1999, **24**, 189–226.
9. Y. Y. Lee, M. Ladisch, M. Holtzapple, R. Elander, B. Dale and C. Wyman, *Bioresource Technol.*, 2005, **96**(18), 1959–1966.
10. N. Mosier, C. Wyman, B. Dale, R. Elander, Y. Y. Lee, M. Holtzapple and M. Ladisch, *Bioresource Technol.*, 2005, **96**(6), 673–686.
11. M. Galbe and G. Zacchi, *Adv. Biochem. Eng. Biotechnol.*, 2007, **108**, 41–65.
12. C. R. Wilke, R. D. Yan and U. von Stockar, *Biotechnol. Bioeng. Symp.*, 1976, **6**, 55.
13. M. Takagi, S. Abe, G. H. Suzuki, G. H. Emert and N. Yata, *in Proceedings of the Bioconversion Symposium, Indian Institute of Technology*, New Delhi, 1977, pp. 551–571.
14. F. Alfani, A. Gallifuoco, A. Saporosi, A. Spera and M. Cantarella, *J Ind. Microbiol. Biotechnol.*, 2000, **25**(4), 184–192.
15. D. D. Spindler, C. E. Wyman, A. Mohagheghi and K. Grohmann, *Appl. Biochem. Biotechnol.*, 1988, **17**(1–3), 279–294.
16. M. E. Himmel, M. F. Ruth and C. E. Wyman, *Curr. Opin. Biotechnol.*, 1999, **10**(4), 358–364.
17. R. A. Nieves, C. I. Ehrman, W. S. Adney, R. T. Elander and M. E. Himmel, *World J. Microbiol. Biotechnol.*, 1998, **14**(2), 301–304.
18. W. Boerjan, J. Ralph and M. Baucher, *Annu. Rev. Plant Biol.*, 2003, **54**, 519–549.
19. US DOE EERE Biomass Program Biomass Feedstock Composition and Property Database, http://www.eere.energy.gov/biomass/progs/search1.cgi.
20. B. S. Dien, M. A. Cotta and T. W. Jeffries, *Appl. Microbiol. Biotechnol.*, 2003, **63**(3), 58–66.
21. S. Zhou and L. O. Ingram, *Biotechnol. Lett.*, 2001, **23**, 1455–1462.
22. L.O. Ingram, F. Alterthum and T. Conway, US Patent 5 000 000, 1991.

23. M. Zhang, C. Eddy, K. Deanda, M. Finkelstein and S. Picataggio, *Science*, 1995, **267**(5195), 240–243.

24. A. Mohagheghi, N. Dowe, D. Schell, Y.-C. Chou, C. Eddy and M. Zhang, *Biotechnol. Lett.*, 2004, **26**(4), 321–325.

25. A. Mohagheghi, K. Evans, Y.-C. Cho and M. Zhang, *Appl. Biochem. Biotechnol.*, 2002, **98–100**, 885–898.

26. N. W. Y. Ho, Z. Chen, A. P. Brainard and M. Sedlak, *Adv. Biochem. Eng. Biotechnol.*, 1999, **65**, 163–192.

27. T. W. Jeffries and Y. S. Jin, *Appl. Microbiol. Biotechnol.*, 2004, **63**(5), 495–509.

28. S. Zhou, L. P. Yomano, A. Z. Saleh, F. C. Davis, H. C. Aldrich and L. O. Ingram, *Appl. Environ. Microbiol.*, 1999, **65**, 2439–2445.

29. S. Zhou, F. C. Davis and L. O. Ingram, *Appl. Environ. Microbiol.*, 2001, **67**, 6–14.

30. W. H. van Zyl, L. R. Lynd, R. den Haan and J. E. McBride, *Adv. Biochem. Eng. Biotechnol.*, 2007, **108**, 205–235.

31. I. S. Maddox, *Biotechnol. Genet. Eng. Rev.*, 1989, **7**, 189–220.

32. D. Antoni, V. V. Zverlov and W. H. Schwarz, *Appl. Microbiol. Biotechnol.*, 2007, **77**(1), 23–35.

33. M. Ozkan, S. G. Desai, Y.-H. Zhang, D. M. Stevenson, J. Beane, M. L. Guerinot and L. R. Lynd, *J.Ind. Microbiol. Biotechnol.*, 2001, **27**, 275–280.

34. L. R. Lynd, P. J. Weimer, W. H. van Zyl and I. S. Pretorius, *Microbiol. Mol. Biol. Rev.*, 2002, **66**(3), 506–577.

35. L. R. Lynd, W. H. van Zyl, J. E. McBride and M. Laser, *Curr. Opin. Biotechnol.*, 2005, **16**(5), 577–583.

36. J. H. D. Wu, W. H. Orme-Johnson and A. L. Demain, *Biochemistry*, 1988, **27**, 1703–1709.

37. J. H. D. Wu, *ACS Symp. Ser.*, 1993, **516**, 251–264.

38. C. R. Felix and L. G. Ljungdahl, *Annu. Rev. Microbiol.*, 1993, **47**, 791–819.

39. W. H. Schwarz, *Appl. Microbiol. Biotechnol.*, 2001, **56**, 634–649.

40. A. L. Demain, M. Newcomb and J. H. D. Wu, *Microbiol. Mol. Biol. Rev.*, 2005, **69**(1), 124–154.

41. Y.-H. P. Zhang and L. R. Lynd, *Proc. Natl. Acad. Sci. USA*, 2005, **102**, 7321–7325.

42. E. Guedon, M. Desvaux and H. Petitdemange, *Appl. Environ. Microbiol.*, 2002, **68**(1), 53–58.

43. T. Reed, M. Graboski and B. Levie, US DOE Report SERI/PR-234-2571, Department of Energy, Washington, DC, 1987.

44. J. P. Longwell, *Fuel Energy Abstr.*, 1997, **38**(6), 445–445.

45. K. T. Klasson, M. D. Ackerson, E. C. Clausen and J. L. Gaddy, *Res. Conserv. Recyc.*, 1991, **5**(2–3), 145–165.
46. A. J. Grethlein and M. K. Jain, *Trends Biotechnol.*, 1992, **10**, 418–423.
47. H. Younesi, G. Najafpour and A. R. Mohamed, *Biochem. Eng. J.*, 2005, **27**(2), 110–119.
48. A. M. Henstra, J. Sipma, A. Rinzema and A. J. M. Stams, *Curr. Opin. Biotechnol.*, 2007, **18**(3), 200–206.
49. H. Baumann, M. Bühler, H. Fochem, F. Hirsinger, H. Zoebelein and J. Falbe, *Angew. Chem. Int. Ed.*, 1988, **27**(1), 41–62.
50. U. Biermann, W. Friedt, S. Lang, W. Lühs, G. Machmüller, J. O. Metzger, M. Rüschgen Klaas, H. J. Schäfer and M. P. Schneider, *Angew. Chem. Int. Ed.*, 2000, **39**(13), 2206–2224.
51. J. Van Gerpen, B. Shanks and R. Pruszko, Biodiesel Production Technology. National Renewable Energy Laboratory, Golden, Colorado, Report Report SR-510-36244, 2004.
52. R.K. Downey, Canola: a quality brassica oilseed. In: Advances in new crops: proceedings of the First national Symposium NEW CROPS, J. Janick and J. E. Simon (eds.), Timber Press, Prtland, OR, 1990 pp.211–215.
53. S.R. Schill, *Biodiesel Mag.*, 2008, March, 64–70.
54. U. Bond and A. Blomberg, in *Yeasts in Food and Beverages*, ed. A. Querol and G.H. Fleet, Springer, Berlin, 2006, pp. 175–213.
55. K. V. Alsaker and E. T. Papoutsakis, *J. Bacteriol.*, 2005, **187**(20), 7103–7118.
56. J. R Mielenz and S. Mickel (CPC International), US Patent 4 493 893, 1985.
57. J. R. Mielenz, *Proc. Natl. Acad. Sci. USA*, 1983, **80**, 5975–5979.
58. J.A. Eratt and A. Nasim US Patent 4 870 014, 1989.
59. R.M. Berka, D. Cullen, G.L. Gray, K.J. Hayenga and V.B. Lawlis, U.S. Patent 5 364 770, 1989.
60. R. D. Coleman, S. S. Yang and M. P. McAlister, *J Bacteriol.*, 1987, **169**(9), 4302–4307.
61. D. W. Goddette, C. Paech, S. S. Yang, J. R. Mielenz, C. Bystroff, M. E. Wilke and R. J. Fletterick, *J. Mol. Biol.*, 1992, **228**(2), 580–595.
62. A. Shaw and R. Bott, *Curr. Opin. Struct. Biol.*, 1996, **6**(4), 546–550.
63. O. Kirk, T. V. Borchert and C. C. Fuglsang, *Curr. Opin. Biotechnol.*, 2002, **13**(4), 345–351.
64. G. Giaever, A. M. Chu, L. Ni, C. Connelly and L. Riles, *et al.*, *Nature*, 2002, **418**, 387–391.

65. J. Förster, I. Famili, P. Fu, B. Ø. Palsson and J. Nielsen, *Metabolic Network Genome Res.*, 2003, **13**(2), 244–253.

66. T. Hirasawa, Y. Nakakura, K. Yoshikawa, K. Ashitani, K. Nagahisa, C. Furusawa, Y. Katakura, H. Shimizu and S. Shioya, *Appl. Microbiol. Biotechnol.*, 2006, **70**(3), 346–357.

67. D. Cavalieri, J. P. Townsend and D. L. Hartl, *Proc. Natl. Acad. Sci. USA*, 2000, **97**, 12369–12374.

68. T. Hirasawa, K. Yoshikawa, Y. Nakakura, K. Nagahisa, C. Furusawa, Y. Katakura, H. Shimizu and S. Shioya, *J. Biotechnol.*, 2007, **131**(1), 34–44.

69. J. I. Castrillo, L. A. Zeef, D. C. Hoyle, N. Zhang, A. Hayes and D. C. J. Gardner, *et al., J. Biol.*, 2007, **6**, 4.

70. Y. Inoue, Y. Tsujimoto and A. Kimura, *J Biol Chem.*, 1998, **273**(5), 2977–2983.

71. N. Hao, M. Behar, S. C. Parnell, M. P. Torres, C. H. Borchers, T. C. Elston and H. G. Dohlman, *Curr. Biol.*, 2007, **17**(8), 659–667.

72. S. Hohmann, M. Krantz and B. Nordlander, *Methods Enzymol.*, 2007, **428**, 29–45.

73. P. K. Foreman, D. Brown, L. Dankmeyer, R. Dean, S. Diener, N. S. Dunn-Coleman, F. Goedegebuur, T. D. Houfek, G. J. England, A. S. Kelley, H. J. Meerman, T. Mitchell, C. Mitchinson, H. A. Olivares, P. J. Teunissen, J. Yao and M. Ward, *J. Biol. Chem.*, 2003, **278**(34), 31988–31997.

74. S. Chen and D. B. Wilson, *J Bacteriol.*, 2007, **189**(17), 6260–6265.

75. A. Aden, M. Ruth, K. Ibsen, J. Jechura, K. Neeves, J. Sheehan, B. Wallace, L. Montague, A. Slayton and J. Lukas, Lignocellulosic Biomass to Ethanol Process Design and Economics Utilizing Co-Current Dilute Acid Prehydrolysis and Enzymatic Hydrolysis for Corn Stover 2002 National Renewable Energy Laboratory, Golden, Colorado, Report TP-510-32438, 2002.

76. R.Wallace, K Ibsen, A.J Mcaloon and W.C Yee, Feasibility study for co-locating and integrating ethanol production plants from corn starch and lignocellulosic feedstock. National Renewable Energy Laboratory, Report TP-510-37092, 2004.

77. K. Sanderson, *Nature*, 2006, **444**, 673–676.

78. J. S. Tolan, *Clean Prod. Processes*, 2002, **3**(4), 339–345.

79. M. S. Krishnan, F. Taylor, B. H. Davison and N. P. Nghiem, *Bioresource Technol.*, 2000, **75**(2), 99–105.

80. T. Han, R. Gonzales, A. Martinez, M. Rodriguez, L. O. Ingram, J. F. Preston and K. T. Shanmugam, *J. Bacteriol.*, 2001, **183**, 2979–2988.

81. M. Sedlak, H. J. Edenberg and N. W. Y. Ho, *Enzyme Microb. Technol.*, 2003, **33**, 19–28.
82. R. Gonzalez, H. Tao, K. T. Shanmugam, S. W. York and L. O. Ingram, *Biotechnol. Prog.*, 2002, **18**, 6–20.
83. J. Zaldivar, A. Borges, B. Johansson, H. P. Smits, S. G. Villas-Bôas, J. Nielsen and L. Olsson, *Appl Microbiol Biotechnol.*, 2002, **59**, 436–442.
84. L. Salusjarvi, M. Poutanen, J.-P. Pitkanen, H. Koivistoinen, A. Aristidou, N. Kalkkinen, L. Ruohonen and M. Penttile, *Yeast*, 2003, **20**, 295–314.
85. A. J. A. van Maris, A. A. Winkler, M. Kuyper, W. T. A. M. de Laat, J. P. van Dijken and J. T. Pronk, *Adv. Biochem. Eng. Biotechnol.*, 2007, **108**, 179–204.
86. M. Kuyper, M. P. Hartog, M. J. Toirkens, M. J. H. Almering, A. A. Winkler, J. P. van Dijken and J. T. Pronk, *FEMS Yeast Res.*, 2005, **5**, 399–409.
87. M. Kuyper, H. R. Harhangi, A. K. Stave, A. A. Winkler, M. S. M. Jetten, W. T. A. M. de Laat, J. J. J. den Ridder, H. J. M. O. den Camp, J. P. van Dijken and J. T. Pronk, *FEMS Yeast Res.*, 2003, **4**(1), 69–78.
88. K. A. Gray, L. Zhao and M. Emptage, *Curr. Opin. Chem. Biol.*, 2006, **10**(2), 141–146.
89. B. Hahn-Hägerdal, M. Galbe, M. F. Gorwa-Grauslund, G. Lidén and G. Zacchi, *Trends Biotechnol.*, 2006, **24**(12), 549–556.
90. J. R. M. Almeida, T. Modig, A. Petersson, B. Hähn-Hägerdal, G. Lidén and M. F. Gorwa-Grauslund, *J. Chem. Technol. Biotechnol.*, 2007, **82**(4), 340–349.
91. T.-A. Hsu, in *Handbook on Bioethanol: Production and Utilization*, C. Wyman ed., Taylor and Francis, Washington, DC, 1996, pp. 179–195.
92. B. Yang and C. Wyman, *Biotechnol. Bioeng.*, 2006, **94**(4), 611–617.
93. M. Galbe and G. Zacchi, *Adv. Biochem. Eng. Biotechnol.*, 2007, **108**, 41–65.
94. J. P. Delgenes, R. Moletta and J. M. Navarro, *Enzyme Microb. Technol.*, 1996, **19**(3), 220–225.
95. A. Martinez, M. E. Rodriguez, M. L. Wells, S. W. York, J. F. Preston and L. O. Ingram, *Biotechnol. Prog.*, 2001, **17**(2), 287–293.
96. S. I. Mussatto and I. C. Roberto, *Bioresource Technol.*, 2004, **93**(1), 1–10.
97. E. Palmqvist and B. Hahn-Hägerdal, *Bioresource Technol.*, 2000, **74**(1), 17–24.

98. E. Palmqvist and B. Hahn-Hägerdal, *Bioresource Technol.*, 2000, **74**(1), 25–33.

99. S. W. Gorsich, B. S. Dien, N. N. Nichols, P. J. Slininger, Z. L. Liu and C. D. Skory, *Appl. Microbiol. Biotechnol.*, 2006, **71**(3), 339–349.

100. Z. L. Liu, *Appl. Microbiol. Biotechnol.*, 2006, **73**(1), 27–36.

101. V. V. Zverlov, J. Kellermann and W. H. Schwarz, *Proteomics*, 2005, **5**(14), 3646–3653.

102. D. M. Stevenson and P. J. Weimer, *Appl. Environ. Microbiol.*, 2005, **71**, 4672–4678.

103. A. Kosugi, K. Murashima and R. H. Doi, *J. Bacteriol.*, 2001, **183**(24), 7037–7043.

104. S. O. Han, H. Yukawa, M. Inui and R. H. Doi, *J. Bacteriol.*, 2003, **185**(20), 6067–6075.

105. S. O. Han, H.-Y. Cho, H. Yukawa, M. Inui and R. H. Doi, *J. Bacteriol.*, 2004, **186**(13), 4218–4227.

106. S. O. Han, H. Yukawa, M. Inui and R. H. Doi, *Microbiology*, 2005, **151**, 1491–1497.

107. N. D. Gold and V. J. J. Martin, *J. Bacteriol.*, 2007, **189**, 6787–6795.

108. K. C. B. Jennert, C. Tardif, D. I. Young and M. Young, *Microbiology*, 2000, **146**, 3071–3080.

109. M. V. Tyurin, S. G. Desai and L. R. Lynd, *Appl. Environ. Microbiol.*, 2004, **70**(2), 883–890.

110. V. Mai, W. Lorenz and J. Weigel, *FEMS Microbiol. Lett.*, 1997, **148**, 163–167.

111. A. J. Shaw, K. K. Podkaminer, S. G. Desai, J. S. Bardsley, S. R. Rogers, P. G. Thorne, D. A. Hogsett and L. R. Lynd, *Proc. Natl. Acad. Sci.*, 2008, **105**(37), 13769–13774.

112. S. G. Desai, M. L. Guerinot and L. R. Lynd, *Appl. Microbiol. Biotechnol.*, 2004, **65**(5), 600–605.

113. F. Teymouri, L. Laureano-Perez, H. Alizadeh and B. E. Dale, *Bioresource Technol.*, 2005, **96**, 2014–2018.

114. S. D. Brown, B. Raman, C. K. McKeown, S. P. Kale, Z. He and J. R. Mielenz, *Appl. Biochem. Biotechnol.*, 2007, **137–140**, 663–674.

115. M. Özkan, E. I. Yllmaz, L. R. Lynd and G. Özcengiz, *Can. J. Microbiol.*, 2004, **50**(10), 845–851.

116. A. A. Herrero and R. F. Gomez, *Appl. Environ. Microbiol.*, 1980, **40**(3), 571–577.

117. T. I. Williams, J. C. Combs, B. C. Lynn and H. J. Strobel, *Appl. Microbiol. Biotechnol.*, 2007, **74**(2), 422–432.

118. J.L. Gaddy, US Patent 6 136 577, 1996.

119. E. R. Weyer and L. F. Rettger, *J. Bacteriol.*, 1927, **14**, 399–424.

120. E. Cornillot, R. V. Nair, E. T. Papoutsakis and P. Soucaille, *J. Bacteriol.*, 1997, **179**, 5442–5447.
121. D. R. Woods, *Trends Biotechn.*, 1995, **13**(7), 259–264.
122. J. Sierra, R. Acousta, D. Montoya, G. Buitrago and E. Silva, *Rev. Colomb. Cienc. Quim.-Farm.*, 1996, **25**, 26–35.
123. E. M. Green, Z. L. Boynton, L. M. Harris, F. B. Rudolph, E. T. Papoutsakis and G. N. Bennett, *Microbiology*, 1996, **142**, 2079–2086.
124. D. R. Woods, *Trends Biotechnol.*, 1995, **13**, 259–264.
125. N. Qureshi and H. P. Blaschek, *Biotechnol. Prog.*, 1999, **15**, 594–602.
126. T. C. Ezeji, P. M. Karcher, N. Qureshi and H. P. Blaschek, *Bioprocess Biosyst. Eng.*, 2005, **27**, 207–214.
127. N. Qureshi, B. C. Saha and M. A. Cotta, *Bioprocess Biosyst. Eng.*, 2007, **30**, 419–427.
128. J. Nölling, G. Breton, M. V. Omelchenko and D. R. Smith, *J. Bacteriol.*, 2001, **183**(16), 4823–4838.
129. T. C. Ezeji, N. Qureshi and H. P. Blaschek, *Curr. Opin.Biotechnol.*, 2007, **18**, 220–227.
130. C. J. Paredes, R. S. Senger, I. S. Spath, J. R. Borden, R. Sillers and E. T. Papoutsakis, *Appl. Environ. Microbiol.*, 2007, **73**, 4631–4638.
131. J. R. Borden and E. T. Papoutsakis, *Appl. Environ. Microbiol.*, 2007, **73**, 3061–3068.
132. http://onlinepressroom.net/DuPont/NewsReleases/June 20, 2006.
133. S. Picataggio, T. Rohrer, K. Deanda, D. Lanning, R. Reynolds, J. Mielenz and L. D. Eirich, *Bio/Technology*, 1992, **10**, 894–898.
134. V. J. J. Martin, D. J. Piteral, S. T. Withers, J. D. Newman and J. D. Keasling, *Nat. Biotechnol.*, 2003, **21**, 796–802.
135. R. D. Perlack, L. L. Wright, A. F. Turhollow, R. L. Graham, B. J. Stokes and D. C. Urbach, US DOE Report OSTI ID: 885984, Department of Energy, Washington, DC, 2005.
136. A. E. Farrell, R. J. Plevin, B. T. Turner, A. D. Jones, M. O'Hare and D. Kammen, *Science*, 2006, **311**, 506–508.
137. S. B. McLaughlin, D. G. de la Torre Ugarte, C. T. Garten Jr, L. R. Lynd, M. A. Sanderson, V. R. Tolbert and D. D. Wolf, *Environ. Sci. Technol.*, 2002, **36**(10), 2122–2129.
138. X. Pan, N. Gilkes, J. Kadla, K. Pye, S. Saka, D. Gregg, K. Ehara, D. Xie, D. Lam and J. Saddler, *Biotechnol. Bioeng.*, 2006, **94**(5), 851–861.
139. C. Luo, D. L. Brink and H. W. Blanch, *Biomass Bioenergy*, 2002, **22**(2), 125–138.

140. S. B. McLaughlin, J. R. Kiniry, C. M. Taliaferro and D. D. Ugarte, *Adv. Agron.*, 2006, **90**, 267–297.
141. S. Hanley, J. Barker, J. Van Ooijen, C. Aldam, S. Harris, I. Åhman, S. Larsson and A. Karp, *Theor. Appl. Genet.*, 2002, **105**(6–7), 1087–1096.
142. P. J. Tharakan, T. A. Volk, L. P. Abrahamson and E. H. White, *Biomass Bioenergy*, 2003, **25**(6), 571–580.
143. R. Ming, P. H. Moore, K. K. Wu, A. D. Hont and J. C. Glaszmann, *Plant Breed. Rev.*2006, **27**, 15–118.
144. F. Chen and R. A. Dixon, *Nat. Biotechnol.*, 2007, **25**, 759–761.
145. M. Sticklen, *Curr. Opin. Biotechnol.*, 2006, **17**(3), 315–319.
146. M. B. Sticklen, *Crop Sci.*, 2007, **47**, 2238–2248.
147. G. A. Tuskan, S. DiFazio, S. Jansson, J. Bohlmann and I. Grigoriev, *Science*, 2006, **313**(5793), 1596–160.

Subject Index